A SHROPSHIRE GAZETT

Credits

ISBN 0 906114 13 6

Published by: Michael Raven
Yew Tree Cottage
126 Jug Bank
Ashley, Market Drayton
Shropshire TF9 4NJ
Tel. 063087 2304

Typeset by: KC Graphics, Shrewsbury

A SHROPSHIRE GAZETTEER

Michael Raven

Front Cover Picture:
This is a photograph of the ceramic tile picture displayed in the premises known as 14-15 King Street, Ludlow, currently occupied by Cotswold Sheepskins and Woollens. The Maypole picture was installed by a previous tenant when the shop was one of a chain of Maypole shops. Nothing is known locally about either the artist or the manufacturer. The building itself was about to be demolished in 1986. However, investigation showed it to be a timber-framed house of the early 15th Century which had been adapted for commercial usage. English Heritage considered it to be a building of national importance and it was restored instead of being destroyed.

To my mother, Marion Raven,
and my father, Leonard Raven (1908-1974)

Preface

This book has taken nearly 2 years to compile, not evenings and weekends, but full time, an average of 10 hours a day, 7 days a week. In all of the thousands of miles I have travelled my constant and ever-cheerful companion has been my dog, Lady, a crossbred border collie.

Within the limitations of space I have endeavoured to make this guide as complete as possible. The gazetteer contains many more entries than any other similar publication and the constituent parts of Telford New Town, in particular, have been well attended.

Almost all the photographs were taken with a hand-held Nikon 35mm camera and a 28-200mm zoom lens. The film used was colour slide material, either Kodachrome 64 ASA or Fujichrome 100 ASA. In excess of 4,000 pictures were taken.

Our labours over the last 2 years have been exhausting and often physically arduous yet we approached each new day with feelings of anticipation. We were hunters of curiosities and beautiful places who never came home empty-handed.

As to the often asked question 'What is your favourite place?' there can be no single answer. Amongst the contenders, though, would be the tranquil dingle of St. Winifred's Well, the handsome Hall of Condover, the empty wastelands of Whixall Moss, the lovely valley of the River Teme at Nether Skyborry, and the lofty beech grove on Linley Hill.

However, if asked to choose somewhere in which to be confined for the rest of one's life, the answer would be: 'On Titterstone Clee Hill'. Gaunt and defaced by man it may be, but here there is an atmosphere that can only be described as spiritual.

Finally, this survey was researched and published without any financial assistance from companies or trusts, tourist organizations or advertising. This independence has enabled us to include information and express opinions without let or hindrance. The piper paid himself and the tune is his own.

Contents

ABDON *10m. ENE of Craven Arms.*
The name is Old English and probably means 'Abba's homestead'. The settlement lies half way up the western slopes of the 1,800 feet high Brown Clee Hill and there are good views from here across Wenlock Edge and into Corvedale. Abdon is most easily approached down lanes off the B4368. Above the village lies Abdon Burf, the gaunt summit of Brown Clee. There are two Iron Age forts on the hill: Abdon Burf, where evidence of occupation from Neolithic to Roman times has been found, and Clee Burf. Both forts are now mere shadows of their former selves having been eroded by mining and quarrying. Coal, ironstone, limestone and dhustone (a hard, black, basalt used as roadstone) have all been dug from the hill. The church of St. Margaret at Abdon stands on a raised circular mound, almost certainly the site of an earlier defensive position. (About forty per cent of the churchyards in Shropshire are circular and were probably the sites of either pre-Christian forts or pagan temples.) Adjoining the church at Abdon are the extensive remains – house platforms and trackways – of a medieval village deserted in the late 13th Century. Medieval ridge and furrow ploughing patterns are also still visible in the area. Some 100 yards SE of the church the site was re-settled in the 16th and 17th Centuries by mine and quarry workers who in turn abandoned the settlement when their employment ceased. Half-a-mile W of the deserted village are the remains of a small 17th Century iron furnace: a wooded dingle with a slag floor, overgrown tracks, the pond bed and weir and the foundations of the furnace itself. Today the church stands alone some distance from the modern development. One mile S of Abdon at **Cockshutford** are the remains of another Iron Age earthwork, Nordy Bank. This was possibly a look-out position for the larger enclosures of Abdon Burf and Clee Burf. In recent times limestone was quarried at Cockshutford and the concrete foundations of the mine buildings still stand beside the road. The rough hill pastures of Brown Clee, which are mostly common land, are grazed by sheep and smallholdings are scattered over the lower slopes. On all sides of the hill are many tracks called 'straker routes' or 'driftways' along which the people of the surrounding villages drove their sheep, and in olden times their cattle, to the common pastures.

ACTON BURNELL *7m. SSE of Shrewsbury.*
A handsome village and one of the most historically interesting in the county. It lies in a wooded valley and is dominated by a group of ancient buildings. The Castle, which is really a fortified sandstone house and probably the earliest of its kind in the whole of England, was built by Robert Burnell (d.1293), Bishop of Bath and Wells and Lord Chancellor to Edward I. He was a wealthy man who had lands in 19 counties and who held 82 manors. Close by the Castle is the Great Barn, or rather 2 stone gables (157 ft. apart) believed to be the remains of a barn. This was probably the meeting place of the first full English Parliament, summoned by Edward I in 1283 whilst he was in Shrewsbury to deal with the 'Welsh problem'. This passed a statute ordering debtors to fix a date for the settlement of their debts and to forfeit their property if they failed to meet the deadline. On the way to the Castle one passes the 13th Century Early English cruciform church of St. Mary. The tower was added in 1888. Inside there is an elaborate tomb to Sir Richard Lee (d.1591), a forebear of Robert E. Lee, the American Civil War hero. The Hall, now the Concord College, is a large but plain Georgian style stuccoed mansion with a giant Ionic portico. It was built by the Smythe family in 1814 and stands in a landscaped park. Attached to the Hall is a Roman Catholic chapel built in 1846 by C. Hanson. In 1939 the house became a convent of the Sisters of Sion and their school was attended by the daughter of Charles de Gaulle. The sisters left about 1970. In the grounds of the Hall are 2 lakes (probably medieval fish ponds), a shell and tile grotto, and a magnificent Cedar of Lebanon that towers above the castle. Just S of the most easterly of the 2 lakes is Sham Castle (SJ.544.016). This is a folly-house with rounded tower-like bays, painted Gothic windows and battlements. It is situated on a rocky knoll surrounded by a field of golden corn which is in turn almost encircled by woods. The 'castle' was built as a music room in the 1780's by the Bruce-Smythe family who then owned, and who still own, the Acton Burnell estate. In the village (which has a population of about 160) most of the domestic buildings are either 17th Century timber-framed farmhouses or 18th Century stone cottages. There is also a red-brick, ivy-clad village stores cum Post Office and some modern housing. In the 19th Century many of the old cottages were clad with rough-casting and the doors and windows were decorated with shallow hoods, thus giving the settlement the appearance of an estate village. It is now believed that before Robert Burnell built his castle and made his deer park the village lay W of Lynall Brook and closer to the route of the old Roman road that runs about 600 yards to the NW of the church. This road ran from Frodesley to upper Cound and on to Wroxeter (Uriconium); it was, in fact, the old Watling Street. A field here used to be called Townend Field. Just 1½m. S of Acton Burnell is Langley Chapel. (*See Langley*.)

ACTON PIGOT *1m. NE of Acton Burnell.*
A tiny, nucleated hamlet. The Anglo-Saxon settlement of Acton passed to William Picot in the 12th Century. The place name can also be spelled Piggott and Piggot. **Golding** Manor lies ½m. NNE. This is a red brick house with quoins, 2 gables and a recessed centre. In the grounds is a brick dovecote.

ACTON REYNALD *3½m. SSE of Wem.*
Lying between Grinshall village and Shawbury aerodrome is the Hall and the park created by the Corbet family in the 17th Century. To make the park complete the ancient and substantial village of Acton Reynald was demolished and all trace of it removed. In 1780 there were 35 farms and houses on Corbet land. By 1810 there were none. The village was not even re-sited but totally destroyed for ever. The Corbet's behaved in a similar fashion on 3 other of their estates, namely Longnor, Moreton Corbet and Adderley. Of their Acton Reynald mansion 3 of the E side bay windows have been dated at 1601 and the fourth at 1625. The portico with 2 pairs of Tuscan columns is also dated at 1625. The rest of the Hall was rebuilt in 1800. Today the building houses a school. Grinshill and its famous quarries lie just across the road. The name Acton means 'the settlement by the oak trees'; 'Reynald' is from Reyner de Acton who held the manor in 1203. Reyner itself is from the German personal name Rainer.

ACTON ROUND *5m. WNW of Bridgnorth down a lane off the A458.*
The hamlet is set in lovely countryside and consists of a few brick houses and an old church with a timber-framed porch and 12th Century ironwork on the door. In about 1750 a N chapel was added by the Acton family. The handsome, early 18th Century brick Manor House has 7 bays, 2 storeys and a hipped roof. It was built as a dower house to Adenham Hall, which lies nearby on the other side of the A458. In the garden there is a modern folly pagoda and in the house is an ape swinging on a chandelier. The mill ceased to work in the mid-19th Century and little of it now remains, though the Mill House is still inhabited, by one of the family which owns the Hall. The village name is possibly derived from a very early circular church dating from the time when the Knights Templar were lords here. The old common, which lay in the Forest of Shirlett, had been cleared and was grazed before being enclosed by Acts of Parliament in 1814 and 1838. Now, much of the area has been re-afforested. Yew trees, often used as direction markers, border the medieval track that led from Shirlett Forest down the hill to the church. Another lane to the church leads from the Hall. If one carries on along this tree-lined track for ½m. the great hollow oak of Acton Round will be found on the right. This is one of Shropshire's largest oak trees with a girth of over 29 feet. It has been pollarded, that is, the branches have been cut back regularly, and it was almost certainly used as a boundary marker. There are splendid views to either side of the track because the route follows the course of a ridge. The lane itself is of interest being metalled but not tarmac-adamed and must be very much like the coaching roads of the 18th Century. Just beyond the great oak tree the track comes to an end.

ACTON SCOTT *3½m. S of Church Stretton.*
Backed by the Caradoc hills and positioned less than a mile off the busy A49 is

Acton Round, two falconers near the church.

the village formerly called Acton Super Montem. It is best known for its Working Farm (established in 1975), a museum of traditional farming methods. This was, in fact, the Home Farm to the adjacent 16th Century gabled brick Hall built by the Acton family and set in a fine park. Within the grounds are an archery butt and a yew hedge 150 yards long. Actons still live in the house. The family is of Saxon, rather than Norman origin, and has numbered amongst its ranks generals, admirals, cardinals, princes and barons. John Acton, the 6th baronet and the Prime Minister of Naples, is reputed to have saved 4,000 Spanish sailors from slavery in an action against Barbary pirates, and was a confidante to Lord Nelson and Lady Hamilton. In recent years the most noteworthy member of the family was Sir Harold Acton who moved in the Mitford and Sitwell circles and who was caricatured by Evelyn Waugh in Brideshead Revisited. In the mid-19th Century the ruins of a substantial Roman villa were uncovered near the Hall. To the west of the A49 is the line of a Roman road. The church of St. Margaret is medieval but was much restored in the 19th Century. There is good brass of Thomas Mytton (d.1577), and monuments to the Acton family. In the churchyard there are, unusually, 5 ancient yew trees, 2 male and 3 female. One tree is self-regenerating by growing a new internal trunk. About 1m. S is the junction of a now dismantled railway with the existing main line which runs parallel to the road. Felhampton Court lies 1½m. SSW of Acton Scott. It is an early Georgian house of 3 bays and 2 storeys built in brick with a large, steep pediment. The name Acton is Anglo-Saxon and means 'the settlement by the oak trees'; *Scott* comes from either Reginald Scot or Walter le Scot, both of whom were medieval land holders in the area. The Acton family took their name from the settlement. Note: The present owner of Acton Hall refused to let us photograph the house, the only such rebuff we have ever received.

ADCOTE HALL *See Little Ness.*

ADDERLEY *2m. N of Market Drayton.*
In medieval times Adderley (*Aldred*'s wood, or clearing in a wood) was a flourishing market town; at the turn of this century it was little more than a hamlet, a decline largely attributable to the Corbet family who destroyed much of the settlement when they landscaped the park of Adderley Hall. The farmer at Adderley Hall Farm used to turn up bricks whilst ploughing the field between the farmhouse and the present, largely modern, village. The field is now laid to pasture but slight earthwork signs of the old village are still visible. Adderley Park, and the parks of Shavington and Cloverley, were all part of the ancient Anglo-Saxon estate of Nigel the Physician in 1086. The medieval church of St. Peter has a Norman font, a tower of 1713 and a chapel – the Kilmorey – of 1637. The rest was rebuilt in 1801 and of this time are the wide, pointed cast-iron windows with perpendicular tracery which have many admirers. Outside the church are a mounting block and a curious, multi-faced stone pillar that appears to be a sundial. The Corbets of Adderley had a feud with the Needhams, viscounts of Kilmorey, who lived at **Shavington** Hall, 1½m. SW. This was largely over the issue of fox-hunting. Kilmorey thought it barbarous and refused to allow his lands to be hunted over by the Corbets. The Kilmoreys were also insulted by not having their own pew in the church. To show his disrespect, Sir John Corbet ordered that a footboy of his who had recently died (in 1633) be buried adjacent to the body of the lately deceased Lady Kilmorey. Only after Viscount Kilmorey had appealed to the Earl Marshall of England was the servant's corpse removed, and only after appealing to the Archbishop of Canterbury did the Kilmoreys get permission to build their own chapel. Shavington Hall was described as being probably the finest brick house in the county. Built in 1685, with Victorian alterations in 1858, it was demolished in 1958. However, the stables, the laundry and some estate workers' cottages remain. The platform occupied by the Hall is about 50 yards to the front left of the stable-block, which is N of the Big Pool. The large, wooded park is now run as an agricultural and sporting estate and there is a plant nursery that is open to the public. (*See Tittenley.*) Note. The remains of the castle of Nigel the Physician – a 21ft. high mound now set between 2 ornamental lakes – are at SJ.665.404. The present Pool House is probably built in the bailey.

ADENEY *1¼m. WSW of Edgmond, near Newport.*
The name is Anglo-Saxon and means '*Eadwynne*'s island'. *Eadwynne* is a woman's name. The settlement stands on a dry sandstone rise in the low lying Weald Moors. The black peat of the drained marshes that surround the hill makes a good, light soil suitable for growing potatoes. Other crops grown by local farmers include corn, wheat, barley and sugar beet. A few sheep graze the meadows but cattle are too heavy-footed for some of the lowland pastures. Adeney (or Adney as it used to be) is a hamlet that consists of little more than a short row of cottages with windows set into recessed blank arcades; the Lodge, which is also a Bed and Breakfast establishment; Adeney House, which has a tall monkey puzzle tree, a landmark for miles around; and the Manor House of red brick with stone dressings, gables, dormers, decorated barge boards, an ivy-clad porch and a small conservatory. Inside it is dark but full of character, and the present owners have a splendid collection of antique furniture. The house must be at least 18th Century and could be older. Outside there is a range of old farm buildings about a sunken courtyard in part of which was once a duck pond. In the garden of the house there is a splendid lime tree and along the roadside are several yews. A track leads down the side of The Lodge to the Weald Moors. It passes over a deep cutting of the now disused Shropshire Union Canal, Newport Branch (SJ.702.181). To the right the canal bed is intact though totally overgrown with bushes; to the left it has been filled in. There are good views from here over the flat moorlands to the Lilleshall obelisk in the SE and The Wrekin in the SW. A pool was the process of being constructed when we were there. In the Middle Ages Adeney was given by King John to Croxden Abbey in Staffordshire. It was later exchanged for Caldon Grange (also in Staffordshire) and became the property of Buildwas Abbey.

ADMASTON *1m. NE of Wrockwardine which is 2m. W of Wellington.*
Admaston (the name means 'Edmund's homestead') is now a place of new houses; it does, after all, fall within the boundaries of Telford New Town. The area around Bratton Road has some character though – older houses in a mixture of styles and a good timber-framed cottage. Admaston Hall is of red brick, 5 bays wide and 2 storeys tall. Social facilities include a Methodist Church, the Pheasant pub, a Community Centre, and a row of 5 flat-roofed shops by a pine tree owned by a gentleman called Bdesha. Amongst them is a Post Office and a general store. To the E, off Shawbirch Road, is **Admaston Spa**. Here were 2 springs, one sulphurous and one chalybeate (iron rich) and around these a small, but fashionable, spa with a hotel and a boarding house was developed during the 19th Century. Only the Clock House remains, all the other buildings, and a small pool, were cleared away when the site was re-developed in the 1980's. The springs were in the area now occupied by the back gardens of the upmarket terraced houses adjoining the Clock House. In the early Middle Ages Admaston belonged to the Manor of Wrockwardine. Before the Norman Conquest it belonged to King Edward and after passed to Earl Roger de Montgomery who held it as part of his personal estate.

ALBERBURY *7m. W of Shrewsbury on the B4393.*
This delightful border village was once much larger than it is today. The substantial remains of the ivy-covered, stone-built keep of the Alberbury castle lie in a walled enclosure about 50 yards in diameter. The fortress dates from the 12th Century and was built by Fulke Fitzwarren. To one side is the lodge to Loton Park, and to the other is the church. Across the road are cottages and modern houses. The church of St. Michael has box pews, a 13th Century tower with a large saddleback roof and splendid arcading. Some people disapprove of the chancel of 1845. The church is set high on a circular mound, probably the site of a prehistoric fort. Loton Park is a red brick mansion with two facades, that of the S being in Jacobean style, and that of the N being Queen Anne. In the 1870's a SE wing was added. The hall, grounds and outbuildings are all very well kept. Sir Michael Leigh-

Sham Castle,
Acton Burnell.

Loton Park,
Alberbury.

Alcaston Manor.

ton lives here. Opposite the hall, on the other side of the road in the park, are the remains of an old army camp. It is here that the well-known motor sport meetings are held. One mile NE are the remnants of Alberbury Priory, founded in 1225 for Augustinian Canons but later transferred to Grandmont. All that is left is incorporated into a farmhouse. The S side of the farm is part of the S side of the nave and chancel. The Chapel of St. Stephen is still complete. Just over the border, in Wales, is the Prince's Oak, so called because in 1806 the Prince of Wales (later to be King George IV) plucked a sprig of its foliage to prove he had been into his palatinate. At the time he was staying at Loton Hall. The tree stands by the main road and has a protective brick wall around it. In 1942, in a field near Alberbury, an important grave was uncovered, the only one of its kind in the county. It contained the skeleton of a Late Neolithic – Early Bronze Age man (2500-2000 BC). His mortal remains are now in Shrewsbury Museum (Rowley House). The name Alberbury is from the Old English and means the fortified place of *Alugerg* (or *Ealhburg*); Both names are those of women.

ALBRIGHT HUSSEY *3m. NNW of Shrewsbury and ¼m. W of Battlefield.*
All there is to mention here is an odd, but charming, house half in timber frame of 1524 and half in brick, dated 1601. It was probably intended to replace the whole house in brick but the work was left unfinished. The premises are now used as a restaurant. There used to be a chapel here but there is now no trace of it. The land hereabouts is very flat. The name is composite: Albright is an abbreviation of the Saxon name Albrighton which means 'the homestead of *Eadbeorht*', and 'Hussey', from Walter Hussey, a Norman who held the manor about 1165.

ALBRIGHTON (near Shifnal) *5m. ESE of Shifnal.*
It lies between Wolverhampton and Newport and is by-passed by the newly improved A41. Albrighton is a pleasant, unpretentious dormitory town for Wolverhampton, but by no means a suburb as some have said. It still retains more than a hint of tree-lined Georgian elegance, but the area around the greens on the S side of the High Street is ruined by the haphazard parking of motor cars. There are several small parades of shops, and there is no shortage of pubs and fish and chip takeaways. The modern approach is most pleasant, thanks to the mature trees of the Hall gardens that line the road. The Church of St. Mary Magdalene is of red sandstone and has a Norman tower with 13th Century lancet windows and Perpendicular battlements. The chancel is Decorated and has an excellent 5-light E window. The nave and aisles are of 1853. Inside there is a notable 13th Century tomb-chest decorated with heraldic shields, a very early example of such ornamentation. There is also an alabaster monument to Sir John Talbot, died 1555, and his wife. The Shrewsbury Arms pub stands opposite the churchyard. It is timber-framed with brick infill. The modern (1954), mostly steel-framed, school in Cross Road is by Jackson and Edmonds. Albrighton is an Anglo-Saxon name and means 'the homestead (or settlement) of *Alubeorht* (or *Aldbeorht* or *Aepelbeorht*)'.

ALBRIGHTON (near Shrewsbury) *3½m. N of Shrewsbury.*
A main road village with some black and white houses, and, unusual in Shropshire, a village green. The church of St. John the Baptist is of 1840, neo-Norman in sandstone with a genuine Norman font and a plain Jacobean pulpit. The 'big house' is now a hotel and the attractive gardens are well kept. The old name of the settlement was Monks Albrighton and it used to belong to Shrewsbury Abbey having been given to that monastery by Alchere prior to 1086. The old pub sign from the inn at Albrighton depicted a man whipping a cat and below were written the lines: 'The finest pastime that is under the sun, is whipping the cat at Albrighton'. This is quite possibly a relic of the ancient pagan custom of sacrificing an animal at what is now called Shrovetide. In other parts of the country cocks were whipped to death during this festival.

ALCASTON *1½m. S of Acton Scott which is 3½m. S of Church Stretton.*
It announces itself with tall metal silos and storage tanks, a torpedo shaped propane gas cylinder, a power cable pylon and mounds of old car tyres. Alcaston is really one large cattle farm with sheds that cover several acres. The farmer's house is modern, made of stone with round-headed windows and a pantile roof. Opposite, in the middle of this 20th Century jungle, is a splendid old black and white Manor House. It is slowly giving up the unequal struggle, neglected and uncared for, the lathes showing like bleached bones now that the plaster has cracked and fallen away. It has 3 gables plus a projecting gabled porch; the right hand gable is a later addition but the house was, in fact, originally larger on that side. The farm lies close by the Byne Brook in the lush valley of Apedale. Wenlock Edge looms just across the stream. The name Alcaston means '*Ealhmund*'s homestead'. In 1066 Edric held the manor. By 1086 it had passed to Helgot and there were 5 villagers, 2 slaves and land for 4 ploughs. It was later owned by the Templars and then by the Fitzalan family, one of the Marcher Lord dynasties.

ALDON *1¼m. W of Onibury, which is 3½m. SSE of Craven Arms on the A49.*
It lies on a hill high above the River Onny, a delightful and unspoiled hamlet of stone-built cottages and farms, ivy covered hedges and cocks a-cock-a-doodle-doing. There are good views of both Titterstone Clee and Brown Clee Hill though from here they do not look at all imposing. Aldon Court lies down the hill a-way, a 5-bay house with a hipped roof, rendered walls and peculiar fenestration – 2 broad bays to the left, and 3 narrow bays to the right. The peace of the settlement is somewhat disturbed by the rumble of large lorries as they grind up the hill to a waste disposal site. The name Aldon is from the Old English *aewell-dun* which means 'river-source hill'. In the woods ¼m. S is the source of a tributary stream of the River Onny. Aldon was a most substantial village in late Anglo-Saxon and early Norman times. In 1066 Siward held the manor but by 1086 most of it had passed to Roger de Lacy. Altogether there were 62 men (one of whom was a bee keeper) registered as living on the manor, plus their families of course. This is an exceptionally high population for a Domesday Book settlement.

ALLSCOTT *2½m. NW of Wellington.*
The early medieval name was Aldescote and the settlement was part of the extensive manor of Wrockwardine. (Aldescote is from the Old English *Aldred*'s cot; meaning *Aldred*'s cottage.) Today Allscott is a pleasant, residential village, a mixture of new bungalows and old cottages. The red brick Manor House is a modest Georgian building with stone dressings, 3 bays, and 2 storeys with a hipped roof. To the rear is a plain, 3-bay extension. The Wrekin District Council want taking to task for letting a good brick house stand empty and boarded-up whilst they use its gardens as their Central Plant Nursery. There is a black and white cottage, a handsome Hall Farm and an old mill, now a house, beside the River Tern. A bridleway to Isombridge passes by the mill and crosses the river on a wooden footbridge. The Mill House has been spoiled by the insertion of aluminium framed windows. On the main road is the Plough Inn and from here the view is dominated by the huge British Sugar factory. Sugar beet is processed in the tall, grey, corrugated iron sheds. Chimneys emit clouds of sickly smelling steam and the great concrete silos are a real eyesore. It is not until one gets close that the size of the place becomes apparant. It is surrounded by security fencing, pipes, reservoirs and scruffy, pre-fabricated buildings. Large lorries race around the narrow country lanes and are a real danger. There is nothing wrong with processing plants being sited in the countryside but things should be in proportion. The Allscott factory is simply far too big. Just up the road is Allscott Inn and opposite the pub are the grey sheds of Staffordshire Farmers, a farmers co-operative.

ALVELEY. *6m. SSE of Bridgnorth.*
It lies in the Severn Valley, an ancient, linear village on the busy A442 equidistant from Bridgnorth and Kidderminster. In the old centre are several listed buildings. The Three Horse Shoes Inn claims to be the oldest hostelry in the county and Oliver Cromwell is said to have stayed here. The Bell Inn contains in its fabric some 20 carved stones variously attributed to the Angles and the Vikings. The Malt House has a monastic origin. Elm Cot-

Atcham, the old bridge over the River Severn.

tage, built in 1672, is constructed of the local red standstone. This stone has properties suitable for its use as grindstones. The old quarry is now a Nature Conservancy Council site of special scientific interest. Alveley Colliery closed in 1969 and the spoil heaps have been landscaped to provide a Country Park. The foot bridge that spans the River Severn was constructed by the colliery to carry 2 narrow gauge coal-tub tracks. The church of St. Mary has a Norman tower, 15th Century clerestory windows, Kempe glass, and a good lychgate, but was over restored by Arthur Blomfield in 1878-9. Inside there is some fine embroidery of about 1500. At the timber-framed Hall Close Farm there have been several sightings of ghostly figures including one of a weeping woman who carries a dagger and a rope in her hands and cries "come with me, come with me, my husband's being hanged". The name Alveley is from the Anglo-Saxon meaning 'the field of the woman *Aelfgyp*'. One mile NNW at a country crossroads is a small cross decorated with Maltese crosses in a solid circular head. This Butter Cross probably marks the site of an ancient open air market. Pool Hall, about ½m. SE of Alveley, off the A442, is a moated house of red brick with a two-storeyed porch. The now demolished Hay House lay 1m. NW and, as its name pronounces, was probably built within the boundaries of a Saxon 'haye', or small, fenced hunting park. In parkland NW of Alveley is Coton Hall, an early 19th Century Italianate stuccoed house with a tower. In the parish of Alveley scrapers made from local pebble flint have been found dating from the Mesolithic (Middle Stone Age), 4000 B.C.

APLEY FORGE AND PARK *See Stockton.*

AQUEDUCT *It lies S of Little Dawley and W of Stirchley (Brookside).*
The aqueduct of the place name still stands. It originally carried the eastern spur of the Shopshire Union Canal over the old Worcester to Wellington road. This length of the canal was opened in 1792 and much of the coal that fired the furnaces of Coalbrookdale was transported along it in tub boats. Later, when the canal was abandoned, the railway was laid in the dry bed. However, the railway had to run straight and isolated bends in the old canal still survive; some are water-filled and used as fishing ponds. The railway track is now a pathway. The aqueduct is single-arched and constructed of grey limestone which in places has weathered badly. On the underside of the arch there are masons marks-triangles and arrows. The road leads downhill through a cutting and comes to a dead end, blocked by the Queensway. It is a delightful little spot embowered in trees. The Silkin Way footpath passes by at the end of the road. Note. The aqueduct is located almost in the centre of the square 8L on the commonly available Telford Large Scale Map, 1:14,000. As to the name of the road there is confusion. The map marks it Aqueduct Road but on the ground the sign says Southall Road. If asking directions locally ask for Chapel Lane; this is a turning *off* the road you wan t. Don't ask for the aqueduct itself; no-one locally seems to know of its existence and you will be directed to all manner of bridges that 'might be it' and become as frustrated as we did. As to the settlement today old houses are noticeable by their absence. Only Tunnel Cottages in Aqueduct Lane have any years behind them. They stood besidc thc Stirchley Tunnel on the Shropshire Union Canal. When the canal bed was used as a railway track by the Coalport branch of the LNWR the tunnel was opened up and became a deep cutting. The railway closed in 1959, and the track is now part of the Silkin Way. There are several groups of modern estate houses to the W of Castlefields Way and in Majestic Way there is a central green with some unfriendly brick blocks with 'half roofs', 'towers' and small windows. These shelter beneath the tree-clad slopes of a spoil mound. The block of flats fronted by Nos. 69 and 79 is a disgrace; it is difficult to believe that someone actually designed this miserable affair. The more recent, private, developments are of traditionally constructed houses in traditional styles. There is a school and a Community Centre but there is also a sense of isolation at Aqueduct. To the E are the open green areas and woods that adjoin Castle Pools and Wilde Waters Pool, that in turn adjoin the more recent developments of Little Dawley. (*See Dawley.*)

ARLESTON *¾m. SE of Wellington.*
Early versions of the name show it to be derived from the Old English *Eardwulf*'s-tun. To the E of Dawley Road there are large new estates that are most unappealing – Flat-roofed blocks of apartments, gardens with broken fences, and streets litered with rusty old cars parked on once green verges. A row of shops faces the Dawley Road to the W of which are more new houses, 2 schools (3 if you count the Old Hall School) and a recreation ground. However, Arleston has one well hidden delight. Take Arleston Lane and follow it S through the estates to open country where sheep and horses graze beneath electricity pylons and suffer the roar of Queensway. Here is Arleston Manor. It is a splendid timber-framed house in traditional black and white livery. Some parts are only facing boards and some parts are painted brick but nevertheless it is a handsome house on a modest scale most suitable for modern living. It has 2 gables, one dated at 1614, the other at 1640. The framing has narrowly spaced studs on the ground floor, lozenges inside lozenges on the first floor, and quatre foils in the gables. Inside there is some good plasterwork. This was long the home of the Griffin family who departed in 1988; it was sold by a son who had a large collection of vintage cars. There are some low, white-washed out buildings, trees in the garden, and a handsome carved stone freize that functions as a wall and features cherubs and swags. The road continues on past the Manor, over Queensway, and downhill to a dip wherein lies Arleston Inn and a few stuccoes cottages. Half way up the hill stop and look to your left. Here is the great hole of the Newdale opencast coal pit. The huge machines that dig and scrape literally make the ground shake and the din is perpetual. Behind the enormous spoil mound is Lawley Common and at the road junction sits the beleaguered Lawley County School.

ARSCOTT *1½m. S of Hanwood which is 3½m. SW of Shrewsbury.*
It lies in undulating country with Pontesford Hill looming large to the SW. At what might be called the centre of the hamlet are several large, red brick farms. On the lane to Hanwood are nicely detached cottages and houses behind hedges in a wooded setting. Amongst them is the Lea Cross Geophysical Co. Ltd. ensconced in a small, white, ivy-clad building. It is no bad thing for high technology companies like this to be sited in the country. They are the cottage industries of today bringing jobs, money and life to rural areas without damaging the environment.

ASH (Magna) *1½m. NE of Prees Heath which is 2½m. SSE of Wem on the A41.*
Ash is a substantial village on high ground with an unspoiled centre, some black and white cottages, an old brick-built smithy and at least 2 old persons' homes. Ash Hall is a delightful, text book 18th Century Queen Anne style house situated on the highest point in the village. It has 5 bays, 2 storeys, a hipped roof, irregular quoins and giant pilasters with finely carved capitals. To the left of the house is a substantial red brick farm. The Church of Christ stands away from the settlement. It is an engaging red brick structure of 1836 by George Jenkin with a chancel of 1901. The tower emerges from the nave and has 2 off-set decorative buttresses. At the entrance to Ash Grange, a late Victorian mansion now divided into flats, a phantom monk in flowing, hooded habit has been seen on at least 3 occasions. The latest reported siting was in June 1972 when a young motorcyclist drove straight through the ghost which glided before him suspended about a foot above the ground. As the motorcyclist hit the apparition it disappeared in a rush of cold air. Half-a-mile SW of Ash is **Brown Moss**. This is a delightful place of large, shallow pools, peat bogs, marshes and some woodland that has been partially tamed and made accessible to the public. There is a tarmaced perimeter road, organised car parking places and lawn like areas of grass that lead to the water's edge. It is still in a sense a wild place with large areas left to nature, thought much of the peat has been drained and converted to acid heath. There are several species of rare plants, such as the orange foxtail, and a prolific wild bird life, which includes more than 30 breeding species. There were thousands of acres of such terrain in North Shropshire before the Duke of Sutherland and others instituted vast drainage schemes and brought them to agriculture. Brown Moss is well sign-posted and lies only ¾m. NNE

Asterley Windmill.

Ashford Manor, Ashford Carbonell.

The Portway, a prehistoric track on the Long Mynd.

of Prees Heath. (Follow the sign to Ash off the A49.)

ASHFORD BOWDLER *2m. S of Ludlow.*
The busy A49 by-passes the village to the W. A lane leads from the main road over a railway crossing to the church and some very old black and white thatched cottages. The church of St. Andrew has box pews, a charming shingled bellcote and Norman doorways and windows in the chancel. The chancel is built on foundations which overhang a precipitous drop down to the dark waters of the River Teme. Opposite the church is Church House, white-painted and 3-storeyed. Ashford Hall, a 7-bay 18th Century brick house with a 3-bay pediment and hipped roof stands amongst trees to the W, on the main road, in a small landscaped park. The suffix Bowdler in the name Ashford Bowdler is from Henry de Boulers, an early medieval lord of the manor. Ashford Bowdler and Ashford Carbonell, just over the river, were collectively called Esseford in Domesday Book.

ASHFORD CARBONELL *2m. S of Ludlow.*
Less than ¼m. N of Ashford Bowdler is the turning off the A49 to its much larger sister village, Ashford Carbonell. It is entered over a brick and stone bridge which crosses the River Teme. Just beyond the bridge, and visible through the trees to the right, is an old water mill, approachable by a right of way through the 5-bar gate. There is a large and quite dramatic horseshoe-shaped weir by the mill, which is in working order but not used commercially. In the Autumn salmon can be seen leaping the weir. Looking down on the village is the church of St. Mary, with its hammer-beam roof, vessica window (oval with pointed ends) and weather-boarded pyramid belfrey. It is Norman and Transitional and is surrounded by 5 even older yews, suggesting a pre-Christian site. Two trees are female and bear fruit. Most of the village houses line the main street, a pleasing mixture of the old and the new. Some old weather-boarded barns are now used as garages and workshops. The name Ashford is self-explanatory. The site of the ford, and the great Ash tree that stood by it, is believed to be a little downstream from the present bridge at Teme's Green along the lane opposite Church Lane. The suffix 'Carbonell' is derived from William Carbonell, Norman lord of the manor. (Note: There is a local tradition that the name 'Ashford' is derived from 'asses-ford', the asses being the pack horses used on the old trade route from the West Midlands to Wales.) The main street used to be a part of an old coaching road that ran from Ludlow to Little Hereford and Tenbury Wells. Ashford Manor lies outside the village on the road to Caynham. It is a charming Elizabethan yellow stone house that has been overlooked by previous authors probably because it is mistakenly called Ashgrove Manor on Ordnance Survey maps. The original house had been made Georgian with sash windows and stucco but in the first half of this century it was reduced in size and remodelled by the present owner. It has attractive gardens which are occasionally opened to the public.

ASTERLEY *1½m. N of Minsterley.*
The name means 'the eastern clearing in the forest'. In the 19th Century there were coalmines in the fields around the village. The spoil heaps are not large and are often wooded so that they are not immediately recognizable. There are several on the land of Church Farm and indeed the mound alongside the farm itself is a spoil heap. Opposite Church Farm is the now empty and crumbling red brick and ivy-clad church; it has no name. Also in the village is a Primitive Methodist Chapel of 1834. It is built of stone and has round-headed windows. Asterley is a sizeable village and there has been some modern development along the main street. The pub is called the Windmill Inn, after the windmill that lies just N of the settlement; the tower still stands and is being converted to a dwelling. Probably the best building in the village is the modern brick and half-timber house opposite the Methodist Chapel. Further along the road is Innisfree, black and white and dated at 1675. A tributary stream of the Rea Brook touches the SW edge of the settlement.

ASTERTON *2½m. NNE of Plowden which is 5m. NW of Craven Arms.*
The road from Plowden to Asterton and thence to Boiling Well on the Long Mynd and thence to either Ratlinghope or Church Stretton is considered by many to be the most dramatic and picturesque in Shropshire. For the first 2½m. the road runs along the lower slopes of the Long Mynd, but at a sufficient height to provide wide views over the lush lowlands of the River Onny to the W and the hills beyond. Against this distant background of grand scenery the lane passes a close landscape of woods, streams, cottages, pools and moorland in ever changing array. Although the map shows the road as a 'broken dotted' track it is, in fact, level and properly metalled. Asterton is a stone-built hamlet in the process of being heavily renovated. The restoration of old cottages is admirable, not so the conversion of barns into blocks of mews houses. There are still several derelict cottages, but not for long one suspects. Modernisation also brings the curse of overhead power cables and telephone wires. One, as yet untouched, building appears to be a 'longhouse', a cottage and attached barn-stable built as one under the same roof. There are 2 red brick houses that look a little out of place and a pub, the Crown Inn. The lanes hereabout have deep ditches alongside for the ground here can be wet. The flat land N of Asterton was once a wild, marshy moor. It is still called **Prolley Moor** though today the land is drained and laid to pasture. In olden times this was where crows in their thousands came to sleep, where witches' covens met and where children chanted "Dead horse, dead horse, Where? Where? Prolley Moor, Prolley Moor. We'll come, we'll come. There's nought but bones." At Wentnor Prolley Moor, 1¼m. N of Asterton, is Robury Ring. This is a prehistoric earthwork of which little is now discernible because a farm has been built in the centre of it. From the entrance drive to the farm there is a good view to the long, narrow fields on Wentnor Hill. The hedges seem to have fossilised medieval strips, complete with the shallow 'S' shape caused by the oxen teams turning. These can be seen even more clearly from the higher ground of Long Mynd. **Medlicott** lies at the head of the Criftin Brook valley 1m. N of Robury Ring. Here there is one pebble dashed cottage, 2 stone-built farms, several old caravans and a great litter of building materials and old machinery. Despite man's carelessness this is still a beautifully situated little place, of some age judging by the gnarled, old hedges. But to return to Asterton. From here the road becomes wickedly steep as it climbs the flank of the Long Mynd to the high moors of its summit. There are marvellous views from the road; the whole of South West Shropshire is laid out before one. Beyond the immense patchwork of fields lie the distant hills of Stiperstones, Linley and Corndon to the W and of Clun Forest to the S. Long-winged, silver gliders soar in the skies above and colourful hang-gliders twist and turn at lower altitudes. They are based at the Midland Gliding Club which was formed in the early 1930's by Charles Espin Hardwick. It is housed in a large Nissen hut-like building and today has over 200 members. The road now follows the route of the pre-historic track called The Portway through heather moors and past numerous tumulii. At Boiling Well one can either head NW to Ratlinghope or SE to Church Stretton. There are wonderful views in every direction and the Long Mynd itself is full of interest and atmosphere. The descent to Church Stretton is along the Burway, a narrow track with a heart-stopping fall into Cardingmill Valley. (*See Church Stretton.*)

ASTLEY *4m. NNE of Shrewsbury, just off the A53.*
The hamlet lies in flat country within earshot of the RAF Station at Shawbury. The church of St. Mary has some Norman fragments and a blocked Norman doorway but is mainly Victorian. Astley House is a stuccoed, 3-bay, Georgian villa with classical Greek overtones and is set around with trees. The name Astley means 'the east *leah*'. *Leah* can mean three things: a) a natural glade, a thinly wooded area in a dense forest where grass grows between the trees; b) a field cleared by man within a wooded area; and c) a wood, the forested area itself, for example Ashley and Oakley. *Leah* is commonly translated as meaning b), but this is often an unsubstantiated conclusion. On the main road, the A53, there are several commercial enterprises: Jubilee Villa, an old persons' home; a pub, the Dog in the Lane; Fairfield, a hotel for cats and dogs; Wendover, a cattery; and Astley Recovery, a 24 hour breakdown service. There is also a small estate of modern houses, some council prefabs and a hand-

Attingham Hall, seen from the bridge that carries the A5 over the River Tern.

ful of substantial Victorian dwellings.

ASTLEY ABBOTS *2m. NNW of Bridgnorth, just off the B4373.*
In Domesday Book the village is called Estleia, from the Old English *east-leah* 'meaning the eastern wood, or clearing in the wood'. The manor once belonged to Shrewsbury Abbey, hence Abbots. The church of St. Calixtus has a Norman nave, a chancel of 1633 and a tower and S porch of 1857. Inside are the Maiden's Garland and other relics of Hannah Phillips who drowned whilst crossing the River Severn on the eve of her wedding day in 1707. Bishop Percy, famous for his Reliques of English Poetry, was curate here from 1752 to 1754. A one-time resident of Astley Abbots was Frances Pitt, the respected writer, photographer and naturalist who did pioneer work in creating a public awareness of British wildlife and its preservation. Stanley Lane skirts the estate of Stanley Hall and from it are good views of the River Severn. Stanley Hall, ½m. E of Astley Abbots, is a good red brick mansion with a black and white gable-end and many features such as the arches below the gable and a side tower with a bell house. It incorporates an earlier sandstone house. The grounds are well-wooded and well-tended. Half-a-mile NW of Astley Abbots is Great Binnal Hall, early 17th Century and better within than without. Half-a-mile W of Astley Abbots, on the B4373, is a fine Elizabethan timber-framed house called Dunvall. It has mullioned and transomed windows, star-shaped brick chimneys and projecting gable wings.

ASTON (near Wem) *½m. E of Wem.*
Aston means 'east town'. The River Roden flows through the hamlet and within yards of the timber-framed Aston Hall (1565). It is unusual in having concave lozenges in the first floor framing and post and pan in the gable. Near the crossroads is Aston House, also timber-framed with 2 gables, a central porch, upright studding and brick chimneys. The Thatches is a handsome, thatched, timber-framed cottage. In such company the row of brick council houses is, to be kind, incongruous.

ASTON BOTTERELL *8m. SW of Bridgnorth and 2m. S of Cleobury North.*
The common Anglo-Saxon settlement name of Aston caused confusion amongst the conquering Normans. The problem was solved by adding the names of the new lords to their manors as a qualification – hence Aston Botterell, Aston Munslow, Aston Eyre. (The same procedure was followed with other common names such as Hope and Albright.) Aston Botterell lies in rolling country on the lower western slopes of Brown Clee Hill. The stone, brick and stucco Manor Farm was the former home of the Botterell family. Botterell is a French name probably from *Boterel* meaning 'a toad'. The right half of the house incorporates a 13th Century Great Hall, and its original exterior doorway is preserved inside. In 1576 a ceiling was inserted and an upper floor created within the old 'earth to roof' hall. Today it looks careworn and neglected and is hemmed around by a farm. For the rest, there is little more than a handful of brick and stone cottages. The church of St. Michael stands on a raised circular mound. The chancel has a Norman N window; a handsome 13th Century Priest's Door; a mid-13th Century S aisle; a porch dated at 1639; and a tower rebuilt in 1884. Inside is a late Norman font and a Jacobean communion rail. Less than ¼m. NE is the track of a dismantled railway.

ASTON EYRE *6m. WNW of Bridgnorth.*
An unremarkable main road village on the B4368. However, the Norman church has what is generally agreed to be the best piece of Anglo-Saxon stone-carving in the county. The tympanum, the semi-circular area between the lintel and the arch over the door, has a representation of Christ's entry into Jerusalem with Jesus on a donkey. The artists were of the famous Hereford School. The church has no dedication and was built as a chapelry to Morville in 1132. Near the church is Hall Farm. The farmhouse appears to incorporate a 13th Century gatehouse and hall. In 1066 the manor of Aston (the eastern settlement) was held by Saxi. In 1086 it was held by Alchere. A later Norman lord of the manor was Fitz-Ayer, hence Aston Eyre. Like so many Shropshire villages Aston Eyre is 'shrunken', that is, its population has been substantially reduced. In 1841 some 130 people lived there; today the inhabitants number about 50.

ASTON MUNSLOW *16m. WSW of Bridgnorth and 4m. ENE of Craven Arms.*
Aston Munslow lies on the B4368 in Corvedale, 1m. SW of its namesake, Munslow. The black and white Swan Inn and a few cottages stand by the main road at the junction with a lane that leads to the village. On the hill is the White House. This is a 14th Century Hall of cruck construction, to which has been added a 16th Century crosswing and in the 18th Century further additions which included a hanging staircase. It has a small country-life museum and is open to the public. Aston Hall is an early 17th Century 'E' shaped house with gables and star-shaped chimneys.

ASTON ON CLUN *2m. WSW of Craven Arms.*
The black poplar is a tree native to Britain but is nevertheless very rare. The most famous in the county is the Arbor Tree of Aston on Clun. It is the centre of the only pagan tree-dressing ceremony to survive in Britain. The tree stands in the middle of the village on raised ground by a bridge over the river Clun. Every year, on the 29th May (Oak Apple Day), it is emblazoned with flags attached to larch poles which are nailed to its trunk. A pageant leads a mock bride and groom to the tree where there is dancing and jollity. The custom probably derives from the Celtic ritual of placing women's clothing on certain special trees to ensure agricultural fertility. This little stone hamlet and its ancient custom are internationally known and the festival is well attended. Aston Hall has 5 bays, a Greek Doric porch and was built in the 1820's. There are 2 round houses and a Gothic pub, The Kangaroo, reputedly the only inn with this name in England. The Forge Garage was once the Blacksmith's shop and indeed is still owned and operated by the old blacksmith. Very recently the main street and earthern house platforms of a deserted village have been found in a field to the rear of the Village Hall. At Little Brampton crossroads, on the edge of the settlement, is an ancient stone signpost with iron arms. The much respected revivalist traditional musicians John Kirkpatrick and Sue Harris live at Aston on Clun.

ATCHAM *3½m. SE of Shrewsbury on the A5.*
Atcham is an abbreviation of Attingham, which has been interpreted as meaning 'the home of *Eata*'s people'. The 13th-15th Century church of St. Eata contains Roman stones from nearby Uriconium. Some have dovetail holes to take a lewis, a Roman lifting machine. The church was harshly treated by Victorian restorers but has good Flemish and German woodwork and a font of 1675. This is the only church in Britain dedicated to St. Eata who was a disciple of St. Aidan. The Rectory is Gothic red brick. The first bridge at Atcham was built by the Abbot of Lilleshall in 1222. The Romans crossed the river lower down at Brompton, where traces of a wooden bridge have been found linking the island opposite the modern Wroxeter. Today the River Severn is crossed by 2 bridges. The modern bridge (1929) carries the A5 but the much finer and older structure (1769-71) by John Gwynne is now for pedestrians only. Gwynne was a friend of Dr. Johnson and a founder member of the Royal Academy. He built his bridge at the site of the original ford. The river was fordable here because it is broad and shallow but this did mean that the bridge had to be long, some 350 feet between abutments. The Mermaid and Mytton hotel was built as a dower house to Attingham Hall. Part of the Hall is open to the public. It is best viewed from the beautiful bridge of 1774 by Robert Mylne which carries the A5 over the Tern, about ¾m. E of Atcham, but this is a dangerous spot with thunderous traffic proving more than a mere distraction. The old village of Atcham was largely destroyed by the owners of the Hall when the grounds were extended and landscaped in the 18th Century. The modern village is a mere remnant. Such callous behaviour by landed gentry was not uncommon. Today the village school is closed and the Post Office has restricted opening hours though the garage and the general store flourish, as much on passing trade as local custom though. The red brick Old Malt House was given to the village by Lord Berwick in 1926 and is very much a local social centre. The ghost of Mad Jack Mytton (see Whittington) is reputed to haunt the hotel every year on the anniver-

Aston on Clun, the Arbor Tree.

sary of his birthday, 30th September. After he had died his body had lain here for a night before burial. Attingham Hall was constructed of Grinshill stone to a design by George Steuart and was built for Noel Hill, first Lord Berwick, in 1783-5 on the site of Tern Hall, most of which was incorporated into the newer building. With its Ionic columns, painted boudoir and Nash's iron and glass picture gallery Attingham is probably the grandest country house in the county. Today it is owned by The National Trust and used as an Adult Education College. In the grounds of the present Attingham Park was located a large ironworks, called Tern Forge, which flourished between 1710 and 1757. Situated on the banks of the Tern it had a mill for rolling brass plates and iron hoops, a mill for slitting iron bars into rods for nailmaking, a wire mill and furnace for converting iron into steel, workshops, and housing for 40 men and their families. All that remains are a few lumps of slag on the bed of the Tern, though some of the ponds may have been used by Repton as part of his landscaping when the Hall was built some 10 years after the Ironworks closed. On the N side of the river between Attingham Hall and Shrewsbury, but hidden from view by a wood, is **Longner** Hall built by John Nash in 1803 to replace a Tudor mansion. The grounds of his pleasing, irregular, Tudor style house were landscaped by Humphrey Repton. Edward Burton (d. 1558), an ardent Protestant, who reputedly died for joy of hearing of the accession of Elizabeth I, is buried in the grounds. He is also portrayed in stained glass in the splendid staircase hall of the house.

BADGER *6m. NE of Bridgnorth.*
Badger is best approached from the B4176. This unites the A442 and the A454, both of which radiate from Bridgnorth. It is a charming and unspoilt village, with cottages of half-timber, red brick and stone set amongst a tangle of wooded lanes. Badger Hall was demolished in the middle of this century but the descending chain of ponds, which were a feature of its gardens, are now incorporated into the village landscape. A Classical temple also remains and is approached along a drive that crosses the road to Albrighton. There used to be a ghostly Lady in Grey who haunted first the Hall, and then, after it had been demolished, the surrounding grounds. She effused a delicate perfume and did no harm to anyone. When the land around the Hall was bulldozed level an ancient jewel box was unearthed which, amongst other things, contained an engagement ring. From that day the spirit has been still. The red sandstone Gothic church of St. Giles (1814) stands alongside one of the pools. Inside there are fragments of old glass in the E window and many monuments by Flaxman, Chantrey and Gibson. Badger was well known to the novelist P. G. Wodehouse (1881-1975). Badger Dingle, a walk in the grounds of the old Hall, appears in his work as Badgwick Dingle. Isaac Hawkins Browne (1705-1760) a poet, M.P., and a leading wit of his age, lived at Badger Hall. The name Badger is derived from the Old English elements *Baegi*, a personal name, and *ofer* which can mean either 'river bank' or 'border' (e.g. of cultivation).

BAGBURY BRIDGE *3½m. N of Bishop's Castle.*
Bagbury is a farmhouse just over the border in Montgomeryshire near Hyssington. Bagbury bridge (SO.321.934) crosses the stream that here marks the border between England and Wales. A former squire of Bagbury was a very wicked man. When he died he turned into a bull, the Roaring Bull of Bagbury. He terrified people on the roads and even entered their houses. Twelve parsons were called to lay the spirit. He was enticed into Hyssington church and using 12 ritual candles the preachers preached and gradually reduced the bull to the size of a sheep. But suddenly he rushed and the candles blew out. He grew even larger than before and cracked the walls of the church. So they started again and finally managed to reduce the bull to a size so small that they could put him in a snuff box. From within his tiny prison the spirit begged to be buried beneath the bridge at Bagbury 'so that every mare that passes over me should lose her foal and every woman her child'. Of course the priests refused and sent the snuff box to Arabia to be buried in the Red Sea. Nevertheless local people went in fear and trepidation whenever they crossed the bridge and the ghost of a big, dark man was seen in the farmhouse many years after the spirit of the wicked squire had been laid.

BAGLEY *4½m. S of Ellesmere and 3½m. N of Ruyton XI Towns.*
The name Bagley is quite common and yet it is not possible to explain its meaning with any certainty. The first element may be from an Old English animal name, probably a wild animal such as the fox. The Bagley described here stands on the minor road that skirts the eastern flank of Baggy Moor (sic). Most of the buildings are built of red brick and the boundary walls of sandstone. There are 2 Georgian houses with hipped roofs, a Primitive Methodist Connexion chapel of 1865 (now a dwelling) and the blue-painted Fox Inn. Half-a-mile NW is **Bagley Marsh** – a handful of stuccoed cottages linked by 2 dead-end lanes both of which are in part 'enclosure roads', that is about 40 feet wide as prescribed by the Act of Parliament but with tarmac only laid on the central 10 feet. The wide verges are hedged and ditched. **Baggy Moor** is not a moor in the accepted sense. It is a one-mile wide, 5-mile long, section of the flat-bottomed valley of the River Perry which is flanked by low, undulating hills. The valley was once a marsh but in 1861 an Enclosure Act was passed and the Moor was drained. The River Perry was straightened and deepened, all fords, weirs and dams were cleared away, and strines dug to carry the water off the land. In all some 1,300 acres were reclaimed. This was the last marsh in Shropshire to be drained. As proof of how treacherous the ground used to be look at the Ordnance Survey Map. For nearly 5 miles not one road or track crosses the river. The land is now both laid to pasture and used for arable crops, though pools are quick to form and linger long after heavy rain. Electricity pylons bestride the length of the moor, not a pretty sight. Two miles SSW of Bagley, near **Wykey,** there is a bridge over the River Perry. The water level is some 20 feet below the surface of the ground. By the bridge is Weir Cottage, indicating that here was once one of the old weirs. The hamlet of Wykey consists of little more than a couple of Georgian farm houses and a few cottages set on higher ground to the W of the river.

BARROW *2m. E of Much Wenlock on the B4376.*
The settlement lies on the edge of Willey Park, a remnant of the Forest of Shirlett. The village of Barrow is no more than a church, a few cottages, a large house, and a farm. It is, however, an ancient place. The Church of St. Giles is raised on a mound, which is probably of Celtic origin. The tower and nave are Norman. The chancel is Saxon, the S wall possibly as early as the 8th Century, and the N wall has a Saxon window. The W doorway to the tower has a tympanum decorated with a rare design of saltire crosses and lozenges. In the churchyard is the grave of Tom Moody (d. 1796) the 'whipper-in' of the local hunt and a horseman of great renown. He worked for Squire Forester at the Willey Hall stables and his ghost, accompanied by a hound, has been seen many times in the countryside hereabouts. Tom Moody has been the subject of numerous songs, poems and paintings and is one of Shropshire's best known characters. Also buried here is Thomas Turner (d. 1809) who was one half of the partnership that established the renowned Caughley Pottery. (*See Stockton and Ironbridge.*) The school and the almshouses were built in 1612, a foundation of John Slaney, and originally stood near Willey Old Hall. They were dismantled and rebuilt in their present position by the first Lord Forester. (*See Willey.*) The school still flourishes and takes pupils from a wide area. The name Barrow is almost certainly not a reference to a burial mound but is most likely to be a conception of the Old English *bearu* meaning either 'a wood' or 'a grove in a wood'. On the left-hand side of the road from Barrow to Much Wenlock is a large brick farmhouse. To right and front of this house is a much older detached stone building, possibly of Tudor origin, with signs of a moat and other earthworks between it and the road.

BASCHURCH *8m. NW of Shrewsbury on the B5067.*
Many a traveller must have been struck by the spacious area next to the level crossing on the main road; something seems to be missing. It is, in fact, the station and the market place that are no longer with us. In 1847 the railway came and with it a cattle auction and a Friday corn, butter and

A tributary of the River West Onny at SO.335.958.

poultry market. These were followed by the construction of the Boreatton Arms, a large inn and posting house with 17 horses, which still stands. Most of the rest of the main road development is prosaic – a garage, a fire station and houses of various ages – but the old centre still retains some character. The stuccoed New Inn lies opposite the church and nearby is the Old Vicarage which is now a guest house. There are a few sandstone cottages and on the Shrewsbury road is Nightingale House which has contrived to retain its ornate iron railings when most of the like were commandeered during the last World War, melted down, and turned into guns and tanks. The red sandstone church stands on a mound and has a 13th Century tower, which survived the wrath of the Welsh who burned the place down in 1404, and an E window by Kempe. Thomas Telford did some restorative work here in 1790 and his bill used to be on display in the church. North-East of the church is the Laurels, a fine timber-framed house with gables. However, the structure of most interest here is The Berth, a pre-Saxon earth fort, possibly the site of Pengwern, the Hall of the Welsh Prince Cynddylan. Cynddylan was killed in the 7th Century and buried in the *eglwysseu Gassa*, 'the churches of Bassa', now Baschurch. It is thought that when Uriconium declined The Berth became the administrative centre for the middle Welsh border country, prior to the rise of Shrewsbury in the 9th Century. The Berth is situated about 1m. N of the town. It consists of a 3-acre enclosure with a 50 foot high mound, surrounded by a rampart. Causeways connect this enclosure to another enclosure and to a deep pond. It is not easy to see from the road and needs to be looked for. It lies to the NW of the Baschurch-Marton Road, beyond Mere House. Finds include a bronze water clock and a 6th Century bronze cauldron. In undulating country 1½m. NW of Baschurch is Boreatton Park. The somewhat austere mansion is now a Peter Lawrence Activity Centre for young children but it is famous because it was here that Dame Agnes Hunt founded her hospital before the establishment of her world famous Park Hall Orthopaedic Hospital near Oswestry. Today happy bands of youngsters troop about singing; there are shire horses and a riding school.

BATCHCOTT *2½m. SSW of Ludlow.*
The name probably means 'the cottage in the valley', though *cott* in a place name can also mean an 'animal shelter' and occasionally 'a manor'. Batchcott is approached off the B4361 down a narrow lane that leads to a wooded valley. It is a tiny hamlet of stone-built cottages and weather-bounded barns with a stream and stony tracks. The road ends here. The ancient custom of dressing the well with flowers on the 14th May was practiced at Batchcott as late as the early years of this century. The church of All Saints lies ¼m. S on the main road to Richards Castle. It is a large, yellow stone affair with an almost detached tower adjoining the S aisle; stately perhaps but unfriendly too and designed to be seen only from one direction. It is of 1891 to a design by Norman Shaw and was paid for by Mrs. Johnston-Foster. Just to the N of the church is the Georgian Vicarage, brick with a Venetian window. Moor Park lies in wooded grounds ½m. to the NE. The unprepossessing house is nicely situated but the grounds are untended, except for the cricket pitch. The main building has 11 bays, shaped gables and giant angle pilasters, but has attracted various unsympathetic modern satellites. It is now a Roman Catholic Preparatory School. The house used to belong to the Salwey family, local landowners of some substance, and in 1861 was almost bought by the Prince of Wales, later to become Edward VII, as a royal residence. He bought Sandringham instead. In 1873 Major Johnston-Foster acquired the estate. When he died his wife built the new church already mentioned. (*See Overton.*)

BATTLEFIELD *3m. NNE of Shrewsbury.*
The NE outskirts of Shrewsbury on the A49 are named after the Battle of Shrewsbury, 1406, when Henry IV won a victory over the rebellious Marcher Lords and in which Harry Hotspur was slain. The King ordered a church to be built on the site with 8 chaplains – secular canons – to pray for the souls of the dead. The building fell into disrepair during the 18th Century and was heavily restored in 1861-2 by S. Pountney Smith. The inside of the church is good. The college that housed the clergy lay S of the church, surrounded by a moat. The red brick vicarage is of about 1862. The church, St. Mary Magdalene, lies isolated in fields N of the built-up area. Heading out of Shrewsbury on the A49 go past the junction with the A53 (on the right and signposted Shawbury and Market Drayton). About ¼m. further on is a lane on the left. This is not signposted. It passes under a railway embankment bridge and leads to the church, where it comes to a dead end. The church at Battlefield replaced that of Albright Hussey, 1m. NW, of which there is now no trace left.

BAYSTON HILL *2½m. SSW of Shrewsbury on the A49.*
Bayston Hill, 'the hill on which stands *Baega*'s Stone' is now a southern suburb of Shrewsbury. The church of 1843 is near the village green but Bayston Hill is now a place of modern housing estates. To the E of the main road is **Bomere Pool** which has 2 ancient legends connected with it. One should remember that although today the pool is a centre for power boating and water ski-ing, it was, until a few years ago, quite isolated, uninhabited and wild with overgrown banks hidden amongst dark trees. The first legend tells of the Great Fish of Bomere. A giant fish with Wild Edric's sword in a sheath attached to a belt around his middle was caught by fishermen using an iron net. But the fish smashed his way free with the sword declaring he would only give it up when the rightful heir to Condover Hall was restored. (Local people believed that at one time the Hall had passed, by deceipt, to a usurper.) The second legend relates to the Bomere Deluge. In Roman times the site of Bomere Pool was occupied by a good-sized city. However, the people returned to worshipping pagan gods. A Roman soldier was sent to re-convert them, but only the daughter of the Governor responded. The Hill of Caer Caradoc exploded with fire, a great flood swamped the town "and the sun in the heavens danced for joy". The Roman soldier was saved but went on the waters of the lake that had formed over the city, searching for the Governor's daughter with whom he had fallen in love. His boat overturned and he drowned. Now, whenever Easter falls on the same day as it did that day, the ghost of the Roman soldier returns searching for his sweetheart. The Pool is also the Sarn Mere of the novel 'Precious Bane' by Mary Webb, who lived at **Lyth Hill** for a time. Lythwood Hall, on the SW outskirst of Bayston Hill was built by George Steuart (who also designed Attingham Hall) for a gentleman called Blakeway who had won a substantial sum in a lottery. It is brick-built but has been much altered and a fine portico has been removed. The last squires, the Halton Harraps, were of charitable disposition. They also buried their dogs and horses in the cellars, now filled in. Today, the house now lurks, ashamed, behind a screen of trees down a short lane at the back of a housing estate. It has been put into multiple occupation and is somewhat scruffy. Lyth Hill is a windswept edge with superb views over Southern Shropshire. Virtually all the hill ranges, with the exception of the Clun Hills, are laid out before one. Here we met a County Ranger mending fences and a young guitarist communing with nature. Mary Webb lived at Spring Cottage, a gabled bungalow with central chimney, from 1917 until her death in 1927. The Old Windmill of 1835 was built to grind corn and also to prepare the flax used at the nearby Rope Walk where ropes were made for local lead mines and collieries. Great Lyth Manor was built in 1638, abandoned in 1948 and only recently restored. It has a reputation for being haunted.

BEARSTONE *See Norton-in-Hales.*

BECKBURY *7m. NNE of Bridgnorth.*
A village just N of Badger with a fat, little pink sandstone church dedicated to St. Milburga. This has a chancel of about 1300, a Georgian nave and tower, a S aisle of 1856 and a N aisle of 1879. The graveyard has many good trees and the cottages opposite make a pleasant group. However, the council has spoiled the village with its bland, red brick council houses. On the hill above the church are several timber-framed dwellings. The Hall is 17th Century and there is a good 19th Century school. The whinnying and clattering hooves of an invisible phantom horse have been heard many times at Lower Hall. It has been suggested that it is the spirit of one of Squire Stubbs' horses,

Battlefield, the church of St. Mary Magdalene, which stands on the mass grave of those who died in the Battle of Shrewsbury, seen over the medieval moat.

possibly ridden by the ghost of the squire himself. Stubbs was a colourful character who, in his later years, lived for the hunt, following the hounds of the South Shropshire pack 6 days a week. He was born at the Hall in 1671 and rode up to 20 miles a day at a steady trot that came to be known as "Stubbs' pace". The name Beckbury is from the Old English *Beocca's burg. Burg*, or bury, means 'a fortified place', and usually refers to a pre Anglo-Saxon earthwork. In 1066 the manor was held by Azor. By 1086 it was held by Roger Hunter from Earl Roger of Montgomery. By the 13th Century it came under the Barony of Wem and then passed to the Prior of Wenlock Abbey.

BEDLAM (Titterstone Village) *6m. ENE of Ludlow.*
It lies half way up Titterstone Clee Hill in the valley of Benson's Brook. The name Bedlam was given to this Victorian mining settlement because there used to be a lunatic asylum here. (Bedlam is a corruption of St. Bartholomew's Hospital). The local people do not like the name because of its associations and want the hamlet called Titterstone Village. So, at the moment confusion reigns: the Ordnance Survey map calls it Titterstone and the road signs Bedlam. There are two rows of terraced houses, built as homes for the workers in the roadstone quarries; a War Memorial; a corrugated iron shack that used to be the Newspaper Reading Room and is now the Village Hall; and a Primitive Methodist Chapel, now a house. (*See Dhustone.*)

BEDSTONE *7m. SW of Craven Arms.*
A small border village with a hillside location in verdant country near the confluence of the River Clun and the River Teme. Manor Farm is an 'H' shaped, timber-framed house with a stone-faced S side of 1775. To the rear of this house are the remains of the Old Hall (circa 1350): 2 large crucks. The Norman church of St. Mary has original windows in the S nave and the N chancel walls and a N window by William Kempe. Bedstone Court is a beautifully situated black and white multi-gabled Victorian mansion of 1884. It was designed by Thomas Harris (author of 'Victorian Architecture') for Sir Henry Ripley, M.P. whose great-grandfather had founded a dye works in Bradford. Bedstone court is a 'calender-house'; it has 365 windows, 52 rooms, and 12 chimneys. The Ripleys left in 1903, though they still own the estate. Today, the house is a school. The village school was established in 1750 but closed in 1947 and is now a dwelling. There is a Post Office and a tuck shop. In the woods of Bedstone Hill, 1m. W of the village, are the earthwork remains of a prehistoric settlement called Castle Ditches.

THE BEECHES *1m. NW of Marton, which is on the B4386, 6m. NE of Montgomery.*
The old name was The Beach and is more correct in meaning than The Beeches. This tiny hamlet lies on the lower slopes of Long Mountain in a steep, wooded valley. Until 1837 the Beach Dingle Wake was held here on Trinity Sunday. An account of the proceedings was given by an old man to Sir Offley Wakeman and reported by Charlotte Burne in her Shropshire Folklore: " . . . some three or four hundred people were wont to assemble about two o'clock in the afternoon, and spend the rest of the day in drinking, dancing, prize-fighting, playing pitch and toss, and so forth. A little below the Beach there is, in the midst of the brook, a stone of several tons weight, called Whirl-stone, which was said to turn round every time it heard the cock crow, but on Wake Sunday it turned round whether the cock crew or not".

BEGUILDY *10m. WSW of Clun on the B4355.*
Just over ½m. NW of the settlement, hard on the border with Wales, is a handsome motte and bailey castle (SO.188.805). The earthworks stand on a ridge above a small stream that abounds it to the NW before turning S to join the River Teme. The motte stands 22ft. above the ditch and the summit is 34ft. in diameter. The bailey is 'D' shaped and extends about 95ft. to the S with a bank about 3ft. high to the SE and the SW. Mary Webb took the name for Wizard Beguildy in her most popular novel, Precious Bane.

BENTHALL *1m. S of Ironbridge.*
The name is from the Old English *beonet*, meaning 'bent grass'. It is a small, main road community which is now virtually a suburb of Broseley. Standing in landlordly isolation from the dwellings of the common folk is Benthall Hall. This is a splendid, stone-built house of about 1583, now in the care of the National Trust but still occupied by as a private house by the donor, Sir Phillip Benthall. The handsome exterior with its star-shaped chimneys, bays and gables contains an equally fine interior – elaborate plasterwork, a carved staircase, 17th Century furniture and a collection of Caughley China. In the porch are discs arranged, it is said, to represent the stigmata, the 5 wounds of Christ. This is possibly the explanation of 'the symbols at your door' in the old song Green Grow the Rushes O. The Benthalls remained Catholic after the Reformation and such symbolism was not uncommon amongst these recusants. The gardens of the Hall abound in crocuses. This is no accident. George Maw, of the famous decorative tile company in Jackfield, once lived here; he was a botanist of international repute and the author of a standard work on the crocus. Close by the Hall is the white-painted church of St. Bartholomew. This was built in 1667 to replace the previous church which had been destroyed by fire in the Civil War. It consists of a nave, chancel and timber-framed belfry with a pyramid roof. The chancel was made Gothic in 1884, and the red brick apse, doorway and arched upper window are of 1893. It has good hammer-beam roofs and is well lit. The pulpit and reader's desk are Jacobean. Today, Benthall looks peaceful enough but this is, in fact, an old industrial area. Coal was mined here as early as 1250 and in 1756 there were 13 coal boats on the River Severn that were attached to the village. By 1770 there was an iron smelting furnace N of the settlement in the valley the Benthall Brook, a tributary stream of the Severn. Originally the bellows of the furnace was water powered but later a steam engine was used. In 1781 the Iron Bridge was constructed and the S access road, still in use and called Bridge Road, came up the Benthall Brook valley, past the ironworks and potteries to join the Broseley-Wenlock turnpike road. By 1784 there was an engineering works capable of making steam engines and in the ironworks as a whole about 700 men were employed. The Harries family, lords of the manor of Benthall, were closely associated with the industry; their partners were the long established ironmasters Banks and Onions. In 1787 Lord Dundonald installed some of his 'new process' coke ovens but these were demolished in 1799 when the operators went bankrupt. The major part of the Benthall furnaces' output was pig iron used for casting. In the 19th Century much of this was exported to the Black Country. The furnace closed down in 1821 but the foundry lasted a few years longer. Supplies of clod coal, suitable for coking, were severley depleted but it seems that the ironmasters found it more profitable to simply export iron ore to the Black Country than to smelt here where communications were bad and markets small. One of the famous sites of the Ironbridge Gorge was the very large cast-iron Benthall Mill water-wheel (dismantled in the 1930's) which features in many photographs and paintings. The mill still stands, derelict but substantially intact, between the Benthall Brook and Bridge Road near the bottom of the hill adjacent to a red brick wall. It is constructed of stone and brick and measures 30ft. by 17ft. Trees and undergrowth are in the process of destroying the mill; surely it must be restored, there is so little that is old in the Gorge. On the other side of the road is a partly exposed tunnel – an adit shaft – one of the many mines with which the hill is riddled. Today coal is still mined at Benthall. Almost opposite the entrance drive to Benthall Hall is another track that leads to an opencast mine. As much grey fire clay as coal is obtained, and is sold to manufacturors all over the West Midlands. Whereas traditional shaft and tunnel coal mines had pit head building complexes that were in place for many years, even tens of years, opencast pits have often come and gone within 24 months and the offices are nothing more than portable sheds. We have found no reference to iron-working at Benthall Hall itself but to the left of the entrance drive there is a shallow pool close to which is a low mound of jet black earth. This is often a sign of iron working.

BENTHALL EDGE *1m. NNW of Benthall.*
The steep, wooded hillsides of Benthall Edge, which extend to about 100 acres, are most easily explored from Ironbridge. The car park adjacent to the Iron Bridge itself, on the S bank of the Severn

Bedstone, near Bucknell, a school cricket match.

Bettws-y-crwyn, a view to the south-west of the church.

Bletchley Manor.

opposite the town, is the logical place to use a base especially as an organized Nature Trail begins here and is marked with numbered posts. Benthall Edge lies W of the Iron Bridge; to the E the slopes are called Ladywood. In Anglo-Saxon times Benthall Edge was part of a great forest but as early as 1250 Phillip Benthall was giving the monks of Buildwas Abbey permission to build a road across the Edge along which they could transport limestone, timber and coal from their quarries. In the 18th and 19th Centuries there were several industries by the waters' edge: boat building, lime burning in the 5 kilns of Bower Yard, and brick making at the White Brickworks near the Iron Bridge. By the mid-19th Century the hills were almost bare, the trees having been taken for charcoal and construction and simply cleared in the course of mining. Most of the dense woods we see today have naturally regenerated since then. Above all, though, Benthall Edge was known for its limestone mines. Like Lincoln Hill on the other side of the river it is riddled with tunnels, shafts and caverns almost all of which are uncharted. The quarries at Pattins Rock were joined to the riverside lime kilns by 2 inclined planes. These kilns were in use until the 1930's. Other, smaller quarries were connected to Broseley by a tramway. The White Brickworks were also served by an inclined plane down which the clay (mined rather than quarried) was brought. Today these inclines are used as pathways. The white limestone, suitable for use in iron furnaces as a flux, was taken across the river to the Coalbrookdale wharfs by Severn trows. This trade was one reason for the construction of the Iron Bridge. In 1862 the Severn Valley Railway came down the gorge on the S, Benthall Edge, bank. At one point the track crosses a most elegant viaduct, with tall, 30ft arches. This is best seen from opposite bank near the Severn Warehouse by the Ironbridge Gorge Museum Centre. Here is a brief guide to the Nature Trail mentioned earlier. The numbers are post numbers. 1. Inclined Plane. The path follows this. It took clay from the adit mines (horizontal shafts driven into the side of the hill) to the White Brickworks. 2. Adit Mine. Now blocked; is about 30ft from post. 3. Ash Trees. Several here, covered in ivy. 4. Alders. They like damp conditions. On the right is a fenced off vertical mine shaft. Not capped. 5. Broseley Fault. The line of it follows the steep scarp slope; clay and coal to one side, sandstone to the other. On the right, a little way on, is a Cherry Tree, with typical striped bark. 6. Sandstone Plateau. Dry ground, better drained. Thin soils. Bilberrys. 7. Wild Service Tree. Near the post are several of these very rare trees which have leaves similar to Sycamores. Service trees are a sign of ancient woodland. At the top of the hill is an old tramway along which tubs of limestone were hauled by ponies. Cross the small stream by the stepping stones. 8. Bower Brook. It flows in an entrenched bed. There are dead elms, woodpeckers, and yellow archangel flowers (also a sign of old woodland). 9. Descend into steep valley. Two valleys meet. Cross the third stream. We are now on acidic sandstone; on the right the stream disappears down a swallow hole – birch scrub with hazel. The open-ness allows guelder rose, fragrant honeysuckle and wild clematis to flourish. 10. Limestone Plateau. Blue violets and wood anenome grow here. The meadow was cleared of trees early in this century. 11. From here there is a good view over Buildwas power station. A grey willow stands to the right of post; to the left is one end of a very long quarry; the whole hilltop has been cut away. The path leads to the right, through a small pit and past 2 small quarries each watched over by handsome beech trees. 12. Pattins Rock. This is the largest limestone quarry on Benthall Edge. It closed in the mid-19th Century but was worked on a small scale for many years after. The vantage point is very dangerous. The fencing is totally inadequate and no warnings are given despite the unexpected sheer drop. Keep an eye on children and dogs. 13. There is a magnificent pollarded beech on the right with a cluster of trunk-like branches. There is some spurge laurel and yew but beech trees dominate the woods here. Limestone Scarp. The path heads steeply downwards and can be very, very muddy. We passed people virtually stuck, covered in grey clay and almost in tears. The lime rich soil is liked by yew trees. Hazel is coppiced here. The tree is cut down and numerous small poles grow from the stump. The wood so obtained was burnt for charcoal and used as pit props. Poles could be 'cropped' at different ages and therefore thicknesses for different purposes. Usual cycle is 10-15 years. Today, coppicing on Benthall Edge is done to provide poles in making small furniture such as chairs, fences and hurdles. Elsewhere, though, it is done to produce a dense, leafy cover favoured by game birds. (Many small countryside woodlands owe their existence to the bloodlust of the local landlord.) 14. Hazel Coppice. Because it is light flowers and wild garlic grow here in the spring. 15. Oaks. After coppicing oak trees sometimes have a short twiggy growth in the lower trunk called epicormic shoots, caused by sunlight on young buds that would normally be shaded. From here there are 2 routes back to the car park: the track of the Severn Valley Railway and the woodland edge trail where deer have been seen. 16. Brick Arch. If you follow the woodland edge the path crosses the inclined plane that runs from Pattins Rock to the lime kilns by means of a brick arch. If you follow the Severn Valley Railway track you will pass over the viaduct. This has low brick walls capped with rusticated stone. It is not obvious that there is a 30ft. drop on the river side. Dogs have died here by leaping over the wall, thinking that the ground level was the same on the other side, and finding out too late that is not. 17. The Bower Brook emerges from the hillside in a culvert, a rusty-red iron pipe; the stream is often only a mere trickle. It passes beneath the railway embankment through a brick culvert and so to the River Severn. At the point where the stream emerges from the iron pipe the railway track crosses a path. If you leave the track and pass under the bridge, heading towards the river, you will find the limekilns in the bank to your left. The brick arches are in various states of disrepair and one of the 5 kilns (the end right) has completely collapsed. Follow the road to Ironbridge and you will pass a wall made of reject pipes from the White Brickworks. Follow the railway track and you will pass under the bridge that carries the inclined plane that you started this wall upon. Whilst walking in the woods there is much wildlife to look out for: bats, of several kinds – Noctules, Pipistrelles and Daubentons (which fly low over the river); grey squirrels; hedgehogs; badgers; fallow deer and muntjac deer; rabbits, moles, foxes, and weasels. These notes were made with guidance from a pamphlet published by Stirchley Grange in 1986.

BENT-LAWNT *½m. W of Hope which is on the A488, 6m. SW of Pontesbury.*
The name is a combination of Old English and Welsh. Bent is from *beonet*, meaning 'bent grass', and *lawnt* is Welsh for 'lawn'. The scattered hamlet stands on the side of a hill. There is a Post Office and general store, a village pump, a row of terraced houses, white-painted cottages, a stone chapel with round-headed windows, holly in the hedges, fields laid to pasture, and good views over Snailbeach and the Stiperstones hills. The distinctive clump of pines on Bromlow Callow lie ¼m. W.

BERRINGTON *3m. SE of Shrewsbury off the A58.*
A place of some character. The red sandstone church of All Saints has a cobbled entrance path; a Perpendicular tower; a 13th Century nave; a Norman font with primitive carvings; and an oak effigy of a late 13th Century cross-legged knight. On Easter Day a Love Feast used to be held at the church. This orgy was already an ancient tradition by 1639 when the Bishop of Lichfield complained about it. The last Love Feast was held in 1713. The timber-framed and gabled manor house is dated at 1658 and stands opposite the church. It has a pool and good, substantial timber-framed barns. The houses of the village are mainly red brick and the half-timbered buildings have brick in-fill. The village hall is a green, corrugated iron hut which stands opposite the Victorian School. In the meadows to the NW are two small meres, Berrington Pool and Top Pool. The name Berrington usually means 'by, or belonging to, a fortified place'. In 1066 the manor was held by Thored (from the Old Danish name Thorth); in 1086 it was held by Azor from Reginald the Sheriff and subsequently passed to Lestrange who held it from William FitzAlan.

BERWICK *2m. NW of Shrewsbury.*
Unsignposted but announced by a pair of stone lodges on the B5067 is the park and mansion built for the Powys family in 1731 by Francis Smith. There is no village, but in the tree scattered grounds are a range of Almshouses (1675) and a Chapel (1672)

Benthall Edge, in the woods near the summit.

with later tower (1731). Berwick House is of brick with stone dressings and though much of it was re-modelled in Victorian Italianate in 1878, it is still a handsome building with terraces leading down to the River Severn. The name Berwick is from *berewic* which is Old English for 'a farm growing corn' but which later came to mean 'an outlying area of an estate'. The present owner of Berwick House is R. E. Angell-James who is currently (1989) the High Sheriff of Shropshire. (*See also Crossgreen.*)

BESFORD *3½m. SE of Wem.*
A tiny hamlet in undulating, wooded country NW of Moreton Corbet. The cottages and farms are mainly red brick but the 3-bay, 3-storey Besford House Farm is black and white, dated at earlier than 1620. The River Roden flows by less than 400 yards to the E. The farming is both arable and pastoral and the landscapes here are most attractive. Besford probably means '*Betti*'s ford'.

BETTON *2m. NE of Market Drayton.*
The hamlet lies in flattish country and consists of little more than Betton House, 2 Halls and 2 farms, all of which lie along the minor road that runs between Market Drayton and Norton-in-Hales. Betton House hides behing a long hedge composed of pink flowering rhododendrons. The house is not large. It has 2½ storeys and the pedimented doorway is flanked by two curved bays, each with 3 sash windows. Behind are mews cottages and behind them is a lake. On the other side of the road, a little to the S, is Betton Old Hall. This is an attractive half-timbered house with upright strutting, a slightly jettied first floor and a gable. Betton Hall stands at the top of the hill hidden from view by a small wood of broadleaf trees. It has been sadly neglected and now lies virtually surrounded by endless rows of black plastic hay bags. The house is square, has 3 bays of 2½ storeys, an Ionic columned and pedimented doorway set to the left, and is clad in grey stucco. The top 'half' windows are imitation blanks. There is a terrace to the garden side and to the right is a tall, ivy-clad arch in rusticated stone that leads to the very substantial stable block and farm outbuildings. This is one of those small country houses that in abandonment takes on an air of sadness that gives it a romantic aura that it never had in its hey-day. There are several well constructed sandstone farm buildings close by and the boundary walls in the hamlet are sandstone also. Nearly opposite the Old Hall is a lane lined with hogweed, by the side of which stands a tiny grey-painted, corrugated iron chapel. It has been abandoned and looks most forlorn. Just S of Betton the road to Market Drayton crosses the Shropshire Union Canal and alongside the bridge is Victoria Wharf where H. Orwell and Son, Coal and Log Merchants, have their premises.

BETTON ABBOTS *3m. SSE of Shrewsbury.*
Betton Abbots is a farm that lies to the W of the Acton Burnell – Shrewsbury road. The lane passes a black and white cruck framed cottage, leads through the red brick farm and on to Betton Pool. Despite the signs this lane is a public right of way. The pool is fringed with rushes and trees and is used for water-skiing, a noisy sport because the skiers are towed by motor boats.

BETTON STRANGE *2m. SSE of Shrewsbury.*
Strange indeed, for the substantial Hall is not even mentioned in some reference books. It lies behind a high brick wall and mature specimen trees. The house has 9 bays, 2½ storeys with dormer windows and a 9-columned entrance porch. It is stuccoed and painted grey with white trim and though now converted to private flats the Hall and its grounds are very well tended. From the back of the house a path leads through a densely planted avenue of yews and other trees for a hundred yards or so to the tiny, red sandstone Gothic church of St. Margaret (1858). The church is totally surrounded by trees and cannot be seen from outside the wood. The atmosphere is positively medieval. The Hall, the path and the chapel make an ensemble that is one of Shropshire's unexpected little delights. Opposite the Hall there is a farm and a house and that is all at Betton Strange. The lane continues to Betton Alkmere where it ends at the large red brick farm. To the N of the farm there is a small mere and to the S is a fragment of a moat. The name Betton is probably from the Old English meaning '*Baega*'s homestead'. The suffix Strange is from the Old French *estrange*; Hamo le Strange was a medieval landowner here (circa 1160).

BETTWS-Y-CRWYN *6m. W of Clun.*
The name is Welsh and means 'the Chapel of Fleeces'. The 'village' consists of a church and a farm on an exposed hill 1,400 feet above sea level, with distant views into Shropshire and the Radnor hills of Wales. As the name suggests this is sheep country and in the church of St. Mary the farm names are painted on the bench ends. The church has a good roof typical of the region, a medieval oak chancel screen, and a Jacobean double-decker pulpit. An old drovers' road passed through Bettwys, and before that a Bronze Age trade route, the Clee-Clun Ridgeway. Below the church is a house with a well in the garden which probably had a pagan religious significance. By the side of a remote hill track at Crossways, 4m. from the Bettws church, is the Cantlin Stone. It marks the spot where in 1691 a pedlar called William Cantlin died. The local churches refused to accept him for burial, but in the end the vicar of Bettws took pity and buried him in Bettws churchyard. This act of kindness was rewarded 184 years later when the Cantlin Stone was accepted as evidence that the land in which it lay belonged to the Bettws-y-crwyn parish, and the parish was awarded several hundred acres by the Forest of Clun Enclosure Act (which disbanded and redistributed the common lands) that it otherwise would not have had. (*See Crossways.*)

BICTON *3m. NW of Shrewsbury on the A5.*
Here is a ruined 18th Century brick church and a new church of 1886. There is little of interest in the village but to the N the River Severn makes a huge meander. The land so enclosed is called The Isle, and here is the site of a moated castle, long abandoned. Part of the moat remains and still holds water. Close by are some derelict black and white cottages, a farm and a large brick house.

BICTON (near Clun) *See Cefn Einion.*

BILLINGSLEY *6m. S of Bridgnorth on the B4363.*
Billingsley lies in the beautiful, rich country of old orchards and wheat fields that lies between the Clee Hills and the River Severn. Dr. Thomas Hyde was born here in 1636. He was the first Englishman to learn the Chinese language. The Norman church of St. Mary was heavily restored in 1875. Inside is a Norman font, a Jacobean pulpit and reader's desk, and an early 14th Century Easter Sepulchre. The first element of the name Billingsley is possibly from an Old English personal name derived from *bylga* meaning 'angry'. The 'ley' ending is from *leah*, which in this case probably means a clearing in Morfe Forest, now long gone. The present (1989) Member of Parliament for the Ludlow constituency, C. J. F. Gill, lives at Billingsley Hall Farm, 1m. N of the village, above the wooded valley of the Borle Brook.

BINGS HEATH *2¼m. SW of Shawbury.*
It lies in the flat lands NE of Shrewsbury on the A53. 'Bing' might be an abbreviation of Billings. If so it could mean that the heath belonged to the Anglo-Saxon people called *Billingas* (*Bill* in Old English means 'sword'.) There is not a lot to say about Bings Heath. It is a main road hamlet of 19th and 20th Century farms and houses mostly of red brick that straggles along the busy A53. 'Elvaston' is a handsome small Georgian-style dwelling of 3 bays under a slate roof with 2-bay windows, a colonnaded porch and iron railings. On the road to Poynton is a clutch of new bungalows and an ivy-clad Methodist Chapel of 1858. To the S of Bings Heath is **Ebrey** Wood and the prehistoric fort of Ebury Hill. Ebury and Ebrey are from the Old English *ea-burh*, 'the fort by the river (or the stream)'. (*See Haughton.*)

BISHOP'S CASTLE *19m. SW of Shrewsbury on the A488.*
A pleasing and unspoilt small town in the hilly SW of the county close to the Welsh border. The site of the Norman Castle lies in the grounds of the 7-bay Castle Hotel (1719). Little remains but the mound, on which is now a bowling green, and a part of one wall. The castle was built about 1127, one of more than 150 such motte and bailey castles in Shropshire. The town was 'planted' (that is, was deliberately

Bishop's Castle, the House on Crutches.

created) on the site of a Saxon settlement in the late 12th Century. By 1285 there were 46 burgesses. The settlement has expanded little since medieval times and maintains its simple grid of streets leading down the hill, away from the castle. The Norman church of St. John the Baptist lies at the bottom of the hill. Only the typically squat border tower is original, the rest was rebuilt in 1860. In the vicarage garden wall is a fine, early 13th Century doorway, which is presumed to have belonged to the old church. Just to the NE of the church is the handsome black and white Old Hall. Near the top of the hill is the brick and stone Town Hall, of about 1765. At the base of it are the 2 small, round windows of the old town Lock Up. Alongside stands the well known House on Crutches which is presently in the process of being renovated. A narrow passage leads to the Market Place. Houses, shops and offices of stone, brick and timber make a pleasing huddle and create a town of great character, though it is a little too quiet for its future to be entirely safe. The oldest pub in Bishop's Castle is the Three Tuns, and here the landlord brews his own beer. The town was formerly the smallest borough in England. It was a 'rotten borough', and until 1832 returned 2 members of Parliament. (In 1820 four candidates each had 87 votes. All were elected.) Like all the market towns of the county – Bridgnorth, Cleobury Mortimer, Clun, Ellesmere, Ludlow, Market Drayton, Newport, Oswestry, Shifnal, Shrewsbury, Wem, and Whitchurch – Bishop's Castle was an 'open village' as distinct from a 'closed village' where the inhabitants were entirely in the service of one local lord, or estate owner. During the 19th Century the big landowners cut down on the number of permanent workers whom they housed on their farms and many cottages were either pulled down or left to become derelict. They increasingly used seasonal labour that was hired for a particular job and then fired. These labourers lived in nearby towns and villages and walked to work each day. These were the 'open villages'. Because these agricultural workers were transient extra houses were not built; the result was gross overcrowding of what accommodation was available. (In Newport, for example, 25 Irishmen are known to have lived in a tiny 2 up and 2 down cottage in Bellman's Yard.) In Bishop's Castle overcrowding was probably worse than anywhere else. The Bishop's Castle Railway ran to Craven Arms between 1866 and 1935 when the company became bankrupt. The route it followed is extremely picturesque. There were stations at Stetford Bridge, Horderley, Plowden and Lydham Heath. In recent years the ancient Friday Market has been revived at Bishop's Castle and a manufacturer of 'animal waste gas extraction plant' has played a major part in rejuvenating the local economy. Note: Bishop's Castle was originally called Lydbury Castle. It was given to the church by Egwin, the Saxon Lord of Lydbury, after he had received a cure at the tomb of St. Ethelbert (d.794) in Hereford Cathedral. The town's name was then changed to Bishop's Castle. St. Ethelbert was the Saxon King of East Anglia. He was to be married to Alfreda, the daughter of Offa, King of Mercia. However, Offa's queen disapproved of the match and had King Ethelbert killed. His chair was placed over a trap-door and he fell into a pit below where he was stabbed to death by the queen's guards. The murdered king was buried at Hereford Cathedral and was later canonised. Within 2 miles to the W of Bishop's Castle is **Bishop's Moat**, a motte and bailey castle with a dry moat, possibly built on the site of a prehistoric fort. This lies on the Kerry Ridgeway, an ancient hilltop track that continues on to Bishop's Castle. One of the most important of the Welsh drover routes (from Montgomery) passed through Bishop's Castle on its way to Shrewsbury and the great livestock markets held there.

BITTERLEY *3m. ENE of Ludlow.*
Bitterley is a residential village of stone cottages and Georgian and Victorian houses in lush country on the western lower slopes of Titterstone Clee Hill. It is most easily approached from Angel Bank on the A4117 Ludlow to Cleobury Mortimer road. The name was formerly *Buterlie* which means 'the place where butter was made'. There is no shop and no Post Office but there is a village hall called the Bitterley Hut and a school of 16th Century origin that had Grammar School status until 1958. The Early English church of St. Mary has a large, squat tower with a shingled turret and a broach spire. In the churchyard is a slender, shafted cross with a 'sight-hole', thought by some to be aligned to a 'lay-line'. Close by is the Georgian facaded Jacobean Bitterley Court. Both the church and the big house stand ½m. E of the main village. The reason lies in the field at the rear of the Court where house platforms and sunken roads mark the site of the original settlement. This was moved to its present position in the 16th Century to improve the view from the big house. Four ancient footpaths, plus the present access road, still converge on the Court and the church, the old centre of the village. Bitterly is the home of the Walcot family. Their coat of arms includes 3 rooks, and is said to have been granted by Henry V to John Walcot who checkmated the king with his rook during a game of chess. In the church there is a fine monument to Timothy Lucye (d.1616) who came from Charlecote Park near Stratford upon Avon. William Shakespeare is said to have been caught poaching by Lucye's grandfather who was portrayed as Justice Shallow in The Merry Wives of Windsor. Near Bitterley, but on the main Ludlow to Bridgnorth road, is **Crowleasowes** Farm. At the end of the access drive is the magnificent Crowleasowes Oak. It is the second largest oak in Shropshire with a girth of over 36 feet. (The largest is at Lydham Manor.) Until recently it had been regularly pollarded, which greatly extends the life of a tree. It is now hollow but still healthy and has been standing here since at least the 16th Century.

BLACK MARSH *8m. N of Bishop's Castle and 1m. WNW of Shelve.*
Black Marsh (SJ 316.995) is little more than a cluster of cottages on the A488. It lies in the broad valley between Stapeley Hill to the NW and Shelve Hill to the SE. This was an area of some significance to Bronze Age man. On Stapeley Hill is the famous Mitchell's Fold Stone Circle (*see Middleton in Chirbury*) and also the remains of the Giant's Cave, about which more later. At Black Marsh is Marsh Pool Circle, another Bronze Age stone circle, smaller than Mitchell's Fold but rarely visited and therefore in some ways more evocative. In the centre of the ring is a large 'King Stone'; there are 37 stones in all. Marsh Pool Circle lies in very rough pasture behind the third cottage on the left on the lane from Black Marsh to Bromlow, Brockton and Worthen. The stones are not large and it is remarkable to think that they have lain undisturbed for 4,000 years. (Some authors have stated that the Marsh Pool Circle, or Hoare Stones as they are sometimes called, is on Stapeley Hill. It is not, it lies on the flat ground of the valley below. North of the circle is **Bromlow Callow**, a hill on which there is a distinctive clump of pine trees, a landmark for many miles around. These trees are reputedly a drovers' mark, deliberately planted to point the way to Welsh drovers journeying to the livestock markets at Shrewsbury. Half-a-mile SW of Black Marsh, on the slopes of Stapeley Hill, is the Giant's Cave. It lies about 30 yards up the hill behind the farm adjacent to the warding Giant's Cave on the Ordnance Survey map. Many people have been misled into believing that this is the name of the farm, a misconception encouraged by the farmer who actually tells sightseers this to dissuade them from tramping over the hill pastures above his house. In fact the cave was really more like a pit and many years ago (but in this Century) a long-horn cow fell into the hole and had to be shot. It was left where it lay and the pit was filled with earth to prevent a similar tragedy. An outcrop of rock, some bushes and pieces of old corugated iron mark the position of the 'cave' which is believed to have been inhabited by pre-historic man but which has not yet been properly investigated. To get to the Giant's Cave turn off the A488 at a small sign saying Lower Stapeley (between Black Marsh and the More Arms at White Grit) and keep right when the track forks. (*For more information on this area see Shelve.*)

BLETCHLEY *3½m. W of Market Drayton.*
The hamlet lies adjacent to the busy A41 which here follows the route of The Longford, a section of Roman road. It is not an attractive place – there are too many scruffy farm buildings – but it does have 2 good timber-framed houses, both of which are visible from the road. To the S, on high ground, is Bletchley Manor with 3 gables and simple upright studding. At the rear the house has been colour washed and it has the worst of the unsightly farm buildings. Across the road

is another, much handsomer, black and white house with a sandstone base and 2 projecting and jettied gables. The left extension is of brick painted to match the original framing. The right bay is quaintly lopsided. The rest of the village lies around this house and alongside the old road which has been by-passed by the higher, new road a few yards away. One mile SE of Bletchley is the Tern Hill crossroads – or rather roundabout as it now is – where there have been several fatal accidents. Half a mile SW of the roundabout are the Clive Barracks and the Tern Hill Airfield, used mainly for training. (*See Tern Hill.*) One-and-a-half miles SW of Market Drayton, on the A53 to Hodnet, is Fordhall Farm, the home of Arthur Hollins and his wife, May. They practise organic farming and Mr. Hollins has written a book entitled "The Farmer, the Plough and the Devil" (published by Ashgrove Press, Bath, BA1 2PW), in which he tells how his father ruined the land with fertilizers and how it has since been rejuvenated. He has invented a machine called the Pulvoseeder which he hopes will be of assistance to other organic farmers. The best of luck to him. Half a mile SW of the farm is an earthwork, possibly an early defensive position connected with the Buntingsdale Hall estate on the hill just over the lake.

THE BOG *6m. NNE of Bishop's Castle.*
It lies in somewhat bleak but nevertheless attractive country at the southern foot of the Stiperstones. It is most easily approached from the A488 via Shelve. The Bog was once a proper little village. It developed around The Bog Lead Mine which opened in about 1740. The miners lived in cottages which had small plots of land attached. These smallholdings were worked by their wives and some of the fruit trees they planted in the hedgerows still survive. Lead and zinc ceased to be mined in the 1880's but barytes mining continued until the 1920's. Up to 200 people had lived here but when the mine closed most of the people left. The school of 1839 shut its door in 1968 and the cottages were demolished in 1970. Between 1983 and 1987 the area was reclaimed by the County Council. The barren waste areas were treated with lime and seeded with grass, the workings were made safe and a walled enclosure was built. In this are informative plaques giving details of the area and suggested circular walks. Nearby is the Stiperstones Field Centre, housed in the old, stone-built chapel. Remains of the mine include: Overgrown reservoir pool, tunnels, spoil mounds, stone foundations and earthwork remains of the old lead mine. The Somme tunnel has been blocked by a grid because it is home to bats, of which there are 2 kinds hereabouts – Brown Long-eared bats and Lesser Horseshoe bats. The lead and zinc ores occur in the Mytton Flags. These outcrop in two S-N bands, one running between The Bog and Snailbeach to the E and the other from White Grit to Batholes in the W. The minerals were deposited in sedimentary rocks by hot gases rising from subterranean granitic magmas. (For more information see 'The Shropshire Lead Mines' by Fred Brook and Martin Allbut, published by Moorland Publishing Company.) Today there is still a scatter of cottages on the hillsides. The slopes are laid to grass with wild areas of yellow flowering gorse, purple heather and scrub woodland. The road S from The Bog, sign-posted More and Linley, leads through the conifer plantations of Nipstone Forest, past the Brookshill Rifle Ranges, and down into the lovely valley of the River West Onny. Here there are views over lush meadows and wooded slopes which are near idyllic. At the bottom of the hill Linley Big Wood faces the traveller. Here one can turn left for Linley or right to Nind and the A488. Along the lane to the latter there are several deserted farmsteads. It has obviously been a deliberate policy to let these places become ruinous. Someone, it seems, wants to keep this lovely valley to himself.

BOMERE HEATH *5½m. NNW of Shrewsbury.*
It stands in attractive, gently rolling country and today it is very much a place of modern houses and bungalows quietly maturing amongst hedges and trees. A few old cottages have survived and the bright red local sandstone gives the settlement some individuality. It is used in the construction of farm buildings, retaining walls, cottages and in both the Presbyterian church and the Wesleyan Chapel of 1836 (rebuilt in 1868). Next to the latter is the Methodist church of 1903 built of hard red brick. There is yet another church, a little Mission Hut on the main road. As to social services there are: 2 general stores; the Red Lion pub; a hairdresser's shop; a Post Office; a garage; and a water pumping station. Percy Thrower, the famous television gardening personality did *not* live at Bomere Heath; he lived at The Magnolias, Merrington, which is ½m. N. (*See Merrington.*)

BOMERE POOL *See Bayston Hill.*

BONINGALE *1m. S of Albrighton, near Wolverhampton.*
This attractive little village is now by-passed by the newly improved A464. The old main road is now lined with flowers and laurels and there is a variety of dwellings: black and white cottages; painted brick terraces; the red brick Georgian style Old Rectory with a hipped roof; some sandstone buildings; and most striking of all, the handsome black and white Church Farm. This has 2 symmetrical gables and decorative timber work, post and pan on the ground floor, and concave lozenges above. The pretty little church of St. Chad stands embowered in mature trees. The nave and chancel are Norman, the S aisle is of 1861, and above there is a weather-boarded belfry with a broach spire. Inside there is a Jacobean pulpit and a plaque commemorating a former vicar who was previously the Archdeacon of Athabaska, Canada. On the main road there is a pub, The Horns; a large and well known plant nursery surrounded by a grim security fence; and, 1m. E, the Manor House. Boningale's most famous son is, without doubt, the scoundrel Jonathan Wild, born here in 1682. He became the leader of organized crime in London. The essence of what he did is quite simple. His thieves would steal goods and Wild would then let it be known that he could, 'through his underground connections', have them safely returned to their proper owner – for a small fee, of course. Every now and again he would betray one of his pick-pockets, burglars, highwaymen or prostitutes to maintain his credibility with the police and with Society. He became enormously rich and had warehouses stuffed full of jewels, paintings and all manner of costly things. Those that were not 'bought back' he shipped to the Continent. Inevitably he was caught and brought to trial. Two hundred thousand people turned out to see him hanged on Tyburn Hill. John Gay used Jonathan Wild as the model for Peachum in 'The Beggars' Opera', 1728. Three-quarters of a mile SW of Boningale is Bishton Hall, a timber-framed house with a stone barn to the S. One mile SE of Boningale is Pepper Hill, an old house of brick and stone positioned on a sandstone cliff.

BORASTON *1¼m. ENE of Burford and Tenbury Wells.*
An attractive little place in undulating country close to the Herefordshire border with some red brick Georgian houses and cottages with trim gardens. At the time of Domesday Book the settlement belonged to the manor of Burford and was called *Bureston*; the name could be derived from *gebur*, Old English for 'peasant', hence 'the home of the peasant'. The church has an undecorated Norman N doorway, now blocked, and some indications of a Norman S doorway; the rest is mostly of the restoration by Henry Curzon in 1884-7. The font is of about 1700 and came from Buildwas.

BOSCOBEL *2½m. ENE of Tong (which is at exit 3 on the M54).*
Boscobel House is a rather ordinary dwelling with panelling and plaster of the early 17th Century. It is internationally known because Charles II stayed here whilst on the run after his defeat at the Battle of Worcester (1651). He stayed first at White Ladies Priory, then in the wood known as Spring Coppice, and then, after a wasted day trying to cross the Severn at Madeley, at Boscobel House. On Saturday, 6th September 1651 Charles and Major Careless hid in a recently pollarded oak tree in Spring Coppice. Whilst there they saw soldiers searching for them. This small story caught the public imagination and the Royal Oak became the most famous tree in England. Hundreds of pubs were called the Royal Oak and souvenir hunters came from far and wide. By 1712 there is a record of a sapling growing alongside the dead tree. Finally, the Royal Oak was dug up and even its roots were used to make souvenirs. Boscobel House originated as a hunting lodge in Brewood Forest and was built about 1580 by the

Giffards of Chillington (in Staffordshire). The name is derived from the latin *boscus bellus* meaning 'beautiful wood', later Italianized to *boso-bello*. White Ladies (more properly the Priory of St. Leonard) was the only medieval nunnery in the county of Shropshire. It was established in 1186 and the nuns were Augustinian cannonesses. The ruins of the priory lie in a field at the side of a small wood and are approached down a rough track. The little sign that marks the turn is often not in place and the track is easily missed. The site is about ½m. SW of Boscobel House on the lane to Cosford.

BOULDON *2m. E of Diddlebury, which is 4½m. NE of Craven Arms.*
A most pleasant stone-built village of some character situated on the western slopes of Brown Clee Hill. It used to have a small, grey, corrugated-iron parish church, but, alas, no more. The water-mill has survived, though, complete with most of its working parts including an iron bucket-wheel cast in Coalbrookdale. There was also an ironworks in the village during the 17th and 18th Centuries. Little is left of this but the tree-clad slag heap behind the mill where once 'the best tough Pig Iron in the kingdom' was made. The furnace probably closed in 1795 when it was sold, one of the last 2 charcoal fuelled works in the county. The first element of the name Bouldon could be either from the Old English *bula*, 'a bullock', or *Bulla*, a personal name, or *Bulwana*, the name of a tribe. The second element is from *dun*, meaning 'hill'. At **Peaton**, 1m. W of Bouldon, there are several weather-boarded houses. These were built by Lord Boyne in about 1965 for workers on his Burwarton Estate.

BOURTON *3m. SW of Much Wenlock on the B4368.*
A mostly stone-built village by a stream in Corvedale on a crossing over Wenlock Edge. The church of Holy Trinity has a Norman S doorway, a weather-boarded belfry and box pews. It is set amongst yew trees and lies a little way up the hill by the Manor. This is an irregular and picturesque mansion by Norman Shaw with a half-timbered gable, tile-hung first floor, impressive stone-arched doorway and 16 chimneys. Bourton Hall Farm is a 17th Century stone house with star-shaped chimneys and a dovecote. The parish of Bourton was once part of a Saxon estate owned by the Wenlock Church in 1066. In the village there are fruit-bearing cherry trees. These were planted in celebration of the Coronation of Queen Elizabeth II, one for each child in the village. Bourton has long been an estate village on the Bourton estate but now, as houses become vacant, so they are sold on the open market. The approach to Bourton from Much Wenlock passes through an avenue of chestnut trees planted to commemorate the Diamond Jubilee of Queen Victoria.

BRAGGINGTON *1½m. W of Alberbury.*
There is no village today, just one farm, Braggington Hall, which is approached down lanes from Halfway House on the A458. The site of the deserted village is ¼m. NE of Braggington Hall. It lies on a N facing slope in the Breiddon foothills, but there is little left to see. In 1963 it was excavated and then levelled. The Anglo-Saxon settlement seems to have been deserted after the Norman Conquest and then re-established some 100 years later. In 1301 it was occupied by 16 Welsh tenants. There was a long-house here from the 14th to the 17th Centuries and there are earthworks by the stream (which here marks the border with Wales) that are probably the remains of a mill. Some medieval ridge and furrow also survives. It is likely that the settlement was abandoned in the early 17th Century, possibly cleared go give privacy to the owners of Braggington Hall. Work on the Hall began in 1650.

BRATTON *2m. NNW of Wellington.*
The Gate pub on the B5063 is the centre of this scattered little community. It is also the northern terminus of the Silkin Way, a 14m. footpath that runs from here, through Telford, to Coalport on the River Severn. To the S of the pub, along the road to Adbaston, is a pleasant line of red brick houses, old and new, and a market garden called Moor Farm that produces vegetables. To the SE is a lane signposted Eyton that skirts the back of the new Shawbirch housing estate. The farm and cottages at the beginning of this lane are also a part of Bratton. Indeed the stream of its name ('the settlement by the brook') passes by the farm. Here, we are on the edge of the endless flat acres of the old marshlands called the Weald Moors. **Long Lane** lies 1¼m. N, on the busy A442 Wellington to Whitchurch road, Here it is the Buck's Head that is the centre of things: a red brick hamlet with a petrol filling station, and a mobile caravan park. On the other side of the road is **Sleapford**, a handful of cottages along a stream – drainage ditch that runs from the Weald Moors to the River Tern, 1½m. W. The name Sleapford meand 'the ford by the marsh'. *Sleap* is Old English and is probably derived from the Old Norse *sleipre* meaning a 'slippery place', or from one of several German words with a similar meaning. **Sleap** itself lies 1½m. N on the A442 by the River Strine and is now considered to be a part of Crudgington. Beside the main road stands a tiny cruck cottage; its inverted 'V' frame can be plainly seen. Between Sleapford and Sleap is Sleap Moor, one of the most extensively drained areas of the Weald Moors. Note: There is another Sleap, near Wem. This has its own article and an alternative entymology is given there.

BREADEN HEATH *4½m. ENE of Ellesmere.*
On the A495 to the E of Welshampton is a delightfully run-down little timber-framed cottage. Breaden Heath lies along the lane opposite, a hamlet of red brick cottages on rising ground. Northwards the country is gently undulating with a scatter of small cottages in fields laid mainly to pasture. **Hampton Wood** Hall is a modest farmhouse clad in stucco with a timber-framed barn amongst the outbuildings. Brook House Farm lies ¼m. SW in a dramatic little valley. The dwelling stands on a promontory with steep slopes to 3 sides. There is likely to have been a homestead here from pre-historic times. To the WSW by ¾m. at **Coptiviney** is a timber-framed farmhouse (SJ.414.368) with 2 rusty-red roofed barns. It stands by a small pool, derelict, deserted and quite woebegone with only a few cream-coloured cows for company. To the N is the steep-sided valley of Spout Wood. The valley bottom is very rough and marshy. Above the N bank is **Northwood** Hall (SJ.412.381), which is marked but not named on the 1¼ inch Ordnance Survey map. It is a substantial farmhouse of red brick with a gable and blue brick diamond decoration. Just to the W of Northwood Hall and running N off the main road is Lion Lane. We have come across several Lion Lanes and have yet to find a satisfactory explanation for the origin of the name. One presumes that lions were involved in some way, either live ones in cages or stone sculptures at the entrances to drives. Private zoos were not uncommon in medieval times. This Lion Lane leads through woods and down to Penley Mill, just over the border in Wales. The old mill is now surrounded by a number of red brick buildings, a factory that makes capacitors, and a garage.

BRIDGNORTH *20m. SE of Shrewsbury on the A458.*
Bridgnorth is a fine hill town perched on a sandstone ridge high above the River Severn. It takes its name from the fact that the previous bridge crossing of the river was to the S at Quatford. Bridgnorth is in fact 2 towns, a High Town, W of the river, and a much smaller Low Town to the E. They are linked by a bridge, built in 1832. Previous bridges may well have used the islands, of which there were 3, though only one still exists. In recent years a new bridge carrying the A458 by-pass was built to the S. The origins of the settlement are unclear. There are records of a Saxon *burh* on the W bank of the river. This could have been on the site of the mound at Panpudding Hill, or more likely on the site of the present castle. There is no mention of Bridgnorth in Domesday Book. It was founded in 1101 when Earl Robert de Bellême abandoned Quatford and moved his castle and church (St. Mary's) to the site of the present town. The castle was in the front line against the Welsh marauders in the 11th and 12th Centuries, but in the 13th Century fell into decline. By the 16th Century it was in total ruin. The town had probably been laid out in the 12th Century to a grid pattern and it was then that a second church, St. Leonard's, was built. The settlement was defended by an earth bank that was partly rebuilt in stone with 5 gates. Only one gate remains, the North Gate, rebuilt and redesigned in 1910. A suburb developed at Low Town over the river, and at one time was as large as the High Town. By 1300 Bridgnorth was only second in importance to Shrewsbury in the county of Shropshire. The prosperity of the town

Bridgnorth, view north-east over the River Severn from Castle Hill.

was solidly based. It was a centre of communications, a river crossing and a port; it had leather tanning and brewing industries and hats, caps and carpets were manufactured here. Little remains of medieval Bridgnorth because most of it was destroyed by fire in 1646, but Bishop Percy's house (1581) at the bottom of the Cartway still survives. The new Town Hall, in the centre of the High Street, was built in 1652. Most of the present buildings date from the 18th Century when the town acquired its dignified Georgian atmosphere. The church of St. Mary Magdalene in East Street was rebuilt in white free stone by Thomas Telford in 1792. Before the modern road was constructed the only vehicular way between the High Town and the river was the Cartway, a delightful, winding lane said to be haunted by a mysterious Lady in Black. Pedestrians have the benefit of several sets of steps and the Cliff Railway. Half way up the picturesque Stoneway Steps is the Steps Theatre which is housed in an old Chapel and also has a ghost, a rather vindictive spirit that punches people! (Bridgnorth is particularly well endowed with supernatural beings – some 21 at the last count.) St. Leonards' church, near the North Gate, has a charming close of mixed almshouses and cottages and lawns. The castle was garrisoned by the Royalists during the Civil War, and after their defeat was dismantled. The Norman keep was blown up but failed to disintegrate. It now leans at an angle of 17 degrees, a greater angle than that attained by the Leaning Tower of Pisa. There are walkways around the cliffs giving spectacular views over the Severn valley in several directions. At the bottom of Cartway are cave houses cut into the river-front cliffs. These caves had extensions to the front, made of brick, and were lived in until late Victorian times. The Severn Valley Railway has found a home in the old railway station. Volunteers run a professional service through lovely country along the banks of the river. In Low Town Raistrick had his foundry, and it was here that the first railway engine was constructed for Trevithick in 1804. (A replica of this engine stands outside Telford Centre railway station.) In fact, there was a considerable amount of heavy industry in the town. The Coalbrookdale Company used Town Mills as a forge, and behind Mill Street are the remains of John Hazeldine's well-known foundry. Much of the Ironbridge gorge trade passed through Bridgnorth: coal from Broseley, crude iron from Coalbrookdale, and agricultural produce from the countryside around. The town remained a major river port until the 1850's and the coming of the railways. The 18th Century River Warden's House still exists with the old wharves before it. There are landing steps at Old Quay, opposite the Stoneway Steps, and there is a continuous iron lath along the river wall to protect towing ropes. The port finally died in 1895 when the last large barge sailed down river. In the 1770's lacemaking was introduced to Bridgnorth by a French refugee, M. de la Motte. In 1781 he was executed as a spy but the craft prospered for some 100 years. It went into decline with the introduction of machine made lace and finally ceased in 1875. The trade was centred on the Cartway. Today Bridgnorth is a flourishing commuter and tourist town. There has been much postwar housing development but this has been mostly in the NE away from the old centre. Francis Moore, of Old Moore's Almanac fame was born here. His predictions, based on astrology, were first published in 1700 as Vox Stellarum (The Voice of the Stars).

BRIDGES *6m. NE of Bishop's Castle, as the crow flies.*
The hamlet lies just off the main road; we say main road but a quieter and more attractive one you will not find. There are 2 bridges: a footbridge over the stoney bedded River West Onny in front of the pub, and a vehicular bridge over the Darnford Brook, at its confluence with the West Onny. The Horseshoe Inn is part stone and part rendered and the barmaid does her knitting during slack periods. There are 2 enlarged and modernised cottages, some mature pines, oaks and beeches, sheep in the meadows, and on the lane to Ratlinghope there is a stone-built house with gables and red and yellow brick dressings that is now a Hugh Gibbins Memorial Youth Hostel. The settlement lies in a bowl surrounded by large but gentle hills. **Overs** lies ¼m. SW, and is little more than a pink painted cottage and a farm on the track that runs to the many modern barns of the cattle farm at **Coates.** On the main road ¼m. SW of Bridges is the **Kinnerton** Methodist Church. Kinnerton itself is a tiny place ½m. W on the road to The Bog and Pennerley. The main road continues

BRINGEWOOD FORGE *3m. WSW of Bromfield which is 3m. NE of Ludlow.*
The River Teme has cut a steep sided gorge here and it is now a most attractive wooded place with rocky cliffs and a spectacular, battlemented bridge. The settlement of Bringewood Forge (SO.454.780) consists of a handful of estate cottages that line the access lane and a large and shamefully derelict house. The Bringe Wood, after which Bringewood Forge is named, is now a conifer plantation on the high ground of Bringewood Chase, 1m. S. There was a furnace at Bringewood Forge that smelted iron using charcoal as a fuel, and a water-wheel to operate the bellows that provided the blast of air, from the early 17th Century. (The people of Ludlow complained about the clearance of the woods on the Chase for use as furnace fuel.) Later, there was a forge-furnace for refining crude iron and a foundry for casting iron goods. Through the 17th Century and most of the 18th Century Bringewood was part of the Knight family empire. They lived at Downton Castle and had widespread iron interests which included furnaces at Bouldon and Charlecotte as well as Bringewood. All 3 were supplied with iron ore mainly from the Clee Hills. There is very little left of the ironworks at Bringewood – a little brickwork, a mound of black earth and a water channel. The upstanding stone building is probably a part of a later tin works. Bringewood was one of the last charcoal furnaces in Shropshire and is thought to have closed in the late 1790's. The Knight family introduced exotic flowers, shrubs and trees in the 19th Century and landscaped the old ironworks area as a part of the Downton estate garden. In this Century these were allowed to grow wild and Bringewood was known for its natural beauty. In the midst of this wilderness stood the most handsome white stone bridge – it still does stand. This was built in 1772 and the Surveyor was the leading architect Thomas Farnolls Pritchard. Below the bridge is a semicircular weir over which the water pours with a great roar. On the downstream side their is a wharf. Today the wilderness has been tamed. Some 2 years ago the spectacular Downton Castle was sold to a Greek businesman and the estate disposed of separately to a French woman, Madame Prima of the family who owns the Perrier Water Company. She, of course, lives in France and the estate is now being exploited for its resources of timber and has been stocked with pheasants to be commercially shot. The river banks are now exceedingly tidy and everywhere there are notices forbidding entry. The local people are seething with indignation at both the loss of the natural beauty (though that will be substantially restored very soon) and the loss of access. Previously they had enjoyed the traditional courtesies and respects of an Old English family, the Lennoxes, who in fact still own some of the estate cottages. Now they have heavy-handed absentee landlords. Bringewood Forge is well worth a visit, if only to see the bridge and its dramatic setting. Now that most of the undergrowth has been cleared there is an opportunity for industrial archaeologists to take a more detailed look at the area than was possible in the past. The Forge is most easily approached down a lane which leads S off the A4113 Bromfield to Knighton road about 2½m. W of Bromfield. In the churchyard at **Burrington**, 2¼m. SSW, there are some tomb covers that were cast at Bringewood. Note: Most of Bringewood Forest is still in Shropshire but Bringewood Forge and Downton are now in Hereford and Worcester.

BROCKHURST *See Lee Brockhurst.*

BROCKTON (near Bishop's Castle) *2m. S of Bishop's Castle.*
Brockton is a village of farms, unusual for its size and completely unspoiled. Unspoiled does not necessarily mean pretty or picturesque; Brockton is a working place and has its share of untidiness and ramshackle buildings. In its way, though, it is quite splendid. A tributary stream of the River Kemp flows through the settlement and is crossed by a vehicular bridge. We used the ford and bottomed the car twice. There are areas of grass between the farms and cottages and several of the buildings have red painted corrugated iron roofs, a characteristic of the area. There is a house of stone with half-timbered upper elevations; a delightful thatched cottage called The Thatch; several stone-built

Bringewood Forge.

Bridgnorth Castle.

Brockton, near Bishop's Castle.

farmhouses; a Methodist church with flaking paint; the big green sheds of Shropshire Highland Seeds Ltd.; stone boundary walls; and though there are few trees we heard birds singing everywhere. There is an almost medieval atmosphere here despite the close proximity of the busy A488. **Lower Down** lies 1m. SE of Brockton. Here there is little more than 2 large, stone-built farms, a pair of semi-detached houses, a row of stone-built terraced cottages with a hipped roof, a telephone box and the remains of a Norman castle. The most southerly farm is quite handsome with its mullioned windows, dormers, central pediment and traditional, weather-boarded barns. The Norman motte is about 9ft. high and 46ft. across at the top but has been damaged by the removal of gravel which was used in the construction of roads. Lower Down is actually quite high up but still low compared to the large pre-historic hill fort of Bury Ditches 1m. SSW. From here there are some of the best views in the county. (*See Clunton.*)

BROCKTON (near Much Wenlock) *4m. SW of Much Wenlock.*
A crossroads settlement in Corvedale on the B4378 with a pub, a school, a Post Office, a cruck cottage and the motte and bailey of a Norman castle. The Post Office is run by the local blacksmithing family who still produce ornamental ironwork. *Broc* or *brock* in a place name usually means either 'a brook' or 'a badger'.

BROCKTON (near Newport) *2m. SW of Newport.*
It lies on the edge of the drained marshland of the Weald Moors. On the A518 is the colourful Red House pub. The lane to the settlement is crossed by a disused redstone, ivy-clad, railway bridge. Longford Grange has attractive, weather-worn, timber-framed barns whilst opposite the red brick Brockton Manor has ugly, galvanized modern farm buildings. In the grounds of the Grange is a large horse-chestnut tree with many branches that have dipped to the ground and rooted, a remarkable sight. There is a War Memorial by a large, 3-storey brick house and a pretty lane leads to Edgmond.

BROCKTON (near Worthen) *¾m. WSW of Worthen on the B4386.*
An attractive and relatively unspoiled little village on the northern edge of the broad Rea Brook flood plain. There is a pub, the white stuccoed Cock Inn; a red brick chapel with round-headed windows; a stone bridge over a fast-flowing stoney-bedded stream; Georgian red brick houses and a few modern bungalows; farms; and an old stone and brick mill disfigured by the insertion of bland, double-glazed windows. On the northern fringe are council houses; some of which are clad in prefabricated concrete slabs. The lane continues uphill to **Hampton Beach**. Here there is a timber-framed cottage with bright red brick infill; a ford (which only carries water at times of heavy rain); stuccoed cottages, white against the wooded bank; scattered cottages and farms and the Hall. Hampton means 'high settlement' and on the highest point is the Hall, an imposing red brick mansion set in a small park with good views southwards to the Stiperstones hills. The central block is 7 bays wide with 2 tall storeys and a parapet; the wings are 2 bays wide and also of 2 storeys, but lower. Beneath the whole is a basement storey. The house was built in 1681-6, on the site of an earlier building, but the centre was altered in 1749 when the end carved bay of the right wing and the arched stone porch were added. On both sides of the porch are circular windows, characteristic of the 17th Century. Despite the rebuilding and the later alterations the hall still lies in the medieval position, that is, not behind the porch but to the side. It is quite possible that some of the internal fabric pre-dates the 17th Century rebuilding. Associated with the house is a large farm. The surrounding country is indulating and most attractive. The lane continues, uphill and down dale, to the hamlet of **Rowley**, a place of scattered farms and cottages with a tiny stone church and a telephone box. The church, now a house, has a clump of marker pines beside it. The fields are all laid to grass; indeed the name Rowley means 'rough pasture'. The settlement lies on high ground and there are good views. To the E is Rowley Hill. **Lower Wallop** lies 1½m. NE of Rowley on lower ground. Here there is a yellow-red brick Georgian farmhouse of 3 bays with a hipped roof with a slightly projecting centre; a red brick Smithy, a pair of semi-detached Council-like houses, boundary walls of stone, a tiny stream and in the hills around a few scattered cottages and farms. The name Wallop is from the Old English *Woell-hop*, which means 'the valley of the stream'. Wallop Hall lies in a valley by a stream ½m. to the W. The earthwork remains of Caus Castle lie on the hill ½m. E. This was once a flourishing Norman town and fortress, the stronghold of the Corbet family. (*See Westbury.*)

BROMDON *10m. SW of Bridgnorth.*
Set amongst the hills and dales between Brown Clee and Titterstone Clee are a few scattered farms, an ill-assorted collection of roadside barns and a caravan site based on a large stone-built farm; this is Bromdon. As to the name: *Brom* means 'broom' and *don* (or dun) means 'hill'.

BROMFIELD *2½m. NW of Ludlow on the busy A49.*
The village lies close to the confluence of the River Onny and the River Teme. A stone and half-timbered gatehouse is all that remains of the 13th-14th Century Benedictine Priory. St. Mary's church is a remnant of the Priory chapel. After the Dissolution the Norman chancel was blocked off and the rest of the building was incorporated into a house, by Charles Foxe, and not restored to church use until 1658. The painting (1672) on the present chancel ceiling has been described as 'the best example of the worst style of ecclesiastical art'. An avenue of 40 yew trees line the E path to the church. There is a centenary memorial to Doctor Henry Hill Hickman, 1800-1830, the pioneer of anaesthesia by inhalation, who was baptised here. A field known as Old Field, a part of the local medieval agricultural system is now incorporated in the golf course. Just to the S of the village is the ancient Oakly (sic) Park, with its comparatively modern red brick Hall of the 18th Century. The park was once famous for its Druid Oaks, but these have sadly nearly all gone. On the edge of the park the River Teme tumbles over a weir, and close by are some Gothic cottages. The Hall is the home of the 14th Earl of Plymouth. The first earl was Thomas Windsor Hickman. The present earl still owns almost all the dwellings in the village and is the largest single employer in the area. Ludlow Racecourse lies ½m. NE, over the railway line. There are 4 Bronze Age burial barrows here at SO.496.776, and a fifth at SO.490.779. They were opened in Victorian times and revealed signs of cremation. In fact these are only a few of the original 20 known barrows that lie in a line on the level gravel terrace that runs NE of Bromfield. In 1966 a quarrying operation revealed a cemetery of 130 shallow pits which have been carbon 14 tested, the results of which shows that human remains were interred here continuously from 1770-1870 BC to 910 BC (Bronze Age). Just to the SW of the pit graveyard is the site of a rectangular Roman marching camp within which is a circular Anglo-Saxon barrow. There is nothing to see of these on the ground; they were only recognized as crop marks on an aerial photograph. (For further information on this necropolis – town of the dead – see The Archaeology of the Welsh Marches by S. C. Stanford, Collins, 1980.) The first element of the name Bromfield is from the Old English *brom*, meaning 'broom', a yellow flowering shrub. Near the River Teme at Bromfield is a meadow called The Crawls where once stood a moated mansion. Once upon a time there lived here a beautiful maiden who fell in love with a penniless knight. Her father said that the only dowry she would receive would be the land she could crawl around on a dark winter's night. She donned leather breeches and astounded everyone by crawling all the way to Downton, 3m. to the SE. Her father was so impressed by her courage that he made her his heiress above her brothers.

BROMLEY *2m. NNE of Bridgnorth.*
It lies just off the A442 on high ground, a charming little place with a wide range of building styles. There are cottages of red sandstone, red brick, variagated yellow brick, half-timber and most noticeably there are shingle-clad walls. These timber tiles look most neat. The 'big house' is symmetrical, has 3 bays and a projecting central gable. Half-a-mile NE is the curious 'resort' of Rindleford on the River Worfe. (*See Rindleford.*)

BROMLOW *6m. SW of Pontesbury.*
It catches the eye for many miles around. Turn a corner and there it is, the distinctive clump of pine trees on the hill called Bromlow Callow. It was originally

Bridgnorth, Castle Terrace.

planted, it is said, as a drover's landmark and we are quite prepared to believe it. Bromlow means 'the broom covered hill', and Callow is from the Old English *calu*, meaning 'bald' and hence 'bare'. It is a very scattered community in a country of small hills and dales. If there is a centre it is the stone and stucco Drum and Monkey, a freehouse inn. The manufacturers of white paint do well hereabouts; most dwellings wear several coats of it. The pastures are rough but the views are splendid. Down the hill to Brockton, ½m. N of the pub, is the motte of a Norman castle, a small 9ft. high mound. There are several of these little earthworks in the area. The mounds, and the wooden towers that stood on them, could be constructed in a few days and were often abandoned after a few months as the Norman troops conquered and moved on to new pastures. At the time of Domesday Book Bromlow was probably a part of the lost manor of Muletune, which could be the place called Munton in later medieval documents. It has been suggested that Munton may have been what we now call Mondaytown, 5m. to the NE, close to Caus Castle. However, may we offer another contender for consideration; namely **Meadowtown**, less than 1m. SW of Bromlow. The narrow lane leads from Bromlow down into the valley of Lyde where there are 2 cottages and up through a small conifer plantation to the bright green fields of Meadowtown. It is a ramshackle little place of brick and stone cottages, cows feeding by scruffy sheds, a derelict red brick chapel with yellow brick dressings and round-headed windows, a black and white dwelling and a derelict house of stone. The lane leads on down a steep hill to Rorrington. The earthwork remains of Castle Ring, a pre-historic settlement-fort lie at the end of a track ½m. to the S of Meadowtown.

BROMPTON *1½m. WSW of Churchstoke.*
The hamlet lies close to the Welsh border at the junction of the B4385, the Montgomery to Bishop's Castle road, and the A489, Newtown to Churchstoke road. It is directly in the line of Offa's Dyke which is broken here by the River Caebitra. Close to the N bank of the river are the earthworks of a Norman motte and bailey castle. The mound is 25ft. high and 28ft. in diameter – a fine specimen. A section of the bailey platform to the SW was removed to construct a mill pool in the 19th Century. At Brompton Hall, adjacent to Offa's Dyke, is a prehistoric burial mound. In 1086 Brompton was part of the manor of Rhiston which belonged to Earl Roger of Montgomery (The Welsh town of Montgomery used to be in Shropshire. It took its name from the Earl who in turn took his name from his home town in Normandy.)

BROOME *2½m. SW of Craven Arms.*
It is very much a place of new houses with more than its fair share of untidy little corners. The Engine and Tender welcomes guests with views of its kitchen and looks more like a semi-detached house than a pub. The settlement increased in size very quickly with the coming of the railway, the line from Shrewsbury to Swansea. There were 2 coal wharves, cattle pens, a timber yard and a warehouse. The warehouse still stands, lost amongst far less handsome satellites. Indeed, the yards are now downright scruffy, littered with scrapped cars, a jumble of makeshift sheds and an engineering works in the process of erecting a modern steel-framed structure. The station is now only a halt. The fences lie broken and bent and the crude iron bridge over the road is painted an ugly black. The oldest house in Broome is the Old Forge. It has 3 pairs of crucks and some wattle and daub panels.

BROOMFIELDS *2m. N of Montford Bridge which is 4½m. NW of Shrewsbury.*
The broom is now gone but the backwoods nature of the area remains. The nucleus of this scattered settlement is at a road junction where there are a few mature red brick dwellings, red sandstone farm buildings and a cluster of tall pine trees. Little cottages, some no doubt originally built by squatters, lurk in odd corners of the landscape beyond. Many are surrounded by clusters of makeshift shelters of corrugated iron and scrap timber. Ramshackle is the word. In the fields are black and white cows and in the hedges are hollies. It is a flat landscape; indeed an old wartime airfield lies just to the S. Just to the N is **Nib Heath**. Nib is from the Old English *hnybbe*, meaning 'a point, a tip'. Just to what this refers we do not know. At the crossroads is a sizeable mere called Cottage Pool with a large resident population of ducks and a transient one of geese. Ancient trees are dotted around the shore and there is an air of happier days past – the rotting, skeletal, landing stage and the derelict boathouse. Above the pool is a black and white cottage and in the pastures are pale tan cows. Just N of the mere is a cluster of houses, bungalows and stone cottages one of which is the lodge to Adcote School. (*See Nesscliffe*.)

BROSELEY *1m. S of Ironbridge.*
The name is Anglo-Saxon and means '*Burgheard*'s forest (or clearing in the forest)'. Broseley is an old town and in the 18th Century, when the industrial revolution was getting underway, it was the only settlement of any size in the Ironbridge Gorge area. It is a place of some substance and character with a long industrial history. The manor was a part of the large Liberty of Wenlock, held by Wenlock Priory and then part of the extensive Wenlock Parish. (Wenlock is now called Much Wenlock.) Coal was mined here in medieval times but in the late 16th Century the industry became organized and developed apace. John Clifford, lord of the manor of Broseley, was a leading figure in the trade. By 1605 there was a 1m. long wooden railway operating between Birch Leasowes, near the church, to the Calcutts on the River Severn at Jackfield. This followed the valley of a small stream and was the first railway in Shropshire and only the second in the whole of Great Britain. By the end of the 18th Century coal from Broseley and Madeley was the main cargo carried down the River Severn – 100,000 tons a year in 1760. By 1756 there were 87 coal boats on the river specifically attached to the Broseley mines. Jackfield, the main port for the town, flourished though it was, from all accounts, a pretty rough place. Hard men lived it up with loose women and liquor. The hardest of these hard men were the bow-haulers who pulled the sailing ships up river against the current. Doing the work of animals they behaved as such. (Not until a tow path was constructed could horses be used.) A complement to the coal trade was brick-making. Clay was a by-product of coal mining. In 1754 the Broseley brickworks were established and sending common bricks and fire bricks by the many thousand to Abraham Darby II for his new Horsehay furnace. (His father, Abraham Darby I, died in 1717 and is buried in the Quaker Burial Ground at Broseley. When he moved to Coalbrookdale in 1708 there were many Quakers in the area and their only meeting place, built in 1692, was at Broseley.) In 1781, after many trials and tribulations, the famous Iron Bridge across the River Severn opened. Several of the shareholders were Broseley businessmen: a maltster, a windmill owner, a grocer who also manufactured soap, and a lawyer. Iron ore, coal and limestone from Broseley and Benthall could now be much more easily transported to markets N of the river. In 1728 a new railway called the Jackfield Rails was constructed, probably following the route of the old line. The first iron furnace in the area was built at Coneybury, ½m. E of Broseley, in about 1786. It was sometimes called the Broseley Bottom Coal Furnace. It produced pig iron and was managed by the established partnership of Banks and Onions. By 1800 there was a large foundry and in 1806 another furnace was established by Thomas Guest. There were good supplies of local ironstone and from 1717 ores were being sent from the Ladywood mines to Coalbrookdale. The clod coal mined all round Broseley – even today the map is peppered with 'old shafts' – was most suitable for converting to coke, the fuel of the iron furnaces. However, supplies began to dwindle and by 1830 all the ironworks had closed down. It was not really the lack of coal that caused the furnaces to close; that could have been brought in from elsewhere. The real reason is almost certainly that it was more economic to export iron-ore to the Black Country where there was a ready market and which had good communications with the rest of the country. The earthenware trades were important locally and bricks and tiles are still made at Broseley, but the industry never grew to any size. Broseley clay smoking pipes were nationally famous and were being made into the 1870's (Roden's was the largest manufacturer) but the industry was an insignificant part of the local economy. There was even at least one 'cotton manufactory' in 1792

Bridgnorth, Bishop Percy's House.

but little is known of this enterprise. When the furnaces 'blew out' there was mass unemployment, the area was extremely depressed, mass emigration occurred and in the second half of the 19th century Broseley was largely derelict and remained a wasteland into the middle of this century. As elsewhere in industrial England cock-fighting and bull baiting were extremely popular 'sports' in Broseley during the 18th and 19th Centuries. Bull baiting, though, was expensive and usually organized by the chartermasters – contractors who supervised the coal mines for landowners and iron-masters. During the years of prosperity there were many squatters in the area but in fact the local landlords were quite agreeable to selling freehold building plots and large numbers of workers took advantage of what, in its day, was a rare opportunity. Today Broseley is prospering once more. Old buildings are being renovated and there is a cheerful atmosphere in the town. The old centre lies at the bottom of the hill by the church of All Saints. This was built in 1845 by H. Eginton and is a large building in Perpendicular style. There is a W window by William Kempe and the churchyard gates are early 19th Century cast-iron. Next to the church, in Church Street, is Broseley Hall, a substantial brick mansion. On the other side of the road is The Lawns, a tall house with a large bow window and dated at 1727. Inside there is a fine staircase and good fireplaces. This was the home of John Wilkinson between 1778 and 1800. His ironworks were at nearby Willey. When he left he rented The Lawns to the master potter John Rose. Today it houses a collection of china-ware and is open to the public. There are several more large houses nearby. Higher up Church Street Nos. 20-22 are gabled and dated at 1663. The Town Hall of 1777 was recently demolished and in its place now stands a garish, modern, brick-red windowed Spar general store. This lies half way up the hill by the old village green and the war memorial. There is a preponderance of Georgian facades to the houses and a good selection of small, individual shops. Inns there are a-plenty and the Gothic blue brick school now houses a library and a health centre. The Methodist Church of 1802, red brick with a pedimented porch and round-headed windows. Most of the 18th and 19th Century housing spreads northwards around a tangle of attractive lanes and down the steep slopes of Bridge Road. This is a delightful highway and is the original road to the Iron Bridge. It follows the valley of the Benthall Brook and it was here that the Benthall furnaces and potteries were situated. The famous mill that was powered by a huge water-wheel still stands in the woods near the bottom of the hill. (*See Benthall.*) These industries caused a nuisance to travellers – the road was sometimes blocked with slag and piles of iron bar – and a new road, the Ironbridge Road, was made in 1828. This cut E through Ladywood and then turned S to join the Broseley-Bridgnorth road. Near the junction is the Lady Forester Hospital, a big rambling Victorian looking building embowered in trees and bushes close to the church. Most of the 20th Century developments have taken place SE of the old town; to the Dunge in the S and towards Folly Farm in the E, an area of old coalmines and clay pits. Adjacent to the modern John Wilkinson School in Coalport Road is a rare little black and white timber-framed cottage, a tiny place at the head of the valley that leads to the Calcutts. One of the oldest buildings in the area stands beside one of the most recent. It is noticeable that whilst the old brick houses are a pale, orangey yellow colour (the colour of the local brick clay) the modern houses are an unsympathetic red.

BROUGHALL *1½m. E of Whitchurch.*
An unremarkable red brick hamlet of farms and cottages on the A525 Whitchurch to Newcastle-under-Lyme road. There is a saddlery and a timber-framed house. The name Broughall is Old English and could mean 'either the hall or the secluded place by either the fort (*burg*) or the burial mound (*beorg*)'.

BRYN AMLWG CASTLE *8m. W of Clun.*
It lies remote, a long forgotten castle close to the border with Wales (located at Ordnance Survey map reference SO.167.846) in the Forest of Clun. Earth ramparts form an oval some 50 yds. by 25 yds. in diameter on a knoll with steep slopes to the N and W and marshes to the S and E. Very little is known about the castle. It was probably built by Helias de Say in about 1142 after he killed 2 of the sons of the Welsh Prince of Maelinydd, Madog ap Idnerth. It then passed to the Fitz Alans of Clun and in the early 13th Century they replaced the timber defences with a stone fort. The castle was purely a military building and was never occupied as a house. The NE corner of the earthworks have been damaged by quarrying and the construction of an access road. The stonework was substantial. To the S are the foundations of a tower 35ft. in diameter with walls 10ft. thick. To the N (adjacent to the quarry) are the remains of a gatehouse with round towers about 20ft. in diameter with rectangular guard rooms to the rear. This gatehouse and the small 'D' shaped towers in the courtyard sides are probably late 13th Century and may have been built by the Welsh Prince Llewellyn (d.1282) who held the castle from 1267 to 1276. Bronze Age implements have been found near the site.

BUCKNELL *4½m. E of Knighton (Powys).*
The substantial and very pretty village of Bucknell lies on the banks of the River Redlake in lovely, lush country close to the Herefordshire border. The church of St. Mary stands on a raised, circular mound, banked at the edges, in the centre of the village. There is re-used Norman masonry in the nave and an early 14th Century chancel but the whole was heavily restored-rebuilt in 1870. The tub-shaped font is either late Anglo-Saxon or early Norman. In the churchyard are some large, old yew trees. In the village there is: the motte of a Norman castle, next to the football pitch; a freestone Gothic railway station of 1860, next to the level-crossing; a garage; a Post Office; a general store; two Victorian Chapels; some timber-framed buildings including the Old School house and the delightful Weir Cottage; a pub, the Sitwell Arms; a fertilizer store; a timber merchant; and a variety of substantial houses in brick and stone. The whole combines to make a most attractive village, one of the most handsome in the county. Even so there are none of the old craftsmen and tradesmen that gave villages much of their human character and most of the residents travel elsewhere to work. The name Bucknell could mean either '*Bucca*'s hill,', Bucca being a personal name, or 'buck-hill', from *bucc*, Old English for 'a male deer'. One mile SE of Bucknell is the wooded hill called Croxall Knoll on which are the earthworks of a large Iron Age fort-settlement. This is one of the possible sites of the last great battle between the Romans and the British, led by Caractacus.

BUILDWAS *2½m. WNW of Ironbridge.*
The River Severn enters its gorge here. The enormous cooling towers of the electricity generating power station dominate the landscape, not unpleasantly. Buildwas Abbey, a Cistercian monastery, is a fine and substantial ruin, now cared for by English Heritage. It was built of Grinshill stone between 1135 and 1137 by Roger de Clinton at a well-used river crossing. The Abbey had an elaborate system of fish ponds and 13 granges (outlying farms, often many miles away). In the16th Century it even had its own ironworks; on the Dissolution of the Monasteries this was taken over by secular operators and was quite possibly the beginning of the huge Coalbrookdale industry. The present Pratt truss steel bridge over the Severn at Buildwas was erected in 1905. It replaced Telford's far more graceful iron bridge which had been made unsafe by movement of the river banks. The banks still move and periodically even the modern bridge has to be jacked up and the expansion plates adjusted. In the village there are 2 school buildings, one Victorian the other black and white; a village hall; a Post Office, and a small housing estate. The church of Holy Trinity was largely built with stone 'quarried' from the Abbey ruins. The fertile land upstream in the flood plain of the Severn has been farmed from the earliest times. Bronze and Iron Age settlements are now being identified by crop marks recorded on aerial photographs. (Plants situated above an old trench which has been levelled by in-filling with top soil are able to get more moisture and nourishment than adjacent plants. The better nourished plant grows greener than its companions. Similarily, a plant that grows over buried stonework has less nourishment and grows less green. These effects are heightened during a period of drought. In practice there are only a few days in the year when these differences are sufficiently marked to be recordable.) The name Buildwas is probably derived

Weir Cottage, Bucknell.

Plaish Hall, Cardington.

The road south of The Bog.

The superb Norman arcade at Buildwas Abbey.

from the Old English *gebyldu-waesse* meaning 'a building in a swamp'.

BURFORD *7m. SE of Ludlow.*
Burford is situated in luxuriant country and lies on the banks of the River Teme which here marks the boundary between Shropshire and Worcestershire. It is situated at an important river crossing and is joined to Tenbury Wells, on the opposite bank, by a bridge. Today Burford is a humble place with little to show that it was once of some local importance. The village is dominated by the Swan Hotel with its large bow window. Between the river and the main road, the A456, is a mound almost certainly the motte of a Norman castle constructed to control the river-crossing. Nearly 1m. W of the village, standing on the site of the Saxon castle, is Burford House. The house has small, modern gardens, which are open to the public, and a large plant nursery made very unsightly by the use of plastic-tunnel greenhouses which are clearly visible from the main road. The National Clematis Collection is held at the gardens and in the Old Stable Block there is an exhibition of the development of the genus from the 16th Century to today. Close to Burford House is the red sandstone church of St. Mary. It has some Norman masonry in the chancel but was much restored by Sir Aston Webb in 1889. St. Mary's is known for its monuments which include memorials to Elizabeth de Cornewayle, c.1370, and Princess Elizabeth, daughter of John O'Gaunt, d. 1426. The name Burford is probably from the Old English *burg-ford* meaning 'the land by the fortified place'. Usually *burg* referred to a pre Anglo-Saxon fort – Roman, or Iron-Age or earlier. In 1066 Burford was held by Richard Scrope (or Scrobb), a Norman who had settled in England before the Norman Conquest. By 1086 it had passed to his son Osbern and became the *caput* of his Shropshire barony. Osbern was also lord of Richard's Castle and had estates in Herefordshire. The barony descended to the Mortimers by marriage. The manor of Burford had many members, including Greete, Tilsop, Stoke, Watmore, Weston, Whitton, Bonaston, Harthall, *Mulla* (probably Court of Mill), *Cromwode* (a lost settlement), and Wn'tone (probably Wootton). Burford itself was a sizeable place in 1086 with 24 smallholders, 12 villagers, 7 freemen, 6 slaves, 3 riders, 2 priests and 2 mills. (It is now believed that the Domesday Book count did not include a large part of the population, namely those who owed no service to the local landlord).

BURLINGTON *3m. NNE of Shifnal.*
A tiny hamlet consisting of 2 large, brick farms and a row of stuccoed cottages in attracttive country just N of the A5 (Watling Street). It stands by a tributary stream of the River Worfe which is crossed here by a ford. The name Burlington probably means the settlement of *Baerla*'s people. *Baerla* may be derived from the Old English *bar*, meaning 'a boar'. There are ferns and old oaks in the hedgerows along the main road and the land is gently undulating.

BURLTON *7m. SE of Ellesmere on the A582.*
The name is from the Old English *burg-hyll-tun*, meaning the 'settlement by the hill-fort'. It is a small main road village built largely of red brick. This area suffered during the border wars and was several times devastated by the Welsh. It is believed that after the Black Death (1349) the village was abandoned and rebuilt about 1m. from its original position. The land is mostly used for mixed dairy farming and there was a local tradition of butter and cheese making. Pigs were an important part of the local economy for they ate the whey. Most villagers kept a pig or two at the end of their gardens. Today the blacksmith, the wheelright, the brewer, the gardeners and grooms are all gone. The Burlton estate was sold in the 1970's and today the village has a sparse look to it. There is a little Sunday School of 1891, a good half-timbered house with ornate chimneys on the road to Loppington and ¾m. further along this road is Burlton Grange, a mock Tudor house which incorporates parts of an older house. **Brandwood** lies 1m. W of Burlton, a scattered farming hamlet on drained ground. The names means 'burnt-wood' a name associated with comparatively late woodland clearance.

BURWARTON *9m. SW of Bridgnorth.*
The name means '*Burgweard*'s homestead'. It is an attractive stone-built village on the B4364 Bridgnorth to Ludlow road below Brown Clee Hill. The splendid park of Burwarton Hall with its fine collection of trees lies on the slopes of the hill. The Hall (1877) was designed by Salvin but has been reduced in size. The occupier of the house and owner of the extensive Burwarton Estate is Lord Boyne, Mr. Hamilton-Russell, the 10th Viscount. The village as a whole is a handsome place with many substantial buildings such as the rock-faced grey stone Vicarage (1893) with its tall red brick chimneys. This was once the Dower House to the Hall but is now privately owned. The chapel-like grey stone school is of 1886. Trees are a feature hereabouts. There is a magnificent hornbeam in the churchyard and along the main road there is an avenue of copper beeches. Opposite the cream-washed Hall Lodge is a rhododendron-lined path leading to the rock-faced church of St. Lawrence (1886), also by Salvin. It was recently de-consecrated and is now being converted to a house. The old, ruined and ivy-clad church is just to the NE across a small brook. The village pub is the Georgian stone-built Boyne Arms. High above the village is the summit of Brown Clee Hill, called Abdon Burf (1792 feet), from where, on a clear day, 15 counties can be seen. There was once a hill fort here but this was destroyed by the Abdon Clee Stone Quarrying company. They left in 1936 and the hill is quiet now.

BURNHILL GREEN *6m. SSE of Shifnal.*
A red brick hamlet with a cream-painted pub, the Dartmouth Arms, which faces the tiny village green. Behind the pub are the shallow earthworks of a medieval moated site. There is a farm with several ugly modern barns and a substantial, gabled, brick schoolhouse erected by the 6th Earl of Dartmouth. 'Burn' in a place name is usually from the Old English *burna* or *burne* meaning 'a stream'.

BUSHMOOR *5m. SSW of Church Stretton.*
It lies on the old Roman road. From Bushmoor to Marshbrook, 1½m. NNE, there is an abandoned stretch of this ancient highway. The A49 follows the Roman route for much of the way from Stretton to Ludlow but here it takes the lower land to the E. Bushmoor is a mature village of old red brick and stone cottages some of which are clad in stucco, and a few new houses and bungalows. There are holly hedges and stone boundary walls and in one garden there are numerous model aircraft perched on poles; in another is the remnant of an orchard. **Leamore Common** is a straggle of cottages that line the old Roman road between Bushmoor and Wistanstow, 1¼m. S. Amongst them are 4 black and white dwellings. At **Felhampton**, ½m. E on the busy A49, is a tall brick house that must have caught many an eye. It is called Felhampton Court, an early Georgian building of 5 bays and 2½ storeys with a steep 3-bay pediment in which is an arched window. Below that is a semi-circular window and below that a Venetian window. Inside there is a fine staircase. This is a country of dips and raises with larger hills dark in the distance.

CALVERHALL *5m. NW of Market Drayton.*
Calverhall is a small village that lies deep in the woods and lanes of rolling pond-ridden country in the north of the County. The 'Jack of Corra' was formerly kept here, at the Old Jack Inn. This was a leather flagon with a silver rim from which any traveller could drink his fill for one penny. In the main street there is a range of humble red brick almshouses of 1724. Some of these were demolished by the architect Eden Nesfield to make room for his new church of Holy Trinity (1872 and 1878). Inside there is a good William Morris and Burne Jones window of about 1875 depicting two women in white. On the other side of the church is a bowling green and a popular one at that. There is no shop and no Post Office but there is a well-used Village Hall. Half-a-mile E of the village is Cloverley Hall, constructed to a design by Eden Nesfield in 1862. A substantial part of the house has been demolished but what is left is still handsome – Elizabethan style red brick with mullioned and transomed windows and a tower. The Hall is set in an attractive, wooded park grazed by cattle and is now used as a Christian Study Centre. (*See Willaston.*)

CANTLOP *4m. SSE of Shrewsbury.*
A hamlet of farms and colour-washed cottages at the end of a dead-end lane in gentle, rolling country. A quarter-of-a-

Child's Ercall.

St. Mary's, Caynham.

View from Fiddler's Elbow, Chapel Lawn.

mile NW is the well known Cantlop Bridge, built in 1812 to a design by Thomas Telford who was County Surveyor for Shropshire at the time. It carried the Acton Burnell–Shrewsbury turnpike road over the Cound Brook. Telford's elegant iron bridge has now been bypassed but stands as a monument under the care of English Heritage. A section of the old road is now a lay-by and picnic area. The name Cantlop may mean 'the settlement in the valley of the Cound Brook'.

CARDESTON *6m. W of Shrewsbury.*
Cardeston is a tiny place just off the A458 Shrewsbury to Welshpool road. The Anglo-Saxon manor was held by Leofnoth in 1066. Under the Normans it became absorbed into the manor of Wattlesborough where they had built a castle. All that remains of the late 12th Century Norman church of St. Michael at Cardeston is a window with a round arch in the S chancel wall; of the church we see today the nave and chancel are of 1748 and the tower is of 1844. The work of 1748 was made Gothic in the 19th Century. Lady Leighton made a contribution of £500 towards the cost of rebuilding the tower. Her miniature engraving of the 'Ladies of Llangollen' (Miss Sarah Ponsonby and Lady Eleanor Butler) was extremely popular in Victorian times and prints changed hands for large sums of money. The church organ is of an unusual barrel design and was made by T. C. Bates, of London, in 1850. Although many of the residents of the village are commuters some are still employed in agriculture. Two of the Smith family have represented Britain in the World Ploughing Championships. The village stands on the edge of a breccia limestone outcrop called Cardeston Rock. This was widely used locally as a building material and the old quarry can be seen alongside the Welshpool Road. The village had a limeworks that operated between 1660 and 1817.

CARDINGTON *4m. NE of Church Stretton.*
The name Cardington is probably from the Old English *cenreding-ton* meaning 'the settlement of *Cenred*'s family (or tribe)'. In 1086 it was held by Reginald the Sheriff and was a sizeable place for its time with 21 families recorded as being resident there. In the early Middle Ages the manor was given to the Knights Templar (who were established nearby at the now deserted village of Lydley) by William FitzAlan and returned to him when the Order was suppressed about 1311. Cardington is situated in the Stretton Hills, lost in a maze of lanes, and is a most attractive village with considerable character. Most of the old buildings are constructed of stone and though there are some modern houses of brick they do little harm. The most recent development of detached stone houses is very tasteful indeed. Some weather-boarded barns, a Victorian school and the occasional black and white house together provide a touch of architectural variety. The weather-boarded barn of the house called Maltster's Tap is dated at 1558 and is an example of a Shropshire Longhouse, few of which remain. Sad to say, the living heart of the village went with the closure of the school, the Post Office, the 2 shops, the bakery, the blacksmith's shop, the pottery and the undertaker's parlour, all now long gone. But the church still stands. St. James' has a Norman nave, an Early English tower with Perpendicular upper parts and chancel windows of about 1300. Inside there is a memorial to Chief Justice Leighton (d.1607) who lived in the splendid **Plaish** Hall, 1m. NE of Cardington. Built around 1540 to an 'H' plan this was the earliest large house to be built of brick in Shropshire. The red brick walls are decorated with blue brick ornaments and have stone dressings. The Judge is reputed to have bargained with Mr. Sherratt, a bricklayer he was about to sentence, that the man could go free if he built him the finest chimneys in the county. The bricklayer slaved for many months and everyone agreed that his work knew no equal. However, the judge broke his promise and had the bricklayer hanged from one of the chimneys he had just created. The bricklayer's ghost is said to haunt the Hall and he has the company of a little grey lady. (About 1916 a skeleton was removed from one of the chimneys and quietly buried in Cardington churchyard.) The name Plaish is from the Old English *plaesc* meaning 'a shallow, or marshy, pool'. On the opposite, northern bank of the valley, is Hoar Edge, famous for the sandstone roofing slates once quarried there. Examples can be seen in Much Wenlock and at Pitchford Hall. Near Gretton, ¼m. E, are substantial earthworks, the remains of a long abandoned manor house. Beyond Cardington, to the W, is The Wilderness, a steep hill that borders Caer Caradoc Hill and its large and dramatic hill-fort earthworks. Chatwell Hall lies 1m. N of Cardington. It stands on the curiously named Yell Bank, a 17th Century stone and timber house once a home of the Corfield family. Chatwell Farm is a Queen Anne house of stone with gables to both sides of a recessed centre.

CASTLE HOLDGATE *See Holdgate.*

CASTLE PULVERBATCH *See Church Pulverbatch.*

CAUGHLEY *See Stockton and also Coalport.*

CAUS CASTLE *See Westbury.*

CAUSEWAYWOOD *2½m. NE of Cardington.*
Causewaywood is a tiny hamlet near a stream on the lower slopes of Lodge Hill just over a mile NE of Chatwell Hall. Charlotte Burne in her Shropshire Folklore writes: "On the road between Acton Burnell and Cardington is a stretch of rude pavement from two to three hundreds yards in length, known as the Devil's Causeway. It once formed part of a Roman road leading from Wroxeter to Rushbury, and in the judgement of C. H. Hartshorne, the local antiquarian authority, was never more than an average thirteen feet wide. But in the summer of 1881, a country woman in the immediate neighbourhood was found to declare that the causeway as it at first existed was a quarter of a mile long, and no less than a furlong (220 yds.) wide. The devil laid it in a single night, and he haunts the scene of his labours still. If you cross the causeway at midnight you will meet him, in the shape of a black man, with cow's horns and hoofs, riding on a white horse. There is no danger in the meeting to anyone with a good conscience, going on a lawful errand. The devil passes by all such persons like a flash of light. But if any man be going on a bad errand, or have lived a careless, godless life, the devil will set upon him, and struggle with him, and leave him half dead". Today, local people will tell you, it is a Headless Horseman who will attack you. The stone road lay in front of Causewaywood Farm (SO.525.986) but has been tarmaced over and is no longer visible. The sandstone bridge, a single arched tunnel, over the stream is called the Devil's Bridge and up the hill, near the first bend, is the Devil's Well. (Not a deep well, but a 'welling' of water from the rocks.) The rest of the settlement lies just N of the farm – stone built cottages and a red brick house. One dwelling has 2 mature yew trees and holly bushes are common here. Holly was used as a winter fodder in olden times. There are stone boundary walls and sheep graze the meadows. The escarpment to the S is called Netherwood Coppice, but is mostly clad with scrub, dead red fern and brambles with patches of rough pasture.

CAYNHAM *2m. SE of Ludlow.*
The name is English and means 'the settlement by the Cay Brook'. Caynham is thought to be one of the earliest English settlements in South Shropshire. (Shropshire was colonised by the Angles, not the Saxons.) With lowland fields close by and the coal, copper, ironstone, limestone and dhu-stone of Titterstone Clee, 2m. away, villagers could be both miners and farmers. The slopes above Caynham are spotted with small villages and single cottages. Centuries before the roads were constructed donkeys and women were used as pack animals. There is a record of coal being sold at Caynham by Wigmore Abbey in 1235. During the Middle Ages Caynham belonged to the Mortimers, one of the powerful Marcher Lord families. The church of St. Mary has an unusual Norman chancel arch, but was heavily restored in 1885. In the churchyard is an old, weather-worn cross about to be preserved by English Heritage. Caynham Camp lies ¼m.N. This is a Bronze Age – Iron Age settlement-fort. Evidence has been found of rock-cut ditches and stone ramparts, indicating a formidable structure. The site covers 8 acres and evidence of occupation from about 1000 B.C. has been found. Today, the area about the church is still pretty but the modern settlement to the W has little to commend it. Here are council houses; a tangle of overhead telephone cables and power

lines; a development of mock-Tudor cottages on the land around the long neglected, large, red brick house called Caynham Court; a weather-boarded village hall painted black and white; a tiny stone-built National School of 1834; a pair of superior Elizabethan style stone houses; and a plaque marking the location of the old Animal Pound where stray animals were kept until claimed by their owners and payment of a fine.

CAYNTON *2m. NW of Edgmond, which is near Newport.*
There is no village as such here but there are 2 mills, an old manor house and the handsome Caynton House, all set in rich Shropshire farming country with big fields and big views to the Wrekin. Caynton House (SJ.704.219), for long a home of the Yonge family, is splendidly positioned at the end of a long drive on a low rise with woods to the rear and great, open cornfields to the fore. It is a symmetrical late Georgian house, 7 bays wide and 2½ storeys high, with 1½ storey wings of one bay with blank arches. The doorway has Doric pilasters and a segmental fanlight. The Manor House lies ½m. SW on the edge of the Meese valley. It is of red brick, 5 bays wide and 2½ storeys high with blocked windows and a later addition to the left. Down the bank, to the W, is Old Caynton Mill, also red brick and equally uninspired as the Manor House. The valley is delightful though, a watery place where the river flows through two widely separated channels. The present mill is of the early 19th Century but there has been a mill here for several centuries. It has been used for crushing linseed, grinding corn and grinding animal feed. In this century the original waterwheel was replaced by a turbine and then by an electric motor. The mill then became derelict and was recently converted into a house. New Cayton Mill lies 1m. N at Howle. There has in fact been a mill here for several centuries also, despite its name. In the early 18th Century iron was worked at a forge powered by the mill. It was part of the empire of Richard Ford (d.1745). His family business went bankrupt in 1756. Iron furnace slag can still be found in the vicinity even though most of it has either been used for road metal or been taken away for re-smelting. The mill building is of red brick on a red sandstone base; the water supply leat was divided into 2 and powered 2 separate undershot wheels. The mill ceased to work just before the First World War. In 1982 a Bronze Age (c.700-700 B.C.) axe-head was found in a field near Caynton Manor. The name Caynton means 'the settlement of *Caega*'s people'.

CEFN EINION *3m. SW of Bishop's Castle.*
Cefn Einion lies about half way between Colebatch (on the A488) and Whitcott Keysett (5½m. SSW in the valley of the River Clun). It is a journey that can be recommended for the splendid scenery and unspoiled hamlets encountered along the way. The traveller leaves Colebatch on a narrow twisting lane that follows a stream into curved, conifer-clad hills and meadows grazed by sheep. The big, many coloured plantation to the left is Blakeridge Wood. A Forestry Commission road leads through this to the steep little valley called Hell Hole. When we visited there was a 'Beatnik' coach parked here with the front wheels removed so that it could not be easily towed away by the authorities. The main lane continues up the hill, past 3 isolated cottages on the opposite side of the valley, over the summit of the pass and down to Cefn Einion. The name can be interpreted as 'the settlement behind the anvil-shaped hill'. There are a few stone cottages clad in stucco and 5 lanes meet here. Our road descends to the valley bottom, crosses the River Unk by way of a stone bridge, and climbs the hill to Shadwell Hall. There are good views of some fine hill country hereabouts. The Hall is an irregular, stuccoed farmhouse with some good traditional stone outbuildings and some ugly modern barns. The name Shadwell probably means 'the homestead near the boundary stream', presumably a reference to the River Unk. To the left of the lane there are lovely views, a patchwork of pink, green and yellow fields dotted with white sheep and black cows. At the crossroads of **Three Gates** there are 2 farmhouses both with tile-hung elevations and more good views. The track descends quite steeply to the grey stone hamlet of **Mardu**. The name could mean 'the place of the black lake'. There is no lake but 3 streams meet here and there may have been a pool or a marsh at one time. Offa's Dyke touches the settlement to the W. Our lane crosses a babbling brook and leads downhill into a gentler, closer landscape. Scattered cottages and farms line the route, amongst them a grey stone chapel with yellow brick dressings and round-headed windows. And so to the village of **Whitcott Keysett**. The name can be interpreted as meaning 'the white cottage by *Caega*'s hill'. It is a delightful and unspoiled little place. The settlement stands about a grassy banked stream: a row of old stone cottages; farms with weather-boarded barns, one of which is used as a public notice board; new houses in buff-coloured brick; Lake House, rendered with 2 projecting gables; a prominent red brick house with a hipped roof; a timber-framed barn, open at the top and stone-walled at the bottom; and alongside one dwelling a collection of very ancient motor cars of some quality covered in moss and quietly disintegrating. There is said to be an 81ft. menhir hereabouts. The lane continues E in the valley of the River Clun to **Bicton**, 'the settlement at the foot of *Bica*'s hill'. A track runs W, past the disturbed ground of a grassed-over medieval moated site, to the River Unk which can be either forded or crossed by a bridge. A black and white house with dormer windows stands on the opposite bank and trees fringe the waterway. This is a tranquil and picturesque little spot that has an air of timelessness. The rest of the hamlet is more mundane: a few stone cottages, a big modern house and a substantial stone-built farm with extensive outbuildings. There are a noticeable number of large arable fields amongst the gently rounded hills. The traveller now has a choice of 2 lanes to return him to the A488. The town of Clun is 1½m. SSE.

CHADWELL *3½m. SE of Newport.*
It lies ½m. W of the A41, a red brick hamlet with 2 substantial houses, one dated at 1731. Chadwell Mill is a renovated working water mill. There is a tea house and guided tours are available. The cottages and outbuildings are variously dated at 1881 and 1850. There are large arable fields hereabouts. The name Chadwell could have been derived from either 'St. Chad's well' or the Old English for 'the cold spring'. In fact St. Chad (d.672) is reputed to have blessed the well, which is at the end of the mill pond furthest from the mill (SJ.786.143). There is a little wooden bridge and a circular, stone walled structure. The spring itself appears to be on the other side of the road.

CHAPEL LAWN *3m. NNE of Knighton.*
Chapel Lawn – a 'lawn' was a grassy clearing in a forest – lies in the lovely Shropshire hill country on the road between New Invention and Bucknall which follows the valley of the River Redlake. North of the hamlet is Hodre Hill, and to the S looms the mass of Caer Caradoc on the summit of which is an Iron-Age fort. This hill is one of several that could have been the site of the Britons' last stand against the Romans. The British leader was Caractacus, hence Caradoc. The earthwork defences of the fort-settlement are quite elaborate and those about the western entrance are noted for their ingenuity. At least 5 hut-circles have been identified and a pit believed to be a grain store is located in the centre of the camp. (For further information on this site and many other forts, castles and moated sites see the Victoria County History – Shropshire, Vol.1.). On the S facing slope of Hodre Hill is Bryneddin Wood. This is an almost pine oakwood but growing high up the slope is a solitary dark green pine. On the shortest day of the year (21 December) between 3pm and 4pm the shadow of the hill fort on Caer Caradoc, which is on the opposite side of the valley, just touches the Bryneddin Pine. The oak wood itself is still common land and the villagers still hold strips that used to be marked by 'meer stones' that ran up the hill. These tenures date back to Medieval times. The church of St. Mary at Chapel Lawn is of stone with lancet windows and was built in 1844 to a design by Edward Haycock. The original chapel was a part of Chapel Lawn Farm (1600). The Woodhouse was formerly a pub, the Woodcocke Inn, and the Quern is an 18th Century stone mill that ceased to operate in 1927. The school and schoolhouse were built in 1866 on what was the village green. **Pentre** lies ½m. NW. It is a tiny hamlet of stone cottages and weather-boarded barns that shelters at the foot of Hodre Hill. It is a charming little place but an ominous planning application notice tells of someone's intention to demolish some barns and build 4 new houses in their place.

Half-a-mile NE of Chapel Lawn is **Pentre Hodre**. It lies on Hodre Hill, high above the river, a handful of stone cottages and barns made ugly by having been repaired with corrugated iron instead of timber. Just up the hill is **Obley**. Smaller than this you cannot get. It is a little unkempt but the splendid hills make up for man's untidiness. Standing all alone is a Primitive Methodist chapel in stone with yellow brick dressings and dated at 1880. **Fiddler's Elbow** lies ½m. NW of Obley. There are no dwellings, only a parking area in the Forest of Black Hill. The name refers to a sharp bend in the road. Just W of here there are splendid views over the South Shropshire hills and dales towards Offa's Dyke. At **Pen-y-cwm**, 1¼m. W, there are 3 cottages and numerous horses, goats and turkeys. In the middle of the moor by a treacherous bog three young lads were playing cricket when we visited.

CHARLCOTTE *1m. S of Neenton which is 6m. WSW of Bridgnorth on the A4364.*
The name Charlcotte means 'the cottage of the free peasants', the *cearls*. At Cinder Farm are the remains of the Charlcotte Furnace (SO.637.861) situated in the gentle valley of the Cleobury Brook. Large mounds of tree-clad charcoal furnace slag surround the site. The stone-built furnace is some 25ft. square and 25ft. high and though neglected is still in remarkably good condition. The interior is especially good and the whole structure ought to be repaired and conserved. Trees and bushes grow on the top of the furnace and these will cause much damage very quickly. Just to the N of the furnace are the foundations of an ore-house and incorporated into the farm buildings some 200 yds. away, are the remains of a paper mill. The furnace mill and the paper mill shared the same water supply channel (leat); both were powered by water-wheels. The furnace water-wheel operated a bellows which blew air into the mix of charcoal, iron-ore, and limestone to make it burn more hotly. It was a failing supply of timber for fuel and competition from the new coal burning furnaces that caused Charlcotte to cease operating in the late 18th Century. The furnace mill was then converted and became a second paper mill. During the 18th Century the furnace was in the control of the Knight family of Downton (Herefordshire) who also owned ironworks at Bouldon, Bringewood (near Ludlow) and Abdon. In a good year Charlcotte produced 400 tons of crude iron. Much of this was taken by cart to Bridgnorth where it was stored before being shipped downstream to Stourbridge. Cinder Farm is a red brick house of 3 bays and 2½ storeys with dormer windows. Sheep graze the fields hereabouts. Charlcotte is most easily approached from Neenton on the B4364. Take the lane signposted to Wrickton and on reaching a pair of semi-detached houses turn hard right down a rough track. This leads to Cinder Farm and the furnace.

CHELMARSH *3m. S of Bridgnorth.*
A village on a ridge with easterly views over the reservoir and its new marina, and across the River Severn to the woods of Dudmaston Hall. The 14th Century church of St. Peter has its admirers, especially of the row of Decorated windows along the S wall. There is a Norman N doorway and the tower was built in 1720. Legend has it that a monk's heart is buried high in the E wall. There is a rare tower clock in the church. Chelmarsh Hall is mostly Victorian but at the back are some blocked doors and windows and inside are some pointed doorways, all of which are believed to date from the 13th Century. It is a most charming building and has great character. By the River Severn is Stern's Cottage, a haunted house that stands on the site of an older building. It was once used for stabling the horses that hauled the river barges and was later lived in by a Severn Valley Railway worker. Phantom figures and voices manifest themselves here, and machinery ceases to function for no apparent reason. Ghostly miners have been seen at the old mine shaft and engine house (which collapsed in 1984) on Chelmarsh Common. Apples for cider and damsons for dye used to be grown hereabouts. In Occupation Lane there was a brick kiln which used local clay. The 'Chel' in Chelmarsh may be from the Old English *cegal* meaning 'a peg', or 'a pole'. It has been suggested that the narrow ridge upon which Chelmarsh stands was nicknamed 'the peg' because of its shape.

CHENEY LONGVILLE *2m. NNW of Craven Arms.*
An unspoiled hamlet that lies along a minor road to the West of the A49. It is a linear settlement of mostly stone-built cottages and farms. Indeed the white and cream painted Georgian house and the weather-boarded barn are conspicuously different from their neighbours. A stream runs alongside the south side of the lane and each house has a little bridge across it. The lane is wide and the 'pavements' are lawned. To the east lies the Manor House, almost hidden from the road; to the west, in the fork of the lanes, is a circular earthwork, probably a prehistoric fort but possibly of Norman origin. The left-hand lane (presuming one is approaching from the east, off the A49) leads to Cheney Longville Castle. This was once a real castle and was besieged during the Civil War. Licence to crenellate was obtained by Roger Cheney of Acton Burnell in 1395. However, today what we see is a substantial 3-bay, 2-storey stone house with dormer windows and an arched carriage doorway in the right-hand side range. In all there are 4 ranges around an oblong courtyard. The complex still has its 14th Century plan and with the exception of the house, the exterior masonry walls are 14th Century also; only the crenellations are missing. The house and its farm buildings stand in a valley across which has been thrown an earthen wall, a bay, to contain a pool. The basin lies in front of the house but is now dry. The pool probably served the dual purpose of fish pond and reservoir to the moat of which it was a part. Once again the prospect has been marred by the erection of ugly corrugated iron barns. The village was probably 'planted' at the time the castle was built but it did not flourish and has remained quite small. The burgage plots run at right-angles to the main street. Today an annual country game-fair is held in the village. In the wooded hills about 1m. W is the prehistoric fort-settlement of Wart Hill. The original name of Cheney Longville was *Longfelle* , from the Old English meaning 'the long clearing'. It became Longville after the Norman Conquest; it is indeed a 'long village'. The Cheney family held the Manor from at least the 14th Century.

CHERRINGTON *See Tibberton.*

CHESWARDINE *4m. SE of Market Drayton.*
The village lies on the edge of a bluff. The church and the old castle lie on the high ground and the village stretches southwards down the hill. The country is pleasant but unremarkable. There has been a settlement here for many centuries and the church probably stands on the site of a prehistoric fort. In 1086 the extensive manor belonged to the Norman Robert of Stafford and was in Staffordshire. In the reign of Henry II it was given to the Le Strange family and the shire boundary was moved so that they could consolidate their lands in one county, Shropshire. It is probably their substantial castle mound and moat that we see today. The moat is filled with dark, grey-black water in which dead trees stand stark. The outside banks and the castle-island are covered in a dense growth of mature broad-leafed trees, brambles and ferns. It is a wild, mysterious place. Moorhens scoot across the turgid waters and many species of wild birds sing in the woods. The moat lies opposite Castle Cottage, about 150 yds. down the lane that passes the church and the Victorian brick-built school. It stands adjacent to the lane but is hidden from sight by a 10ft. high earthen, wooded bank. The handsome church of St. Swithin has a good Early English chapel, and a Perpendicular tower. Below the bell openings are carvings of a Talbot dog and the Staffordshire Knot of about 1470. On the SE buttress is a benign, carved lion and on the NE buttress a very good dragon. The rest of the church was restored in Early English style by J. L. Pearson in 1888 and includes stained glass by William Kempe. Opposite the church is a good group of houses which include the sandstone and brick school, built in 1738 and restored in 1877. The Old Vicarage is a large red brick house with a tiled roof laid to a diamond pattern. In the main street there are 2 pubs – the Fox and Hounds and the Red Lion – a Post Office, a newsagents, a butcher and a grocer, who also bakes "Billington's famous gingerbread" which has been made on these premises since 1817. The village is by no mcans pretty and there is much 20th Century housing both in the main street and on small estates adjoining it. Cheswardine Hall was originally called Hill Hall and in fact lies at Chipnall, 1m.

Cheswardine, the castle moat.

NNE of Cheswardine. The present Hall, built in 1875 by the Donaldson-Hudson family, is a neo-Elizabethan mansion with a Tuscan-columned porch. Since the 1940's the Hall has been a Catholic College, a school for 'naughty boys' and is currently a home for the elderly. From 1946 to the early 1970's there was an Artificial Insemination Centre in the stables of the Hall. Though the bulls are now gone there is still an administrative office here, in a new building of 1981. The county boundary with Staffordshire lies just over 1m. E. It has been suggested that the name Cheswardine means 'the cheese-farm on the river bank'.

CHESWELL *1¼m. NW of Lilleshall, near Newport.*
The old name was Chrestill which could mean either 'Christ's Hill' or 'the hill with a cross'. The settlement is overlooked by Cheswell Hill, a rocky, sandstone edge that has been quarried in the past. There are a handful of newish houses, a range of old farm buildings, a marshy pool and a manor house. Two dirt tracks connect it to the outside world and in wet weather there is mud a-plenty here. Such quiet, remote hamlets are few and far between these days; motor cars look totally out of place. The Manor House is 5 bays wide and 2 storeys high with dormers. Like all the buildings here it is of red brick though there is sandstone in some of the farm buildings. Cheswell is on the edge of the Weald Moors. When these were drained a new prosperity came to the area and most of the old sandstone and timber-framed houses were rebuilt in fashionable brick. The boundary walls are still stone though. In recent times the Manor House has been eclipsed by Cheswell Grange. This is a large brick house with gables and tall chimneys and a large number of modern barns. Although you can see it from the hamlet there is no longer a direct vehicular route. The Grange has its own access lane off the Newport to Dorrington road. The estate was owned by the Whitfield family but some years ago the houses were sold off to owner-occupiers and the farmland was purchased by the Prudential Insurance Company. If one leaves Cheswell by the northern lane one passes over a causeway which looks as though it may once have been a pond bay. To the left is a tree-fringed, watery looking hollow; to the other is a huge arable field in which land drains were being installed when we visited. To the NW is the windmill tower of Longford Mill Farm. (*See Longford.*)

CHETTON *4m. WSW of Bridgnorth.*
The village lies on the northern slopes of Brown Clee Hill at the end of a dead-end lane that leads off the B4364, the Bridgnorth–Ludlow road. As one approaches the settlement Watsbatch Manor is passed on the left. This is a handsome red brick house of 2½ storeys and 3 bays with single storey 2-bay side wings. The pedimented porch is supported by Doric columns and the house is screened from the road by a hedge of conifers, an alien feature in an English landscape. The village lies on high ground. There are brick cottages; a row of dull council houses; a corrugated iron barn; a few modern detached villas; a stone-built farmhouse with weather-boarded barns; a pub, the Old Inn, of brick and stone, parts of which are said to be 800 years old; a modern rectory; and the church of St. Giles. The church is hidden by a screen of trees, amongst which is a most handsome yew. The Chancel is 13th Century – see the N and S windows, doorway with shafts, roll mouldings on the arches, and quadrant moundings on the Chancel arch. The nave was rebuilt in 1788; the tower was added in 1829 and the whole was restored in 1891. The village is beautifully situated with wide views over undulating country-side. There is a shallow seam of coal in the area and villagers used to collect it from stream beds and by digging small pits. Before 1066 Chetton belonged to Lady Godiva the wife of Earl Leofric of Mercia (1018-1057). After the Norman Conquest the manor passed to Earl Roger (de Montgomery) and he held it himself. It later passed to the de Broc family. The name Chetton is probably from the Old English *Ceatta*'s-*tun*, *Ceatta* being a personal name and *tun* meaning 'homestead', or 'settlement'. **Criddon** (SJ.663.912) lies ½m. N of Chetton as the crow flies but over 4m. by road. This tiny hamlet was a separate manor in Norman times and belonged to the Fitz Alan family. The name means '*Crida*'s homestead on the hill'. Crida was the founder of the kingdom of Mercia in about 586 A.D., but this is more likely to have been the home of a namesake.

CHETWYND *1¼m. NW of Newport on the A41.*
A most attractive area. The old main road bends around the rocky, tree-clad ridge called The Scaur (from the Old English, *scearp*, meaning 'rugged and steep'). Here we have seen more than one car leave the highway and its occupants come to a nasty end. Below the hill is the 23 acre lake of Chetwynd Park, though the big house stood on the other side of the road ½m. N near the brick Dovecote (which has 700 nesting places) and the Home Farm buildings. The Elizabethan Hall, for long the home of the Borough family, was demolished in 1961 and replaced by a much smaller modern house. King Charles I is reputed to have spent 3 nights at the old Hall in May 1645 whilst en route to Naseby. A leat from the River Meese leads to the old laundry where the remains of an undershot waterwheel are still in situ. It was used to both pump water and to generate electricity. Talking of water, close to the nearby Puleston Bridge (1812) is a Severn-Trent Water Authority bore-hole and pumping station (1975) that produces 400,000 gallons a day! In 1981, during dredging operations in the river, a 10ft. section of a prehistoric dug-out canoe was found and is now at the National Maritime Museum in London. In 1988 the Deer Park, the lake, 2 attractive lodges and much of the agricultural estate of the Hall were sold off to a speculative development company. They immediately auctioned off the lodges and began to decimate the woodlands, much to the chagrin of local people. Mind you, the estate had already been savaged by having been cut through by the Newport by-pass. This was undoubtedly needed, but it did not hve to run through an ancient and most attractive deer park. The church of St. Michael (1857) lies near the intersection of the old A41 and the new road. It is very good Victorian in rock-faced sandstone with a broach spire by Benjamin Ferrey. Inside there are round marble piers with extravagantly carved capitals. The churchyard is said to be haunted by the ghost of Madam Pigott, the mistress of the old Elizabethan mansion in the 1770's, who died in child-birth. It is said that when doctors told her husband only one could survive he replied: 'lop the root to save the branch.' In fact both died. The ghost of Madam Pigott has been seen many times in the Deer Park, sitting on a rock combing her baby's hair. Her favourite place is Cheney Hill which used to be steep and high-banked. Here she would alarm travellers on horseback by leaping up behind them and clinging on until they came to running water which she refused to cross. Cheney Hill is said to have got its name because horse drawn waggons had to have their rear wheels chained to stop them skidding down the steep, loosely surfaced incline. The name Chetwynd is from the Old English *Ceatta-gewind*, *Ceatta* being a personal name and *gewind* meaning 'a winding path up a hill'. In Anglo-Saxon times the manor was owned by Lady Godiva. After the Norman Conquest it was given by Roger de Montgomery to Tarold (de Chetwynd). His family held it until 1381 when it passed to the Peshalls who lived there until the end of the 15th Century when it came into the possession of the Pigott family. They held it for 11 generations. In 1803 the estate was bought by Thomas Borrow, who came from a family of Derbyshire ironmasters. He changd his name to Borough and by the 1880's the family was a major landowner with estates in Shropshire, Derbyshire and Staffordshire.

CHETWYND ASTON *1m. SSE of Newport.*
Along the main road there are modern houses, a fencing firm, a few old cottages and the Wheatsheaf Inn. The main settlement, however, lies off the road along a lane to the E. It comes as happy surprise to find a small, almost untouched farming community. Almost, for The Grange is a modern bungalow, out of place amongst the red brick farms and cottages and the Georgian charm of Stoneleigh House. (To the rear of the latter is a half-round tower with a steeply pitched conical roof.) At the end of the lane is the rather strange looking Field Aston Manor. This is a mature, brick-built, 3-bay house with big mullioned windows of awkward proportions and a cluster of 4 chimneys in one off centre block and another at the gable end. There are stone walls in some of the old barns and stone boundary walls around some of the cottages such as those that

Chirk Bank, the viaduct and the aqueduct.

front Thistlewood, dated at 1701. The group of buildings adjacent to the manor house stand around a small green in a rough circle, the classic early Anglo-Saxon settlement pattern. Such groupings are quite rare nowadays. To the E is a rectangular reservoir pool and a farmers' access bridge over the new A41T, the Newport by-pass. The name Aston means 'eastern settlement'. In 963 Church Aston and Chetwynd Aston were collectively called *Eastun* in a land charter granted by King Edgar to his thane Wulfric. Later, the 2 settlements, which are both in the parish of Edgmond, were called *Magna Aston* (Chetwynd Aston) and *Parva Aston* (Church Aston). Magna means large; parva means small. Presumably the manor of *Magna Aston* came into the posession of the Chetwynd family of Chetwynd, just N of Newport, and became Chetwynd Aston.

CHILD'S ERCALL *2½m. SW of Hinstock which is 5m. S of Market Drayton.*
A red brick village that stands amidst flat land, close to a wartime aerodrome. The church has a 12th Century chancel doorway, an arcade of the 13th Century, an aisle of the 14th Century and good stained glass windows by Kempe. There is, or was, a pool hereabouts with a mermaid. She enticed two farmers into the water with the offer of a golden treasure. Her intention was to drown them but as one farmer reached for a lump of gold he cried 'Jesus, this is a bit of luck', at which the mermaid screamed and dived deep into the water. It was the use of Jesus' name that saved the men from a watery grave. Local people today are not clear as to the location of this pool, and many think that it is now dry. The name Ercall is thought to be an old Celtic name, pre-dating the Anglo-Saxon period (when the majority of settlement names were established). *Child* was a medieval term meaning 'a knight'; there are traditional epic ballads entitled Child Waters, Child Owlet and Child Maurice.

CHIPNALL *1¼m. NNE of Cheswardine which is 5m. SE of Market Drayton.*
At the time of Domesday Book it lay in Staffordshire, as did nearby Cheswardine. In the reign of Henry II both manors were given to the Le Strange family and the shire boundary was moved so that they could consolidate their land in one county, Shropshire. Chipnall is an attractive little red brick hamlet on a rise in well-wooded country just S of the lovely valley of the Coal Brook. On this stream, 1m. NE of Chipnall, is Chipnall Mill Farm. The mill had 3 pairs of stone used to grind corn and an overshot iron wheel. It last worked in 1889 and was demolished in 1963 though there are foundations and grinding stones still in situ. It is said that when the mill ceased operation hordes of rats knawed their way through the doors of the farmhouse seeking succour and shelter. At Chipnall Hall Farm is a field called The Bloody Breech, a folk memory, perhaps, of some ancient battle. Cheswardine Hall lies S of Chipnall on the road to Cheswardine. (*See Cheswardine.*) The name Chipnall was formerly Ceppecanole (in Domesday Book) and Chippeknol (in 1260); it means either 'the knoll where logs were obtained' or '*Cippa*'s knoll', where *Cippa* is a personal name. To the N and E the hills are forested, a coniferous replanting of the ancient oak woods of the Forest of Blore where the Bishop's Lichfield used to hunt and which supplied fuel for the glass furnaces established here by the Bishops in the 16th Century. The reconstructed remains of one of these furnaces can be seen just over the border in Staffordshire at SK.760.312. (*See Broughton in 'Staffordshire and the Black Country', by M. Raven, 1988.*)

CHIRBURY *2½m. ENE of Montgomery.*
Two miles W of Chirbury is Offa's Dyke which hereabouts still fulfills its original prime purpose, that is, to mark the border between England and Wales. Six roads meet at Chirbury. It lies at the feet of the mountains of Wales with views across the mighty Berwyns, home of Owain Glyndwr, the Welsh national hero. Legend has it that he never died and will one day emerge from the mists and make his country free again. There are some black and white houses, such as the Old Schoolhouse of 1675, but mostly the buildings are of red brick. The present school is in a 17th Century grey stone building and is still a going concern with accommodation for 30 children. The church of St. Michael is large with the massive tower to be expected in border country. The nave and aisles are from the early 13th Century church of the Augustinian priory that once stood here; the tower is Perpendicular, the blue brick chancel is of 1733 and the whole was restored in 1871. There is stained glass by both William Kempe, and Tower and Powell. Chirbury is the *Cyrigbyrig* of the Anglo-Saxon Chronicle. In the field called Castle Field, or sometimes King's Orchard, are the slight earthen remains of Chirbury Castle. They lie just to the W of the village, N of the Montgomery road and to the E of, and about 25ft. above, the stream. This is probably the site fortified by Queen Ethelfleda in 915. What with the Anglo-Saxon castle and the Norman monastery, it comes as no surprise to learn that the name Chirbury means 'the fortified place with a church'. Lord Herbert of Chirbury was an Elizabethan historian, philosopher and diplomat. He left a large library of chained books to the village but these are now kept in Shrewsbury at the County Library. The Herbert Arms pub was named after Lord Herbert. His brother was George Herbert, the poet. **Marrington** Hall, 1m. SE, just off the A490, is an extremely handsome black and white house of 1595, though much of what we see today is of the 19th and 20th Centuries. It stands by Marrington Dingle, a dramatic, wooded gorge with public footpath access. South of the Hall are the earthworks of 2 prehistoric forts. They face each other at the point where the River Camlad makes a semi-circular loop. The Camlad is reputed to be the only river to flow from England into Wales.

CHIRK BANK *6m. N of Oswestry.*
Chrik Bank is a small settlement on the Shropshire–Denbighshire border ½m. S of Chirk. The A5 swoops down into the valley of the River Ceiriog. Just to the W is the impressive aqueduct built by Thomas Telford to carry the Ellesmere Canal over the valley. (The Ellesmere Canal is now a part of the Shropshire Union Canal.) Running parallel with, but at a higher level than, the aquedect is Henry Robinson's railway viaduct. Both structures are in full working order. The village of Chirk Bank lies mostly along the road to Weston Rhyn. At the top of the hill is the motte of an old castle and a sizeable settlement of red brick houses and stuccoed cottages with a Post Office – General Store close to the canal. On the main road itself there is a garage; a pub, The Bridge; and Seventh Heaven, a company that deals in antique beds. The name Chirk is thought to be an Anglo-Saxon version of Ceiriog, which is a very ancient British name the meaning of which is not known. (Note: the splendid Chirk Castle lies 2m. W of Chirk, set in a huge park close to Offa's Dyke. At the park entrance are a set of superb wrought iron gates, the masterwork of the Davies brothers of Wrexham.)

CHOULTON *4m. E of Bishop's Castle.*
The name means '*Ceol*'s homestead'. In 1066 it was held by Gunfrid. (He also held the land on which was to be built Wigmore Castle, in Herefordhsire, and which was to become the home of the great Marcher Lord Mortimer family.) By 1086 Choulton had passed to Robert Corbet but was later split, half going to the Botterell family and half to the Fitzherberts. Today it is a remote little farming hamlet on high ground above the valley of the River West Onny. There is a part stone and part timber-framed farmhouse; a stone, gabled cottage; a white painted, stone-built house; a junction of 2 lanes; some sheep on the pastures and little else. **Eaton**, 1m. N on the A489, was part of the Domesday manor of Choulton. The name Eaton means 'settlement on a river' and it does, indeed, lie close to the lovely West Onny. (*See Horderley.*) The country here is gentle and rolling but to the E is the looming mass of the Long Mynd, at the foot of which are the handsome white houses of Myndtown. At Eaton there is a grey stuccoed farmhouse with double end gables, a weather-boarded barn at the road junction, and a modern bungalow by the stone-built bridge; that is all.

CHURCH ASTON *1m. SSW of Newport.*
Aston means 'eastern settlement'. The first church was built in 1620 but was only a 'chapel of rest'. The present church was built in 1867 by G. E. Street. It has an unusual lead flèche over the nave and a stained glass window by Morris & Co. The old village has been overrun by modern houses and is now to all intents and purposes a suburb of Newport. The Hall was built in 1830 by Ralphe Leeke. In 1851 Mr. Ormsby Gore M.P. took residence here. He was Groom in Waiting to the Queen and Deputy Lieutenant of

Combermere Abbey.

Childe's School, Cleobury Mortimer.

Mawley Hall, Cleobury Mortimer.

All Saints', Little Stretton.

Shropshire. There were limestone pits near The Last Inn on the A518 but in 1861 they were flooded and abandoned. (*See Lilleshall.*) The Shropshire Union Railway from Stafford to Wellington passed through the village; it opened in 1841 and closed in 1963.

CHURCH PREEN *5m. WSW of Much Wenlock.*
Hidden in the hedges and woods you could easily miss it. The 13th Century church of St. John the Baptist is unusual in being long (70ft.) and narrow (13ft.), with nave and chancel joined without demarcation. The church originally belonged to a small monastic house based at Much Wenlock. It adjoins the Manor House which is a Victorian extension to an older building by Norman Shaw. All except the ground floor has been demolished. At the crossroads between Church Preen and Hughley is the handsome school, of 1872, also by Shaw. However, Church Preen is best known for its magnificent and very old Yew Tree. It dominates all around it, including the church it stands beside. Like most very old yews it is hollow. These hollows were sometimes used as pulpits by early Christian preachers. Without any doubt the tree pre-dates the church. The yew was revered by the Celts, and Christian churches often took over their holy places. The iron bands placed around these trees, like that at Church Preen, do very little good, and often do harm. Yews can live as long as 2,000 years; their secret is regeneration. The old tree can virtually die, become hollow and then spring to life again through the old, outer shell. Because their heartwood has long decayed, it is not possible to tell the age of the tree by counting growth rings. There are modern, timber houses belonging to the Forestry Commission on the path down to the church. From the lanes around there are good views over the valley to tree-clad Wenlock Edge. As to the name Church Preen, the first element is self explanatory but the second is from the Old English *preon* menaing 'a pin' or 'a broach' but used here to describe the shape of the high ground on which the settlement stands.

CHURCH PULVERBATCH *9m. SW of Shrewsbury.*
A small, unremarkable village recorded in Domesday Book as Polrebec, the name by which Mary Webb alluded to the settlement in her unfinished novel 'Armour Wherein He Trusted' (1926). Church Pulverbatch is situated in the northern foothills of the Long Mynd and has a population of about 300. There are 2 pubs – the White Horse and the Woodcock – a shop, and a community hall. The village lies on high ground on the road that runs from Belle Vue in Shrewsbury, through Nobold, Longden, Pulverbatch and Bridges to Bishop's Castle. This road passes between the Stiperstones to the NW and the Long Mynd to the SE and is a route of some character; the attractive landscape and the villages through which it passes are as yet largely unspoiled. The church of St. Edith at Church Pulverbatch lies within a circular churchyard which is quite probably the site of a prehistoric fort. The first known priest here was 'John the Chaplain' who served the community prior to 1193. The chapel was a daughter church to the church at Pontesbury and throughout the Middle Ages was 'a benefice without the care of souls'. In the early 15th Century it was destroyed by Welsh raiders but had been rebuilt by 1521. In 1773 it was made Classical and the tower was erected, and in the 1850's the nave and chancel were rebuilt in Decorated style by Edward Haycock. There are box pews, a door and a window with Gibbs surrounds and a Jacobean chair. Sukey Harley lies buried here; his story features in the book 'More than a Notion', (1964) by J. H. Alexander. Close to the church is Lower House Farm, brick with 17th Century back and Georgian front. Georgina Frederica Jackson (1824–1895), who compiled the authoritative Shropshire Word Book, and upon whose collection of folklore Charlotte Burne drew so heavily, was born at Pulverbatch. At **Castle Pulverbatch**, ¾m. SW of Church Pulverbatch, is the motte of a Norman castle. (Most Norman castles were originally built of timber and were never replaced by stone structures. When the timber decayed all that was left was the mound, and sometimes the ditch but this is often filled in by farmers for the safety of their livestock.) Here the castle mound is 28ft. high and is still almost encircled by its moat. It stands on a rocky promontory 100ft. above the stream. The original horse-shoe shaped bailey adjoins the mound to the NE and at a later date a second horse-shoe shaped bailey was added to the W. The first element of the name Pulverbatch is probably the name of a stream which itself was derived from the Norwegian *puldra*, 'to gush'; the second element simply means 'stream'. Just to the SW of Castle Pulverbatch is **Cothercott**, high, hill farming country. Once there were mines here and copper and barytes were extracted from the ground. There was a narrow gauge railway down to the brook where the ore was washed and from the mine on Huglith Hill there was an aerial cableway to Malehurst. Spoil heaps and old shafts can still be seen.

CHURCH STRETTON *13m. S of Shrewsbury on the A49.*
In late Victorian times Church Stretton developed from a small market town to a holiday and retirement spa. 'Little Switzerland' it was called and the town is indeed less important than its dramatic mountain setting. The characteristic building style is brick or stone with black and white half-timbering on the upper floors and gables. There are few buildings of note but nevertheless the town has a certain charm. Detached villas dot the lower slopes of the hills – Woodcote and Scotsman's Field are especially good – and it is these hills that are the great attraction: Lawley, Caradoc, Bowdler and Ragleth to the E and brooding Long Mynd ('long mountain'), with its ancient track called the Portway, and beautiful Carding Mill Valley to the W. The valley between the hills in which Church Stretton lies was caused by a double geological fault. On the rocky summit of Caradoc Hill is the prehistoric earthwork fort-settlement of Caer Caradoc. The church of St. Lawrence in Church Stretton is large and cruciform with a Norman nave and Perpendicular upper part of the crossing tower. The rest is Early Englsh except for the W aisles and transepts which date from 1867 and 1883. Hesba Stretton (1832–1911), the incredibly popular Victorian novelist whose story Jessica's First Prayer (1866) sold several million copies, lived in Church Stretton and there is a 'Jessica window' in the church. Also in the church, but now relegated to a place of relative obscurity above the external opening of the N doorway, is a badly worn 21 inch high carving of a pagan fertility symbol: a woman blatantly exposing herself. There are similar 'sheela-na-gigs', as they are called, in the churches at Holdgate and Tugford. They are widespread in Ireland and are probably of Celtic origin. In the hedgerows of the lanes around Church Stretton are examples of the comparatively rare Service tree (Checkers Tree). This tree often grew in old forest woodlands and tends to only be found in areas that have been continuously wooded since medieval times. But here the hills are king. One word of warning though. The Burway, the steep, narrow road from Church Stretton to the top of Long Mynd, is very, very dangerous. In fact, it is hair-raising and sooner or later there must be a disaster. It has an unwalled precipice to one side and traffic is allowed, unbelievably, to travel in both directions on what is no more than a metalled track. Indeed, the Long Mynd has something of a reputation as a taker of mens' lives. One of the many hollows and 'batches' is called Dead Man's Hollow and the November Fair at Church Stretton was nicknamed Deadman's Fair because several people died in the hills on their way home after attending the market. Running parallel to the busy A49 is the B4370 from Church Stretton to **Little Stretton**. It is well worth the short deviation. It passes through an unexpected valley, with the tree-shrouded Brockhurst Castle to the left. Little remains of this castle now, but in its day it was a large stone-built fortress comparable to Ludlow Castle. The road continues to the pretty little village of Little Stretton. It has a 20th Century black and white timber church of 1903 with a thatched roof and several much older timber-framed buildings. These include the Manor House of 1600, now heavily restored, the Manor Cottage, the Tan House (an old home of Derwent Wood) and the Malt House. North of Church Stretton is **All Stretton**. The Hall is built of brick with stone dressings and hipped roof and is now a hotel. Opposite is the Yew Tree pub with its very considerable car park. The stone church of St. Michael was built in 1902 by A. E. Lloyd Oswell. However, it is for its water that All Stretton is famous. Between All Stretton and Church Stretton is Cwm Spring, the waters of which are bottled by Wells (Drinks) Ltd. Their bottling plant occupies an old lemonade factory. There is a roadside tap at which

Church Stretton, the 'sheela na gig' over the north doorway of the church of St. Lawrence.

anyone may fill there own containers free of charge. Behind the factory, up the valley, there is an underground reservoir and a most attractive Italian style gazebo with a wrought iron, domed roof. Just N of the spring is the Church of England Primary School, a flat-roofed modern thing, and opposite that is the Police Station. North of Cardingmill Valley, near the circular earthwork of Bodbury Ring on Bodbury Hill, is the Church Stretton golf-course, one of the highest in the country; the 14th green is 1,200 feet above sea level. The **Long Mynd**, on which it is situated, is 10 miles long. The hill is owned by the National Trust but is mostly common land. The 20,000 or so sheep that graze the bracken covered slopes belong to commoners, but are virtually wild. Some meet untimely deaths, savaged by dogs, frozen in winter snows or trapped in mountain bogs. There are about 25 Bronze Age burial mounds on Long Mynd and most of them are located along the prehistoric Portway track. Note: the name Stretton is Old English for 'a settlement on a Roman road'. The 'All' in All Stretton is probably from a personal Anglo-Saxon name, most likely either *Aelfred* or *Aelfgyp* (a woman's name).

CLAVERLEY *5m. E of Bridgnorth.*
Claverley – the name means 'the clearing where the clover grows' – lies in lanes off the A454 Wolverhampton to Bridgnorth road. It is a very pretty village in red sandstone country. White washed cottages and pubs face the handsome 15th Century timber-framed former vicarage, which is flanked on one side by the 12th Century church of All Saints and on the other by the old sandstone school. In the churchyard is one of the oldest and largest yew trees in the county with a girth of 29ft. It is hollow, not well cared for and is almost certainly older than the church. Recent research suggests the tree is 1,500 years old. When restoration work was carried out on the church in 1902 the remains of two humans and a small animal were found buried North-South (Christian burials are always East-West), under the floor of the nave where the pulpit now stands. The churchyard is mounded, is the central point of the village and lies near a stream, further evidence that points to an earlier history as an Iron-Age religious centre. Indeed, in Claverley Parish have been found scrapers and other implements made from local pebble flint, which date from an even earlier period – from the Middle Stone Age (Mesolithic), about 4,000 B.C. The present church was founded by Roger de Montgomery (d.1094). In the nave are very rare, early 13th Century wall paintings of fifteen fighting horsemen, who represent the battle of the Vices and the Virtues. There is also an alabaster memorial to Sir Robert de Brooke, d.1558, who was the leading lawyer of his time, a Norman tub font and a Jacobean pulpit. A short distance downhill is the fine black and white Pown Hall. The White House is a picturesque black and white cottage in the High St. of the village. It has a magnificent and much photographed English country garden which overflows with colour in the summer; pictures of the garden were used by Cadbury's for chocolate advertising some years ago. The couple who created it used to be farmers in the farm across the road, before it was sold to a housing developer. The extremely handsome Ludstone Hall lies 1m. NNE of Claverley. It stands behind a high brick wall but can be seen through the entrance gates. It is a Jacobean house built by John Whitmore in brick with stone dressings and has a recessed centre with a semi-circular bay, mullioned and transomed windows and a moat. The Hall was restored in 1832 and today both the house and the gardens are kept in immaculate condition. Farmcote Hall, 1m. SW, is a red brick Georgian house that was the home of Catherine Glynne who was courted here by Gladstone, later to be the Prime Minister of Great Britain. He stayed at Pown Hall whilst visiting her. Other famous visitors to the area include Queen Mary, who stayed at the 18th Century Dallicote Hall, and Princess Alexandra who was a guest at Chyknell. King's Barn, about ½m. SW of Farmcote Hall, is partly brick and partly timber-framed. Woundale Farm, just over 1m. W of the village, is a good timber-framed house with a fine porch. High Grosvenor, ¼m. NW of Woundale, is another good timber-framed house with narrow uprights and concave lozenges in the W gable. Two miles S of Claverley, standing in its substantial park, is Gatacre Hall, a red brick Georgian house on a site where the Gatacre family are said to have lived since about 1050. Gatacre Park is a neo-Elizabethan house of 1850 approached off the Bridgnorth to Stourbridge road.

CLEE HILL *See Doddington.*

CLEE ST. MARGARET *6m. NE of Ludlow.*
The settlement lies in the lanes between the B4368 and the B4364 on the western slopes of Brown Clee Hill. Somewhat disconcertingly, the Clee Brook fills the main road of the village for 50yds. before disappearing over a steep bank. The church of St. Margaret has an early Norman chancel wih herringbone masonry, an original N window, and Jacobean chancel seats and pulpit. The impressive Iron Age earthworks of Nordy Bank stand ¾m. E of the village. Local tradition has it that if a pregnant woman sees a mountain hare on Brown Clee Hill she must tear her dress or her child will be born with a hare lip. Fairies are said to live hereabouts.

CLEESTANTON *7½m. NE of Ludlow.*
A ramshackle little place of farms and cottages, several of which are in disrepair. A tiny red brick chapel is now a house and sheep and horses now graze the rough pasture that covers the site of the deserted medieval village. This lies behind the partly excavated hill at the crossroads where a large barn burned down recently. To the SE looms the great hulk of Titterstone Clee Hill.

CLEETON ST. MARY *9m. ENE of Ludlow.*
The settlement lies on the NE flank of Titterstone Clee Hill and is most easily approached from the S by leaving the A4117 at Doddington. A lane leads off the road to Farlow and Oreton, and passes through Catherton Common to the Victorian hamlet of Cleeton St. Mary. Here is a red brick school with a bellcote; a pair of stone cottages; a terrace of 8 yellow brick houses; a few pine, yew and broadleaf trees; and a handful of new houses. Above the hamlet is the summit of Titterstone Clee with the radar domes and aerials of the Civil Aviation Authority standing in silhouette on the horizon. However, it is not the modern settlement of Cleeton that is the attraction here; it is the old, deserted village and its splendid moated site that lie on lower ground ¼m. N. The lane from Cleeton drops down a steep slope (at the bottom of which are some grassy banks by a stream where people often picnic), passes over a cattle grid at Cleeton Gate Farm, then bears right and down another hill to a grey rendered house (SO.605.793). Opposite this house is a grassy and often muddy track. This sunken lane leads to the ruined brick and stone chimney breast of the old timber-framed Cleeton Manor Farm that was burned down about 1955. The cellars are still intact. Beyond the ruin are some willow trees and a brook, babbling over a stoney bed, and beyond the brook are the ruins of stone barns, now covered in banks of ferocious nettles. The path continues as a 'holloway', a sunken track. Between the holloway and the right-hand field boundary 5 house platforms have been identified – 3 alongside the track and 2 behind them. This is the site of the old village and, indeed, the field is called House Meadow. The track leads down the slope, bears left at the field boundary and passes through a gateless gap. Modern man now takes a hand and the path passes over a dam wall, made this century to create a fishing lake. This pool has been allowed to silt up and nature has created a most charming spot – bullrushes, reeds and flowering water plants inhabit the waters' edge and broadleaf trees and a plantation of conifers provide a backcloth. The walker should now turn back and retrace his steps until he reaches the ruined stone barns covered in nettles. Turn left here and move uphill through a grassy field that was once an orchard and in which some fruit trees still remain. At the top of the slope is a broad ditch about 6 feet deep. This is the moat of the Medieval manor house. To the right is a causeway that leads on to the moated mound. This is flat, circular and has a diameter of about 50 yards. A stream is divided so that it runs around the mound filling the moat with water. The ditch is much deeper on the stream-divide side, about 12 ft. ditch-bottom to mound-top. This earthwork is considered by some to be the finest moated site in Shropshire. The field W of the moat is called Bottom Meadow. It is a delight, full of wild flowers and several kinds of grasses, like meadows used to be. Cleeton Court lies on the lane, W of the deserted village. It is

Claverley, the sculptor Anthony Twentyman. The bronze that he holds is in the possession of the author.

a substantial timber-framed house of brick construction with the framework of the original Great Hall exposed in an upstairs bedroom and the downstairs lounge. At a later date the building has been faced with stone and that is how we see it today. To the front it has 2 large gables with a narrow recessed centre. The windows are a motley crew.

CLEOBURY MORTIMER *11m. E of Ludlow.*
Cleobury Mortimer is a charming small town on the A4117, the road from Ludlow to Bewdley. The main street is lined with terraces of Georgian houses and shops, and on the high ground there is a most attractive area where the pavement is raised and lined with bollarded trees. The brick-built Manor House is of about 1700 and lies at the W end of the town; it has 7 bays, a hipped roof and Doric pilasters. The road descends to the River Rea, passing the church of St. Mary with its sturdy Norman tower of pale green sandstone and tall twisted spire. Opposite the church is the handsome mid-18th Century Childe's School, stone with a hipped roof. The school is on the site of the castle built by the medieval Marcher Lord Ralph de Mortemer (d.circa 1104), hence the French suffix to the Anglo-Saxon name of the town. One of the Mortimer family went on to become King of England as Edward IV and his 2 sons were the ill-fated 'Princes in the Tower'. Cleobury (pronounced Clibbery) has an old industrial tradition. In the late 16th Century Robert, Earl of Dudley, built a charcoal furnace for smelting the locally mined iron ore. Iron, coal and paper making were important in the area well into the 19th Century. A mile out, in the direction of Bewdley, just off the A4117 is the red brick Mawley Hall, visible from the road, standing in its well-wooded park on a hillside above the River Rea. Built in 1730 for Sir Edward Blount, the exterior has weathered badly but the interior is the most splendid of its date in the county, with excellent plasterwork and wood carving. The staircase handrail is in the form of a serpentine snake. The architect was probably Francis Smith. Half a mile out of Cleobury, on the same road and in the same direction, is Castle Toot, a substantial Norman motte now the happy setting for a modern house. At nearby Wall are the remains of a Roman Fort, one of a series established on the edge of the Civil Zone; a farm now occupies the small square earthwork and the road cuts through its ramparts and ditches. It is generally accepted that William Langland (C.1331–C.1400), the poet whose vision concerning Piers Ploughman is one of the masterpieces of medieval literature, was born in the Western Midlands. Cleobury Mortimer claims the honour but so do Ledbury and Great Malvern. At Hopton Bank, W of Cleobury on the A4117, is the Clee Hill Bird Breeding Centre where over 400 birds and animals can be seen. The Centre specializes in endangered species. There is a large collection of wild fowl and a Children's Corner. Reaside Manor Farm lies 1m. S of Cleobury Mortimer. It is an early 17th Century stone-built house with many gables, a porch on the W front, and star-shaped brick chimney stacks. Inside is a good staircase and some good plaster work. Claybury Mill, Pinkham, has a breastshot water-wheel and is open to the public.

CLEOBURY NORTH *7m. SW of Bridgnorth.*
A small, main road settlement on the B4364 Bridgnorth to Ludlow road that has a most attractive village shop. This is housed in a part of the Old Hall. Until recently Cleobury North was almost entirely owned by Lord Boyne as a part of his Burwarton Estate. Today, as properties become empty they are sold off on the open market. The stone-built church of St. Peter and St. Paul has a Norman tower with a top storey of brick, a chancel largely rebuilt in 1890, and good roofs of the 16th and 17th Centuries. Opposite the church is the handsome Cleobury Court, for long a home of the Hamilton Russell family. The Village Hall was originally a granary with sheep-pens beneath. The only new buildings in the village are the old people's bungalows. A little curiosity is the toilet at Mill Farm, a hut built over a stream. Adjoining are fisheries that used to belong to the Burwarton Estate but which are now privately owned. There are few obvious signs of it now but coal was being mined all around here at least as early as 1260, when Walter de Clifford granted land for assarting (the act of clearing woodland) in Cleobury North, with a licence to "dig coals within the Forest of Clee". Small, collapsed 'bell pit' coal mines are scattered over the hills. They look like small shell craters – mounds with depressions in the centre. The people here usually mixed mining with farming. The route of a dismantled mineral railway, that ran from Abdon to Cleobury Mortimer, passes through the village. Close by the well-wooded churchyard there runs a small stream, the Cleobury Brook, a tributary of the River Rea. The settlement lies at the foot of Brown Clee Hill. Clee is probably an ancient Celtic name and its meaning is unknown. In the early Middle Ages the King of Almain held a part of the manor for a time before giving it to the Knights Templar. A witch by the name of Prissy Morris once lived at Cleobury North; she had the power to stop a horse dead in its tracks.

CLIVE *3m. S of Wem.*
A most pleasant, airy, stone-built village, with a church spire that is a landmark for many miles around. It lies on the slopes of Grinshill Hill; indeed the name Clive is from the Old English *clif*, meaning 'a cliff'. At Grinshill village, ½m. SE, the hill is still quarried for building stone. (The stone surround to No. 10 Downing Street was taken from Grinshill.) The Elizabethan Clive Hall – some say it is 14th Century – has been isolated from its outbuildings, which are now across the road and badly converted to houses. The Hall is timber-framed with herringbone infill but has been totally ruined by the addition of stucco and the insertion of modern windows., It stands in the main street to the E of the village at the fork in the road and has been converted to flats for old people. It has been suggested that Clive of India's family came from here, before moving to Styche Hall at Moreton Say near Market Drayton. One certain previous occupant was William Wycherley, the Restoration dramatist, who wrote somewhat coarse plays for the somewhat coarse court of Charles II. His best known work is 'The Country Wife'. His bones lie in the village churchyard. All Saints is a Norman church that was rebuilt in 1885-94. The S nave masonry and the nave doorways are original. The S nave doorway has decoration characteristic of the Norman period – shafts with stiff leaf capitals and zig-zags in the arch. The tall spire has already been noted. The school of 1873 lies above the church from which it can be reached by a steep and stoney path called 'The Glat'. The traveller would never know it but there is a network of tunnels beneath the village, part of old copper mines which were worked from Roman times until 1886. The main shaft has been enlarged and is now used as a cistern for water which is pumped out to the cottages and farms of the **Sansaw** Estate. The house and park of Sansaw lie ¾m. SSW of the village. The Hall is a Queen Anne revival house in red brick with a hipped roof, stone dressings and a stone portico. It is 3 bays wide, 3 storeys high and has projecting pavilions to either side. The art dealer and critic John Ireland lived here. **Yorton** lies ½m. ESE of Clive. There is a main line railway station where one can catch trains to Shrewsbury and Crewe, and a pub to pass the time whilst waiting.

CLUDDLEY *1½m. SW of Wellington.*
A strange little place marooned in a cul-de-sac at the point where the old Holyhead road joins the new M54. The Old Mill is quite delightful: a small, red brick house with a white stucco porch and charming outbuildings – some with stone dressings, some weatherboarded – and the Windmill Tower, all embowered in trees. There is a notice at the entrance that reads: 'Conundrum. The Directors of Cadburys have consulted the author Don Shaw who has advised that a casket is NOT buried on this property which is privately owned.' The occupiers of the house had been invaded by great numbers of at times belligerent 'treasure' seekers. However, this charming property has for company 2 yards wherein are parked great numbers of Unigate milk delivery vans and lorries, not a pretty sight. For the rest there is little more than a modern dwelling called Arleston House; an old, large, red brick house; a white painted farm building, and some black and white cows grazing in far from peaceful meadows adjacent to the M54 embankment. The old name for Cluddley was Clotleye, which is from the Old English *Clate-leah*, 'the clearing where burdock grew.' In the Middle Ages it belonged to the manor of Wrockwardine which was for long held by the Lestrange family. One mile S of

Cluddley is the great hill of The Wrekin. The road from Cluddley to Huntington passes between The Wrekin and its long foothill, **The Ercall**. There are deciduous woods and several small quarries, now overgrown. At the foot of the pass is Forest Glen, a ramshackle building of timber construction that was once a tea-house but now appears to be a cabaret club. Between here and Wellington, to the N, is a pretty, forested lane called Ercall Road along which are some spaciously set modern houses and the popular Buckatree Hall Hotel, now extended and stuccoed and with a pool decorated by the presence of white swans.

COMBERMERE *3½m. ENE of Whitchurch.*
Combermere Abbey lies in Combermere Park beside Comber Mere. The house and the mere are in Cheshire but the park stretches over the border into Shropshire. The country is undulating and the topsoils are a red, clayey, drift which was deposited as the glaciers of the last Ice Age melted and moved north. The entrance to the Park is off the A525, the Whitchurch to Nantwich road. Beside the stone lodge are 2 stone pillars on which are perched sculptured birds of prey. The long drive winds through pastures dotted with many ancient oak trees, craggy old fellows with broken limbs and hollow bodies. We have never seen so many old trees in one place. Combermere Abbey lies in a broad bowl facing the reed and rush fringed mere which wanders around like a broad river. It is an idyllic setting. The abbey was founded in the early 12th Century for Benedictine 'black' monks. Amongst the properties owned by the Abbot was the town of Market Drayton and it was the monastery that developed the settlement. Leland describes a time of great excitement at Combermere – the day the hill disappeared. Overnight a small hill collapsed into the ground leaving a hollow. What had happened was that an old underground salt pit had collapsed. The hollow filled with brine and the monks began to produce salt by evaporating the water in heated pans. Local, lay producers complained, though, and they had to cease. After the Dissolution of the Monasteries the estae passed into private hands. The bells of the Abbey are now at Wrenbury church. When they were removed they were ferried across the lake. One fell into the water and a workman cursed aloud, the last sound he was ever to make, it is said, as a huge, demon-like figure reared out of the pool and dragged him down to a watery death. The house we see today was built on the ruins of the Abbey by the Cotton family. It is an irregular brick pile, partly rendered in grey cement, with fat pointed-arched windows – a house of some character but by no means elegant. There is a large walled garden and a good coach-house and stable block which has 2 little lead capped towers and cast-iron latticed windows. Opposite the house, across the lake, is a very tall obelisk monument, a local landmark of some note. There are woods on the northern shore of the mere and in the pastures to the W are acres of rampant worms. Whole fields are totally covered in little mounds of their pipe-like extrusions. In the early 19th Century the leading soldier of Cheshire lived at Combermere Abbey (as it is still called). He was Sir Stapleton Cotton, the commander of Wellington's cavalry in the Peninsula Campaigns. He was Le Lion d'Or to the French because of his swashbuckling character and sartorial flamboyance. It is said that a little girl haunts Combermere. Whenever she appears it signals a bereavement in the family of the owner. The present incumbants are Lord and Lady Lindsey. The name Comber is from the Old English personal name *Cumbra* which might have a connection with the Welsh name for Wales, Cymru, or the Celtic tribe, the Cambrians.

COMLEY *2½m. NNE of Church Stretton.*
It lies on the lower, northern slopes of Caer Caradoc in the gap between that hill and The Lawley. Both hills have prehistoric fort-settlements on their summits. The green sandstone rocks that outcrop here have been much studied by geologists, from Murchison in the 1830's to Dr. Cobbold in the 1930's. There is a small black and white cottage, an irregular house of brick called The Wood, a row of stone cottages, a stream, deciduous woods, holly in the hedges and ponies in the paddocks. If one turns E at the road junction and then NE at Folly Bank, and through Enchmarsh, the high ridge road of Yell Bank is reached. It is narrow and sunken and must have been a highway (literally) from time immemorial.

CLUN *5½m. SSW of Bishops Castle on the A488.*
The lovely valley of the River Clun runs W to E between Newcastle and Aston on Clun. Moor-topped hills line its banks and as the old rhyme puts it: "Clunton and Clunbury, Clungford and Clun, Are the quietest places, Under the Sun." At the town of Clun the river is crossed by a stone saddleback bridge with 5 segmental arches and triangular breakwaters, which continue to parapet level to provide pedestrian refuges. It was built about 1450 and is almost completely original, very rare for a bridge of this age. The town lies on both banks. North of the river is the Norman 'plantation'. This lies below the substantial ruins of the Norman Castle, and here are the Town Hall, the Court House and most of the shops. To the S the road climbs the hill, lined with painted houses, which leads to the typical border church of St. George. It has a sturdy Norman tower topped by a truncated pyramid roof. This was the old, pre-Norman centre of the town. The church itself appears to have been built on the site of an Iron Age fort and there was an important Saxon manor here. Indeed, the name Clun is British and is related to the Gaelic *clon* and the Welsh *llan*, both of which mean 'church'. It is therefore quite probable that Clun was an important Celtic religious centre. Tucked away on the NE fringe of the town is a surprise: the charming stone-built Hospital of Holy Trinity, founded in 1618 by Henry Howard, Earl of Northampton. It consists of a quadrangle of one-storey accommodation for old people, around a square of lawns and flowers. To the front it has been extended by the addition of a school wing with a veranda on round arches; this balances the forward extension of the old stable block on the other side which has been converted to provide further accommodation. The Hospital still fulfills its original function. (A hospital was originally a place of hospitality, not a place to practice medicine.) The whole has a friendly, cloistered atmosphere. In the dining room hangs a 'Sword of Peace'. In the town centre, near the crumbling Court House of 1780, is the Buffalo Inn where Sir Walter Scott is reputed to have stayed whilst writing his novel, 'The Betrothed'. The hills around were quite densely populated in pre-Roman and Roman times. The evidence lies in the numerous tracks, hill forts and burial mounds found on them. One of the largest and best preserved of the hill forts is Bury Ditches (SJ.327.837), sometimes called Caer Ditches, 2½m. NE of Clun. In the Middle Ages Clun was at the centre of the Forest of Clun. Today there are quite possibly more trees than in Norman times. In medieval terminology a Forest (with a capital F) was an area under Forest Law, an area subject to rules restricting rights of grazing and taking wood etc., and not specifically a wooded area. By the 17th Century much of the uplands had become common land. Most of this was enclosed by Acts of Parliament in the 19th Century and the majority of the fences, hedges and stone walls date from this time. Farming in the area is based on livestock; cattle for beef and Clun Forest sheep for both wool and meat. Clun Castle was built shortly after the Conquest of 1066 by Robert de Say. Like most of the border forts it had a comparatively short active life and by 1272 was in a state of disrepair. It is larger than it looks from close quarters and is best seen from a distance. There is a good view from the Newcastle road. Clun Castle is almost certainly the Garde Doleureuse of Scott's novel 'The Betrothed' (1825). Today it is owned by the Duke of Norfolk. The village of Clun was called Oniton by E. M. Forster (1879-1970) in his novel Howard's End. The area called Welshman's Meadow probably refers to a place used by the Welsh cattle drovers as a stopping place on their way to the great border market at Shrewsbury.

CLUNBURY *4m. WSW of Craven Arms.*
It is one of those quiet, unsung little places that have no great beauties, no great place in history, no famous people or intriguing curiosities and yet is in some way perfect because of this. Clunbury is a grey stone village of Anglo-Saxon origin with some good Elizabethan and Jacobean timber-framed houses, a sturdy Norman church, a tiny Post Office, cottages covered in bright red ivy, an old, stuccoed, Mill House by the chattering River Clun and high above the looming, bracken-clad mass of Clunbury Hill. There are some delightful spots by the river. Willows

drape themselves along the banks and gardens run to the water's edge. The grey stone school lies above the village and, happy to say, is alive and well. Beyond is the long, ridge-like summit of Clunbury Hill which was a part of the famous Clee-Clun Ridgeway, a trading route used by prehistoric man traced by a trail of flint chippings. These hill-top highways were used for thousands of years until the Romans built their lowland, metalled roads. It is a friendly hill with grazing sheep; an ancient stone used as a parish boundary marker; a scatter of solitary trees and clumps of tall pines standing stark on the skyline; green, angular fields on the lower slopes; whinberries and sphagnum moss (once used for dressing wounds) and in the autumn a cover of golden red on the higher reaches. The venerable church is friendly too, despite its strong, stubby and battlemented tower of the 13th Century. It is dedicated to St. Swithin and the chancel and the nave are Norman. The word *Clun* is ancient British and its meaning is lost in the mists of time though it might be related to the Welsh *llan*, meaning 'a church' or 'religious place'. The hill country W of Clunbury is quiet, unspoiled, and more than a little mysterious. Partaking of this mystery is Chapel Farm just across the river from Clunbury village. This used to be a chapel but strange sounds were heard to emanate from it at night. Local people attributed these noises to the Devil himself and refused to continue worshipping there. Since it became a farmhouse the old chapel has been demon-free. Ida Gandy lived at Clunbury from 1930 to 1945. Her husband was the local doctor. She spent much of her time exploring the countryside and wrote an affectionate and evocative account of these wanderings called 'An Idler on the Shropshire Borders' (Wilding & Son Ltd., 33 Castle Street, Shrewsbury 1970).

CLUNGUNFORD *3m. SW of Craven Arms.*
Here the River Clun has turned and heads southwards towards gentler and more wooded country. The name Clungunford is from: 'Clun', the River Clun on which the settlement stands; 'gun', from Gunward the pre-Norman Conquest holder of the land; and 'ford', a river crossing. Some local people call the village Gunnas. The church of St. Cuthbert and its accompanying Priests' House are both 13th Century but have been much restored. The church replaced the previous timber-built Anglo-Saxon chapel. The ancient Easter Day evening custom of the Love Feast was held at Clungunford church until 1637 when the Bishop banned this orgy from his house of God. Public opinion was such that he had to consent to its continuance in the Parsonage. Just NE of St. Cuthbert's is the motte of a Norman castle. The church and castle lie close to the river whilst the village straggles up the hill eastwards, a mixture of rendered council houses and good detached villas. Near the top of the high ground is Clungunford House, an attractive Georgian brick mansion of 1828, hidden from the road by the trees of its small park. It has a symmetrical front of 3 broad bays 2 storeys high; the centre bay is recessed and fronted by a columned stone porch and the outer bays have pediments. The house was built by the Rocke family and stands on the site of the old Manor House which was demolished to make way for it. To the rear, alongside the road, are some good stone barns. The river was first crossed by a bridge in 1657. The present structure replaced it in 1935. Stones of the original ford can still be seen in the river bed. Abcott Manor, a well restored house with some good timber framing lies, ½m. W of the village. Ferney Hall lies 2½m E, on the road to Onibury, via the stone-built hamlet of Shelderton. The old house was burned down in 1856 and replacing it in its William Kent park there now stands a symmetrical, neo-Jacobean mansion by either John Norton or Pountney Smith. Set in a wall of trees, 1½m. WSW of Clungunford, is Broadward, an irregular brick house which has been given battlements and then been partly rendered. It has a turret, a red brick tower and a glazed porch. The whole, somewhat bizarre, edifice has subsided slightly in several different directions. In the grounds of the house, near the banks of the River Clun, are two mounds: one appears to be a Norman motte and the other is a circular platform some 125 feet in diameter at the top. In 1867, during drainage operations, a large hoard of Bronze Age weapons was found some 5 ft. below ground level. The hoard contained parts of at least 46 spearheads (20 being barbed), 7 fragments of spearhead sockets, 5 ferrules, 2 bugle-shaped objects, a tanged chisel, 11 leaf-shaped sward fragments and a great number of animal bones. It is likely that these were offerings to water gods and that the level of the land was subsequently raised by alluvial earth deposited at times of flood.

CLUNTON *2¼m. E of Clun.*
A disappointing main road village in the lovely valley of the River Clun. Bury Ditches lies 1½m. NNW in the midst of a large Forestry Commission plantation. This magnificent Iron Age fort stands on the summit of Sunnyhill and can only be approached on foot. The path starts at the car park (SO.334.839) on the road between Clunton and Brockton. The fort is ellyptical in shape and has 2 entrances, the SW entrance being of an unusual design. Some of the best views in Shropshire can be obtained from Bury Ditches.

COALBROOKDALE *See Ironbridge.*

COALPORT *See Ironbridge.*

COCKSHUTT *4m. SSE of Ellesmere.*
A small, main road village with an attractive red brick church of 1777, a handful of timber-framed cottages and much new housing. Just to the N are the low-lying peat lands of Whattal Moss. This was a mere, or small lake, that drained naturally. A dug-out canoe, circa 700 B.C. (late Bronze Age, early Iron Age), was found in the bottom layers of the peat. It is 10ft. long and the side walls are one inch thick. The boat is currently on display in Shrewsbury Museum (Rowley House). Adjacent to Whattal Moss are Cross Mere, a popular fishing pool, and tiny Sweat Mere. Cockshutt House Farm, on the A528, is an 'open farm' with a conservation trail and a display of small-scale farming techniques. To the SW of the village are several old houses. Shade Oak is ¾m. SW of Cockshutt but the old oak of its name was unfortunately blown down in a gale in 1968. To the rear of Petton Hall, now a school, is a moat isolated in a field. Stanwardine Hall, large and somewhat ungainly, was built about 1588 probably by Robert Corbet, brother of Sir Andrew Corbet of Moreton Corbet (d.1578), and the shaped gables of Stanwardine do have a resemblance to its more illustrious relation. Wycherley Hall is half-timbered black and white with attractive gardens. The name Cockshutt is problematic. 'Cock' may be from a personal name such as *Cocca* but could also mean a wild bird; *Shutt* is an old Shropshire word meaning 'a closed passage'. Perhaps there was a tunnel-net bird trap here in Anglo-Saxon or Medieval times; the area was rich in wild fowl attracted by the meres and mosses.

COLD HATTON *6m. S of Hodnet.*
'Hatton' is from the Old English *haep-tun*, meaning 'settlement on a heath'. The heath in question is Hine Heath, which may mean 'the heath of the monks'. The Domesday manor was held by Gerard of Tournai-sur-Dive but it later passed to Lilleshall Abbey. The settlement stands on the A442 close to the confluence of the River Meese and the River Tern. The Duke of Sutherland was busy here, building red sandstone cottages, 2 up, 2 down and all facing S, from material quarried at Cliff Rock, a wooded gorge of the River Meese. He also sank wells for water and planted hawthorne hedges around the fields of his estates (which were sold in 1912). It was the custom in this area that the landlord gave his tenants a walnut tree on the coming of age of a son and heir. Descendents of some of these trees still exist. The Seven Stars was frequented by drovers who grazed their cattle overnight on the common. Today there is a modern motel, a garden nursery, a boarding kennels for pets, and by the meadows still stands the 'Chutor house', a large brick kiln that was used for drying flax that had been retted in the River Tern.

COLD WESTON *6m. NNE of Ludlow.*
The settlement is about 1m. from Stoke St. Milborough and is approachable only across fields at Ordnance Survey map reference SJ.552.828. A farm, a church and good views over Corvedale; on the face of it that is all. But in fact this is the site of a deserted village. The restored Norman church of St. Mary is surrounded by sunken roads and house platforms, all that remain of a village which was in decline before the Black Death. A common reason for rural depopulation was the change from arable to pastoral farming. Less labour is required to look after sheep

Clunbury, the River Clun.

and cattle than is needed to grow crops. In this case the village was also badly sited, lying 800ft. up on a cold, north-facing slope. Today, there are only 4 houses in the whole parish: Cold Weston Court, 2 cottages and a farm. With the abandonment of the village the 15th Century saw the open fields enclosed to form one pastoral holding. At Cold Weston there is a medieval cattle road. Cattle are heavy and their hooves quickly churn up unmetalled roads. For this reason they were moved from one place to another along special cattle tracks that often ran parallel to the ordinary highway, and they always skirted villages. These roads became sunken through wear and tear. At Cold Weston such a drift-way or straker, as these lanes are sometimes called, can be seen running alongside the ordinary village road. There is a cottage in the parish that lies on the border between Cold Weston and Stoke St. Milborough. It is said that in the early 19th Century this served as a retreat for unmarried mothers who, by moving from one side of a room to another, could claim financial assistance from both parishes.

COLEBATCH *1½m. S of Bishop's Castle on the A488.*
A main road village of mainly stone-built cottages situated in a broad valley surrounded by rounded, wooded hills. A charming little lane leads off the main road and follows the course of a tiny stream. Here is a variety of dwellings including some black and white cottages, old weather-boarded barns and the motte of a Norman castle. The timber-framed house in the middle of the village is 14th Century and is believed to have been the kitchen to the old Manor House, now no longer with us. On the other side of the road is a house called Brookside. This used to be the Bishop's Castle Estate Office. There were formerly 2 pubs in the village, The Barley Mow and The Fleece. The name Colebatch is from the Old English *col-haece* (or *bece*). *Col* can mean either 'coal (charcoal)', and hence 'black', or 'cool' (of temperature) or an abbreviation of *Cola*, a personal name. *Baece* means 'a stream', or sometimes 'a valley'. Several interpretations are therefore possible: 'the valley of the Cole Brook', 'the black-river', 'Cola's Stream', etc. A 12th Century lord of the manor was Lefwyn de Colebech. He took his name from the village, not vice versa.

COLEMERE *2½m. SE of Ellesmere.*
The hamlet of Colemere lies near the southern shore of the lake Cole Mere. The first element of the name might be an Anglo-Saxon personal name or be from *col*, Old English for coal and hence a discription meaning 'black'. Inland lakes surrounded by trees often do look black. In the village are 5 most attractive thatched cottages, and a group of 20th Century farm workers' houses of brick with timber-clad gables and steep pitched roofs. The lake is the centre of a Nature Reserve. There is a car park, a toilet block and a picnic area. A daisy-strewn meadow and a sandy beach edge the placid waters and there is a nature trail in the woods that surround the rest of the mere. The lake itself is used for sailing and unfortunately the boats are kept within an unsightly wire-net compound, the only blemish on a picturesque and tranquil scene. The handsome church of St. John, embowered in trees, overlooks the lake from its position on raised ground. It was built in 1870 to a design by Street and is constructed of stone with decorative red bands. The stone was brought from Cefn Mawr down the Shropshire Union Canal which skirts the Mere to the N. Close by the church and almost hidden by trees is the Old Vicarage, rock-faced with cast-iron windows in which are diamond shaped lights. West of the church is an avenue of mature chestnut trees. This leads to the hamlet of **Lyneal**. The name might mean 'the secluded place in the limewood'. Most of the buildings are of red brick but there are some half-timbered cottages – The Island, and Rose Cottage which is also thatched. The tiny school of 1880 has a simple charm and has so far survived the education cuts. Half-a-mile NE of Lyneal is the attractive Fernwood Caravan Park. It is mostly hidden from the road but what can be seen does no harm to the country at all. Farming produce of the area includes potatoes (for crisp manufacture), milk, sugar beet and hay/silage.

CONDOVER *4m. S of Shrewsbury.*
The village lies in well-wooded, undulating country. Condover Hall is the finest Elizabethan house in the county. It is constructed of pink sandstone and was built for Thomas Owen, who died in 1598 with the mansion not quite finished. The mason was probably the famous Walter Hancock. The impressive entrance gates frame the avenue of clipped box and yew trees. Beyond stands the Hall with its towers, gables, transomed and canted bay-windows, chimneys and pedimented, columned doorway. At the rear of the Hall is the Park through which runs the Cound Brook. The Hall has had several owners and Clive of India rented it for a time. Whilst it was in the possession of Reginald Cholmondeley (pronounced Chumley) Mark Twain was a guest (in 1873 and again in 1879). Mary Cholmondeley, the novelist, lived here for a few months in 1896 before her father sold the Hall and moved to London. Today Condover Hall is a school for the blind, owned and run by the National Institute for the Blind. There is a legend connecting the Hall to the Great Fish of Bomere. (*See Bayston Hill.*) The village has a green, not usual in these parts. In medieval times Condover was the administrative centre of the Condover Hundred, and there seems to be a connection between this and its possession of a green. Recent research has shown that in Shropshire there are many old timber-framed houses clad in a later skin of brick. In the village of Condover 6 such houses have been identified – early ones, at that, having a 'cruck' construction. Rural depopulation is evident here and is typical of the county as a whole. Of the 61 hamlets of the Condover Hundred mentioned in Domesday Book, 8 have been deserted and 31 have shrunk to either one or 2 farms. The pink sandstone church was largely rebuilt after the tower crashed through the nave in 1660. Inside are some notable monuments to local families – the Scrivens, Owens and Cholmondeleys. Richard Tarleton was born in Condover, the son of a farmer. He became Court Jester to Queen Elizabeth I and was the most famous comedian and actor of his day. Despite being short and hump-backed (and very ugly) he was a fine swordsman. However, it was for his quick wit that he was most applauded and, on occasion, feared. He was a great favourite of the Queen and it is probable that Shakespeare had him in mind when he wrote of Yorick in Hamlet. Tarleton died in 1588 and was buried in Shoreditch. A collection of his jokes and stories was published entitled 'Tarleton's Jests'. Either these are hack imitations or the Elizabethans were unbelievably easy to please. It could be, of course, that it was Tarleton's delivery and physical presence that were the essence of his humour. There are some comedians today who can make one laugh without saying a word. If 'Tarleton's Jests' are a true record of his material this must be the explanation of his popularity. Tarleton was born in the area called Pyepits, where the present school of 1880 is situated. Indeed, the school yard is located on the site of the old clay pits from which material was extracted to make pipes – pipepits becoming Pyepits. The name Condover is probably derived from *Cound*, from the Cound Brook on which it stands and *ofer*, Old English for 'a bank'. Less than 1m. N of Condover on the road to Bayston Hill is Norton Farm. Between the farm and Bomere Pool is an ARC sand and gravel pit in which the bones of 4 mammoths were found in the mid–1980's. The substantial remains are of one adult Mammuthus Primigenius and 3 infants. The bones have been radio carbon dated at 12,700 years old and are on temporary display at the Cosford Aerospace Museum. When the exhibition ends the County Council hopes to find a permanent home for them in Shrewsbury.

COALBROOKDALE *1m. NW of Ironbridge.*
To the W of Ironbridge lies the industrial heart of Telford. In the steep valley of the Coal Brook Abraham Darby took over an already existing iron works. By his perfection of the method of smelting iron using coal, as coke, the Ironbridge Gorge area was to become the foremost producer of iron in the whole of the country during the 18th Century. Let it be said straight away that this was not the 'Birthplace of Industry'. This is merely a slogan dreamed up by the advertising department of the Telford Development Corporation. (If such a laurel lies anywhere it is with the fibre mills of Derbyshire and Lancashire. It was the organization of factory labour that was the most important development, for without it inventions and new processes could not have been utilised in any effective manner.) The Quaker, Abraham

Rose Cottage, Coalbrookdale.

Cound Hall.

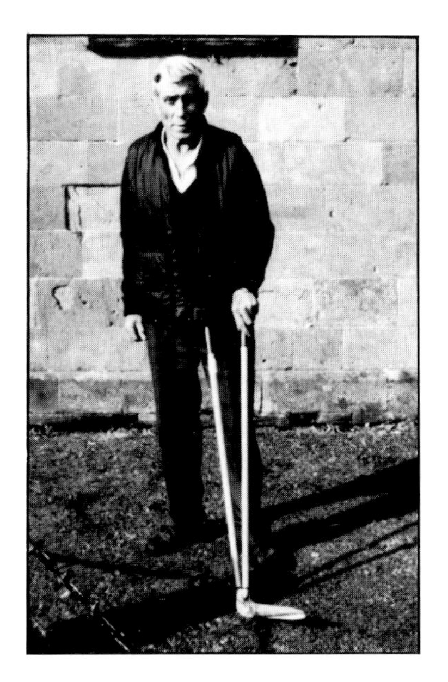

A gardener at Condover Hall.

Boring Mill Pool, Coalbrookdale.

Darby I, came to Coalbrookdale from Gloucester in 1708 and in 1709 he first used coke to smelt iron ore at the Old Furnace. (In 1954 this was excavated from beneath 14,000 tons of slag and is now covered in a modern glass protective pyramid.) It took 10 years of experimentation before production was of any sizeable consequence. Darby made important developments, not only in smelting, but also in casting and his innovations sparked off an expansion of the trade. His work was continued by his sons and grandsons and above all by the Reynolds family. New works were opened in Horsehay, Ketley, Little Dawley and Madeley Wood. Darby laid railway lines (the wagons were drawn by horses) to connect the iron ore fields at Dawley with the coal field and the furnaces at Coalbrookdale. By 1785 the Coalbrookdale Company controlled 8 furnaces, 9 forges and 20 miles of railway lines. About this time John Wilkinson started up in competition at Snedshill, Willey, Hadley and later at Broseley. At the outbreak of the Napoleonic War South Shropshire had double the number of furnaces found in any other county in the country. After 1850 the industry declined. Today, there is much to see in the Coalbrookdale valley. There is Darby's second home with what were the company's offices beside it; the Quaker Burial Ground (the Meeting House has gone); the museum at the Coalbrookdale works; the still functioning casting factories; the gaunt, blue brick Literary and Scientific Institute (1859), now a Youth Hostel; several rows of late 18th Century workers houses; and of course the valley itself which still retains its rural character. Special mention must be made of the delightful little museum at Rose Cottages, the oldest building in the valley. The left-hand cottage contains a small collection of articles made of wood, including coracles and displays relating to the management of woodlands. Behind the cottage is a workshop where various articles are manufactured. The whole is organized by the Greenwood Trust, a charity supported by the Ironbridge Gorge Museum. There is no admission charge. For many years the whole iron-working system at Coalbrookdale was dependent on water power. Water-wheels worked the great bellows that blasted air into the furnaces and lifted the great hammers in the forges. The little Coal Brook could not manage by itself. Five ponds were dug and when the water had fallen to drive the wheels it was pumped back up to repeat the cycle. A sixth pond, the New Pool, was the highest and was used as a reservoir. This was built about 1700. It is now about a half of its original size. New Pool lies just S of the car park on Cherry Tree Hill. The next pool in the sequence is the most famous. Upper Furnace Pool lies adjacent to Darby Road and provided power to Darby's original furnace. A wooden bridge leads across the now heavily silted pool to Green Farm. The third and fourth pools are now lost beneath the modern Glynwed ironworks. Lower Furnace Pool was opposite the church and Upper Forge Pool (one of the very early 16th Century pools) was opposite the picnic site and car park adjacent to Rose Cottages. (Near Upper Forge was a timber-framed house called White End, the home of Abraham Darby I before he moved to Madeley Court. White End was destroyed in 1939 during road widening.) Only the floodgate of the Upper Forge Pool remains. This is on the other side of the road by the red brick pillars. The stream is divided here, half continuing on downhill and half taken under the road to Boring Mill Pool which is behind the black and white Rose Cottages. The Boring Mill is a simple brick shed with a pitched roof and large black doors. The Upper Forge – a humble brick building now painted white – lies N of the car park-picnic area; Rose Cottages and the Boring Mill are S of it. At Upper Forge there was once a small community – houses and workshops. It didn't have a pool but the water-wheel was operated by water brought in through a pipe. Later, a Boulton and Watt steam engine provided the power. This is now the only 18th Century industrial building left in Coalbrookdale. Early puddling experiments were carried out here. The Boring Mill pool is now much reduced in size and overgrown but still holds water. It provided power for both the Boring Mill (where the cylinders of steam engines and the barrels of canon were bored) and Middle Forge. Water from this pool was returned up the hill to the Old Furnace Pool by the steam engine called Resolution (1781) which stood near the Old Furnace. Today a large, red brick house with round-headed windows in the upper storey stands on the site. Resolution Cottage and Engine Row stand beside it. A ½m. long tunnel led from Boring Mill to Resolution where it connected with a 120 ft. shaft. Quite an operation. The lowest pool, the 6th, was at Lower Forge. It has almost disappeared but lies on the opposite side of the road from the police station, between Dale Road and Paradise. From here the stream passes under the roads to join with the River Severn by the Dale End car park. Note: the furnaces and forges had much ancillary equipment that was worked off the same water-wheels – grindstones, lathes, circular saws, etc.

COALPORT *2m. ESE of Ironbridge.*
Coalport was planned and created as a canal-river interchange port by the ironmaster William Reynolds. In 1793 he leased some low grade farmland on the banks of the River Severn, S of Ironbridge and built quays, warehouses, cottages and a porcelain works (1796). There was a soapworks, adjacent to Ludcroft Wharf, a timber processing plant and in 1798 Benjamin Edge established his famous rope and chain works. In 1814 Gilbert Gilpin left the employ of the Botfields and also set up making chains to a new design that enabled them to replace ropes in collieries and heavy engineering works – an important technical innovation. Coalport is a red brick place especially well-known for its very collectable porcelain. The Coalport works were taken over by John Rose, who had learned his trade at the renowned Caughley pottery. Rose lived at Hay House Farm, on the hill ½m. W. This is a striking early 18th Century brick house of 7 bays with projecting 2-bay wings and bold, blank-arcaded chimneys. It was at Coalport that Rose adopted and made famous the 'Indian Tree' pattern. The factory is now closed but there is a small museum, 2 bottle-shaped kilnhouses, a footbridge over the river and the remarkable Hay Incline, which ceased operation in 1884. Canal boats were hoisted on to wheeled cradles and hauled up, and lowered down, the Incline between the Severn at the bottom and the Shropshire Canal 207ft. above at the top. The Incline is 350yds. long with a gradient of 1 in 3 and a steam engine was used to haul the boats up the hill. At the base of the Incline is the Tar Tunnel. This was begun as a wagon way tunnel by William Reynolds about 1786. He intended to connect the Severn with the lower workings of coal pits at nearby Blist's Hill. However, the walls of the tunnel exuded a tarry petroleum at the rate of 1,000 gallons a week, an unexpected bonanza. It was processed at the mouth of the tunnel and exported all over Europe. By 1893 the flow had much diminished and by 1901 the tunnel was being used for its original purpose, coal being transported through it to be loaded on the barges on the Severn. Nowhere in the Ironbridge Gorge are the woods more evident than along the road from Coalport to Ironbridge. Also evident is the state of the road itself. Settlement and landslip make it impossible to keep the road in good repair. Downstream of the Coalport factory is the Coalport Bridge. The first bridge here was built in 1780 of timber and was called Preen's Eddy Bridge after the S bank place name. This was replaced in 1799 by the elegant, single-span iron bridge that we see today. The date 1818 on the midspan panel refers to repair and strengthening work carried out in that year. Note: 'Coalport' is now only a trade name. Since 1926 Coalport china-ware has been made in Stoke-on-Trent by the Wedgwood group of Companies.

CORELEY *4m. WSW of Cleobury Mortimer.*
The hamlet of Coreley lies on the SE slopes of Cleehill in pleasant but unspectacular country. In 1255 the tenant-in-chief was the splendidly named Roger de Mortuo Mari. The church of St. Peter has a squat 13th Century, tower topped by a later shingled broach spire, and a brick nave and chancel with Gothic windows of 1757. The pulpit is Jacobean. The first element of the name Coreley is thought to be from *cran*, or *cron*, Old English words for the bird 'the crane'. Hence, the name meant something like 'the wooded slope where cranes are found', or 'the clearing close to a crane's resting place'.

COSFORD *8m. NW of Wolverhampton.*
Cosford lies just to the NW of the village of Albrighton on the A41 and just S of the Tong junction on the M54. The old settlement has been completely taken over by the RAF who have built what

The Hay Incline, Coalport.

amounts to a small town here. The main road 'village' is a mess and the camp buildings themselves are charmless. The railway line from Wolverhampton to Shrewsbury passes through the camp and there is a stark public station here. Cosford is famous for two things: the Indoor Sports Arena and the Aerospace Museum. International athletics meetings are regularly held here and most are televised. The museum is one of the largest in the country. It was founded in 1979 and includes the British Airways collection of civil aircraft and what is said to be the most comprehensive collection of missiles in the world. A recent acquisition is an Argentinian Pucara captured during the Falklands 'War' of 1982. There are also, of course, examples of military aircraft, old and new. One of these is a World War Two Lincoln bomber (RF 398) which was brought to the museum to be restored in 1977. Strange and unexplained phenomenon have occurred in this aeroplane: whistling noises; discrepancies in temperature within and without the aircraft; a phantom airman wearing a white polo-necked sweater and bomber jacket; and mysterious faces appearing on photographs. It has been suggested that the apparition is that of pilot Major Miller who knew the aeroplane well and who died in an air accident near Codsall. Finally, the name Aerospace Museum may mislead some into thinking that 'space vehicles' are represented in the collection, but 'mooncraft' and the like are not here; far from it, at the time of writing there is an exhibition of Mammoth bones that were found in a quarry near Condover. (*See Condover.*)

COTON *2m. WNW of Prees which is 6m. S of Whitchurch.*
A hamlet on the road between Prees and Whixall in the flat country of North Shropshire. It is dominated by Coton Hall, a large red brick mansion with some black and white work and a tower. In this century it has been the home of Lord Hill 'of Hawkstone', as the locals called him. He is fondly remembered tootling about in his little Austin A40 motor car and embibing liberally at the Bull and Dog, a handsome old half-timbered pub. The next squire of Coton was a less traditional character. He was Mr. Mellings, a scrap metal merchant from Manchester, who commuted daily in a large chauffer-driven American car. The Hall is presently being renovated and developed by its new owner, Mr. Swainbank. Just past the entrance to the big house is a black and white thatched cottage that was once 2 and lived in by the milkmaids who worked at Rookery Farm across the road. (Note: there is another Coton Hall near Alveley.)

COUND *6m. SE of Shrewsbury, off the A458.*
Cound Hall lies near the main road. It is not signposted but the 2 entrances have lodges, on the right-hand side of the road if approaching from Shrewsbury. It has been empty for some 10 years now, the windows boarded and the grounds unkempt. It is owned by a family of retail shop owners who seem to be holding it as an investment. The house is an imposing red brick house of 1704, with stone dressings and fluted Corinthian pilasters and was built for Edward Cressett by John Prince of Shrewsbury. The village lies ½m. off the road, S of and behind the Hall. It is divided by the Coundmoor Brook. Cound has the church of St. Peter, mainly 13th Century with a Victorian N aisle and chancel, a 'sanctuary knocker', a Norman font and Georgian memorials which include a hanging monument to Dr. Edward Cressett, Bishop of Llandaff (d.1755). Upper Cound hides amongst its gardens, trees and flowers with black and white cottages, both old and new. On the main road is the substantial Cound Lodge Inn, late 17th Century brick with 2 gables. The original front faces the river and the old road on which the railway track was laid. It is the back of the inn that we now see from the new road. There was also a small port here which especially flourished in the 18th Century as a coal shipment point. But times change; now the port and the railway are long gone and the pub turns an embarrased back to the busy modern highway. A mile to the S of Stevenshill is a sizeable earthwork fort, and ½m. W is the course of a Roman road which crossed the Cound Brook near here on its way to Uriconeum (Wroxeter).

CRAVEN ARMS *6m. S of Church Stretton on the A49.*
It is an old crossroads but a young town with few buildings of note; a functional settlement with light industries connected to agriculture and farming, and unremarkable housing. The Roman road from Wroxeter (Uriconeum) to Leintwardine (Branogenium) passes ¼m. W of the town; in fact, from Church Stretton to Craven Arms the modern A49 follows the general route of this Roman highway and in parts replaces it. At Craven Arms the A49, the main N–S road, and the B4368, the main E–W road of the area, meet and form a junction. However, it was not the roads that made Craven Arms but the coming of the railways. All that originally stood at the crossroads was an inn, the Craven Arms, named after the Earls of Craven. The present inn was built in 1841 but there was a hostelry here long before that. The old settlement was the hamlet of Newton, ¼m. W of the inn. Then came the railways; first the Shrewsbury to Hereford line, in the late 1840's, followed by the Knighton line that linked Central Wales to Craven Arms, and the Buildwas line to the Shropshire coalfield that travelled through Apedale along the northern foot of Wenlock Edge. The town was now a busy junction. All this activity brought a gleam to the eye of the Earl of Craven. He decided to build a town. Much as the Normans did he 'planted': streets were laid out in a grid – Market St., Newton St., and Dale St – and divided into building plots; a bridge was built across the Onny; an auction market was started and houses were built. The Church of England ignored the town, however, and Craven Arms is still in the parish of Stokesay, 1m. S, and the town is still without its own Anglican church. But the Baptists are here (1871) and the Methodists (1880). Remarkably, one of the rows of houses built then fossilised part of the pattern of the medieval open field system of Newton. Albion Terrace, off the Clun Road, curves away at an unusual and apparently uncalled-for angle. It does, in fact, run exactly parallel to one of the old open field boundaries. At the beginning of this century it seemed that Craven Arms was destined to become a town of considerable size. For no good reason this just did not happen. It could simply be that it has a somewhat bleak and unfriendly feel and is almost totally lacking in character. Every Autumn one of the largest sheep sales in the county is held at the Craven Arms livestock yards. Less than a mile SW is the impressive earthwork of Norton Camp. It lies at the southern tip of Wenlock Edge, 1,000 ft. above sea level, and covers 10 acres. The area around Craven Arms was, without doubt, an important communications centre in Iron Age times, just as it is today. The inhabitants of Norton Camp defended these trade routes and quite possibly extracted tolls. Today, the deep ditches of the settlement form field boundaries, and even the parish boundary of Culmington follows its contour. The earthworks of the fort have made the ground too irregular for cultivation and the area has become tree-covered. Eight footpaths and rights of way still converge on the Camp. Near **Horderley**, 3m. NNW of Craven Arms on the A489, is a hill at the top of which is a boulder called the Devil's Stone. As late as the early years of this century children who collected wild fruits, such as wimberries, were careful to always leave propitiary offerings in holes in the stone, a remarkable survival of an ancient pagan custom.

CRESSAGE *7m. SE of Shrewsbury.*
Cressage lies on the A458, Shrewsbury to Much Wenlock road. It is famous for the reputed visit of St. Augustine who is said to have preached under the Cressage Oak in 584 AD. The old name of the village was Christesache, 'Christ's Oak'. The original tree is believed to have grown where the present War Memorial stands. It was part of the medieval Long Forest, the wooded areas of which had been cleared by 1616. A cutting from this original tree was planted in an open field near Lady Oak Leasowe, between the A458 and the course of the now dismantled railway (SO.582.046). The original road passed close by the tree but was resited when the railway was constructed along its route in 1861. Only a fragment of this tree now remains but a new, young, third generation tree is growing out of the hollow centre of the old oak which died in the frosts of 1982 at an age estimated to be 700 years. Between the War Memorial and the old oak tree is a stone marking the position of the old church dedicated to St. Samson which was pulled down in 1841 because it was subjected to constant flooding. The new church of Christ Church was consecrated in the same year. It has a Norman font, a gallery supported by cast

Cressage, Christ's Oak and The Wrekin.

iron columns and some good stained glass. The Old Hall Hotel is locally well known. It is partly timber-framed, of the 17th Century and stands by the River Severn. Castle Mound is the motte of a Norman castle which commanded the ford close to the present bridge. This reinforced concrete structure was built in 1913 to replace the previous timber toll bridge of about 1800. Mr. Price of Oak Tree Farm is now retired from farming but in 1987 he and his dog Davy won both the national and international Supreme sheep dog trial championships. There is much new housing in the village which no doubt pleases the landlord of the Eagle pub and provides new blood for the local football team. Six roads meet at Cressage. One mile SE is Belswardine Hall, said to be of 1542, but it was much added to and altered in the 19th Century. All that remains of the original house are two large chimney breasts.

CRIFTINS *See Dudleston.*

CROESAU BACH *3m. WSW of Oswestry.* It is 'bach' indeed, a handful of stone and stucco cottages, a farm and a small disused quarry at a crossing of roads. One mile NW is the hilltop lake of Rhuddwyn and the prehistoric fort of Coed-y-gaer. To get to **Llyn Rhuddwyn** take the road to Llawnt and Rhydycroesau and bear left after about ¼m. This lane leads to 2 farms, Pentregaer Uchaf and Pentregaer Issa. The lake lies at the top of the hill at SJ.232.288. Dark coniferous woods now surround it but there are paths through the trees and around the waters' edge. This is a wild place. Coppiced alders and other brushwood fringe the pool which is now much overgrown with reeds and bullrushes though there are still some areas of clear water. The woods have been used for breeding game birds and are littered with corrugated-iron shelters, blue and yellow plastic bags and milk churns on stands raised about 1ft. above the ground. The centre of these works is a compound surrounded by wire netting with entrance traps. There was no sign of life though. Near this compound is an abandoned and roofless 2-roomed stone building about the size of a small cottage. The oval earthworks of the pre-historic fort stands just N of the lake. To the NW there is a very steep slope – 70 degrees or more with a fall of 600ft. – which we fought our way along for nearly a mile by mistake, having lost our bearings. Much of the original rampart of the fort has disappeared. What remains lies at the N and the SE and varies in height from 2 to 3ft. To the SW there was a ditch but it was much disturbed by picnic parties in the early years of this century, before the conifers invaded and took command. The paths are very wayward, coming to an end at thorny thickets or young, dense plantations. The name of the earthwork, Coed-y-gaer means 'the wood by the fort'. The old deciduous wood is gone, only a handful of scattered oaks, beeches and birches remain. These trees, and a few grassy glades provide a welcome relief from the ominous gloom of the endless ranks of conifers. To the NW there are several old quarries now covered in undergrowth and hedged around by trees and wire fences. To the NE, at the bottom of the precipitous slopes, is the River Cynllaith which here forms the border between England and Wales. At Bwch y Rhiew there are rifle ranges. The shooters fire from Wales into the butts in England. There was constant firing whilst we were there and more than once I was concerned for the safety of my dog who wanders far and wide in such country. To conclude it should be mentioned that the Celts, who almost certainly occupied the fort of Coed-y-gaer during the Iron Age, worshipped natural objects such a stones and trees. Circular pools on high places were of especial significance to them and it is not unlikely that they made propitiary offerings to the spirits who occupied the waters. Llyn Rhuddwyn is almost circular. **Craig-llwyn** lies ½m. W of Croesau Bach. Between the 2 settlements are Ashbank, a dog and cat boarding kennels attached to a stone farmhouse, and Pant Hir, a 7-bay white and black painted house set well back from the road behind a horse paddock. The signs of Shirecraft Fireplaces direct one along a rough track to a disused quarry with rusty-red grading and crushing plant and abandoned motor cars. There are many such old quarries hereabouts. Just to the W, on the main road is the hamlet of Craig-llwyn which consists of little more than a few trees, a farm, the old smithy, a modern bungalow and a disused chapel. To the W by ½m. the road makes an acute angle bend (SJ.227.274). There is a place to park a car here and take in the view which is splendid indeed, a great, wide valley with hills ranging into the distance.

CROSSGREEN *2½m. NNW of Shrewsbury.* In 1988 Mary Webb's last and best known novel, Precious Bane, was filmed as a television play. It was broadcast during the Christmas holiday and received great popular and critical acclaim. The Sarn Mere of the novel was based on the brooding, legend-ridden Bomere Pool (*see Bayston Hill*) but the film was made at Alkmund Park Pool (SJ.480.160). This lies hidden from sight in a depression between the B5067 at Crossgreen and the railway, a ¼m. to the E. It is a lovely spot. The wind-rippled waters are fringed with reeds, woods and meadows, a place previously only known to locals, fishermen and those who shoot game birds in Pool Coppice and Park Wood. These lands are a part of the Berwick Estate which is centred on Berwick House, 1m. SE. The timber-framed cottage seen in the film was located at the southern tip of the lake but was only a polystyrene shell and has since been removed. (Some filming was also done at the County Showground.) Alkmund Park Farm lies ¼m. S of the pool. This is a 3-storey red brick house with a gable and good traditional outbuildings. It lies on a public footpath that stretches from Coton Hill in Shrewsbury, past the mere, to a minor road that links the B5067 and the B5124. The hamlet of Crossgreen lies at the junction of this road and the B5067. Here there is a substantial brick farmhouse that has been painted and boarded to make it appear to be timber-framed; a range of redbrick farm buildings; a pair of stuccoed, semi-detached houses also painted black and white; a Victorian black and white school with a gable and a bellcote; a few scattered cottages; and most striking of all, Upper Berwick Farm, a most handsome 'Queen Anne' house of about 1700 set a ¼m. W off the main road. This is a red brick mansion of 6 bays and 2 storeys with irregular quoins, a 2-bay pedimented porch, a hipped roof with a steep 2-bay pediment containing a round window, a walled garden and several satellite cottages. The country is most pleasing with numerous dips and raises, woods and meadows, and in the distance the 'blue remembered hills' of South Shropshire. **Leaton Knolls** lies ½m. N of Crossgreen. Here there is a farm with a good, red brick stable quadrangle, a few cottages, a nursery in a large walled garden, some greenhouses and mixed woodlands which reach down to the banks of the River Severn.

CROSS HOUSES *5½m. SE of Shrewsbury on the A458.* The old hospital dominates the village. It was built in 1793 as the Atcham Union Workhouse and was enlarged in 1871 and 1903. During the First World War the building was converted to a Military Hospital. After the cessation of hostilities it became a General Hospital and remained so until its closure in 1987. The railway line to Shrewsbury was dismantled in 1964 but the old trackway can still be seen. The oldest building in Cross Houses is probably The Fox. In the other pub, The Bell, a harvest festival is held every year. The charming Italianate house called **Cronkhill** lies ¼m. NNW of the village on the road to Atcham. It is believed to stand on the site of a prehistoric defensive position and was built for the land agent who administered Lord Berwick's Attingham estate. It is said to be the earliest house of its kind in the whole of England. John Nash was the architect and the designs for it were almost certainly those of 'the villa in Shropshire' that were exhibited at the Royal Academy in 1802. It is a building of great character and has a square tower, a colonnade with a parapet, round-headed windows and batteries of chimneys. It stands on a hill with a farm adjacent.

CROSSWAYS *3m. NW of Newcastle, which is 3m. WNW of Clun.* Crossways is the nearest habitation to the Cantlin Stone, but is unmarked on many road maps. Crossways is most easily approached thus: take the road heading NW out of Newcastle. After ½m. bear left at Caldy Bank. Follow this lane, ignoring any right or left turns, for 1½m. and you will come to a cattle grid. Cross over this and 1m. further on you will come to a plantation of coniferous trees and 2 cottages. This is Crossways. The road for that last mile (from the cattle grid to Crossways) marks the 'boundary of cultivation'. To the right, downhill, is fenced

above: *Alkmund's Park Pool, Crossgreen.*
left: *Old canal bed, Muxton Bridge, Donnington Wood.*
below: *Upper Berwick Farm, Crossgreen.*

pastureland. To the left, the hill top, is Rhos Fiddle, a rough common area covered in bracken, gorse and tough bogland grasses. This is something like the post-forest natural habitat of these high, moorland hills. Three quarters of a mile N of Crossways is the Cantlin Stone. To get there turn right at Crossways on to the forestry road, pass through the double gates and when the forestry road swings right, downhill, leave it and keep straight on down an unmade track. It is best to park here and walk the remaining 200 yds. After going through a five-bar gate the Cantlin Stone will be found on the left of the track in a small fenced compound. The Stone is now broken but the writing is still legible. William Cantlin was a pedlar who had a round which included isolated hill farms. On one of his trips, in 1691, he collapsed and died in this bleak and remote area. The stone marks the spot where he passed away. The crudely engraved inscription reads: "W. C. DECSED HERE BURIED 1691 AT BETVS." There was some disagreement between the parishes of Bettws-y-crwyn and Mainstone as to who should be responsible for the cost of Cantlin's burial, he having died a pauper. The stalemate was broken by the Vicar of Bettws who showed Christian compassion and buried poor Cantlin in his churchyard. Some 200 years later this act of charity was duly rewarded when the Cantlin Stone was accepted by the Enclosure Commission as proof that the land in which it lay belonged to Bettws, and they therefore awarded the parishioners several hundred more acres than they would otherwise have received. Close to the Cantlin Stone is the remains of a cross erected by Beriah Botfield of Ludlow. The road from Crossways to Mainstone, 5½m. NE, is to be recommended for its splendid views over the Forest of Clun hills. On leaving Crossways, and having travelled the 1m. back (SE) to the cattle grid whence you came, turn left and follow the lane straight to Mainstone. As you near your destination you will see a church on the left-hand side of the road. In it is a curiosity. (*See Mainstone.*) Note: a Forest, with a capital F, is a medieval term for an area under Forest Law and is not necessarily wooded. Indeed, most Forests were kept as hunting grounds and included large areas of heath and moor, much more suited to men hunting on horseback than heavily wooded country.

CRUDGINGTON *4½m. N of Wellington on the A442 to Market Drayton.*
Here is situated a creamery which manufactures under the Dairy Crest name. The factory and its tall, silver chimney are a landmark for many miles around. It began as a farmers' co-operative but in 1935 was taken over by the Milk Marketing board who still have their Research and Development Establishment here. Some 300 people are employed at the complex. Some may feel that a factory in the countryside is incongruous, but this one does process a local product and was placed where it is to be central to its suppliers. (Historically most industry was in the country; iron works, paper mills, woollen manufactories etc., were all country based. Only in the last 150 years or so have we come to associate industry with towns.) The hamlet of Crudgington stands on a small rise near the confluence of the River Tern and the River Strine at the crossing of the Market Drayton – Wellington road and the Newport – Shrewsbury road. In 1086 there were 4 fisheries and a small village. In later times it became a part of Lilleshall Estate and the Duke of Sutherland improved the land with drainage schemes. These required resident estate workers to keep the ditches clear. The Duke built several brick houses nearby, and a line of agricultural workers' cottages stretches E for over 1m. along a lane, called Crudgington Green, that runs parallel to the River Strine, the main artery of the drainage system. As to the rest of the village there are several black and white cottages, including one of cruck construction, and a few modern houses. Although the former Methodist chapel is now a dwelling, the church of St. Mary (1863) was re-opened in 1970 after having been closed for some years. (*See Kynnersley.*)

CULMINGTON *5m. NNW of Ludlow on the B4365.*
It lies in the valley of the winding River Corve. A pretty place of grey stone and half-timbered cottages, surrounded by rich meadows. The church of All Saints has a 14th Century tower and a nave and chancel all in one. The spire is a curiosity. The original was left uncompleted and was capped by a lead make-do, to which, in recent years, was added an aluminium frame. Some 300 yds. NW of the church are the earthwork remains of a Norman motte and bailey castle. At Callow Hill (bare hill), 2½m. NW, is Flounder's Folly, a tower erected in 1838 to mark the spot where the estates of 4 local landowners meet. It has been suggested that the River Corve was once called the Culm which is derived from the Welsh *Cwlwm* meaning 'a knot', hence 'loop', hence 'winding', hence 'winding river'.

CWM HEAD *2m. NNW of Wistanstow, which is 2m. N of Craven Arms.*
A couple of cottages, a modern bungalow and a church stand at the head of a pretty, wooded valley on the B4370. Such a tiny place as this would not warrant a priest of its own so a coach house and stable stand close by the church for the benefit of the visiting vicar. St. Michael's was built in 1845 to a neo-Norman design by H. C. Whitling. The apse, nave and tower are all of rock-faced stone and the coach house–stable is made to match. Further down the valley are some attractive cottages positioned between the road and the stream.

DAWLEY *1m. SW of Telford Town Centre.*
Dawley is now a part of Telford New Town, with its garish plastic-covered factories and spartan, treeless housing estates. However, the old Anglo-Saxon village of Dawley still retains a little individuality. The area was originally forested and the name means 'the wood, or the clearing in the wood, of *Dalla*'s people'. Dawley is situated on the Shropshire Coalfield and it developed quickly in the 19th Century. Industries based here included the manufacture of bricks and pipes, coal-mining and iron-working. In 1801 the population of Dawley was 3,869; by 1871 it had risen dramatically to 9,503 (census figures), but, as the Potteries and the Black Country continued to develop, so Dawley and the area around it declined. By the middle of this century it had become a desolate place, surrounded by an industrial wasteland; a wasteland that was returning to nature with scrubby undergrowth covering old derelict buildings and waste tips – almost a romantic scene. Then came the new town, originally called Dawley New Town, before the name was changed to Telford. Things have improved but it will be many years before the disparate parts of the New Town become integrated. Today, Dawley itself consists of little more than one long pedestrianised street. There are several Nonconformist chapels and in Dawley Magna is the large, stone-built Victorian church of Holy Trinity which has an unusual Norman font. The stone-built castle at Dawley was held by the Royalists during the Civil War. It fell to Cromwell's men in August 1645 and in later years it was 'quarried' for hardcore before being buried under slag from the Castle Furnaces. It lay about 260 yds. S of the church. Captain Matthew Webb, the first man to swim the English Channel (1875), was born at Dawley in 1848. He died trying to swim the rapids below Niagra Falls.

DETTON *3m. N of Cleobury Mortimer and 1m. ESE of Oreton.*
Here there is a deserted medieval village in the valley of the River Rea just to the NE of the Hall. Only a single cottage remains facing the old village High Street. There was formerly a chapel and the field in which the settlement is situated is now called Chapel Field. The village houses were of timber and had been deserted by 1300. The population was probably between 40 and 50 and they occupied the Manor House, at least 2 other large farms and about 6 small crofts.

DEUXHILL *3m. S Bridgnorth on the B4363.*
An almost deserted village. The remains of the church lie at the rear of a Georgian farmhouse a few yards up the road to Eudon George. Hall Farm is dated at 1601; it is half-timbered and has been recently restored. There are good views over the rolling, slightly mysterious countryside to the S. Deuxhill is pronounced Dukeshill.

DHUSTONE *4½m. E of Ludlow.*
The hamlet is named after the hard, igneous dolerite that caps the Carboniferous limestone rocks of Titterstone Clee Hill, on the middle slopes of which the settlement stands. This dark stone is called dhu-stone, from the Celtic for black-stone. It has been quarried for use as roadstone for many generations. The

Crudgington Moor from the B5062.

lane to the hamlet climbs uphill off the A4117, the Ludlow to Bridgnorth road. It is a surprise to see Victorian terraced houses half way up a barren, windswept hill. Washing flutters in the ever present breeze and unsightly sheds adjoin rough pastures where sheep graze. Spoil heaps are everywhere, though many are now grassed over. The lane crosses the track of a dismantled mineral railway by means of a causeway, and continues past the houses to a huge car parking area around which are the numerous gaunt ruins of abandoned quarry buildings. The views from here are most impressive. This is a big, strong hill and man has been here since time immemorial. He has made his home here and mined its coal, limestone and dhustone and grown crops here and grazed his animals. Marks from many ages are etched into the hill. The most recent are the dishes and aerials of the Civil Aviation Authority's National Air Traffic Services' radar installation. Their equipment tracks and guides civil aircraft and provides meteorological information. The big silver dome (like a golf ball on a tee) is the meteorological radar and the 2 dishes are Secondary Surveillance Radars. One thing they do not do here is track satellites and the Ordinance Survey legend 'Satellite Earth Station' is quite simply wrong. The sheep that inhabit the hill belong to the commoners, local people who have rights of free grazing.

DIDDLEBURY *4½m. ENE of Craven Arms on the B4368.*
The name is from *Duddela*'s-*burg*; *burg* meaning a fortified place and *Duddela* being a personal name. Today the main road developments will deceive the traveller. The modern houses have thankfully been built alongside it and the real, old village has been allowed to slumber undisturbed. There are lovely stone houses, a stone bridge, a stone church and a stoney-bedded stream that runs parallel with the main street before heading south to join the River Corve. The charming Glebe Farm has one stone gable and one black and white gable and is now a guest house. The church is very old with a Saxon nave and N wall entirely of herringbone masonry. The chancel is Norman and the tower is a problem; the experts cannot agree, but it seems likely that the lower part is Saxon and the upper part Norman. A charming village. Nearby are several fine halls. Delbury Hall (its name is an abbreviation of Diddlebury) is constructed of red Georgian brick and stands in a park that stretches back to the main road. About 2½m. NE is Broncroft Castle, most of which is an 18th Century addition to a 14th Century tower (the lower of the two main towers). It is a handsome, well-kept place. A mile or more to the SW is Elsych; moated, of stone and half-timber framing and probably Elizabethan. It is approached down a lane off the main road in the direction of Craven Arms. At Corfton, ½m. SW, are the moat and earthworks of a Norman motte and bailey castle. The shadowy earthwork remains of another castle, Corfham Castle (SO.525.849), are situated in a field a mile to the ESE of Diddlebury, alongside the road to Peaton. Here was once a great stone castle but as early as 1550 only one tower remained. Corfham was the home of one of Britain's most famous beautiful women, the Fair Rosamund, beloved of poets and the mistress of King Henry II. Her father, Walter de Clifford, was granted lands at Corfham by King Henry. The first element of the name Corfham is from the Old Englsh *ceorfan* meaning 'a cutting' and hence, 'a pass'; the 'pass' here is the valley of the River Corve which name is itself derived from *ceorfan*.

DITTON PRIORS *7m. WSW of Bridgnorth.*
It lies on the lower northern slopes of Brown Clee Hill, a small, comparatively isolated village of stone, half-timber and red brick cottages, farms and terraced houses. Most of the people who lived here were connected with the quarrying of Brown Clee Hill for its coal, limestone and road-stone. A mineral railway used to run down to Cleobury Mortimer. The industrial tradition of this country village is continued on the Trading Estate on its eastern fringe. This was originally built as a Naval Camp and was later used by the American Army as a supply depot. The Church of St. John the Baptist is mostly of the 13th Century with a weather-boarded broach spire which was rebuilt in 1831. It is noted for the trussed rafter roof of the nave and chancel. There are yew trees in the churchyard. In 1066 Earl Edwin held the manor but after the Conquest it passed to Earl Roger (de Montgomery) who held it himself. In 1086 it was a sizeable settlement with at least 20 villagers and 8 smallholders (and their families) and 10 slaves. The manor also owned a salthouse in Droitwich. The name Ditton is Anglo-Saxon and probably means 'the settlement of *Dudda's* people'; 'Priors' refers to Wenlock Priory to which it was given by Henry II.

DODDINGTON *3m. W of Cleobury Mortimer on the A4117.*
The settlement consists of a scatter of buildings along the main road. Between here and **Clee Hill** village, which is ½m. E, there are many parking places to enable motorists to stop and take in the spectacular views. On sunny summer days it almost becomes a holiday resort. The hills above the road are scarred with old quarries and pits. The spectacular quarries at Clee Hill are still worked by ARC who mine the dhustone. (*See Dhustone.*) The ashlar church of St. John (1849) at Doddington lies opposite the turn to Earls Ditton. To the E of Doddington is the metalled track (signposted Farlow) leading to **Catherton Common**. Here are heather and bracken moors grazed by the sheep of freeholders, and spotted with the crater-like remains of 'bell pits', the result of small-scale, open-cast mining, some of which date back to the 13th Century. Many of the cottages hereabouts were built by squatters.

DONINGTON *It adjoins Albrighton (near Shifnal).*
The church of St. Cuthbert has a 14th Century chancel and a holy well. St. Cuthbert's Well is just below the church, off a hilly bend in the road and down a stepped bank in a dark, wooded glade. The waters form a pool and are reputed to relieve and cure complaints of the eye. It is an attractive spot and well cared for. The name Donington is from *Dunningtun*, 'the settlement of *Dunn*'s people'. If we had not known the early medieval name of the settlement we might have been confused by the present first element of the name *don* which is an old name for 'water', especially as there is an ancient spring here. The *in* or *ing* middle element indicates that the place was more than a single homestead because it usually means 'of the tribe', or 'the people of', in this case *Dunn* or *Dunna*. Donington is no longer a separate settlement but has become a residential suburb of Albrighton.

DONNINGTON *1m. ESE of Wroxeter.*
A tiny hamlet just off the B4380. Richard Baxter (1615-91), the Puritan divine, and Richard Allestree (1619-81) both attended the free grammar school here. The patrons of the school were the Newport family of Eyton-upon-Severn. A later master was the Welsh writer Goronwy Owen (1723-69). The hamlet of **Charlton Hill** lies ¼m. SE. Here is a good brick house of about 1660 with gables, dormer windows and brick freizes. Charlton Castle is 2¼m. NNE, on the other side of the A5, but only the earthern mound remains. Close by is a moated site.

DONNINGTON *5m. SW of Newport.*
Now a part of Telford New Town, this is one of 3 Donningtons in Shropshire but without any doubt this is the best known because of 2 disastrous fires at the huge Army Central Ordnance Depot which adjoins the village to the N. The last fire, in 1988, caused an estimated 150 million pounds worth of damage. Donnington used to be a farming hamlet but by the mid-19th Century there was an iron works, a railway station. and coalmines all about. Today it is little more than a suburban sea of estate houses. In the Parade there is a small shopping centre and a pub, the Champion Jockey, a tribute to 'local lad' Sir Gordon Richards (born 1904), who was champion jockey 26 times between 1925 and 1953. There are 2 schools, a modernistic Roman Catholic church, and some distinctive dormer-windowed Duke of Sutherland cottages. (Donnington was on his Lilleshall Estate, sold early this century.) **Donnington Wood**, 1m. S, is a wilderness of old mines and spoil heaps which is currently in the process of being reclaimed and landscaped. The church of St. Matthew was built in white stone to a design by Sir George Gilbert Scott in 1845. (*See Donnington Wood.*) The name Donnington is Old English and means, in this case, 'the settlement of *Deora*'s people'.

DONNINGTON WOOD *It adjoins Donnington which is 1¼m. NNE of Oakengates.*
There were once a few scattered cottages and at least one small village here but most of the dwellings have been destroyed by modern mining operations. In medieval times this area was part of the hunting park of the Leveson family of Lilleshall, but as early as the 16th Century there were ironworks here. In the 18th and 19th Centuries the area was developed at a furious pace and hundreds of acres were turned into a wilderness as the huge resources of coal, iron ore and clay were ripped from the ground. Often all 3 minerals were found in the same pit or quarry. In its natural state the land was either flat or gently rolling; today it is hilly country. Now that the barren slopes are clad in grass and scrub woodland it looks natural once again. The last mine to produce coal in the area was the Granville pit. This closed in 1979, the last deep mine in the Shropshire coalfield. That pit is now the centre of a grand reclamation scheme by the Wrekin District Council. With the aid of a grant from the European Development Fund and the expertise of Budge contractors the spoil mounds of the mine are being processed to recover small coal and are then to be landscaped as part of the Granville Country Park. This is a major project covering some 360 acres stretching from just N of St. Georges and just S of Muxton that have been made safe by the capping (or filling) of about 200 old mine shafts. (Local people believe that there are many more uncharted shafts with only flimsy covers of timber or corrugated iron which could collapse at any time.) Paths have been made to link areas of archaeological or natural interest and it is, indeed, a most intrigueing area. There are several entrances and there are car parks near the Donnington Wood Roundabout and at **Muxton Bridge** (which is at the end of Muxton Lane, entered at Muxton). The bridge used to cross the canal but when this was abandoned the bridge was removed and the void filled with earth. Adjacent to the car park are the remains of the Muxton Bridge Colliery: the tree-clad spoil mound; the capped mine shafts – circular and outlined with bricks; the large and handsome, winding-engine house of about 1880, ruinous but still substantially upstanding; part of the track of the plateway complete with a length of iron rail and a wharf wall; the remains of the pumping engine; and the bob wall of an enclosed beam engine complete with fixing bolts. The path leads SW and follows the now dry bed of the Donnington Wood Canal, which ran from Donnington Wood to Pave Lane, 1½m. SSE of Newport on the A41. This was completed in 1768, one of the earliest canals in Britain (the earliest was built in 1760) and the first in Shropshire. Today it is pleasantly overhung with trees and is virtually an arboreal tunnel for about ⅓m. The 2 pairs of semi-detached cottages across the fields are farm workers' dwellings and have nothing to do with the mine. At the end of the 'tunnel' the path rears right up a short incline, then left along the handrails and over the course of the canal, which dissappears here. Keep straight on when the rails go right. To the left is the spoil mound called White's Bank which in the Autumn is covered in cotton plants of 2 species. An old railway cutting runs parallel to the path. Go through the car park gates cross the car park (White's Cottage is on the left) and turn right on leaving. This is Lodge Road, a limestone metalled track. On the left is Lodge Pool (actually a canal basin and wharfage). Turn left to the pool and the spectacular ruins of the **Old Lodge Furnaces** are revealed. The great stone walls are part of the 3 charging ramps which in total measure some 120ft. long by 40ft. high. The mound against which the ramps stand is artificial, made of old spoil mound material and therefore unstable. To hold back the earth a great retaining wall was built and the ramps project forward from this. Large parts of the ramps are missing. Within them there were workshops and store rooms. In front of the ramps are the remains of the 3 furnaces. These are circular, about 15ft. in diameter and constructed of fire brick; the inside surfaces are coated with a dark grey molten metal and brick layer. The furnaces were originally very tall, taller than the ramps. The iron ore, coal and limestone (a flux that collected impurities and was drawn off to solidify as slag) were brought in along a railway that ran along the ramp and were loaded into the furnaces at the top. The mixture was set alight and the flames fanned by a blast of cold air pumped in by an engine. The engine house stood to the right of the ramps. Once fired a furnace could burn continuously for up to 8 months. The molten, crude iron was tapped twice a day at the bottom and run into sand molds. These took a shape that was like a mother pig feeding her piglets and so the iron blocks were called 'pig iron'. This crude 'cast iron' was sent to forges for refining – to be re-heated and beaten by great hammers to remove impurities and make the iron less brittle. This wrought-iron was much preferred to cast-iron. The pool below the furnaces was a basin on the Donnington Wood Canal, along which we walked at Muxton Bridge. The canal entered from the N under a bridge that carried Lodge Road. This has now been removed and the canal is blocked by an earth bank. The basin has been dredged and the wharfs have been restored. South of the canal basin is a high dam wall. This was built to greatly extend a small natural pool and its purpose was probably to act as a reservoir for the canal. At the base of this earth bank-dam are some very rare coking ovens – the furnaces burnt coke, not coal – and in the same general area (to the left front of the furnaces) are the remains of the offices and the stables. Note: there were 5 furnaces altogether but 2 were built later (in 1859) than the others and were probably little used. When the works was closed in 1888 it is likely that they were dismantled and the expensive fire-bricks removed by the Lilleshall Company for use elsewhere. They were fed not from ramps but directly off the extended retaining wall and stood to the left of the 3 original furnaces. The whole complex is most impressive and evocative. Here it stands, in a scrubwood wilderness, surrounded by hills and for all the world looking like a lost city in a South American jungle. In front of the furnaces and to the right is a jumble of huge blocks of stone which measure about 5ft. x 3½ft. x 2ft., Each stone has 4 holes about 4½ inches in diameter. The Old Lodge Furnaces are often simply called the Lodge Furnaces and in modern times the Granville Furnaces. This last name is confusing and should not be used – look for it in any histories or documents and you will not find it. The Lodge Furnaces were built in 1825 and abandoned in 1888; they were known for the high quality of the 'cold blast' iron they produced. Granville Park has many other areas of archaeological interest. These include the remains of the Barnyard Colliery; the Waxhill Colliery and the Waxhill Barracks, workers houses complete with chapel; the Old Hunting Lodge of about 1679 of Leveson family, which gave its name to the Furnaces, and now lies burried beneath the Lodge Mound spoil heap; and the likely remains of at least one early industrial site near the canal basin. The canal was constructed long before the Lodge Furnaces were built and it may well have served an earlier furnace, forge or foundry here as well as being used to ship cargoes of iron and coal elsewhere. For futher information on Granville Park interested persons can contact either the Wrekin Council Action Centre, Chapel Lane, St. Georges, Telford, or the Ironbridge Institute, Coalbrookdale Museum, Telford. The Institute has produced 2 reports on the site – Research Papers Nos. 24 and 32. The Lodge Furnaces were not the only ironworks in Donnington Wood. There were several early sites, as yet, unidentified and the large, brick-built Donnington Wood Furnaces constructed by William Reynolds and Joseph Rathbone in 1783. These became part of the Lilleshall Company and in 1858 were closed down when the 2 additional furnaces were built at the Lodge Works in 1859. South-west of the Granville Colliery (SJ.726.121) site is Grange Colliery (SJ.721.113) where both coal and ironstone were mined. This is no longer in production but the rusty-red headgear still stands proud, high amongst above the litter of caravans that are stored around its feet, but itself dwarfed by the old, tree-clad spoil mound. This colliery only lies ¼m. N of the A5. Locally it is called The Windings.

DORRINGTON *6m. S of Shrewsbury.*
An old, mainly red brick village on the A49. The black and white house that catches the eye is the Country Friends' Restaurant (formerly the Old Hall), parts of which are dated at 1673; adjacent is the Old House of 1588. There is a village hall, a pub, the Horse Shoes Inn (1734), and at Grove House antiques are sold. There used to be a station here, on the Shrewsbury Hereford Railway (1852), and there were regular sheep and cattle sales at the Railway Inn (now the White House). The

line is still open but the station closed in 1958. Most of the houses are 19th Century or earlier but there are some council flats and a few developments of modern houses. The trim little church of St. Edward was built in 1845 by Edward Haycock to service the newly created parish. It has a certain elegance and is faced with ashlar. Opposite the church is the Church of England Primary School in Victorian red brick with ornamental bands and arched windows. This replaces the old school built by Thomas Allcock. Richard Allestree (Provost of Eton) and his contemporary Richard Baxter (the 17th Century Puritan divine) were *not* educated here as stated in some reference books; they attended the school at Donnington under the Wrekin in the parish of Wroxeter. John Boydell, engraver and publisher, was born in Dorrington in 1719 and became Lord Mayor of London. Half-a-mile S of the village is the entrance to **Netley** Hall which stands amidst the landscaped acres of its wooded park. This is the home of the Hope-Edwards family, benefactors to the village. There are good views of the Stretton Hills from here. The red brick house was built in 1854-8 by E. Haycock, who also designed the church in the village. Netley Hall has 5 bays, 2½ storeys, a porch of paired Tuscan columns, an ashlar-faced centre and a topmost balustrade. Bed and breakfast can be obtained here.

DUDLESTON *3m. NW of Ellesmere.*
Dudleston, 'the settlement of *Duddel* and his people', is a quiet, red brick hamlet on a hill in rolling country near the northern border of the county. The Manor House is large but was built to an unremarkable design in red brick with stone dressings and yellow brick ornament. The church of St. Mary stands on an ancient site and has old stone masonry and therein lies a puzzle, for in 1799 the building was described as being timber-framed and roughcast. Perhaps the render covered the masonry and the earlier timber-framing has since been removed. The upper parts of the octagonal tower are dated at 1819. Inside is a handsome oak chest bound in iron, a Jacobean pulpit, and a stained glass E window by Kempe. There are several mature yew trees in the churchyard; one, adjacent to the tower, is of some antiquity and is now hollow and girded about with an iron-band, yet still very healthy. The church appears to be sited within an old defensive earthwork, parts of a raised, roughly circular bank being in evidence. A quarter-of-a-mile SE of the village is Plas Iolyn, a 5-bay Regency house in ashlar stone with pedimented porch suported by two pairs of Ionic columns. Half-a-mile N is Kilhendre Hall, a house of about 1800 set in a landscaped park. Dudleston Hall lies 2m. SSE of the village. It is a red-brick Georgian farmhouse with a 3-bay front, longer 3-bay sides and a hipped roof. At Coed yr Allt, 1½m. NW of Dudleston, are wooded cliffs from which there are superb views over the River Dee (the Afon Dyfrdwy). **Dudleston Heath**, also called **Criftins**, lies 2½m. SE of Dudleston. It is a scattered settlement of some considerable size. In the 18th Century squatters built cottages on the common before it was enclosed. Local clays were found to be suitable for brick-making and there was once a flourishing industry here. A great part of the heath was a moss-land with a light cover of oak trees but the land has now been drained. Most of the more recent development has occurred along the main road, the B5068, where there are 2 service stations, a Methodist church, the Fox Inn, a Post Office–General Store, and a rash of council houses and bungalows. On the lane to Gravel Hole is the church of St. Matthew, built in 1874 of red brick with stone dressings to a design by W. G. McCarthy with E windows by William Kempe. Inside, a feature is made of the exposed brick walls. Half-a-mile N of the main road is the hamlet of **Gadlas** where there is a moated site, some 300 yds W of Gadlas Hall. In the summer of 1940 the Hall was rented by the wife of General de Gaulle, who was in Dakar at the time. On his return de Gaulle made several visits to Gadlas. At **Old Marton**, 2½m. S of Dudleston but only 1m. S of Dudleston Hall, is Old Marton Hall, a good timber-framed 15th Century farmhouse with close set studs (vertical uprights infilling the main frame).

EARDINGTON *1m. SSE of Bridgnorth.*
A small village on the road to Chelmarsh which here runs parallel to the Severn Valley Railway. Eardington House, Grange Farm and May Farm are the only noteworthy buildings. Less than a mile S of the settlement is a surprise. If heading southwards take the first turn on the right, signed Astbury. The road descends into a narrow wooded valley. This is Upper Forge. There is a range of old red brick buildings on the left. Just past these is a small bridge over the Mor Brook. Walk to the right, a few yards into the woods, and you will find a splendid waterfall. It is in fact a concrete dam, although there must have been a natural waterfall here before. It is part of a large iron-working complex. Hidden in the trees, now moss and ivy covered, are considerable structures of brick and stone, with rusted iron pipes emerging, all built into the bank. In the meadow on the other side of the road are pools, and in the far corner is a tunnel. Water power was used to operate the bellows that pumped air into the furnaces, and to work the hammers of the forging mills. A quarter-of-a-mile upstream are the remains of a disused mill. A little further along the road, up the hill, is a row of terraced cottages, the old homes of the iron workers. They are still occupied. It is strange to think of the noise and bustle that was once here. Now all is silent and nature is healing her scars. The Eardington Forge was built in 1777-8 by John and William Wheeler. By 1803 it was the property of Pemberton & Stokes and was taking 100 tons of refined iron every quarter year from the Horsehay ironworks. In 1809 the complex was acquired by John Foster and John Bradley. It closed in 1889. There is another mill, a fully restored working corn mill, N of Eardington at **Knowlesands**, where the Highley road crosses the Severn Valley Railway. It is called Daniel's Mill and is signposted off the highway. The buildings are brick, painted cream, on a sandstone base but the main attraction is the great iron water-wheel, cast in Coalbrookdale in 1854. At 31ft. in diameter and weighing 31 tons it is reputed to be the largest working water-wheel in Great Britain. There has been a mill on the site since Domesday, but the buildings we see today are largely the work of the George family who have been here since about 1700. The Georges still live in the mill; it was they who recently restored it to full working order. Daniel's Mill is open to the public on weekends during the summer and wholemeal flour, stone-ground on the premises can be purchased at the tea shop. North-west of Eardington, ½m. along the lane that joins the village to the B4363, is **Moor Riding** Farm. This is a pretty black and white building set well back from the road, but still visible from it. There is a curious custom connected with the farm that is still solemnly enacted in London each Autumn by the Queen's Remembrancer in the Law Courts. The ceremony commenced in 1281 when Edward I gave land to Earl Roger. In return the Earl had to provide a horse, a suit of armour and weapons to guard the King on hunting trips on the Earl's estates. To remind the Earl of his duty he had to deliver to the King each year an item of war, usually a pair of daggers. In time this became a billhook and hatchet – tools used for hedging – more becoming to the farmer who inherited the obligation. In the Autumn of each year the ceremony of handing over is still held, though the owner of the farm no longer has to attend personally. There are only two other such ceremonies in the country. Near the farm are the kennels of the Wheatland Hunt. The name Eardington means 'the settlement of the people of *Eandred*', or a tribal leader with a similar name.

EARDISTON *6m. SE of Oswestry.*
A stone-built hamlet in attractive, undulating country most easily approached off the A5 at West Felton. Between the A5 and the settlement is the lake and the wooded park of the Pradoe estate. The house of Pradoe was built by the Reverend Pritchard in the last half of the 18th Century. The cost – some £5,000 – ruined him and he never lived there. An apparition dressed in a black frock coat with a white frill about its neck and black breeches and stockings was seen many times by the very sane and sensible mother of the present owner-occupier, Colonel Kenyon. The supposition is that the Reverend Pritchard occupied in death what had been denied him in life. The church stands on the edge of the park and was built as the private Chapel of Pradoe in 1861 by Rhode Hawkins. The hymn writer Bishop Walsham How is commemorated in a glass mosaic reredos of 1899. Four hundred yards N of Eardiston is Tedsmore Hall, a brick mansion in a substantial wooded park. A public road leads through the park to the strangely

Daniel's Mill, Knowlesgate, Eardington, near Bridgnorth.

named hamlet of **Grimpo**. The first element of this name could be from the Old English personal name *Grim*, which is often associated with ancient earth works, or from the Old Norse *grimr* which means 'a masked person', one who hides his identity', as did the god Odin with whom the word is often associated. Grimpo used to be called Grimpool and is in origin a squatter settlement that developed on the edge of unenclosed marshland.

EASTHOPE *5m. SW of Much Wenlock on a lane off the B4371.*
An attractive village in a secluded dip in the hills of Wenlock Edge. There is a Post Office but no school; children are bussed to Brockton and Much Wenlock. The village used to belong to the Lutwyche Estate but this was broken up and sold in 1952. There has been a church here from at least the 13th Century but the present building is of 1927. The windows in the chancel are of about 1300. In 1333 the vicar of St. Peter's killed the church's patron, John de Easthope, and the vicar's ghost has been seen many times since; his spirit is earthbound as punishment for his crime. There are more ecclesiastical spirits at Manor Farm. This building originated as a cell for monks attached to Wenlock Priory in the 13th Century. Two of the bretheren quarreled in the kitchen, came to blows, fell down the cellar steps, and subsequently died of their injuries. Their spirits are said to return to haunt the house from time to time. It is thought that their earthly remains lie in the churchyard, beneath an old yew tree, where there are 2 tombs without names, just simple crosses within circles. The Manor House was extended in the late 15th Century; there are some unusual round-headed windows and a fine plaster ceiling by the same craftsmen who worked at Wilderhope Manor. The Malthouse has a cross-wing construction, the only building of its kind in the county, it is said. The earliest part is a hall of the early 14th Century to which was added a second hall in 1450. The timber frame is infilled with red brick. These bricks were made in a kiln in the meadow above the large trout pool, the clay being dug out of the ground on the spot. To the SW of the church, 400 yds. away over the pool, is Mogg Forest. In these woods is a prehistoric fort with 2 earth ramparts. Local tradition has it that the Druids were active here. **Lutwyche** Hall lies ¾m. SW of Easthope in the woods of Wenlock Edge. It is a brick mansion of 1587, built originally to an 'E' plan, but the area between the wings was infilled with a mid-18th Century Hall which was made neo-Jacobean in Victorian times by S. Pountney Smith. Inside there is good plasterwork and a fine staircase. The house has been used variously as a private residence, a school and a hotel but at the time of writing it is unoccupied. The novelist Stella Benson (1892-1933) was born at the Hall. Her mother was the younger sister of Mary Cholmondeley of Hodnet who was also a novelist.

EATON CONSTANTINE *2½m. SE of Wroxeter and 7m. SE of Shrewsbury.*
It lies at the foot of the wooded lower slopes of the Wrekin, overlooking the Severn Valley. Richard Baxter, the 17th Century Puritan writer and preacher, spent his early years here. The pretty black and white Baxter's House of about 1645 still stands. The name Eaton means 'the settlement by the river'; the River Severn flows by ½m. to the SSW. In 1242 the manor was held by Thomas de Constentin, who took his name from Contentin, Normandy. Today there are some modern houses in the village but the pub, The Castle, is long gone as are the Post Office and general store. **Upper Longwood** lies 1m. NNE of the village. The hamlet has an old water pump and a pair of semi-detached cottages called The Warren. These may well have originally been occupied by a warrener, a man who bred rabbits which were an important food in early medieval times. At the foot of the Wrekin, ⅔m. W, is Neves Castle, a sizeable mound that could be the motte of a Norman castle but about which very little is known.

EATON MASCOTT *7m. SSE of Shrewsbury.*
An unspoilt brick-built hamlet with some sandstone and a little half-timber in gentle, hilly country close to the Cound Brook. Several cottages are painted cream with green woodwork. The essence of the place is 2 farms – North Farm and South Farm. The Anglo-Saxon settlement was held by the Norman Marescot family in the early Middle Ages.

EATON-UNDER-HAYWOOD *4m. SE of Church Stretton.*
Commonly called simply Eaton. It is a tiny place by the Eaton Brook at the foot of Wenlock Edge in Apedale. The church of St. Edith of Wilton (an Anglo-Saxon Saint) lies embowered in trees and is highly regarded. It has a Norman nave and 13th Century chancel in one, with a squat tower topped by Perpendicular battlements. The floor slopes gently upwards towards the altar and in a recess in the chancel N wall is an early 14th Century oaken effigy. The pulpit and reading desk are Jacobean, as are the panelling and the Communion Rail in the chancel. One mile to the SW is New Hall, a farmhouse of Elizabethan half-timber and Jacobean brick. In a first-floor room is a mid-16th Century wall painting of a bearded man, 4 ft. high, stabbing a deer to death. One mile to the NW is Ticklerton Hall, a 17th Century house with 2 dovecotes. The name Eaton has 3 common meanings: 'the settlement by the river'; 'the island in the river'; and 'the settlement to the east'. Here the first probably applies. Eaton belonged to Wenlock Abbey both before and after the Norman Conquest. Today the population of the parish is only 180 compared to 546 in 1871. The major employer in the village is a factory poultry farm. The National School of 1860 closed in 1926.

EATON-UPON-TERN *4m. S of Stoke-upon-Tern which is 5m. SSW of Market Drayton.*
One tends not to pass through Eaton-upon-Tern by accident. It is lost in the lanes between the A442 and the A41 and the residents must delight in the quietude here. Eaton is a most mature and attractive village with a variety of dwellings – Georgian, Victorian and modern – set behind walls, hedges, and iron railings and embowered in a variety of trees and bushes. The 'big house' is the Grange, a handsome, stuccoed mansion of 2 storeys and 3 broad bays with a canted bay, a gable and a group of 3 narrow round-headed windows above the entrance. In the garden are some tall and extremely handsome trees and to the rear substantial ranges of disused farm buildings. There are several other large houses and this is, indeed, a neat, trim and well-heeled little place. There is no shop, pub, Post Office or church but there was once an RAF wartime aerodrome. This adjoins the settlement to the NE and the flat, green arable fields are still littered with large black hangars and crumbling concrete runways. The village is a good ¼m. E of the River Tern. Mill Lane leads off the 'high street', past the grey sheds of Shelley Signs, to a small wood in which are the remains of the old mill. A dry pond, some cut sandstone walling and a small, derelict, red brick building stand near the sandstone and concrete weir. This is positioned on a bend in the river and has a central cutwater and a footbridge. On the downstream side the force of the water has created a turbulent pool and a small sandy beach. The track comes to an end at a pair of old mill cottages one of which is called Green End. The name Eaton is from the Old English *Ea-tun*, 'settlement by a river'. At the time of Domesday Book it belonged to the manor of Stoke-upon-Tern.

EDGE *2¼m. N of Pontesbury.*
It does, indeed, lie on an edge of a kind but we would be happier describing it as a hill-ridge. The medieval name was *Eggett*, from the Old English *ecg* which can mean either 'a hill' or 'an edge'. At the time of Domesday Book the settlement belonged to the manor of Ford which was part of the personal estate of Earl Roger of Montgomery. Edge is a farming hamlet; most of the fields are laid to pasture and grazed by black and white cows. The 'big house' is the Red House, a Georgian red brick building 2½ storeys high and 3 bays wide with a pedimented doorway. To the left is a remnant of the previous dwelling with a black and white gable-end wall. Elsewhere in the hamlet are farm buildings of stone, an ageless and dateless material. That cannot be said of the machine-made bricks of the few 20th Century houses that have sneaked in through some planner's back door.

EDGEBOLTON *¾m. NE of Shawbury on the A53.*
The early medieval name of Edgebolton was Erchaldinham. This has 3 elements: Erchald-in(g)-ham. *Erchald* could be a personal name, possibly derived from Ercall which is believed to have been the ancient Celtic name of a wide area that included High Ercall and Child's Ercall;

Eardington, the waterfall at Upper Forge.

ing, which is from the Old English meaning 'belonging to the people of'; and ham, from *hamm*, Old English for 'a meadow (usually beside a river)'. The name might therefore mean 'the riverside meadow of Ercald's tribe', or perhaps, 'the riverside meadow of the people of Ercall'. It was part of the manor of (Great) Wytheford which after the Norman Conquest was given to Reginald the Sheriff. By the 13th Century the overlordship had passed to the Fitzalan family. Today Edgebolton is a small village that straggles along the A53 between Shrewsbury and Market Drayton. It lies close to the River Roden on land that rises above the flood plain. It is a mature little place with a part timber-framed barn; a terrace of red brick and stuccoed cottages, a substantial farmhouse in Georgian style; a black and white cottage; walls built of large sandstone blocks; a pine tree; and a petrol service station. Opposite one of the few new houses, a bungalow called Sunnyside, is the site of a brick and stone cottage reputed to be the oldest dwelling in Shropshire. It was dismantled a few years ago when the road was widened. All the stones are numbered and lie in piles nearby awaiting re-erection close to the original site. Triangle Cottage it is (was) called. From here there are view's northwards to Shawbury aerodrome and the magnificent ruins of Moreton Corbet Castle.

EDGMOND *1¼m. W of Newport.*
It lies on a slight rise in flat country, a most pleasant residential village with a wide variety of housing. There is a general store, a Post Office, a hairdressing salon, a newsagent's and 2 pubs, The Lion and The Lamb. The nationally known Harper Adams Agricultural College and the Allied National Poultry Institute are important to the local economy, providing both work and a demand for student accommodation. At the end of the 11th Century the Norman Lord, Roger de Montgomery, gave the manor of Edgmond and adjoining land to his Abbey of St. Peter in Shrewsbury. The most attractive sandstone church of St. Peter at Edgmond has an embattled 15th Century tower with pinnacles, a Norman font and pre-Rafaelite glass. The ancient, and once widespread, custom of 'clipping the church' was revived here in 1867 and still continues. The congregation, choir and clergy encircle the church and sing the hymn 'We love the place oh God'. Up until 1939 this was followed by the Edgmond Wakes, a time of fun and frolics. Just W of the church is the old rectory now called the Provost's House, after Prebendary Talbot, who was Provost at Denstone College (near Rocester in Staffordshire). It is a fine 14th Century building which originally had a 'great hall' with, on one side, the solar (living room), bedrooms and a kitchen, and on the other a private chapel. The interior accommodation has been altered since, and to the exterior have been added Georgian curved bays and a Victorian timbered gable. It remains a handsome structure in well-wooded gardens and grounds. Opposite the Provost's House, across the road, is Edgmond Hall, a modest Georgian house in good gardens. Charlotte Sophia Burne (1850–1923), whose Shropshire Folk Lore (1881) is the standard work on the subject, lived in Edgmond for several years. Her book was based on the notes of Georgina F. Jackson to whom she was an assistant. Several folk carols were collected in Edgmond by Cecil Sharp at the turn of this century. The name Edgmond probably means '*Ecgmund*'s hill'.

EDGTON *3½m. NW of Craven Arms.*
It lies on a slope in the lovely hill country between Bishop's Castle and Craven Arms, a land of valleys, streams and woods. In medieval times Edgton was part of the Barony of Clun. Six footpaths and 5 roads converge here, a sure sign that it has been of some local importance in times past. The village houses and farms are mostly stone and half-timber. The Victorian red brick school has bands of blue and yellow brick decoration and now accommodates a craft pottery. This is a welcome sight; the countryside could do with more craft industries for all too often villages are becoming mere dormitories. The parish church is dedicated to St. Michael, a nomination that often indicates antiquity. However, it was almost entirely rebuilt in 1895 – nave, chancel and bellcote – with only the 13th Century windows surviving from the old church. The benches are partly of 1631. Next to the church stands the gaunt, stone Manor House, complete with skinny, white-painted, modern window frames. The old farm buildings have been converted into accommodation for the retired and externally are not a joy to behold. There are dormer windows, glassy porches, and brown painted frames inserted between the old, bleeched oak posts. Just down the road, by way of contrast, there is a simple little timber-framed cottage near a small chestnut tree that is quite charming; and at the bottom of the hill is a lopsided farm, black and white and most ancient. Rarely does one see such a higgledy piggledy yard so littered with chickens and tools, old sacks and machinery in such charming and delightful disarray. A picture book place, but real not just pretty. Such as these are treasures and it is a tragedy that an unholy alliance of developers and planning officers are robbing us of something that once gone is gone forever. All too often today 'restoration' means 'rebuild and replace in the style of'. Old things have ambience because they are old and new things do not because they are new.

EDSTASTON *2m. N of Wem, just off the B5476.*
One is greeted by a large, empty house standing forlorn in a field. The Norman church of St. Mary looks very good. It has 3 Norman doorways, and that on the S side, leading into the nave, is one of the best in the county. The doors and their ironwork are also probably original. The bellcote is 19th Century. The church lies on a mound and the road curves around it, indicating that the site is quite probably prehistoric. George Bernard Shaw was a frequent visitor to Edstaston and local people have memories of him, both at the church and walking the lanes. On the northern side of the village is an old canal warehouse, now well converted into a house. The canal is filled in but the stone bridge remains. Both the bridge and the warehouse were designed by Thomas Telford. The land is flat and abounds the 'Moss Country' (marshes now drained by ditches) to the N. The name Edstaston probably means 'the homestead of *Eadstan*'.

ELLERDINE *5m. S of Hodnet.*
Ellerdine is a scattered farming community in flat, rich country. Six footpaths converge here, a significantly high number which suggests that the settlement was of some importance locally in times past. The Domesday Book holder was a man called Dodo; later, the manor was part of the lands given by Henry II to Iorwerth Goch, as reward for his services as interpreter in negotiations between the Welsh and English during the border wars. The estate remained intact until 1930 when it was broken up and sold. Ellerdine school, schoolhouse and village hall stand in the corner of a field called Nightly Brockles. Grandfather Clocks were once manufactured at Oakgate Farm, **Ellerdine Heath** where there is a pub, the Royal Oak (The Tiddlywink to locals) and a red brick Methodist Chapel. The name Ellerdine is thought to be from *Ella*, an Anglo-Saxon personal name, and *worpign*, a word that can mean many things but most often 'a homestead'.

ELLERTON *1½m. ESE of Hinstock which is 7m. S of Market Drayton.*
A tiny agricultural hamlet 1m. N of Sambrook. The settlement is centred on Ellerton Hall which stands on a rise between 2 streams, Wagg's Brook and Goldstone Brook. The house is constructed of red brick with sandstone dressings and has 9 bays, gables and an offset porch. To the rear are mews cottages. The Hall was built in 1836, on the site of a previous house, for Robert Masefield. In 1906 Colonel Masefield sold it to Colonel Lawrence. The route of a Roman road passes either through or very close by the Hall. This road ran from Stretton (on the A5 near Gailey Island) to Whitchurch. It crosses the Goldstone Brook on the causeway-dam of Ellerton Mill Pool. It is most attractive here. The old water-wheel is still in place, though only the metal skeleton remains, the wooden buckets and other parts having rotted away. The water came to the wheel above axle height but flowed underneath it, a not very common arrangement called 'pitchback'. The mill ground corn until the 17th Century, fell into disuse and then became a paper mill between 1698 and 1722. In 1789 it was burnt down, was rebuilt as a corn mill in 1795, and ceased to operate in 1909. Below the dam is a turbine fed by a culvert. This was used to generate electricity for the Hall and other estate buildings. The 5 small estate cottages that stand opposite the mill were converted into one house in 1976. There is

Easthope on Wenlock Edge.

Eudon George.

Eyton, near Plowden.

a curious Round House at Ellerton. It is as if an old, stone-built tower windmill had been cut down in height and a pitched roof fitted over it. The corners of the roof are supported by posts and a chimney has been installed on the right-hand side.

ELLESMERE *16m. NW of Shrewsbury, on the A528.*
Ellesmere is the capital of Shropshire's 'Lake District', an area of glacial drift – deposits of sand, clay and gravel left as the ice melted and retreated northwards. Depressions in this drift have filled with water forming the meres (small lakes). Some old meres have silted up and become peat-bog marshes, now largely drained and used as pasture land. Other meres were drained quite deliberately by landowners from the 16th Century on. Still, a glance at the Ordnance Survey map shows that many meres remain: Cross Mere, White Mere, Cote Mere, Blake Mere, Newton Mere, Kettle Mere, and of course the Mere on the shores of which lies Ellesmere town. These lakes were important fisheries in the Middle Ages, and most are still fished though now only for sport. There was a town here in Saxon times, probably on the site now occupied by the parish church. (The name is from Elle, or Ella, an Anglo-Saxon personal name). The Normans built a castle and 'planted' a new town. The castle was originally held by Roger de Montgomery but passed to Llewellyn, Prince of Wales, when he married Joan, the illegitimate daughter of King John. The extensive Honours of Ellesmere, Clun and Oswestry, together formed the major part of the Shropshire Marcher Lordships. The lords of these lands had considerable power. In return for constraining the Welsh they were given authorities and rights normally only held by the King himself. These Lordships were only technically made a part of Shropshire as late as 1536 (The Act of Union with Wales). Today, the motte and bailey castle at Ellesmere is the site of a bowling green. The 116 acre Mere is used for pleasure boating and has a permanent stock of ducks and other wild fowl. There is also a heronry. On summer evenings and weekends Ellesmere takes on the character of a seaside resort. According to local legend the lake was formed when an old woman refused to give water to a neighbour whose well had run dry. The spirit in the old woman's well caused the waters to rise up and overflow and drown her. The Mere and the adjoining Cremorne Gardens were given to the public by Lord Brownlow in 1953. The Mere extends to 114 acres and the Gardens cover about 12 acres, part of which are formal and the rest left as open parkland. The Moors are a strip of common land that lie between the road and the pool. The town centre is on higher ground to the W of The Mere. Here are the shops and a good mixture of half-timbered and Georgian houses. The old ashlar-faced Town Hall of 1833 has a very prominent pediment which overhangs the street. Cross St, opposite the Town Hall, narrows like a funnel, typical of roads originally used as market places. In Birch Row and Talbot St., are timber-framed houses of note. The church of St. Mary stands on higher ground to the E of the town, overlooking the Mere. The old church was largely rebuilt in 1849 by George Scott, but the following are original: the crossing tower, Early English in the lower part and Perpendicular in the upper; the S chancel chapel, which has a superb early 14th Century roof; the N chapel which is late Perpendicular and the N transept door which is Early English. Oteley Hall lies on the opposite side of the Mere to the town and can be glimpsed through the woods of its park. The modern house of 1963 replaces a neo-Elizabethan house of 1826-30. The old mansion was haunted by a ghost, a White Lady, and when the new house was built a chimney from the old dwelling was left as a home for the spirit. The poet Francis Kynaston (1587–1642) was born at Oteley. He attended the court of Charles I and wrote an heroic romance, Leoline and Sydonis, which achieved some acclaim. Ellesmere is probably as well known for its canal as its mere. The first section of the Ellesmere Canal was constructed by Thomas Telford and William Jessop and opened in 1795. It ran from Chester to Netherpool on the banks of the Mersey. Netherpool was re-named Ellesmere Port. In 1805 Ellesmere was connected to Chester. The Canal Basin at Ellesmere is in good order and is being further improved. It is an area of some character with an old warehouse, and a dairy that was once the Bridgwater Iron Foundry. Up until the First World War there was an annual cheese market at Ellesmere. One mile S of the town is Ellesmere College (1879), a fine red brick building with stone dressings constructed to an 'H' plan. There is now an extension block to the rear, well done but nevertheles an intrusion on the overall design. To the side of the main building is an Arts Centre which stages concerts and shows which are open to the public. A mile or more SW, set in its park, is The Lyth, a house of 1819 with an attractive trellis veranda of iron. This has been the home of the Jebb family for many generations. Eglantyne Jebb (1876–1928) was a co-founder, along with Dorothy Buxton, of the Save the Children Fund, in 1919. At **Lee**, 1½m. S of Ellesmere, is Lee Old Hall of 1594, a most splendid timber framed-house which can just be glimpsed from the road.

ENGLISH FRANKTON *5m. SW of Wem.*
English Frankton is a scattered, red brick village in pleasant country close to the Shropshire Lake District. There are a noticeable number of derelict buildings hereabouts. The most grand of these belonged to Frankton Grange, on the lane to Colemere ½m. NW of the village. The roadside has been planted with a splendid collection of mixed broadleaf trees and beyond them is landscaped parkland. At the top of the rise are the ivy-clad, brick-built stables of the Old Grange (ST.447.307). The big house is long gone. The name Frankton probably means '*Francas*'s field'. There is another settlement called Welsh Frankton.

ENSDON *6m. NW of Shrewsbury.*
Ensdon lies just off the A5 in gently undulating pastoral country. To the N are the wooded slopes of Nesscliffe and to the W is the hulk of Breiddon Hill. The hamlet is centred on Bower Farm, a brick and stone ivy-clad house with some old sandstone barns and shrub be-decked boundary walls. At the cottage called Ensdon Lea collie dogs are bred. Dominating the hamlet today are 2 blocks of semi-detached houses with hipped roofs. A low, tree-clad, hill stands beside the settlement. This is very much a landscape of rocky hillocks and watery hollows. Across the main road is Ensdon House: red brick, 2½ storeys, 3 bays, dormer windows and a central pediment. Close by is an unexpected crescent of 9 orange brick houses. At **Little Ensdon**, ½m. N, there are only 2 dwellings, a brick farm and a stone cottage, both of which are painted white. In the Middle Ages Ensdon was a member of the manor of Shrawardine, a seat of the Fitzalan family. The name used to be Edenstone, which can mean several things. We suggest '*Eada*'s settlement by the stoney hill'. Six footpaths converge on the hamlet, an indication that in the past it was of some local significance.

EUDON GEORGE *3½m. SSW of Bridgnorth.*
A farming hamlet in the lightly-wooded valley of the Borie Brook. There are 2 excellent half-timbered houses here but neither can be seen to their best advantage from the road. North Eudon Farm lies amongst brick and stone barns, a work-a-day place guarded by vociferous dogs. Eudon Grange lies ½m. SW at South Eudon. This is quite superb. It is an irregular house of medium size with gables and a porch and is dated at 1618. Between 1985–8 it was most excellently restored by the current owner and he is to be congratulated. The roadside buildings, however, are somewhat grim – large, modern, corrugated iron constructions that give no hint of the treasure they obscure. There is another, less grand, part black and white house in the same ensemble. One feels that in the past Eudon George was probably a substantially larger place than it is today. The name Eudon George is explained thus: Eudon means 'Yew-tree hill' and George is from the holder of the manor in 1242 who answered to the name of William de Sancto Georgio.

EYTON-UPON-SEVERN *5m. SE of Shrewsbury.*
Just over 1m. S of Wroxeter is a scatter of black and white houses and red brick farms, and a charming building now used as a 'summer let' called The Tower. This is Eyton. Lord Herbert of Chirbury was born here at the now demolished Hall. He was a distinguished diplomat, philosopher and historian. South of the village, approached down a rough track which passes The Tower, is the Eyton-upon-Severn National Hunt racecourse. A derelict farm overlooks wide vistas and the River Severn meanders through lush meadows and fields of corn. To the W the

Clee Hill, an old quarry.

Wrekin looms large, a striking sight. The name Eyton can mean either 'a settlement on an island in a river', or 'a settlement on the banks of a river'. (Ey is from the Old English *eg* or *ieg*.) The lane from Eyton leads to **Dryton**, a small hamlet with an attractive black and white cottage. The lane continues on to meet the B4380 at Lower Longwood crossroads. On the way, atop the hill, it passes a magnificent oak tree, the Dryton Oak. Just N of the **Lower Longwood** crossroads is an area called Watch Oak. There used to be a very large and very ancient oak tree of that name here. Longwood is not marked on the Ordnance Survey map, but is signposted on the ground. It consists of a few houses along the B4380. There were several small coal pits here that provided fuel for the tile works near Eyton and the brick works sited in the clay pits on the hill on the lane to Upper Longwood. The chimney and kilns of the brickworks still stand close to the roadside, rapidly being claimed by nature.

EYTON UPON THE WEALD MOORS *2m. N of Wellington.*
A most pleasant little place, only approachable down a single lane off the A442. The village lies on a low rise on the SW edge of the Weald Moors. This was marshy, inhospitable country covered in thorns, alders and willows. From the 16th Century onwards this 'morass' was drained and enclosed. The removal of the water caused the land to sink, by as much as 3ft. in places, leaving roads standing like causeways above ground level. The area is still obviously 'wet'. One should take care when pulling off the road in a car because the verges are liable to be very soft. The village of Eyton has a red brick church of 1743 with a tower topped by a quaint little pyramid roof. The apse is polygonal and was added in 1850. There are a few houses and a handsome white stuccoed Hall. The course of a now disused canal passes by to the NW. (*See Wappenshall.*) The fields are large and flat and mostly used for arable crops such as potatoes and wheat. However, only ½m. W, on the other side of the A442, are the ever-growing estates of modern houses that have already enveloped Shawbirch and made it a suburb of Wellington. The name Eyton upon the Weald Moors is explained thus: Eyton means 'the homestead on an island', usually in a river, but in this case in a marsh; Weald Moors means 'wild-moors'.

FARLEY *1¼m. NNW of Pontesbury.*
The Domesday Book name was Fernelege, from the Old English *fearn-leah*, meaning 'the clearing covered in ferns'. In 1066 the manor was held by Ernwin and he continued to hold it after the Norman Conquest from his new overlord Roger Corbet of Caus Castle, which lies some 3 miles to the W. Today it is a tiny red brick hamlet situated on the high ground above the flood plain of the Rea Brook (pronounced 'Ray') where the land is mainly laid to pasture. Less than ½m. SW, on the road to Asterley, is the spoil mound of an old coalmine. There are many more around Asterley itself.

FARLOW *3½m. NNE of Doddington, which is 8m. E of Ludlow on the A4117.*
This is a pleasant, pastoral and arable country, a landscape full of unexpected little hills and dales. The village itself is torn in two. The church and the school are on a hill (the name Farlow means 'fern hill') and in the valley deep below is the Post Office–general store. They are joined by a very steep, winding lane which must be lethal in winter. The houses, farms and cottages are of brick and stone. The church of St. Giles has a nave, a chancel and a bellcote, all in yellow stone. It was rebuilt by R. Griffiths in 1858 and all that remains of the old, Norman church is a fragment in the S doorway which has a zigzag ornament. The Vicarage is large, and smothered in orangey-brown render. There are good views in all directions. The church stands in part of a field called Green Meadow, once a venue for cockfighting, locals say. Hillhead Farm, next to the church, was formerly the Maypole Inn. The only pub in the area still in service is the Gate Hangs Well. This is a mile or more from the Post Office. Behind the pub, in a maze of lanes, are Hill Houses, the cottages and attached smallholdings of an old coal, iron and limestone mining community. Farlow was in Herefordshire until 1844. In medieval times it belonged to the extensive composite manor of Leominster, of which it was an outlier. A mile or so to the E is the village of **Oreton**, with which Farlow is often associated. This is also in 2 parts; the white-painted New Inn and a few cottages lie ¼m. from the Post Office and the majority of dwellings. The line of a dismantled railway follows the valley of the River Rea hereabouts. In the past there was a long established limestone quarry and lime works; 'Oreton Marble' was used in the construction of the Roman Catholic church at Ludlow. In Factory Lane there is a car repair shop but the track gets its name from an old cotton mill that stood by the Rea Brook. The raw material was brought here by mules from Prescott. The Post Office-cum-general store is also, unusually, a bakery. At Lower House there are the earthworks of a Bronze Age settlement.

FAULS *6m. SSE of Whitchurch.*
Tucked away off the main lane of the hamlet is Moat Farm, a timber-framed cottage, parts of which are thought to be of the 11th Century. In recent years it has been sympathetically restored and a small extension added to the left-hand side. The moat is now filled in but its position is marked by a low ha-ha. In the field to the left of the cottage is a hollow in which a spring emerges and flows underground to a small pool in front of the dwelling. This formerly fed the moat. Today Moat Farm is a cheerful place, a smallholding full of horses, ducks, dogs and other domestic creatures. As to the rest of the settlement, this largely consists of modern houses and bungalows, pleasant enough and quietly maturing. The cruciform church of Holy Immanuel has a bellcote and is built of red brick with blue brick decoration and stone dressings. The Old Vicarage stands behind, large and irregular with mullioned and transomed windows and clusters of tall chimneys. Nearby is the ship-boarded Village Hall. Down the hill, ½m. N, is **Darliston**. A group of mostly old houses straggle along the lane. The black and white Bainton Cottages were once 3 dwellings but are now one. The right-hand parts are 16th Century, and the left-hand part is 17th Century. Down the road is a large farmhouse with rendered walls and a patterned, tiled roof, beneath which is an old timber-framed house. Set back on the opposite side of the road is The Grange, another timber-framed house of the 16th Century, which was enlarged in the 18th Century and now stands anonymous in its new guise. The hamlet adjoins the A442 and near the junction is a brick-built Wesleyan chapel of 1861.

FELTON BUTLER *7m. NW of Shrewsbury.*
Felton (Feltone) has a Domesday Book entry and was at one time owned by the King of Almain. Felton is from the Old English *feld-tun* which means 'the settlement in a large clearing in the forest'; Butler is from a medieval tenant, Hamo fitz Buteler, who held the manor from the Baron of Clun, c.1165. If one approaches the hamlet from the A5, Breiddon Hill looms on the horizon behind the settlement. The tall, green silo tower of Manor Farm also looms large and is something of a landmark in this flat country. Felton Butler has a certain charm and a sense of space – wide grassy verges and big skies. The Manor House stands on a slight rise. It is a timber-framed building with brick infill, a gable and a dormer window. Beyond the substantial farm buildings is a substantial duck pond, more a small mere in fact. A few cottages and another farm, with red painted barns, a few Jacob sheep and some cows, yew trees and long views over a quiet, pastoral landscape and that is Felton Butler.

FITZ *5m. NW of Shrewsbury.*
The hamlet lies between the River Perry and the River Severn. The red brick church of St. Peter and St. Paul was built in 1722 and restored by Sir Aston Webb in 1915. Inside there are 13th Century tiles by the font, a gallery on cast-iron columns and a barrel organ. In the churchyard is the grave of Owen Edward Fitz Tudor, died 1986. He was of a long established local family who presumably took their name from the village. The settlement name was once Fittesho. The first element could be from a personal name such as *Fitta*, or from the Old English *fitt*, meaning 'fight'. The second element is from the Old English *hoh*, meaning 'a spur of land'. The Old Schoolhouse is constructed of large blocks of red sandstone and stands opposite the red brick Old Vicarage. The Old Manor lies behind the church. To the front there is a half-timbered central section which has white rendered parts to either side. At the back it is all half-timbered but much of this is a Victorian facade. Fitz is a small, unremarkable but

also totally unspoiled village. The children of Fitz now attend **Grafton** School, 1¼m. to the NW. This was originally the living and recreation quarters of the RAF personnel stationed at Forton Airfield during the Second World War. Mytton Hall lies ½m. SW of Fitz. It has a stuccoed, 3-bay, late Georgian facade which incorporates a porch supported by unfluted Ionic columns.

FORD *5m. W of Shrewsbury.*
An ancient settlement just off the A458 Shrewsbury to Welshpool road. The traveller is recommended to approach the village from the westerly access, by the cream-coloured, hip-roofed Cross Gates pub. The easterly access takes one through a row of red brick council houses, flats, bungalows and a mundane modern, box-like school. How the Atcham Rural District Council could inflict this development on an historic and most attractive village is quite beyond understanding. The old part is quite charming. The ford has been replaced by a bridge and by its wooded banks are 2 cottages: Brookside is of white-washed stone and Brook House is timber-framed. High above them is the church of St. Michael. As a result of a drastic restoration in 1875 it looks too new for its obviously ancient setting. Little more than the Norman S doorway and the priest's doorway remains of its predecessor. Close by the church are 2 large red brick mansions. The handsomest is the 18th Century Ford House, which has 5 bays, 2 storeys and a Tuscan porch. There are substantial outbuildings, presently being developed, and in the grounds are tennis courts and mature broadleafed trees. The Mansion House is also 18th Century (1779) and has 7 bays, 3 storeys and a 3-bay pediment with garlands. There are some Jacobean parts inside. Over the stream, on the other side of the valley, is a little more of the old village. In New Street, opposite Lower Farm, is a derelict timber-framed cottage. When asked if it had a name the farmer who owns it replied: "I calls it, 'Tumbledown', but 'Burn it Down' is more like." In fact he is offering the cottage free of charge to anyone who will dismantle it and take away the component parts. Apparently the authorities concerned with conservation have agreed to its removal and re-siting. There are some stone and brick cottages with soft sandstone boundary walls, a Post Office-cum-general store, a Georgian house or two and the Ford Private Clinic, embowered in mature trees. The course of a dismantled railway lies just to the SW of the River Severn which passes sluggishly by ½m. N. In 1066 Ford was, by the standards of the day, a very substantial place with at least 84 families and a mill. The manor also had a half share in a fishery. This was probably the fish weir marked on the Ordnance Survey map that lies between Ford and Montford. (Note: A fishery in Domesday Book always means a structure of some kind, not just a place where fishing took place.) The Norman Lord Earl Roger of Montgomery held it himself. Prior to the conquest the Anglo-Saxon Earl Edwin had held it. In the early Middle Ages Ford became the *caput*, the head, of the old Rhiwset Hundred which was re-named Ford Hundred. The manor passed through the hands of Reginald the Earl of Cornwall, the Crown and, for most of the 13th, 14th and 15th Centuries, the Audley family.

FRODESLEY *7m. SSE of Shrewsbury.*
Frodesley is a pleasant, unspoilt, one street village of stone and red brick houses. It lies just off the old Roman road that runs from Wroxeter (Uriconeum) to Leintwardine (Branogenium). The church of St. Mark was built in 1809 and the N aisle was added in 1859. Inside is a 'primitive' carved Norman font and box pews. In the churchyard is a splendid and rare Monterey Cypress. Sir Herbert Edwardes was born at the Rectory. He was instrumental in keeping Afghanistan neutral during the Indian Mutiny, thus saving much bloodshed. The village street leads eastwards for a mile before coming to a dead-end at Frodesley Lodge. This is a gaunt, grey-stone, gabled Elizabethan house of 1591 with mullioned windows, tall chimneys and a roughly semi-circular tower. It stands on Lodge Hill with a coniferous plantation to its back and wide views across the Shropshire plain to the front. In fact the setting has been greatly damaged by the destruction of the old oaks and Spanish chestnuts that once graced the slopes of the hill. The Frodesley Oak has a girth of 27ft. This grand old tree is barely alive and it will be a sad day when it finally dies. The house was probably built as a hunting lodge within the boundaries of Frodesley Park, the pale of which can still be traced in the present day field fences and hedges. South of Park Farm the stone wall, built in 1609, that surrounded the park is still in good order. There are fields within the wall called Big Deerhouse Leasow and Little Deerhouse Leasow. The open parkland was enclosed into fields in the late 18th Century. The name Frodesley is taken to mean either 'the clearing in the forest of *Frod* (a personal name)', or 'the forest of *Frod*'.

GARBETT *4m. NW of Knighton (Powys).*
It lies in the South Shropshire hills N of the River Teme, a tiny stone-built place down a dead-end lane. Despite its small size Garbett is known to many because it is situated on Offa's Dyke and the Offa's Dyke Long Distance footpath. Garbett Hall is a part-rendered farmhouse, somewhat larger than it looks, but still of no great size. Three-hundred yards S, at the entrance to the lane, is Selley Hall. This is a surprisingly large house for a remote country place. It is 3 full storeys high with a 3-bay rendered front and has substantial extensions to the rear, a not unhandsome house but spoiled by the modern double glazed, aluminium-framed windows. The lanes hereabouts are deep and narrow, winding their way through hills grazed by sheep.

GARMSTON *½m. NW of Leighton which is 4½m. WNW of Ironbridge.*
New houses proliferate and the farm buildings of Garmston House have been converted to dwellings. Yew Tree House is timber-framed and dated at 1622. The 2 thatched cottages opposite the top of the steep hill called the Rudge were formerly occupied by the blacksmith and the thatcher, Mr. Percy Hopcroft. Examples of Mr. Hopcroft's work can be seen in the White House Museum at Abbey Foregate, Shrewsbury. The name Garmston is derived from *Garmund*'s-*tun*. At the time of Domesday Book it was a part of the manor of Leighton which was held by Reginald the Sheriff, whose wife was the neice of Roger de Montgomery.

GLAZELEY *3m. S of Bridgnorth.*
A hamlet on the B4363 Bridgnorth to Cleobury Mortimer road. There is a farm with large red brick barns, a couple of cottages and the solid, rock-faced parish church of St. Bartholomew which was rebuilt in 1875 to a design by Blomfield. There is a brass engraving of Thomas Wylde (d.1599), a stained glass E window by William Kemp, and a Norman font in the churchyard. The Vicarage and a small memorial garden stand close by the church. The settlement lies on the southern slopes of the Borie Brook and overlooks the mature woods of Chelmarsh Coppice. The country hereabouts is most pleasant and unspoiled. Glazeley (Gleslie) has a Domesday Book entry. In 1086 the Lord of the manor was Azor who held it from Earl Roger de Montgomery. It later went to the Fitz Alans who sub-let it to Lestrange. The first element of the name Glazeley might be from the Welsh *glais* or the Old English *gleis*, both of which mean 'stream'.

GOBOWEN *3m. NNE of Oswestry.*
A work-a-day place on the A5. There seem to be more Welsh accents here than English. The railway passes through the south-western fringes of the town and the Italianate station (1848) has achieved some fame. We found it not altogether impressive and certainly most run down. In 1987 repairs were in hand, but had come to a halt for lack of funds. The church of All Saints is of 1928 with a tower of 1945, all done in Perpendicular style. The Square is the main shopping centre. Other facilities include 3 garages, a working man's Club, a library, 2 pubs – The Cross Foxes and the Hart and Trumpet – 2 Nonconformist Chapels and a Roman Catholic church.

GOLDSTONE *5m. SSE of Market Drayton.*
A tiny hamlet of little more than 3 dwellings in pleasant, gently undulating country. Goldstone Hall Hotel is a brick house with some old timber framing exposed inside. The mature grounds have some majestic copper beeches and a curtain wall of limes. In the back garden there is a small copper sundial dated 1649, and a modern, steel man-hole cover caps an old sandstone-lined well which at around 100 feet is reputed to be the deepest in Shropshire. The 'Manor House' is an unpretentious little brick building. There are those locally who say

that the site of the original manor house is behind the Hall.

GRAFTON *3m. N of Montford Bridge which is 4½m. NW of Shrewsbury.*
The name is from the Old English *graf-tun* which means 'the settlement in or by the grove of trees'. It lies in gently undulating and most attractive country just S of the River Perry and adjoins the village of Yeaton. Grafton is a red brick hamlet of farms and cottages. Grafton House is a large, irregular building part timber-framed and part brick. There are yew trees, weather-boarded barns, stone boundary walls, and many of the houses still have woodwork painted in the dark green so popular in Victorian times. The area by the river and the bridge is very pretty with the cottages and copper beech hedges of Yeaton House clustered on the bluff, and the old mill beneath them beside the Perry. In the early Middle Ages Grafton was a part of the Manor of Fitz which at the time of Domesday was held by Robert (Picot) de Say. It later passed to Isabel de Say, Baroness of Clun.

GREAT BOLAS *6m. WNW of Newport.*
An unpretentious village in easy-to-get-lost-in country drained by the River Tern and the River Meese. The 17th Century church had a new brick tower and nave in 1726–9, designed by John Willdig, which is capped with urns instead of pinnacles. Inside are box pews thought to be from the previous 14th Century church. On the fringe of the village is a white-painted cottage with a delightful garden. This used to be the 'Fox and Hounds', the pub that is associated with the sad tale of the 'Cottage Countess'. Briefly, the story is as follows: As a young man, the Earl of Burleigh lived incognito in Great Bolas and he fell in love with a local girl called Sarah Hoggins. They married and lived happily at Burleigh Villa. This is a substantial Georgian House with good grounds, 1m. W of Great Bolas, and still stands. On the death of the 9th Earl (of Exeter), Lord Burleigh inherited the family estates in Lincolnshire. He and his wife took up residence in the great Burleigh House, built by Queen Elizabeth's treasurer, the first Lord Burleigh; alas, Sarah was snubbed by polite society and her life made intolerable. Guests, for example, would speak in French amongst themselves in order to exclude the Shropshire miller's daughter from their conversation. The story is related in the documentary novel 'The Cottage Countess', and in a poem by Tennyson, though he exercised considerable poetic licence. The name Bolas might mean 'the forest in the wet place where wood for bows is obtained', or 'the wood by the bend (of a river?) in the wet place (or marsh)'. A medieval Lord of the manor was one of the Foresters of the Forest of the Wrekin, and in the late 13th Century there was a gallows and a twice yearly court to try cases of murder and theft.

GREAT CHATWELL *4m. SE of Newport.*
A charming little place most easily approached off the A41. The country all around is most attractive despite the over-large fields and the wooden huts of a factory farm, a blot on the landscape if ever there was one. The name Chatwell might mean '*Ceatta*'s well, or stream', *Ceatta* being an Old English personal name. (*See Chadwell.*) The rendered and white-painted detached cottage was once the local telephone exchange and is reputed to have been the last manual exchange in England. Opposite the cream-painted Red Lion inn is Chatwell Lodge. This also used to be a pub. A stone wall at the rear of the lodge is thought by local people to have been part of a monastery. The imposing stone archway close by the Lodge marks the entrance to Chatwell Court, a modest red brick house surrounded by some good mature trees, including copper beech and cedar. In the grounds, and visible from the road is a black and white house 'The Old Barn', that once served as a chapel. To the left of the Court is a lane that leads past a substantial 4-bay Georgian farmhouse down to a stream which has been dammed to form a small lake. The track continues on, and sitting in a field all alone is a brickworks. The Lodge was built of bricks made here. It is an interesting complex. There are 3 buildings: a) The Drying Shed, a 1-storey structure, 20ft. x 60ft., of brick with a patterned tiled roof and a large square chimney of 2 pairs of blank arches, 1 on either side of the door. This building is now becoming ruinous and the roof has partly collapsed. The bricks were dried here before being fired in the kilns. b) A red brick, 2-storey building, 20ft. x 40ft., which stands at right angles to the shed. This is in 2 parts: To the left is a cottage; to the right is a kiln room with a cone-shaped kiln reaching to the roof. It originally reached higher, above roof level. It is open to the front and has 6 small fireplaces around the base. You cannot see the kiln from outside. c) If one stands with one's back to the cottage-kiln house there is a view over a pool (constructed in 1983 as a fishing pond), to the right of which, hidden amongst the trees that line a small stream, is a furnace kiln. This is circular, about 15ft. in diameter, and has a domed roof, about 10ft. high. There is a low arched entrance and tall arched entrance. The building is constructed of brick and around the top, outside, is an iron band. The walls are 2½ft. thick and the whole is in sound condition. The ground is very marshy roundabout. On the higher ground there are some mature oak trees. The kilns, house and workshops constitute a very rare rural industrial complex and surely ought to be protected. It may not be very old but it is unusual. All too often we value things when it is too late. The clay, by the way, was dug from the bank behind the cottage-kiln and drying shed. There was also a seam of limestone. Note: Yes, we know Great Chatwell is in Staffordshire, but it is close to the border and through an oversight we omitted it from our 'Staffordshire and the Black Country' volume.

GREAT HANWOOD *See Hanwood.*

GREAT NESS *See Nesscliffe.*

GREAT OXENBOLD *2m. E of Shipton, which is 10m. W of Bridgnorth.*
Oxenbold was an Anglian manor held in 1066 by Edric and Siward and in 1086 by Helgot. A later holder was Robert de Girros. The diminutive settlement lies at the end of a track off the B4368 Bridgnorth to Craven Arms road. Wenlock Priory had a grange (an outlying farm) here and the 13th Century Chapel has survived though the windows are now blocked up. To the W of the chapel is the Hall, parts of which are of the 16th Century. There are some good examples of medieval monastic fish ponds here also; they are now dry but the earthworks are still evident. The name Oxenbold means 'the place where oxen are kept'.

GREAT RYTON *1¾m. S of Condover, which is 5m. S of Shrewsbury.*
It is sometimes called simply Ryton (as on roadsigns) but this leads to confusion with the Ryton 3m. W of Albrighton (near Shifnal). Great Ryton is a charmingly ordinary village. It lies on a rise in undulating pastoral country and is of some antiquity. The name is Anglo-Saxon and means 'the settlement where rye is grown'. Several of the houses are of 16th Century origin. Yew Tree Cottages were once Ryton Manor and Pinfold Cottage was once a shepherd's cottage. Ryton Grange, a low part-timbered and stuccoed farmhouse, is most fetching. There are a few modern dwellings; a couple of black corrugated iron barns; a little red brick mission church; a rebuilt blacksmith's shop – now a house; the cream-coloured Old House, home for 300 years to the Atkis family and reputedly visited by Dick Turpin; stands of pine trees; and a delightful brick cottage with striking, cast-iron windows made at Coalbrookdale. The village pub is the Fox Inn, a freehouse, built of brick with stone outbuildings. The pub,the Old House and the cottage with cast-iron windows are technically in **Little Ryton**, but today Great and Little are really one settlement. **Wheathall** lies ½m. E. The northern approach road runs along the stone boundary wall of Condover Park and beside large arable fields drained by deep ditches. Park House is a modest red brick country house of 5 bays with a 1-bay extension to the left. Wheat Hall is a modest red brick farmhouse with gables, and a paddock to the left full of donkeys. Wheat House is part timber-framed with mostly upright studding. There was a sign for a ford but we didn't see one. On the lane back to Ryton we passed a field full of feeding rooks and were reminded of the local superstition called 'telling the rooks'. Any local matter of importance, should be told to the rooks. If not they would fly away and take their good luck with them.

GREAT WYTHEFORD *2m. SSE of Shawbury which is 7m. NNE of Shrewsbury.*
It lies in flat country S of Shawbury. The manor of Great Wytheford (*Wicford*) was

An old farm at Edgton.

Moat Farm, Fauls.

Burleigh House, Great Bolas.

The Cottage Countess (see Great Bolas).

held as 3 units in Anglo-Saxon times. After the Norman Conquest 2 of these 3 parts, which constitute the majority of the land, passed to Reginald the Sheriff. The third part went to William Pandolf. In the 13th Century the manor passed to the Fitzalan family and much later to the Charltons. At the time of Domesday Book there was a mill on the River Roden. This was quite probably on the same site as the Wytheford Forge, a furnace-forge that worked iron. This was operated by the Dorsett family who were tenants of the Charltons of Apley Castle (near Hadley, Telford) to whom they were also land agents. The forge was close to the present miserable modern concrete bridge that crosses the river adjacent to Forge Cottages; there is 'black earth' and areas of disturbed ground. The cottages are a block of 4, constructed of red brick under a hipped roof with attractive, white painted, cast-iron window frames containing small, diamond and crystal-shaped lights. If one stands on the bridge Wytheford Hall can be seen, through the trees behind the cottages. It has 5 bays, 2 storeys, a hipped roof and a stone porch of columns supporting a 'broken' pediment. The Hall is part of a very substantial farming complex. There are 2 other large red brick houses, a modern bungalow, ranges of traditional red brick and stone farm buildings, a small green with an ancient oak tree and many modern 'factory-farming' sheds. It is, to all intents and purposes a hamlet in its own right. On the main road there are a couple of new dwellings, a stuccoed cottage, and a small caravan site. **Little Wytheford** lies ½m. NW, as the crow flies, on the other side of the river. This is also a farming community. There is little more than a red brick and part-timbered farm, an attractive cottage, and the white-rendered Wytheford Grange. As to the name, Wytheford means 'the ford by the willows'; but, the Domesday name is Wicford, which means 'the dairy farm by the ford'. (*Wic* is Old English and can in fact mean many things but most commonly means dairy farm.)

GREETE *5m. SE of Ludlow.*
A red brick hamlet set amidst hills. The church of St. James is of stone, rendered at the rear. It has a heavily restored Norman doorway, Early English lancet windows in the chancel and the nave, and a late Perpendicular window with timber tracery in the nave also. The church is hemmed around with nasty corrugated iron barns – some farms look more like factories than factories these days – and one of the least attractive large houses we have seen – a farmhouse with blocked windows and heavy chimneys at each corner. Opposite the church is Greete Court which was the original moated manor house and at one time was a 3 to 4 storeyed hunting lodge, one room square, with 4 gables. The present house is brick with some timber-framing and has a priest hole in a chimney stack. This was reputedly used by Charles I who gave the owner, Thomas Edwards, a memorial ring with his portrait which is now held at Netley. The Old Rectory of 1848 is now a private house. Less than a mile to the NW is Stoke Court an imposing Tudor house to which, in 1702, were added new red brick fronts with shaped gables to the W and S. The house has a walled garden and a farm, 'Stoke Court Organic Farm Enterprises'. The name Greete may mean 'gravelly soil', or perhaps 'gravelly river bed'.

GRINDLEFORGE *2½m. S of Shifnal.*
It lies just N of the unremarkable hamlet of Grindle (meaning 'green hill') in the well-wooded valley of the River Worfe. A handful of cottages nestle on the banks of the flood plain above the sandstone bridge. The bridge also seems to have acted as a dam. It is very pretty here but the rash of poles carrying electricity supply lines and telephone wires do much to detract from the view. The old iron forge was built against the N bank of the stream by the sandstone cutting. Upstream, alongside the road, is the renovated Hinnington Mill. It is now a dwelling but was formerly, amongst other things, the premises of Hinkesman's Hinnington Brewery. It is a slightly odd building in red brick with dormer windows and a timber-clad gable end.

GRINDLEY BROOK *1½m. NW of Whitchurch.*
The name is from the Old English and means either 'the green wood' or 'the green clearing in the wood'. It is a main road settlement that straggles along the A41. Most of the dwellings are small, brick cottages but there are some larger houses such as Brook House, a 2½ storey Georgian farmhouse with camping and caravanning facilities, and Grindley Brook Hotel, looking a little Continental with its louvred shutters. The red brick and corrugated-iron mill is now an upholsterer's workshop. Next door is a little shop that sells the unlikely combination of books and herbs, and opposite is an unsightly electricity sub station. Down the road is a transport cafe, a service station and a pub, the Horse and Jockey. Grindley Brook is probably best known for its sequence of canal locks. The Shropshire Union Canal passes under the A41 and there are locks on both sides of the crossing. The South Cheshire Way passes the service station. It leads to the canal and a pretty area by a hump-backed bridge. Here, if you are in luck, you will chance upon one of the most attractive narrow boats you are ever likely to see. "Venetia" is resplendent in black livery and decorated with traditional castles and roses. The towpath follows the Sandstone Trail which leads past a house called the Land of Canaan. The country around Grindley Brook is undulating and mainly laid to pasture.

GRINSHILL *3m. S of Wem.*
The area is famous for the quarries that supplied stone used in the construction of some of the county's finest buildings. It is called white stone but weathers to a pale, green-grey. The Romans used Grinshill stone at Uriconeum (near Wroxeter), and George Steuart employed it at Attingham Hall. Other minerals, including copper, were mined at Grinshill. Today, most of the quarries are overgrown with small trees and dense undergrowth. At the top of the hill is Corbet Wood where there is a public amenity area and a car park. Polecats have been seen here. The wood is approached off the road between Preston Brockhurst and Clive. Next to the car park is a working quarry and dressing yard where the stone is sawn into blocks before being sent to masons all over the country. The stone is quarried by drilling holes into the rock-face. These holes are filled with a powder that expands overnight. This cracks the rock into huge blocks weighing 10 tons or more. The attractive village lies at the foot of the hill. It has a rather garish neo-Norman red sandstone church of 1839, a sturdy 3-bay Manor House of grey stone dated at 1624, the early 18th Century red brick Higher House next to the church, and the early Georgian Elephant and Castle Hotel. The village hall was formerly the school (1862). To the E is the fine Stone Grange. This was either built by, or bought by, the Shrewsbury School in 1617, as a retreat from the county town in time of plague. It is known locally as the Pest House, and overlooks the cricket pitch. On the W flank of the hill, only ½m. away, is the village of Clive. There is pleasant country all around.

HABBERLEY *1½m. S of Pontesbury, which is 7m. SW of Shrewsbury.*
Habberley lies on high ground in a shallow, wide bowl amongst the hills. It is unspoilt and the whole area is most attractive. Somewhat hidden amongst trees, but marked by its tall, star-shaped chimneys, is Habberley Hall. The oldest parts date to 1595. It is timber-framed, with a stone gable extension to the left and a red brick gable extension to the right. William Mytton, the 18th Century historian, lived here. The church of St. Mary lies on a raised mound. It is Norman, but has been largely rebuilt using the old masonry. Two original doorways remain. 1¼m. WSW of Habberley is Vesson's Farmhouse (SJ.391.020), near Eastridge Coppice at the northern end of the Stiperstones. In 1895 the flags on the floor of the house were seen to be broken, damaged by the hooves of the horses that dragged in the Christmas Brund. This was a large log of oak, holly or crab-tree which was carefully tended so that it burned throughout the festive season. This custom was widespread locally, and is a relic of Pagan fire worship. The first element of the name Habberley is thought to have derived from an Anglo-Saxon personal name, perhaps *Heapuburg*.

HADLEY *1¼m. ENE of Wellington.*
Hadley is one of the older settlements on the South Shropshire coalfield that constitute Telford New Town. The name is from the Old English *Headda*, a personal name and *leah*, either a forest or a clearing in a forest. During the early years of the industrial revolution it was one of the very few social and shopping centres of the area. Coalmines, brick factories and iron-

works surrounded it at Haybridge, Ketley, Hadley Castle, Trench Lock, and Wombridge. The modern centre of Hadley is quite small but not without its attractions. It is approached from the island at the end of High Street. Here is the Bush pub; a supermarket; an Indian restaurant; a wall painting of 2 cricketers, one white and one black; a nice group of stuccoed cottages by the new Cyril Maynard Court; the handsome Castle Farm Community Centre in mature red brick with mature trees in the garden; a Methodist church with a yellow brick facade, round-headed windows and red and black brick decoration; and so to the shopping centre, painted brown and cream and built around a pedestrianised square with shops to 2 sides, the nicely matured magnolia painted brick King's Head pub and a modern sculpture called Chassis Plant (1981) by David Moloney – tall, grass-like spikes 'planted' in a bed of natural shrubs. Here, too, are the police station, the library and the modern health centre which is fronted by yew trees, a nice attention to detail. The parish church of Holy Trinity lies E of the centre in Church Street. It was built in 1856 to a design by Owen of Southsea and is constructed of red and yellow brick with stone-dressed lancet windows, plate tracery and a turret. The Old Vicarage is of brick with stone dressings, a pointed-arched doorway and stands in a wooded garden adjacent to the church. Also in Church Street is Dove Cottage, a typical, solid middle-class Victorian house. Castle Street also has some attractive old houses. However, there are also tall, dull-red and cream tower blocks and estates of unremarkable modern houses. South of the centre are schools and playing fields, a gymnasium and a swimming pool. Manor House Farm was a timber-framed house with gables, narrowly spaced studs and lozenges. It was demolished in the 1960's. A mulberry tree from the grounds of the old house now stands opposite the new flats called Mulberry Court. Adjoining Church Street to the NE are the Blockley brick and tile works. At Kearton Terrace, opposite the Valens, one can see into the storage yard where there are bricks of many colours – yellow, red, orange, buff and dark brown.

HADLEY CASTLE *1¼m. ENE of Wellington.*
If one leaves Hadley and travels up the Hadley Park Road towards Leegomery Roundabout signs on the right will direct the traveller to Hadley Park Farm in the area now called Hadley Castle. There are several interesting things to see here. First are the Hadley Park Lock 'guillotine gates'. They were cranked up and down vertically between the posts of a tall scaffold, a bucket of heavy stones acting as a counterweight. The canal closed in 1944 and the gates have lain here untouched since then. Remarkably they are all complete, the only ones of their kind to have retained all their working mechanisms. But they are in a bad way and should sureley be protected, if not actually renovated. The lane crosses the canal by a bridge adjacent to the lock gates and continues on to Hadley House, a handsome red brick building of 2½ storeys and 3 bays with a hipped roof and a projecting porch with a 'broken' pediment. This was once the home of John Wilkinson, the great ironmaster. The farm buildings lie to the left whilst to the rear is the Hadley Park Mill. This is a red brick tower mill with battlements built about 1750. It was designed as a windmill, rare in Shropshire, but was later converted to operate by water. From a distance it looks as though it could be part of a castle. In 1871 a firm of Birmingham screwmakers, Nettlefold and Chamberlain, set up business nearby and sold their products under the 'Castle' trade-mark, hence their works became the Castle Works. The first ironworks here were built in 1804, the New Hadley Furnaces of John Wilkinson. Today it is the site of the mamoth GKN Sankey factory which covers some 37 acres and in its hey day (as Joseph Sankey & Sons) employed more than 6,000 men. Car, truck and tractor body parts are made here. The windmill, as a watermill, was in use until the early 1900's; today a white barn owl nests atop it. Hadley Farm itself is a working dairy farm but the owners have been very enterprising. They make their own ice cream which they sell in the Farm Shop along with a variety of other products and have organized a Farm Trail. One learns that far from simply turning their cows out to pasture a dairy farmer often grows crops such as barley, maize and fodder beat, all to be used as animal food. Pigs are kept here, too, for they can be fed on skimmed milk. (With strict quotas on milk production dairy farmers are having to find 'on farm' uses for their produce.) There are also numerous small animals about the farmyard – ducks, sheep, goats, horses, hens and a donkey. These are not 'working' animals and are kept simply as an attraction, and an attraction they certainly are; most of the day most dairy farms are like ghost towns – not a soul nor a creature to be seen.

HADNALL *5m. NNE of Shrewsbury.*
The village straggles along the main road, the A49, in flat country. General Viscount Hill was born at Prees but for most of his life his home was Hardwick Grange, Hadnall. The house has been demolished but the full size replica of the windmill at Waterloo that the General built in the grounds still stands. It can be seen from the A49 just N of the village. The Old Stables are now an industrial estate but were it not for the signposts you would never know. General Hill served in the Peninsula War with Sir John Moore and was Wellington's right-hand man at the Battle of Waterloo. He was made a Knight and his statue stands atop a huge column opposite the County Council offices in Abbey Foregate, Shrewsbury. In the churchyard at Hadnall is buried Charles Hulbert (1778–1857), the author of several books of local interest. These include The History of Salop and A Manual of Shropshire Biography. He married the daughter of Thomas Wood, the founder of the Shrewsbury Chronicle. The church is Norman with 2 original doorways remaining, but was largely rebuilt in Victorian times. It has glass by Kempe, a font from Malta and a good set of tubular bells. Near the crossroads is the moat of Hadnall Hall. The hall is now gone, the stone used for the church it is said. The Princes in the Tower are reputed to have hidden in the old house. There is a village shop and Post Office and a new school but the tailor, the carpenter, the wheelwright and the undertaker have long gone, together with the mill and the station. At **Haston**, 1¼m. NW of Hadnall, is a very fine half-timbered house called Black Birches. It is a late 17th Century mansion with an extension to the left of 1880. At the back there is a brick tower and much stonework.

HALESFIELD *2½m. S of Telford New Town Centre.*
There was a coalmine here once. Some of the brick-built pit buildings still stand and likewise the now tree-clad spoil heaps. Today this is an industrial estate; no-one lives here. There are companies of many kinds and though some attempt has been made to let lamb sleep with lamb and lion with lion there are still some strange bedfellows. The smaller firms suffer the most, somewhat crammed and cramped in non-purpose-built sheds. Their bigger brothers presumably pay for the luxury of being nicely detached and set about with lawns. It is surprisingly quiet here and people are noticeable by their absence. Very few buildings are built to last; steel frames and cladding proliferate. The industrial archaeologist of tommorrow will be faced with a litany of names, ghostly bodies with no flesh: Billiard Cues of England, Safety Glove, Shropshire Forklift, Avant Electronics, Sigas, Peaudouce, C. S. Carpets, Bulldog Glazing, SWK Testing Lab., Aerotronic Controls, Cobra Supafoam, Redland Ready Mix, Morgan Windscreens, Budget Gas, Grange Fencing, Eurogrid Forging, Windy Baloons, Newport Non-Ferous Foundry, Alycast, Ben's Party Packets, Leabank Office Equipment, Centre Switchgear, Reliance Electric, Piccadilly Precision Engineers, M + M Tools, Alcan Ecko, Secal, Telmark Meats, Tesa Metrology, Hoover Ball and Bearing Co., Darby Housewares, Myson Pressure Castings, Plastic Omnium, Link 51 and the Plastic Processing Industry Training Centre, etc.

HALFORD *½m. NE of Craven Arms.*
It lies just across the River Onny from Craven Arms, a small place of grey stone buildings with a church, a mill and the Old School of 1876. The church of 1887 has a Norman S doorway, a chancel of 1887 and a nave restored in 1848. High above the village is Flounders' Folly, built by Benjamin Flounders in 1838. It is said that he raised it to to view his ships in the Bristol Channel, but finding his view impeded by some inconsiderate mountains he died of sorrow. The modern name of Halford is a contraction of the medieval Hawkeford which could mean either 'the ford where hawks were commonly seen'; or 'Hawker's ford', where Hawker could be a personal name or a follower of that sporting trade.

HALF WAY HOUSE *11m. W of Shrewsbury.*
It lies on the A458 approximately half way between Welshpool and Shrewsbury. Most of the village has been by-passed by the new, improved road. There are 2 pubs, the Half Way House and adjoining it the Seven Stars Inn, and a petrol station. Amongst the old cottages and the new houses is a rickety, old brick and corrugated iron barn of many rusty coloured hues. On the road to Marche is a neat little stone-built farmhouse with a range of outbuildings connected to it, almost a 'longhouse'. Across the road from the pubs is a farm called Bretchel. Close to this is the motte of a Norman Castle. The mound is some 6ft. high and 16ft. long. 'Bret' is probably from the Old English meaning 'Briton', that is a Celt. To the SW by a quarter of a mile is the splendid Marche Manor. (*See Marche.*)

HALL OF THE FOREST *3m. WNW of Newcastle on the B4368.*
A solid, grey stone farmhouse stands in a field near a wood, a remnant of the old forest. This has a romantic association as the house where the heroine of The Lady of Bleeding Heart Yard found peace. Ida Gandy outlines her sad story thus: "The lady, Frances Villiers, when only fifteen was forced by her grim stepfather, Sir Edward Coke – a so-called 'vindicator of national liberty' – to marry Sir John Villiers, brother of the Duke of Buckingham. To gain his purpose, Sir Edward tied her to the bedpost and beat her daily till she gave in. From then onwards one storm after another swept over her. Her husband went mad; the Villiers family disputed the fatherhood of her baby; she was excommunicated and imprisoned; condemned to walk barefoot in a white sheet from St. Paul's to the Savoy (a penance which she spiritedly refused to perform), was hunted from one hiding place to another by her implacable relations, till finally, after a hair-breadth escape, disguised as a page, she was carried away by her lover, Sir Robert Howard, to this remote house on his Border property. Here she probably stayed five years, much of that time alone except for her servants. A romantic story witha romantic setting".

HAMPTON LOADE *5½m. ESE of Bridgnorth.*
It lies opposite Hampton on the other side of the River Severn. Loade is from the Old English *lod*, 'a water course', but usually indicates the presence of a ferry. There has been a pedestrian ferry here for many years. There was also an ironworks. In 1796-7 John Thompson built forges powered by both steam and water at the confluence of the Paper Mill Brook and the River Severn. They changed ownership several times but survived quite late, until 1866. Today Hampton Loade is a working man's weekend retreat and picnic place with a grassy car park and a pub beyond the wood. At **Hampton** there are a few old cottages, another pub some makeshift, riverside chalets, a station on the Severn Valley Railway line, and a camping and caravanning site.

HANWOOD *3½m. SW of Shrewsbury.*
Sometimes called Great Hanwood; the name is from the Old English *hana* meaning 'a cock' or 'a wild bird'. In medieval times the manor belonged to the Corbet family as lords of Caus. In the fields to the E are two separate moated sites. The substantial modern village lies along the busy A488 Shrewsbury to Bishop's Castle road. The main line railway crosses the road here, re-crosses and crosses it yet again. In the past there have been some varied industries in the parish – brick manufacture, a soap factory, lime kilns, a brewery, a coal mine; and a mill (Hanwood Mill) that in its time has ground corn, lead ore and barytes, processed flax and made paper. The red brick church of St. Thomas was rebuilt and enlarged in 1856 and has a polygonal apse with stained glass by David and Charles Evans, a Norman font and a weather-boarded belfry. Along the main road there is much 20th Century housing, some of it with back gardens facing the main road, never a good idea. The White House is a large, timber-framed guest-house and the church of England Primary School is yet another modern brick and concrete flat-roofed thing. The population of Hanwood today is about 1500; it has more than doubled since 1970.

HARDWICK *4m. NE of Bishop's Castle.*
It lies 1m. NNW of Eaton, which is on the A489. The name Hardwick used to be Heredwyke, which is from the Old English and probably means 'a sheep farm'. In the Middle Ages it was a part of the extensive manor of Lydbury North, for long the property of the Bishop of Hereford. If one approaches from Eaton the lane runs parallel with the River West Onny for a short distance. The 'terrace' on the opposite bank is a low embankment of the old railway that ran from Craven Arms to Lydham. Hardwick Hall has a plain, stone back but a delightful black and white timber-framed front with dormers and gables. The little projecting porch is especially attractive. The wing to the left of the porch is post and pan but elsewhere there are diagonal struts and concave lozenges. Behind the farm, on the other side of the road is the motte of a Norman castle, a tree-clad mound some 9ft. high and 50ft. across at the top. By the road is a small pool embowered in trees, the king of which is a huge, old oak.

HARLESCOTT *2½m. NNE of Shrewsbury town centre.*
It is now a suburb of Shrewsbury. The name probably means 'the cottage of *Heorulaf*'; *Heorulaf* might be of Scandinavian origin. In the Middle Ages there was a substantial moated house here called Harlescott Grange. All but the W side of the moat still exists, sitting uneasily in the midst of a modern housing estate. It enclosed an area some 42 yds by 31 yds, was 17 yds wide and the walls were stone lined; a formidable obstacle. Today, Harlescott is known throughout the Welsh borderlands for its large Livestock Market. This has attracted a variety of business and service industries: agricultural suppliers, meat processing companies, banks and building societies, garages and pubs. There are also 2 Retail Parks on which are located warehouse-shops such as Queensway, B&Q, Saverite, Leo's Co-op, Texas, MFI, Payless DIY and Allied Carpets.

HARLEY *2m. NW of Much Wenlock on the A458.*
A pretty hillside village with some black and white cottages. It is a peaceful place now that a by-pass carries the main road traffic. The tower of the church of St. Mary is medieval but the rest was rebuilt by S. P. Smith in 1846. There is stained glass by Evans of Shrewsbury which includes scenes based on famous paintings. The main road to Much Wenlock climbs Wenlock Edge through a cutting reputed to have been dug, or widened, by French prisoners during the Napoleonic Wars. To the E a narrow lane leads off the main road to Wigwig and Homer. The road sign is much photographed. At **Wigwig** there is a concrete surfaced ford and a pair of old brick cottages, inhabited but in colourful decline. Homer is a place of new houses in surprisingly large numbers considering the rural and most attractive countryside in which they are set. They benefit from their surroundings but contribute nothing in return. At **Yarchester** Cot, in a field near the A458 due S of Wigwig, fragments of Roman pottery and tiles have been found lying on the surface of the ground.

HARMER HILL *8m. N of Shrewsbury on the A528.*
The settlement lies at the junction of the A528 to Ellesmere and the B5476 to Wem. Modern houses and bungalows dominate the scene but a few old sandstone cottages and Georgian brick houses still stand. There are 2 pubs, the Bridgwater Arms in bright red sandstone, and the Red Castle, clad in stucco. The small, red stone church has grey stone dressings and a 6-sided chancel; adjacent is the attractive brick and stone cottage called Langley's. **Shotton** Hall lies just to the E, an agglomeration of stuccoe, timber cladding, red brick, and sandstone doing its best to hide from the world behind a high stone wall. The large farm down the road has fields laid to both arable crops and pasture. Pine trees are a feature here. They are also an important feature of the landscape to the W. The Ellesmere road skirts the wooded escarpment that overlooks the flat, low lands of Baschurch. Opposite the Presbyterian Chapel of Wales (1920) and the adjacent village hall is a car park. From here one has access to the cliffs and woods that stretch 1½m. NW to Myddle. This is a most attractive area which in places is everybody's idea of a romantic medieval landscape – craggy rocks weathered and worn, and mature Scots pines standing above them with their Spartan heads raking the horizon. Hidden amongst the trees are many small quarries, now overgrown and most picturesque. Probably the best way to see all this is to take Lower Road out of Harmer Hill. This descends to the once marshy plain and follows the red, rocky

Grotto Hill, Hawkstone.

The Livestock Market, Harlescott.

General Hill's windmill, Hadnall.

'The Nest', Webscott, near Harmer Hill.

cliffs all the way to Myddle. There are a few cottages and farms along the road and one cottage in particular is absolutely beautifully situated. It is about 1 mile along the lane at **Webscott** and is called The Nest (SJ.476.229), though the name is not displayed. Perched amongst the cliffs with wooden walkways over the rocks and surrounded by pine trees and lawns it is quite delightful. A little further along 2 old red sandstone cottages stand near the deeply entrenched stream that drains the land hereabouts. These old marshlands and the Heremeare (Harmer Moss) were drained by Sir Andrew Corbet and Mr Kelton at the time of Richard Gough, famous for his History of Myddle. Opposite the road junction are dramatic quarries with tall, sheer walls still bearing the marks of the chisels and levers used by the old miners.

HARNAGE *¾m. SE of Cound, which is 6m. SE of Shrewsbury.*
The hamlet lies between the Coundmoor Brook to the SW and the River Severn and the A458 to the NE. Harnage House is of the late 17th Century. It is built of brick with stone quoins, and has decorative brick hands, 2 symmetrical straight gables and wooden mullioned and transomed windows. The staircase is probably Jacobean. Harnage Grange lies 1½m. SSE of the settlement. It was built to replace a Queen Anne mansion which was demolished about 1878. Of the present large house the W end is of the original pre-Reformation grange and the E end, with its 2 good, stepped gables and star-shaped chimneys, is of Jacobean brick. Stevenshill Fort, a substantial, traingular shaped earthwork on a hill above the Coundmoor Brook, lies ¾m. to the SW.

HATTON GRANGE *2½m. SSE of Shifnal.*
The main entrance to the park of Hatton Grange is off the A464 Shifnal to Wolverhampton road, but there is a back way along a lane leading from Evelith Manor. A medieval grange – an outlying farm – was established here by the monks of Buildwas Abbey. The present red brick house was built in 1748. It has canted bay windows, a Tuscan-columned porch and a 3-bay pediment to the main front. The house is on high ground and overlooks the woods and lakes of its small park. The lakeside gardens are most attractive, especially in early summer when the rhododendrons are flowering. The 3 pools are in fact medieval fish ponds established by the monks. The gardens are open 2 days a year to raise funds for a nurses' charity. The name Hatton is from the Old English *haep-tun* meaning 'the settlement on the heath'.

HAUGHMOND *3½m. NE of Shrewsbury on the B5062.*
The early monasteries in Shropshire were built in isolated woodland clearings but never too far from a main road. Haughmond Abbey was built in 60 acres of cleared forest, but within 4m. of Watling Street (the A5); a kind of colonial outpost. In the early Middle Ages the monastery acquired extensive and far-flung estates, including the Manor of Boveria, and grazing rights for its horses on the Long Mynd. Granges, that is farms on the fringe of cultivation, were planted and there is reason to believe that the monks encouraged farmers to rent lands on their estates by giving them loans. Haughmond Abbey was founded in about 1135 by William Fitz Alan for Augustinian monks. All that remain of this original building are parts of the foundations of the church chancel. In 1154 Henry II came to the throne and his old tutor became Abbot of the Abbey. The monastery was rebuilt to a magnificence befitting the abode of a close friend of the King. Additions were made in the 13th and 14th Centuries, and some rebuilding occurred in the 15th Century. In 1539 the monastery was dissolved and the church and dorter were demolished. The Littleton family acquired the property and converted it to a secular dwelling. It was they who added the bay windows and fireplaces. During the Civil War the house was burnt down. The site was then left derelict and local farmers pillaged much of the stone. In the latter half of the 19th Century a farmer occupied the ruins. He made a house in the kitchen and stabled his horses in the chapter house. He left in 1933 when the site was purchased by the National Trust. The Chapter House is the most complete part of the Abbey and has Norman arches and statues between the pillars. There are several fish ponds scattered all around the Abbey. The large pond, downhill to the W, may have been used as a cess pit as all the sewerage channels run in this direction. If this is so it would not have been used for breeding fish. Despite its huge size the Abbey only had 14 monks in permanent domicile. Today, the site is in the care of English Heritage. To the WNW, by ½m., is **Sundorne** Farm which contains the castellated red brick gatehouse, curtain wall and chapel of Sundorne Castle, built by George Wyatt in the early 19th Century. This big house has been demolished, the chapel is used as a barn, and the great lake is now much diminished since the collapse of the reservoir dam. Before the lake was made the Sundorne Brook passed under the B5062 through a ribbed arch bridge of stone probably of medieval monastic origin. The little arch is still in place. On the other side of the road from the entrance to Haughmond Abbey is the large wooded area of **Haughmond Hill**. This was, without doubt, occupied by prehistoric man. It commands good views over the Severn plain to Shrewsbury. The conical hill of Queen Eleanor's Bower, and a fort to the S of Haughmond Hill (near Uffington), and Ebury Fort to the N (about a mile from the Abbey), have produced evidence of Iron Age or earlier settlement. Extensive quarrying has quite probably destroyed other remains. In ancient times this hill must have been quite densely populated. It is situated in the middle of a river complex with the Severn to the W, the Tern to the S, the Roden to the E, and a system of streams to the N. Haughmond Hill is therefore surrounded by a roughly circular water system with a diameter of about 4m. A glance at the 1¼ inch Ordnance Survey map shows this quite clearly. At the top of the hill are old, now disused, quarries. The rock here is pre-cambrian, the oldest known, and it was taken for use as roadstone. A ready – mixed concrete company now operates from the old workings, but all their raw materials are brought in from elsewhere. There is a large lake in the lower levels of the quarry with cliffs around, making a dramatic scene. Access is gained from the main road, on the opposite side to, and higher up the hill from, the entrance to Haughmond Abbey. The name Haughmond is problematical, but it has been suggested that it might mean 'the hill upon which haws grow'; haws are fruit of the hawthorne.

HAUGHTON *1½m. W of Roden which is 7m. NE of Shrewsbury on the B5062.*
Haughton is a scattered hamlet of red brick and stuccoed cottages and farms. The attraction here lies on the low-domed and wooded rise called Ebury Hill, just NW of Haughton Farm. Part of the summit of the hill has been quarried and low cliffs now encircle the NW corner where there is a deep, water-filled pool. The quarry lies within the perimeter defences of a substantial pre-historic fort-settlement. This is roughly circular and has a diameter of about 200 yds. The embankment and ditch are best preserved in the SW, especially the S where the bank rises to about 6ft. inside the enclosure and about 10ft. outside, into the trench. The whole site is now a mobile caravan park. It was empty when we visited in December but is obviously a well organized concern with toilet and water facilities within low timber palisades, a childrens play ground, and concrete roads. The area is well-wooded with small birch trees and the like, and there splendid views to the E and the S. The name Haughton can mean several things but most commonly 'the settlement on the spur'. In this case it is likely that the spur was an area of dry, slightly raised land that projected into a marsh. The fields hereabouts are drained by ditches and there are numerous glacial ponds called kettle-holes, always a sign of a clayey soil that holds water.

HAWKSTONE PARK *12m. NNE of Shrewsbury by the village of Weston-under-Redcastle.*
The sandstone craggs and the wooded hills of the Hawkstone ridge are a landmark in the countryside. Probably best known today for its golf course and motor-cycle scrambling meetings, Hawkstone has been a pleasure park from the late 18th Century. Indeed, the golf club is centred on the hotel built to cater for 19th Century visitors. It is an extensive Parkland and not all the parts are easily accessible from each other. To the N, and with its own separate entrance from Marchamley, is the red brick Hall, a handsome, grade 1 listed building of mixed architectural ancestry. The central part was built in 1722 for Sir Richard Hill and the wings were added by his nephew, Sir Rowland

Hill, about 1750. Behind the house is a church, built in 1934 by The Redemptionists (who had bought the property in 1926). The Redemptionists are a Roman Catholic missionary order and they still own the Hall. The courses held here attract Catholics from all over the world. As a condition of a grant by English Heritage for roof repairs the house must be open to the public for at least 28 days in the year. The well-known Hawkstone motorbike scrambling course also has its own separate access near the West Midland Shooting Ground which is well signposted off the A53 Market Drayton to Shawbury road. The track includes a very steep hill and the area is well-wooded. Part of the course actually crosses the embankments of Bury Walls Camp, a large 20 acre Iron Age hill-fort which has triple earthen ramparts. Below the hill, next to the scrambling track, is Hermitage Farm, an Elizabethan house recently listed. In the rock-face above it is the cave in which a hermit was paid to live in the 18th Century as part of the Hawkstone Park attractions. It can be seen quite clearly from the road between Hodnet and Lee Brockhurst. The part of the park accessible to the public centres on the hotel at Weston with its well-known golf course. It is possible to walk from here to the secluded Hidden Valley, with its trees and cliffs and to the Castle Rock where are the scant remains of a genuine 13th Century stone castle – the Red Castle. This is the hill nearest the Hotel. The Red Castle is said to be the scene of the legendary fight during which Sir Lancelot slew the giants Tarquin and Taruinius. Here, too, is the cave where there used to be a stone lion. The next hill, adjoining Castle Rock to the N, is Grotto Hill where is the once spectacular Grotto and the Victorian Folly Arch that stands clear against the sky on the cliff edge. A track leads from the Lodge at Weston through the golf course, past the hills just described, and through a Tunnel, cut in 1853, to the Hall. There were once all manner of other sights to be seen in the park. Some are still visible but others are either uncared for or in decay. They include: the Natural Arch; the Obelisk, erected in 1795, which is really a Tuscan column 112 ft. high that was originally topped by a statue of Sir Rowland Hill, Lord Mayor of London in 1355; a red sandstone Summer House with battlements and pointed windows; a grey Circular Tower with some original medieval work; an Elysian Hill; a Scene in Otaheite; a Scene in Switzerland; an Awful Precipice; a Menagerie, which had live animals but which now lies in ruins on the fringe of the woods on the opposite side of the road to the reed-filled Menagerie Pool. The Menagerie Pool lies alongside the Hodnet-Weston road ½m. E of The Citadel. The Citadel was built as the dower house to the Hall in 1790. It has three circular towers connected by walls with canted bay windows and a higher stair-turret on the middle tower. North of the Hotel is a real Windmill with some of the wooden working parts still in place. Stretching NE of the windmill for 1½m. is the long, narrow Hawke Lake which was made in 1790 and is now used for sport fishing. Altogether there were 10 miles of walks within the Park. In 1774 Dr. Johnson visited Hawkstone and found it a place of 'terrific grandeur'. Today no attempt has been made to either restore or upkeep the man-made attractions of the Park but, little matter, it is still a place of great natural charm and the adjacent country likewise. Anyone interested in exploring Hawkstone ought to do so with the aid of a guide. One such is Jack Jones who conducts regular tours. He can be contacted by telephone at Lee Brockhurst 291.

HEATH *4m. E of Diddlebury, which is 12m. SW of Much Wenlock.*
Situated high up on the western slopes of Brown Clee Hill, Heath is not so much a village as a scatter of houses and farms. It is to see the Norman chapel (SJ1557.856) that so many come. One of the simplest, most original, and most remote of all Norman churches in Shropshire, it is all that is left of a deserted village. In the fields around it are extensive earthworks, the remains of roads, fish ponds, house platforms, fields and even a stone-lined well. There is some ridge and furrow with the characteristic inverted 'S' at the ends, indicating a medieval origin. The chapel is a simple sandstone, barn-like building with a chancel and arched doorway. It possesses neither tower nor belfry. There are slits high up in the walls and the interior is painted white. The furniture consists of box pews, a double-decker pulpit and a Norman font.

HENGOED *2½m. N of Oswestry.*
A scattered community. The church of St. Barnabas (1849–53) was designed and largely paid for by the vicar, the Reverend Albany Rossendale Lloyd. Alas, his handiwork is no more; it was demolished in 1985, though not without local opposition. At Upper Hengoed is the pub, The Last (meaning a cobbler's last), and the Post Office and village store.

HIGH ERCALL *7m. NE of Shrewsbury on the B5062.*
Ercall is probably a Welsh name for the general area. High Ercall is an attractive village at a busy road junction. Close to the crossroads is the stone and brick Hall of 1608, built for Sir Francis Newport by Walter Hancock. Although still a substantial building it has been reduced in size. In the back garden is a curious arcade of 4 arches on round piers which was probably part of an open loggia of the previous Hall built by the Arkle family. There was also a moat and drawbridge but these have long gone. To the N of the Hall is a substantial mound, very noticeable when viewed from distant, higher ground. In the farmyard that adjoins the Hall is a black and white timber-framed building, possibly older than the Hall itself. During the Civil War High Ercall and Ludlow were the last garrisons in Shropshire to hold out for the Royalists. The Earls of Bradford used to be resident here and it was they who founded the Almshouses in 1694. These are made of brick and set around a 3-sided courtyard. The Norman church, with its massive tower, survived the Civil War almost intact, despite having been described in 1646 as being demolished (perhaps for political reasons). The church is set to the rear of the Hall. Today there is a school, a shop, a boarding kennels, a pub and at the Gospel Oak local youths still meet, though most have things other than religion on their minds. During the war there was an airfield close to the village. In the last Century, at Walton, ½m. N of Ercall, there lived one Thomas Leigh who was reputed to be a Wizard. He was benign and only used his powers to the good, such as curing sick animals and charming robbers to return stolen goods. His death is said to have been caused by a rival Wizard called Jack o'the Weald Moors. One mile W of High Ercall, over the River Roden on the road to Shrewsbury, is the village of **Roden**. The large house is now an old person's home and most of the rest is connected with the CWS market garden nurseries. There is a neglected timber-framed building of some years; a small modern housing estate; and many acres of greenhouses – a rather bleak, spread-about little place.

HIGH HATTON *2¼m. S of Hodnet.*
High Hatton lies on a slight rise in flat country. Opposite the old, rock-faced school of 1876, now a house, is the black, corrugated iron village hall. For the rest there is little more than a few Council houses, some cottages and a water tower. High Hatton Hall is of red brick, 2½ storeys high and 3 bays wide with a pyramid roof, a service wing to the rear and an adjacent farm with some timber-framed barns. Before 1066 *Hetune* was held as 4 manors by Alric, Wulfheah, Wulfgeat and Leofric. In 1086 it passed to Earl Roger of Montgomery who gave it to Reginald the Sheriff (Reginald Balliol), the husband of the Earl's niece. The name Hatton is from the Old English *haep-tun* meaning 'the settlement on the heath'. The heath here is Hine Heath. Hine is thought to be from *hiwan*, 'the heath of the monks'.

HIGHLEY *5½m. SSE of Bridgnorth.*
Highley is a large village with a population in excess of 3,000. It lies on a ridge on the steep W bank of the River Severn, opposite Alveley. Coal was mined here until 1969 but now that the spoil heaps have been landscaped, the unknowing stranger would find little evidence of the village's once major employer. In fact, the view from the main street bears the greatest witness. Running off at right angles are several rows of neat, red-brick, terraced miners' houses. At Woodside is a typical inter-war council estate. On the main road is an especially striking detached house in white and green livery that would look quite at home in the South of France. The church of St. Mary has a Norman nave and chancel, and the Church House is half-timbered. On the southern fringe of the village is a lane signposted 'To the River'. Near the bottom of this lane is Highley Railway Station. A prettier station there never

was. It is on the enthusiast-run Severn Valley Railway Line and has appeared in innumerable films. With a steam train in the station it is as if time stopped 40 years ago. The lane continues on to the river bank. Here are a few quaint little summer chalets, colourful and flimsy, but not really objectionable. It is said that the stone for Worcester Cathedral was quarried hereabouts. The medieval name for Highley was Hagelei (and versions of this) which is thought to be from the Old English *Hugga*'s-*leah*, *Hugga* being a personal name and *leah*, meaning 'a wood' or 'a clearing in a wood'. Rudolf de Huglei, an early lord of the manor, took his name from the settlement, not vice versa.

HILTON *5m. ENE of Bridgnorth.*
Hilton is small village on the busy A454, Bridgnorth to Wolverhampton road. The countryside around is undulating and most pleasant; the rocks are red and the soils likewise. The settlement lies in the valley of a tributary stream of the River Worfe and is a mature little place best known for its main road pub, the Black Lion. There are several good black and white houses; Georgian dwellings in red brick; sandstone cottages; and some modern houses which range from detached villas to the flat-roofed blocks adjacent to the flat-roofed Hilton Water Supply Works. (Water is pumped out of a bore hole here.) The Manor House is a substantial building of 5 bays and 2½ storeys, and has a pedimented doorway and irregular side wings. Little stone lions guard the wrought iron gates. The Post Office has found an unlikely home in one of the semi-detached Council houses. Sandpit Lane leads to a sandpit just S of the village. The name Hilton is from the Old English *hyll-tun*, meaning 'the settlement on the hill'. The 'hill' is the slope of the river valley side.

THE HINCKS *1½m. WNW of Lilleshall.*
A 3-mile road, which for much of its length is no more than a dirt track, runs from the A518 West of Lilleshall to Kynnersley through the Weald Moors. The lane dog-legs in 3 dead straight sections and the surface undulates considerably due to movement of the timber raft on which it 'floats'. The ditches only drain the water as deep as they are cut, and in many places the peat is much deeper than the streams and so remains waterlogged below the surface. The Hincks lies along this road, a scatter of brick houses and Duke of Sutherland Cottages. It was the Duke, as simple George Granville Leveson Gower, who organized the drainage works on the Weald Moors. The farms and fields were later sold off and are now owned by many small farmers and pension funds. Hincks plantation stands adjacent to the settlement. This is a wood of scrub bush and mature broad-leafed trees which is commercially managed. Recently, the remains of a Spitfire that crashed here during World War II were retrieved from the peat in this wood. It lay 14 feet down, the Belgian pilot still at the controls. The engine is now at Cosford Aerospace Museum. When the gas pipe line was laid across the Moor a section of it disappeared overnight in an especially waterlogged area just NW of the sewage works. A small plantation of poplars (SJ.697.162) has been planted there in an attempt to dry the land. In periods of wet weather huge lakes appear on the moors hereabouts. One mile S, on the edge of the Weald Moors, is the Army's Central Ordnance Supply Depot.

HINDFORD *2m. ESE of Gobowen and 3½m. NE of Oswestry.*
A red brick hamlet that stands alongside the Shropshire Union Canal. These lands were once wet but are now drained to the River Perry which flows by ¼m. S. There is a good Georgian house of 3 broad bays and 2½ storeys, and a pub now called 'Mad Jack Mytton's Restaurant'. Mad Jack was a colourful local squire who lived at Halston Hall, 1m. S of Hindford. (*See Whittington.*) The first element of the name Hindford could be a reference to either 'a female deer', in particular a red deer 3 years of age or older from the Old English *hind*, or 'a skilled farm worker' who had his own cottage and was of high standing, from the Old English *hina* and the Middle English *hine*.

HINSTOCK *7m. NW of Newport.*
A mainly red brick village that straggles along the busy A41 in good, rolling farming country. One has the feeling that in the past the settlement was a lot livelier in the past than it is now. The present church of St. Oswald is mostly early 18th Century with a S aisle of about 1850. In 1086 the manor was held by Saxfrid from William Pandulph, who in turn held it from Earl Roger de Montgomery. Later, the village came under the Barony of Wem. The Domesday Book name of Hinstock was simply Stoche, from *Stoc* meaning originally 'a place' but later indicating a religious connection. Today there is a mushroom farm (the major employer in the village); a Post Office and general store; a village hall of 1961; a Wesleyan Chapel; 3 pubs; a school of 1839; 2 building contractors; several market gardens and in the country around farms that produce milk, potatoes, barley and sugar. Hinstock Hall was a large house built in 1835 by Henry Justice on the highest point of the village. It was supplied with water from a small reservoir at Bearcroft (from *bere* meaning 'barley'). During the Second World War it was occupied by the Fleet Air Arm. In the 1950's the main building was demolished but one wing and the stable-block have survived and are used as a farm. The 7-acre wood and quarry close to the new road are now in the care of the Shropshire Conservation Trust. It is a Nature Reserve which, because of its dense rhododendron cover, is much favoured by birds.

HINTON *1¼m. NNE of Pontesbury.*
A most attractive, well wooded little place set on a hill in undulating country where the fields are mostly laid to pasture. The rambling, red brick Hall hides behind a high, red brick wall and tall pine trees. Sitting by itself in the middle of a meadow beside the house is a small timber-framed stable with accommodation for 2 horses and a hay loft above. Opposite the Hall is the Grange, stuccoed with 'applied framing', but genuine framing in the outbuildings to the right. A white-painted cottage stands stark against a red-painted barn and at the back of the Grange is a group of new houses nicely done. The narrow lane from Hinton to Edge is flanked by ancient hedges with oaks and holly and many other plants. The name Hinton can mean either 'the settlement on high ground' (from O.E. *Hean-tune*), or 'the settlement belonging to the monastery' (from O.E. *Higna-tun*). As the hamlet stands on a hill and has a Grange (an outlying farm of a monastery) it is anyone's guess as to the original meaning.

HODNET *16m. NE of Shrewsbury.*
The A53 and the A442 cross here, and, indeed, for a short stretch become one. Hodnet is a substantial village nationally known today for the gardens at Hodnet Hall. The name is almost certainly derived from the Celtic *hodnant*, meaning 'a peaceful valley', an indication that there was already a village here when the Romans arrived in Britain. The Normans built a substantial motte and bailey castle (circa 1082) and the now overgrown earthwork remains can still be seen in the grounds of the Hall, SW of the church. In 1275 Sir Odo Hodnet (Norman lords often adopted the name of their new lands) constructed a deer park and re-routed 2 roads to make his boundaries complete. The old village lies around the church and the Hall at the top of the hill and there are many black and white houses, both here and on the road to Market Drayton. Stretching down the hill, southwards, are newer red-brick houses. This development probably started in the reign of Henry III when the town was given a charter granting a weekly market and an annual fair. The large sandstone church of St. Luke is built on a mound (probably the site of prehistoric fort) on the highest ground in the village. It has the following points of interest: a 14th Century octagonal tower, the only example in the whole country; a Norman wall to the right of the entrance porch with Norman arches over the doorway and the windows; and 3 churchyard gates, a lychgate, a wedding gate and a christening gate. The church was extensively restored in 1846. Inside there are monuments to members of prominent local families – the Hills, the Vernons and the Hebers. The present neo-Elizabethan hall was built for the Heber-Percys in 1870 to an 'E' plan design by Anthony Salvin. The Heber family includes, amongst its numerous ranks, the hymn writer, Henry Heber, the author of 'Holy, Holy, Holy, Lord God Almighty', and 'From Greenlands Icy Mountains'. He was rector here before going to India and becoming Bishop of Calcutta. Richard Heber, his brother, is reputed to have had the largest private collection of books in Europe and was a founder member of the Atheneum Club. (There is a small collection of rare books in the church including a Nuremburg Bible of

Hawkstone, the disused windmill by the lake.

1479.) The famous Gardens at Hodnet Hall were laid out in this century by Brigadier Heber-Percy. They are arranged around the lakes that lie in a gentle valley below the house, the 'peaceful valley' of the settlement's name. They are worth seeing at any time of the year and are open to the public. Adjoining the Gardens is Home Farm which has, without any doubt, the finest timber-framed barn (1619) in Shropshire. The frame is laid out in square panels with weather-boarding to the upper parts and later brick infill to the lower. Later, different useage has led to some alterations to the rear. The Old Vicarage, built by Richard Heber, stands in splendid isolation embowered in trees at the end of a long drive to the NW of the church. It is of red brick with stone dressings, 5 bays, 2 storeys, tall chimneys and a sandstone entrance porch with a Tudor arch. In this house was born Mary Cholmondeley, (pronounced Chumley). She wrote several novels, the best known being Red Pottage. Her father inherited the superb mansion of Condover Hall from Reginald Cholmondeley (who had entertained Mark Twain there). He and Mary moved to Condover but only stayed a few months before selling the Hall and moving to London. To the side of the Old vicarage at Hodnet is a folly, a pseudo-Classical ruin that can be seen from the Whitchurch road. (The Vicarage itself is approached off the Market Drayton road down a lane opposite the red brick Lyon Memorial Hall of 1913). The Bear Inn at Hodnet has a resident ghost called Jasper, the spirit of a 16th Century merchant, who, being down on his luck, was unable to pay his bill. Though it was a bitter winter's night and snow lay heavy on the ground, the landlord threw out the merchant who perished with cold. As he lay dying the merchant cursed the landlord and that same night the inn-keeper died of a heart-attack. The new school was opened by the Duchess of Gloucester in 1986 and the Old School was recently sold for conversion to a private house. Leaper's Hill, on the road to Hawkstone through Paradise Valley, is thought to be a corruption of Leper's Hill.

HOLDGATE *2m. S of Shipton, which is 7m. SW of Much Wenlock.*
A shrunken village that lies below Brown Clee Hill in Corvedale and was originally called Holegot Castle (Castrum de Holegot). By 1086 the Anglian village had declined and the area was described as 'waste'. It is presumed that the Normans reinstated the village when they built the castle and the church, which were erected contemporaneously. Immediately to the E of the church is a steep-sided circular motte (mound) with a shallow ditch around it. In the old, timber castle that stood hereabout, Fitz Holgot entertained Henry I in 1109. In 1080 Robert Burnell, the Chancellor of England, built a new stone castle in the bailey of the old fort and it must have been very fine indeed judging by the quality of the masonry in the one remaining tower, which now lies hidden behind a 17th Century farmhouse. The church of Holy Trinity has a red-roofed tower that is a landmark for many miles around. It also has some excellent stone carving, probably executed by the Hereford School of Romanesque Architecture which commenced work in the 12th Century at Shobdon church (now demolished). Their best work is now at Kilpeck church. Not only were their own buildings and carving designs of exceptional quality, but, perhaps even more importantly, they influenced and provided inspiration for the many who came after them. Traces of their work can also be seen at Stottesdon, Aston Eyre, Tugford, Uppington and Much Wenlock. Holdgate church has an example of another school of carving. Overlooking the churchyard and placed beside a chancel window on the S side is the 18 inch high, carved relief figurine of a woman sitting so as to indecently expose herself. This is a 'Sheela-na-gig', almost certainly a pagan fertility symbol of at least Anglian Age. These are rare in England but common in Ireland. The village itself is small and largely built of sandstone. The disturbed earth around would indicate that it was originally much larger. Three quarters of a mile SW, as the crow flies (but 1½m. by road), is Thonglands, a moated manor house, part stone and part timber-framed with a gable to the front. Here a phantom choir is said to sing at midnight when the mood takes them.

HOOK-A-GATE *3m. SW of Shrewsbury.*
It lies on the Shrewsbury–Longden road, a small village with some character. There is a little red brick chapel; some old brick cottages; a 2½ storey Georgian house, rendered and painted grey and black; The Cygnets, a cedar-boarded pub whose roadside car park does nothing for the townscape; and a surprisingly busy air for so small a place. The New Inn welcomes the traveller from the S and the Royal Oak from the N. In Skimhouse Lane there was a soap-making factory in Victorian times. The tallow was skimmed off the boiling liquid and used to make soap and candles. The school of 1864 closed in 1975 and the building is now the Shrewsbury Teachers' Centre. The Rea Brook passes by the back of the village in a steep little valley and sheep graze the hills all around. Just to the N is **Redhill** and here is the for-long derelict 17th Century Seveon's Mill. There is a lay-by on the main road and a sturdy stone bridge over the river. The railway parallels the road at this point. Just to the S of Hook-a-Gate is **Welbatch**. There was a flourishing clog-making industry here until the end of the 19th Century. Today there is little more than one good Georgian house with an elegant porch. Down a dead-end track, NE of Welbatch, is Whitley Hall (1667) which was built by the Owens of Condover. It stands on the site of a much older manor house. The interior is Jacobean and better than the gaunt exterior might lead one to expect. The name Hook-a-gate could mean several things. The Old English *hoc* meaning 'hood' is sometimes used to describe geographical features such as a 'bend in a river', or 'a headland'. The Old Norse *gata*, meaning 'a road', and the old Englsh *gate* could be the origin of the second element of the name. The road crosses the river here at a bend in the river. There are other explanations of the name.

HONNINGTON *½m. SW of Lilleshall, near Newport.*
A most pleasant little place on the A41, the old Newport to Donnington road, at its junction with the lane that leads to Lilleshall Abbey. The road makes an apparantly unnecessary bend but is in fact skirting an old pond. This is not visible from the road but stands embowered in mature beech trees opposite the farm. The water leaves the pool over a shallow cascade. The farm is Honnington Grange, a squat, red brick house of 2 storeys with a 3-bay centre and one-bay wings. In the walled garden is a fine pine tree and in the yards to the rear are numerous red painted barns. A Duke of Sutherland cottage and 2 pairs of bright red brick semi-detached houses complete the inventory of buildings at Honnington. The name might be from the Old English for 'the homestead where honey is made'. (*Hunig* means 'honey'.) It is more than likely that in medieval times the farm had connections with Lilleshall Abbey, 1m. SE.

HOPE *4m. SW of Pontesbury on the A488.*
It lies in the heavily-wooded Hope Valley on one of the most attractive roads in Shropshire. The church of Holy Trinity was built in 1843 by Edward Haycock. It stands below the highway in what amounts to a small, landscaped park in which there are 35 species of trees. It is approached over a little white footbridge and the setting is truly delightful. Near the school there is a natural rock arch. The whole area is riddled with old lead mines, and the glistening white spoil heaps are a very visible feature of the valley landscape as many have been deposited alongside the main road. The miners often had small plots of land attached to their cottages on which they could graze a cow, keep a few chickens and grow potatoes and vegetables. The rough pastures on the hilltops were originally common lands until they were enclosed by mutual agreement of the users (as distinct from enclosure by Act of Parliament). Hope Common, only 110 acres, was divided into 24 smallholdings. The mines in the Hope area have been worked since at least Roman times and probably before that. The Roman open excavation can still be seen, a great gash in the hillside above the ruined 19th Century mine workings which stand alongside the E flank of the A488 at **Gravels**, 1 mile SW of the church at Hope. On the opposite side of the road are enormous mounds of white spoil. This is so poisonous nothing will grow on it. The workings can be overlooked from the steep lane that leads to the hamlet of stuccoed cottages and modern bungalows called **Gravelsbank**. At the top of the hill is a dwelling called Freshwinds (SJ.333.002) from where there is also an excellent view of the Stiperstones and the Devil's Chair outcrop of quartzite rocks, one of the most famous of

all Shropshire landmarks. The Romans also used deep mining methods for in the 1870's miners broke into old workings 50 ft. below the surface and found pottery and tools of Roman age. The Roman Gravels, as these mines are called, were commercially developed in the 1870's and were second only to Snailbeach in output. They closed in 1895. The name Hope is from the Old English *hop* meaning valley. (*See Hopesgate and Shelve.*)

THE HOPE *2½m. N of Ludlow.*
A scatter of houses and cottages on a hillside above Stanton Lacy. The Hope is almost encircled by plantations of coniferous trees with open areas of scrub and gorse. The dwellings are of red brick, stone and half-timber, with roofs of red tiles and thatch. From the back track that leads to the settlement there are wide views over Ludlow Racecourse. In the valley are smallholdings with sheep grazing in small orchards by a stream. Near the top of the hill is a brick-built Primitive Methodist Chapel of 1882, still in regular use. Downton Hall lies ½m. NE from the chapel, straight across the cross-roads, past the peculiar little lodge, and down a private track. (*See Hopton Cangeford.*) The name Hope means 'valley', from the Old English *hop* (which can also mean 'enclosed dry land in a marsh').

HOPE BAGOT *5m. N of Tenbury Wells. Best approached off the B4214 Tenbury-Clee Hill road.*
This peaceful, well-wooded little hillside village has, as a sign of its ancient pagan origins, a spring issuing from beneath an old yew tree on the roadside edge of the churchyard. It is near to the hillside entrance to the church, not that adjacent to the car park. Christian churches were often built on the sites of Celtic temples. The Celts worshipped both water and trees, and the yew was especially esteemed. There are several path-marker yews alongside the lane leading to the church. Recently a long line of old marker yews was chopped down. These lined an ancient 'holloway' that leads from the church to Monk's Field. Here, local tradition has it, there was a monastery, the ghostly bells of which are heard from time to time. The small church of John the Baptist has a Norman nave and chancel, a tower of indeterminate age and an interior that has escaped restoration. Hope Court and the Old Vicarage are Georgian, the black and white cottage near the church is 16th Century, and the Pot House (on the road to Whitton) was the brewhouse for the 17th Century Whitton Court. The village is somewhat scattered and the only social facility is the Village Hall, which is shared with Whitton. As to the name Hope Bagot; Hope, from *hop* meaning 'a valley', is a common Anglo-Saxon place name and Bagot is a corruption of Baghard. In the Middle Ages Robert Baghard held the manor from Roger de Mortuo Mari.

HOPE BOWDLER *1¼m. SE of Church Stretton.*
This is one of the estates held by the Anglo-Saxon warrior lord Wild Edric at the time of the Norman Conquest. The English name of the village was simply Hope, meaning 'a valley', to which was added the name of the Norman lord of the manor, Baldwin de Bollers, who was granted the Honour of Montgomery by Henry I. The village of Hope Bowdler is small and built of stone and lies in hilly, wooded country. The church of St. Andrew and its square, squat tower was rebuilt in 1863 by S. Pountney Smith and in the 1880's some good William Kempe stained glass was installed. The approach to the church is lined with Irish yews. The school of 1856 closed in 1948 and today the children go to Church Stretton. The village hall, as in many Shropshire villages, is the social centre of the community. Between Hope Bowdler and Hope Bowdler Hill, which rises above the village to the NE, is an ancient Celtic Field System; 1½m. N is the great Iron Age fort of Caer Caradoc.

HOPESAY *1m. N of Clun, which is 2½m. WSW of Craven Arms.*
The settlement has a population of about 60, none of whom is native to the area. Outsiders have been attracted to this lovely place, and little wonder. It lies secluded in a sunny valley surrounded by hills – Burrow Hill, Aston Hill, Hopesay Hill and the distinctive Wart Hill. On 2 of these, Burrow Hill and Wart Hill, there are prehistoric fort-settlements. A feature of the village is the presence of several large Victorian houses. These have pleasure gardens well stocked with mature trees though some hide behind high stone walls. Until recently almost all the occupants of the red brick and stuccoed cottages were in service to the big houses. The oldest dwelling in the village is probably Hopesay Farmhouse, part stone and part timber-framed. In fact Hopesay was an important Norman manor and had several attached 'vills' separately mentioned in Domesday Book. Later, it came under the Barony of Clun. The name Hopesay is from *hop* meaning 'valley' and *Say* from the de Say family, lords of the manor and descendants of Picot de Say. The squat church of St. Mary is mostly 13th Century with a nave and chancel of about 1200. The excellent, chestnut-panelled nave roof is 15th Century. The woods of Burrow Hill reach out and almost touch the church. Adjacent is the gaunt, stone Old Rectory. There is no shop, pub or Post Office in the village which is so quiet as not to be true. Just to the S is a track leading E to **Perry Gutter**. This is a delightful little valley in which there are 2 cottages, one of which can only be reached by foot. This is a shrunken village where once there were more cottages, a shop and a blacksmith's forge. Further E again is **Oldfield**, another old village shrunken to one stone barn that has now been converted to a house. North-West of Hopesay, on the lane to Round Oak, is a hamlet called **The Fish**, which takes its name from an old ale house. This has now also been converted to a dwelling and stands adjacent to Oakham House, itself a conversion of 2 cottages formerly known as Fish Cottages. The lane continues ever upwards to **Round Oak**, 1¼m. NNE of Hopesay. A cluster of stone cottages stands alongside the road and alongside almost all of them, it seems, stands a horsebox. The people here make the most of the riding on the National Trust owned Hopesay Hill and Common. Horses roam free here, too; we have seen a small herd of prancing white horses frisking in the evening sun, an evocative sight. The view southwards from here over the Hopesay valley to the hills beyond is quite splendid. The road to Newington (Craven Arms) curls around the steep, conifer-clad slopes of Wart Hill and its Iron-Age settlement. As the road starts to drop there is a weather-boarded barn and near this is a large circle of mature trees, a place of some significance to someone at sometime. There are more spectacular views, northwards this time, to the Stretton Hills and Apedale.

HOPESGATE *⅓m. N of Hope which is on the A488, 6m. SW of Pontesbury.*
A scattered community on the high ground above Hope. At the top of the steep hill is the flat-roofed, orange brick Church of England School. Adjacent to this is the Hope Village Hall and adjacent to this a small detached cemetery. The stone-built Stables Inn stands on the road to Ladyoak and Ploxgreen. Just E of the pub is the long track that leads to Hope Hall. The present building is a modest stuccoed farmhouse painted cream with pale blue bargeboards (pale blue is a popular colour hereabouts). It stands on a pine and conifer-clad knoll and has superb views in all directions: S to the Stiperstones, N to Aston Hill, E to Snailbeach and W to Bromlow Callow and the hillside cottages of Bent-lawnt. To the NW, alongside the road, is the small, tree-clad mound of a Norman motte, 9ft. high, oval and 15ft. long at the top. Beyond is the many-gabled red brick pile of Leigh Manor abounded on 3 sides by mixed woodland. The fields are all laid to pasture.

HOPTON CANGEFORD *5m. NE of Ludlow.*
Also called Hopton-in-the-Hole the tiny settlement stands at the junction of 3 valleys on the lower SE slopes of Brown Clee Hill. It has a stream, the Hopton Brook, and a red brick church of 1776 which was made redundant in 1983 and is now a craft pottery. Before it was sold the family crypt of the Rouse-Boughton's, the local landowners, was bricked up. There is a cottage called the School House, but there appears not to have been a school here and to confuse the stranger further the Old Vicarage is now called The Gables. About 1m. SW, as the crow flies, is **Downton** Hall. Hopton Cangeford lies on the Downton Estate and is owned by Miss M. F. Rouse-Boughton. She lives at the Hall, an 18th Century house of brick and stone with early 19th Century alterations and a Tuscan porch. Inside is an excellent circular entrance hall with Ionic columns, a honeysuckle freize and a roof of glass. Today, it languishes, the front

boarded up though the well-wooded gardens are tended. (Many of the mature oaks and beeches in the country hereabouts were planted by an 18th Century squire.) The western approach to the Hall is guarded by a Lodge with an amusing facade – a shaped gable and pointed windows on the ground floor, and an ogee arch over the first floor window. Behind the facade is a plain red brick house which is now derelict. The Old English name, Hopton, means 'the settlement on an area of dry land in a marsh'. Cangeford is from Cangefot, an old French–Norman nickname.

HOPTON CASTLE *6m. ESE of Clun.*
The settlement stands at the meeting of 3 valleys, with wooded hills all around and a stream by which stands the 14th Century stone keep of the 12th Century Norman castle. In 1644, during the Civil War, 24 Roundhead soldiers, commanded by Colonel Samuel Moore, were besieged in the castle by Cavaliers. Several of the Cavaliers were killed whilst attempting to enter the castle on scaling ladders. Because the attacking force had superior numbers and equipment, and would therefore in time inevitably win the engagement, the Roundhead defenders had violated the Rules of War by not surrendering to prevent unnecessary loss of life. After three weeks the Roundheads did surrender but it was adjudged that they should have done so earlier. Each of them had an arm or a hand cut off and they were then all thrown into a muddy pit. If they attempted to climb out they were beaten back with heavy stones. There they were left until they died. Today the village consists of the castle ruin, the church of St. Mary (1871) by T. Nicholson, the black and white former vicarage and a handful of cottages.

HOPTON WAFERS *2½m. W of Cleobury Mortimer.*
The tiny village lies just to the N of the A4117 in a well-wooded valley. The ashlar church of St. Michael was rebuilt in 1827, though the old tower was retained. Inside there is a carved and gilded representation of the Royal Coat of Arms of Queen Victoria, a large relief of Thomas Botfield, d. 1843 (by E. H. Bailey who also sculpted the figure of Nelson in Trafalgar Square, London), and some fragments of 15th Century stained glass. Opposite the church is a charming group of old cottages in stucco, stone and half-timber. At the back of these and attached to them is Hopton Manor, a brick house of the late 18th Century. A quarter of a mile N, hidden in its well-kept park, thought to have been landscaped by the illustrious Humphry Repton, is Hopton Court. The present large, red brick house of 5 bays and 3 storeys was built by Thomas Botfield in 1726 to replace an earlier farmhouse. In about 1812 the verandah was added, probably by John Nash. In the garden, against a high brick wall, is a rare Coalbrookdale cast-iron framed greenhouse which is shortly to be renovated. When we visited Hopton Court there was a most jolly atmosphere; young people played tennis and weekend guests ambled in the grounds, very much everyone's idea of an English Country House on a summer's evening. South of the village is a red brick castellated building standing in the middle of a field. This is a water station on the pipe that links the Elan Valley reservoirs with Birmingham.

HORDERLEY *4m. NE of Craven Arms on the A489.*
The A489, between its junctions with A49 near Wistanstow in the E and the A488 at Lydham in the W, is a relatively busy but most attractive road. It follows the course of the lovely River West Onny which here has cut a gorge through the toe of the Long Mynd. This is believed to have been made at the same time and in the same way as the Ironbridge Gorge. Both are believed to have originated as overflow channels from the immense Lake Lapworth, a body of water trapped between the glacier front of the last Ice Age and the high ground of the Long Mynd-Wenlock Edge hills. From Wistanstow the road leads through rocky cliffs and mixed woodlands to the handsome stone houses of the hamlet of **Glenburrie-Lower Carwood** and on to Horderley. In 1066 Horderley was held as 2 manors by Algar and Dunning but by 1086 was held by Odo who soon after gave it to Shrewsbury Abbey. Today the settlement consists of 2 houses; one is a rendered 3-bay farmhouse with stone out-buildings called Heather Nursery, and the other is a red brick bungalow with gables that was an old railway station. (The track of a now disused railway from Craven Arms to Lydham also passes through the gorge.) Just NW of Horderley the valley opens out and there are pastures on the flood plain with moorland hillsides to the N and the conifer plantations of Plowden Woods to the S. A few cottages lie alongside the road which now rises above the river. Hillend Cottage and Hillend Farm refer to the end of the Long Mynd. This is old looking country with oaks, ferns and bracken, many coloured trees, sheep in the pastures and a great variety of ever-changing scenery. The road descends to Plowden. (*See Plowden.*)

HORDLEY *2½m. SSW of Ellesmere.*
In 1086 the manor was held by Odo but a few years later he gave it to Shrewsbury Abbey. The hamlet lies in undulating country on the edge of Baggy Moors, reclaimed marshland in the valley of the River Perry. To the N is the Shropshire Union Canal over which are a surprising number of bridges. The recently (1967) restored church of St. Mary has a Norman nave and chancel, a later timber-framed red brick belfry, a patterned slate roof and a 17th Century pulpit. There are 2 yew trees in the churchyard and mature broadleaf trees, including horsechestnut, in the area around. The red brick barns opposite the church are Tudor and belonged to the old Hordley Hall, a home of the Kynaston family. There is a timber-framed house with brick infill and Hordley House is white painted with pediments. The lane to Lower Hordley passes through flat arable land. Here are stuccoed council houses, a Post Office and red brick farms; here, too, is the factory-abattoir of City Meat surrounded by security fencing. This is where the contented cows who graze the peaceful pastures hereabouts are slaughtered.

HORSEHAY *3m. SSE of Wellington.*
Horsehay lies on high ground 1½m. N of Coalbrookdale. In 1754 there was little more here than a farm and an adjacent old corn mill which stood beside a small pool. In that year Abraham Darby II rented the property and entered into complicated and difficult engineering schemes that involved the creation of 2 more pools, the extension of the original pool (Horsehay Pool as we know it today) and the construction of a blast furnace. When this went into production in 1755 it was the first furnace to conclusively prove that coke was superior to charcoal in the reduction of iron-ore to iron. The water of the pools was required to operate the bellows that fanned the furnace flames. All the materials – limestone, coal and iron ore – were brought in tubs that ran on a system of wooden rails. The pig iron produced was similarly transported down to Coalbrookdale for refining. Later, Horsehay had its own forges. By 1756 there was a brickworks here and in about 1790 the famous conical Roundhouse pottery kiln was constructed. This well known landmark was pulled down in about 1970. In 1838 the Coalbrookdale Company purchased the Pool Hill estate. Much to their delight they found a seam of clay suitable for use in the manufacture of ceramic pottery and lost no time in engaging in this trade. In 1857 the railway arrived, but only to witness the end of the industry. When the decline in the iron trade came it came early to Horsehay. The first of the new works to be opened was one of the first to be closed. The blast furnaces 'blew out' in the 1860's and the forge hammers rested in 1886. Many workers were made redundant and there was great hardship in the area. In recent times there has been extensive open-cast coalmining to the W and the land is in the process of being landscaped. In the village there are 3 pubs, the Forester Arms, the Labour in Vain and the Traveller's Joy; a Post Office; a shop; a Methodist church near which is Simpson's Pool, embowered in trees; and a mixture of old and new houses. It is not a large place and is dominated by the old premises of A.B. Cranes, which occupies the site of Darby's famous ironworks, and Johnston's Pipes. Johnston's have a large quarry, now water-filled and surrounded by a dramatic, wooded cliff. This is in the process of being pumped dry and filled with earth. There are 2 factories: one makes concrete pipes, the other glass reinforced plastic pipes (GRP). A.B. Cranes closed down in about 1983. Much of the site has been cleared and what is left is now a small industrial estate. In Bridge Street, opposite the old works, is a railway locomotive shed now maintained by the Telford-Horsehay Steam Trust. The boundary wall of the A.B. Cranes site (on the other side of the road) is partly built of

Hopesay, from Wart Hill.

below: The porch of Hardwick Hall.

above: Hopton Court, Hopton Wafers.

left: Motte of a Norman castle at Hopesgate, Hope.

blue slag from the old iron furnaces that once stood near here. The road runs along the top of the dam that holds back the waters of Horsehay Pool, the main reservoir pool of Darby's ironworks. This is now a delightful, tree-fringed lake with water birds galore. Along the W bank is a row of 27 terraced houses; how happily situated they are. This is Old Row, built by the Coalbrookdale Company to house its Horsehay workers. It is now called Pool View. This lane becomes New Row, a later development of 12 houses by the Company. New Row leads to **Spring Village** where a few old cottages have attracted an 'infill' of modern houses that is not unattractive. There are 2 small wooden bridges over little inlets from the lake in the area of the springs. Many of the unfortunates interned on Woodside (Madeley) would give an arm and a leg to live in this unexpectedly cheerful little corner of Telford New Town. Perhaps Horsehay's best known character was 'John Bull, the greatest man in the world'. This was William Ball (1795-1852) the grandson of a migrant from the Potteries. Ball began work at the age of 8 as a puddler's assistant at the Horsehay Forge and later became a shingler. He had great physical strength and is reputed to have lifted a piece of red hot iron weighing 9 hundredweight from the ball furnace to the forge hammer. In the 1840's he gave up work after injuring an eye; by 1850 he weighed 36 stone and dressed as John Bull made 'appearances' in several large cities including London. Adjoining Horsehay to the SE is **Doseley** which appears to consist of little more than a few Victorian houses with some modern infill, the Cheshire Cheese pub near the railway viaduct, a scrappy little industrial estate that lies alongside the Johnston's quarry and the old brick church of St. Luke which is now a dwelling (though you would never know just by looking). Johnston's offices stand opposite the church. The name Horsehay is Anglo-Saxon and probably means 'the enclosure where horses are kept'. (A 'hay' can also be either a part of a forest or a clearing in a forest.) North of Horsehay is Lawley Common, a large unfenced area of sloping grassland on which cows and sheep graze. At the end N of the Common is a huge mound of earth that looks for all the world like a pre-historic fort; at the time of writing it was being beseiged by a small army of gigantic yellow earth-movers. This is Newdale and the open cast pit lies beyond the spoil mound. (*See Newdale.*)

HORTON *3¼m. NNW of Telford Town Centre.*
It lies on the edge of the Weald Moors and the country is cut through by deep drainage ditches. These old marshlands are now intensively farmed. Along the lane from Leegomery Roundabout on the A442 (Queensway) are several groups of spaced out cottages. One group faces the Moors and another turns its back to them. The village of Horton is as yet unspoiled. There are a few brick cottages, a couple of new bungalows, some farms, the Queen's Head pub, Lille of Telford (maker of dungarees), a boundary wall made of iron furnace slag, cows and a few ducks beneath larger than usual skies. The lane that runs NE to the Humbers passes the unmarked entrance drive to **Hoo** Hall, the 'big house' of the hamlet of Hoo. This is a substantial half-timbered house dated at 1612 which has some good carving beneath the windows by the porch. The approach is now lined first with fast growing conifers and then by a phalanx of 8 marblesque statues. To the left-rear of the house are medieval fish ponds. The lane continues eastwards past the Donnington Garrison Stables, a fruit farm (see Lubstree), scattered brick and stucco cottages and scrub woodland to **The Humbers**. Here are army houses, a row of pine trees, some Duke of Sutherland cottages and the main gate of the huge Army Central Ordnance Supply Depot. In the last few years this has suffered 2 major fires both of which resulted in multi-million pound losses. South of Horton, and in another world, is **Hortonwood**, an expanding industrial estate of the more prestigious kind. There is space between the factories and warehouses, trees and grassed areas. Firms already here include Horton's Ice Cream, Brass Farm, Bischof and Klein (polythene film packaging), Omron, and Edros Emyos. How long, one wonders, before the old village of Horton is submerged in a sea of steel-frames and plastic cladding?

HOWL *5m. NW of Newport and ½m. W of Standford Bridge.*
The village houses are mainly red brick and there is a little Methodist church. Howl Manor is a large farm; the white rendered house lies mostly hidden amongst tall trees. Between the Manor House and the village is Howl Pool Farm which stands opposite a delightful mere. This small lake is man-made and the road forms a dam. The slender, cone-shaped, sandstone tower on the other side of the road to the pool used to house a water pump. It is said to have been built in 1845 as a windmill and used to have a sail until the 1920's. The name Howl is thought to be derived from either an Anglo-Saxon or a Scandinavian word meaning 'hill', i.e., 'a place on or near to a hill'. Howl lies on a slope. To the N and W of Howl is **Ercall Heath**. Before the First World War there was a 900 acre forest here, mostly coniferous with some oaks to the S. At the outbreak of hostilities the woods were requisitioned and for many months a force of about 500 men laboured to fell the trees which were dragged away by teams of horses to Crudgington station. Two cock-pits were discovered during the clearance: circular depressions lined with turf where cockerels were fought and men bet on the result. After the war the land was sold in small lots at low prices. Along the lanes wooden bungalows were built but much of the interior remained unsold and became a heathland of brambles, bracken and scrub woodland. When the Second World War broke out some of the heath was cleared and with great difficulty, and a great deal of time, was brought into agriculture. This process continued after the war and today potatoes, sugar beet and cereal crops are grown quite successfully.

HUGHLEY *4m. SW of Much Wenlock. Most easily approached along a minor road off the B4371.*
'The vane of Hughley steeple veers bright, a far known sign'. A. E. Housman. Everyone delights in pointing out that Hughley has a bellcote, not a steeple, so we will desist. (Apparently, Housman wrote his poem with another church in mind.) The church clock, with its octagonal face, was presented to the parish by the Earl of Bradford when his horse won the Derby in 1892. In the chancel are some fragments of 15th Century stained glass. Hughley Brook runs through the village which lies in lovely country below the wooded scarp of Wenlock Edge. The Edge is white limestone; the valley is red sandstone. The name Hughley is from the personal name Hugh (Sir Hugh de Lega was an early 12th Century landowner here) and *leah* which is Old English and means either 'a wood', or 'a clearing in a wood'.

HUMPHRESTON HALL *¼m. N of Albrighton (near Shifnal).*
It warrants its own official signpost off the A41 near Albrighton but is in fact a single large farm with an Elizabethan manor house. The house cannot be seen from the road but a public footpath crosses a field to the rear and the delightful mansion stands revealed across a substantial remnant of the old moat and waterside trees. The centre part of the house is timber-framed and there are white-painted brick extensions to either side. The tall chimneys are of brick and there is a large sandstone fireplace housing. There are cows in the fields and the country around is flat.

HUNTINGTON *¾m. NNE of Little Wenlock.*
It lies on high ground 1½m. E of the tree-clad mass of The Wrekin. There is a farm, with pastures grazed by cows and Jacob sheep; a large and very stoney arable field; a handful of old cottages now improved and extended; a few modern houses with pony paddocks; and an old, derelict stone building. To the S lies the **Coalmoor**, aptly named for here coal was opencast mined. Today there is a coal washing plant and the great holes in the ground that are left from the old workings are now used as refuse tips. Action Waste Limited are indeed active here. The Smalley Hill Tip (SJ.663.073) is a wilderness of dangerous pools of sludge and slurry, banks of spoil and a great black hole surrounded by security netting. However, there are still a few woebegone farm houses and cottages here and much of the land is still laid to pasture. **New Works** is ½m. NE of Huntington. Here is a brick farm partly clad in grey cement render that some 150 years ago was used as a religious meeting house. Opposite are old opencast pits in the process of being landscaped. Down the hill are large numbers of modern houses and bungalows

Humphreston Hall, Albrighton, near Shifnal.

with the ocassional old cottage fighting to breath in between. The area to the N is well wooded, though much of this is due to be destroyed by yet another monsterous open cast pit. There appear to be no social facilities here, but then Lawley lies less than ½m. E.

IGHTFIELD *4m. SE of Whitchurch.*
A pleasant little place in undulating pastoral country. There is a Post Office and general store, a War Memorial, a timber-framed cottage, a school and the church of St. John the Baptist. The church is Perpendicular (1350–1550) and was restored in 1865 when the chancel was rebuilt. Inside is a significant brass depicting Dame Margery Calverley. Between the church and Ightfield Hall, which lies ¾m. NE of the village, is a long avenue of trees. Legend has it that they were planted in a single night by the Devil (some say a Knight). There is also a ghost. A previous owner of Ightfield Hall, driving a coach and four horses, returns home occasionally. The Mainwarings used to live here, including Arthur Mainwaring, the 18th Century poet. He was the editor and a contributor to The Medley and the first number of the Tatler was dedicated to him. Today, the Hall is a somewhat untidy farm with a brick facia hiding the old timber framing. The moat is very wide and in part still water-filled. It is used today as a duck pond. A half-timbered barn, possibly older than the house, is in a sorry state but still standing. The first element of the name Ightfield is possibly derived from the Welsh *eithin*, meaning 'furze'–a spiny, evergreen shrub with yellow flowers like gorse.

IRONBRIDGE *10m. SE of Shrewsbury.*
On the banks of the river at Ironbridge, just below Severnside Steps, is the rough wooden hut of a coracle-maker. These little craft were once a common sight hereabouts. They are of an ancient Celtic design and remind the traveller that man has been in this dramatic, wooded gorge for time unrecorded. It is believed that the Ironbridge Gorge was created thus: originally the River Severn flowed northwards and entered the Irish Sea at the estuary now occupied by the River Dee. Then, during the last Ice Age, the ice sheet dammed this exit and a huge lake was formed between the ice front and Wenlock Edge. This is called Lake Lapworth after the man who initiated this theory. These waters escaped by carving a new, S-W route, what we call the Ironbridge Gorge, now occupied by the River Severn. As the River Severn cut through the carboniferous rocks it exposed seams of coal, iron-ore, limestone and clay. These exposures were easily mined and from at least early medieval times there has been an iron-working industry in the area. (*See Buildwas.*) The town (if one can call it that) is famous for its iron bridge, the first such structure in the world. This was designed as though it were to be made of timber, using carpenters' joints and strutting, and was cast at the Coalbrookdale Company foundries in 1778. It opened to the public in 1781. There were a few half-timbered houses in the area before the bridge was built but the settlement we see today developed because of it. A small, planned centre was constructed at the N bridgehead. Facing the Market Square is the former Market House of about 1790 with its 5 distinctive segmental arcaded arches; the Tontine Hotel of about 1785; and 2 red brick buildings each of 5 bays. There used to be a row of stuccoed shops with balconies opposite the Tontine Hotel adjacent to the bridge but these have been demolished. There are shops a-plenty though, both in the Market Square and facing the river down Tontine Hill. There are also several public houses, a garage and a cafe. On the hill above the square is the yellow brick church of St. Luke, designed by Thomas Smith of Madeley and built in 1836. The people who came to live here ranged from customs officials to miners and company managers. In true democratic style they built their houses cheek by jowl; substantial dwellings in Gothic and Classical style adjoin crudely built cottages of brick and stone. Church Road, Belmont and Hodgebower are, in fact, quite delightful little lanes and most of the houses have superb views over the gorge. Car parking is a problem, though, and there are some steep climbs that must be daunting to the ill or infirm. By 1840 the population of Ironbridge was about 4,000. The business of the town was trade, not industry itself. Having said that the most famous furnaces in the gorge lie just to the SE, namely the Madeley Wood Furnaces or Bedlam Furnaces as they are sometimes called after an Elizabethan mansion of that name which once stood here. These furnaces were built in 1757 by the Madeley Wood Company in which the lord of the manor, John Smitheman, was a partner. They were later bought by Abraham Darby and finally closed in 1832 when the ironworks were moved to Blists Hill to be nearer the coal supplies. It is the Bedlam Furnaces that feature in the most famous painting of the area 'Coalbrookdale by Night' by Philip James de Louthbourg. The furnaces are alongside the road and are an 'open site' that anyone can visit free of charge. It was the 2 mile stretch of river between Ironbridge and Coalport that caught the imagination of many contemporary artists – the flames and smoke, the red night sky, the hustle and bustle of boats on the river, the roar of the furnaces and the clanking of engines all confined within the cliffs of the gorge. On the steep slopes above the industry was, and still is, Lloyds Coppice, a mixed woodland that has been carefully managed over several hundred years, ever since it was first used as a major source of charcoal. To conclude these notes on Ironbridge, mention should be made of the Madeley Wakes which in the 19th Century were actually by far the most celebrated in the gorge. Bullbaiting was especially popular. The bull was baited by dogs 3 times a day on 3 consecutive days – Monday, Tuesday and Wednesday – first at the Horse Inn, Lincoln Hill, then in front of the Tontine Hotel and finally at Madeley Wood Green. To the NW Ironbridge runs into the Coalbrookdale at **Dale End**. Here is the Severn Wharf and the Severn Warehouse. This piece of mid-Victorian jollity in red and yellow brick has two embattled towers and was built by the Coalbrookdale Company; the river barges were loaded here. Today it is an information centre and a good place to start a visit to the area. Ironbridge was, in fact, a port of some consequence. Barges, the 'Severn Trows', took heavy cargoes of iron and coal down river to Bridgnorth, Gloucester and Bristol. In the 1830's there were, at any one time, up to 150 vessels on the river between Coalport and Ironbridge. This trade was damaged with the coming of the canals and was finally killed when the Severn Valley Railway opened in 1862. In the middle of this century a new trade came to the area, namely the manufacture of teddy-bears. Merrythought Ltd., the largest manufacturers of cuddly toys in the U.K., have their works at Dale End and TV cameras are no strangers here. Just W of the Severn Warehouse and the junction of Dale Road and The Wharfage is the Dale End Riverside Park. It is probably best described by following the official walk which is waymarked with numbered posts. 1. To your left, across the river, is the 10-arched viaduct of the Severn Valley Railway, built in 1862 and closed in 1963. Head upstream towards the cooling towers of the Buildwas power-station. 2. The buildings you pass on the right, between posts 1 and 2, were part of the Severn Foundry of the Coalbrookdale Company. All the furnaces and forges of the ironworks in the valley were powered by water-wheels. To regularise the supply 5 pools were made by damming the Coal Brook that enters the Severn here. 3. The vegetation on the river bank is dominated by 2 alien, Japanese plants: the knotweed and the giant knotweed, which is comparatively rare in Britain. 4. The Buildwas power-station dominates the view here. It is coal fired and can generate 1,000 megawatts of electricity. Millions of gallons of water a day are used to cool the steam that drives the turbines. (The pure water used to make the steam is turned back to water by cooling and is then re-used.) 5. We are now at the Ironbridge Rowing Clubhouse, a nondescript cedarwood shed with a green felt roof. The club was founded in 1870 and featured in society life of the time. Their regatta is held in June. Canoeists favour the rougher waters of the Jackfield Rapids, downstream. 6. We are now at the foot of the Albert Edward Bridge (dated 1863) which was cast at Coalbrookdale to a design by John Fowler. It has a single 200 ft. span. The bridge originally carried the Wellington to Much Wenlock railway but is now used only by the coal trains that supply the power-station. We now turn from the river and walk back through the centre of Dale End Park. 7. The unprepossessing orange-red brick Valley Hotel is 2½ storeys high and 6 bays wide on the garden side where there is a pedimented entrance porch supported by columns. It was once the home of the Maw family, the founders of Maw's Tile Works at Jackfield. (The

Pool Lane, Horsehay.

Buildwas, from The Lodge.

The Lodge, Ironbridge.

works are now a craft centre.) There is a display of tiles in the house entrance. Next, on our left, is Eastfield House, the former home of a local doctor called Webb. His son, Captain Webb, achieved international fame when, in 1875, he became the first man to swim the English Channel. Later, from 1880, the managing director of the other great Jackfield tile factory, Craven-Dunnill, took up residence in Eastfield House. Rawdon-Smith loved the Rowing Club and his garden. It was he who planted most of the large specimen trees that are now a part of the park. Amongst them are: copper beech, yew, Indian beam, locust, weeping ash, tulip tree, pencil cedar, Chinese stewartia, Persian ironwood, red oak, variegated holly, grand fir, Californian redwood, Austrian pine, Bhutan pine and walnut. 10. The park bench is a good example of the decorative cast-iron ware produced by the Coalbrookdale Company. Continue back to the car park. There are several such planned walks and town trails in the Ironbridge area. As one looks around the area today it is difficult to realize that until recently this was a dead place. There were rows of empty, derelict houses and overgrown ruins. What is more it had been this way since the late 19th Century. When the output from the ironworks of Coalbrookdale declined, and the centre of industry moved to Birmingham and the Black Country, Ironbridge fell into decay. Nature began to reclaim the land. Then, in the early 1960's, the historical worth of the gorge was acknowledged. Millions of pounds were spent in grant aid to revive the area, and in 1968 the Ironbridge Gorge Museum was founded. These activities were helped by the creation of the New Town of Telford, which lies just to the N. The proximity of the gorge gave romance and respectability to the sprawling, modern industrial desert. Today, Ironbridge is most picturesque and at holiday times takes on the air of a resort. And yet, for all of man's endeavours in the area, there is a sense of the wilderness here – the steep slopes of the gorge, the fast flowing and forbidding river, the damp, dense woodlands, the landslips that regularly buckle the roads and tip buildings into the Severn – nature almost reclaimed the place for her own once, and will no doubt try again. Note: Curiously, one rarely finds mention of **The Lodge**. This stands aloof from the industrial houses that scramble up the hillside above the Iron Bridge. It was quite probably the first building of any substance to be constructed in the gorge and stands on the highest point of the hill, near Beeches Hospital. The access lane commences opposite the White Horse pub on Lincoln Hill. The lodge was built about 1530, probably by Sir Robert Brooke of Madeley Court. It would then have been quite isolated and may have served as a hunting lodge. It is built of local limestone to an 'L' shape with walls 2 ft. thick. At the angle is a gable with a small, circular window; it is thought there was originally a short tower in this position. There are several chimney stacks of mature red brick and the windows were once probably mullioned. Old photographs show one mullion in place and though this window is now partly blocked the bar can still be seen. There are 2 priest's holes and Charles II is said to have surveyed the Severn from here in hopes of crossing it during his flight after the Battle of Worcester. From the cellar there leads a now bricked up tunnel. Over the years brick additions have been made to The Lodge and today there are 3 cottages attached. The hill on which the house stands, **Lincoln Hill**, was extensively mined for limestone and is riddled with tunnels, shafts and enormous caverns. The ground below the road has been made safe but the cost of attending to the whole hill is so great as to be prohibitive. Benthall Edge, across the River Severn, is similarly honeycombed. The current owner of The Lodge is Paul Tozer who, with his partner Mike Onions, has a workshop in the arcaded, blue brick, Old Police Station (1862) in Ironbridge. Here they fashion high-quality, hand-made, acoustic, steel-strung guitars that have a most excellent reputation.

ISOMBRIDGE *3m. NW of Wellington.*
In 1066 Ulf held the manor. By 1086 it had passed to Ralph Mortimer of the great Marcher Lord family. Shortly after Domesday Book, however, the manor became the property of the holder of the office of the King's Royal Forester of Shropshire. In time the settlement declined in importance and became a part of the scattered manor of Great Bolas. An early form of the name is *Esnebrugg* which means 'the bridge of the servant', *esne* being the Old English for servant. Isombridge stands close to the River Tern but there is no sign of a bridge there now. Today, Isombridge (the locals pronounce it Isumbridge – the Is as in the word 'is') is a tiny, isolated hamlet in flat country. What there is here is this: 2 mid-20th Century houses; a gabled red brick farmhouse; a row of 3 half-timbered cottages now rendered and painted grey; a small orchard, a rare sight these days; 3 detached barns; and a sand pit quarry with 2 pools in the abandoned workings. It is a somewhat cheerless little place with a litter of scrap cars in the field beyond the grey painted cottages. Decline from its days of importance it certainly has, but then many Shropshire villages have declined to the point of extinction. Allscott lies ¼m. SSE and can be reached by a bridleway that crosses the River Tern by a footbridge. The Wrekin stands clear on the horizon but the view is marred by the steaming hulk of the British Sugar factory.

JACKFIELD *1¼m. NNE of Broseley.*
Jackfield is a somewhat neglected area of the Ironbridge Gorge and yet it is one of the most fascinating. There is not much to see of its early industries; the 2 large tileries that dominate the settlement today are both mid-Victorian. There is much archaeological work to be done at Jackfield. The settlement has a long and colourful history as both a manufacturing area and a port. It began as a port. By 1605 James Clifford, lord of the manor of Broseley, had constructed a wooden railway from his coal mines to the River Severn at Jackfield. This was the first railway in Shropshire and the second in the whole of Great Britain. The track followed the valley of a stream and it was around its confluence with the Severn that Jackfield developed. There was a pottery here from at least 1634 and the area was known for its drinking mugs. Small stone cottages were built and at least one grand timber-framed house (of 1654) which was later converted to a pub and called the Dog and Duck. On the stream above the settlement there were at least 2 corn mills which were still standing in the 1790's. In 1713 the Thursfield family arrived from Stoke on Trent and set up a pottery. By 1750 they were making their celebrated 'Jackfield Ware', a highly vitrified black earthenware decorated with gold flowers and figures. Their 'works' consisted of kilns built on to the end of cottages E of the stream confluence with the Severn. But it was still coal that was king here. Jackfield was a port. By 1756 Broseley had 87 boats and Benthall 13 boats engaged in transporting coal down river to Stourport, Gloucester and Bristol. The boats were captained by men referred to as 'owners' who took the coal on credit and paid for it when they had sold it. The boats, the Severn Trows, were hauled back up-river by bow-haulers. These men led awful brutish lives and not unnaturally behaved in awful brutish ways. They thieved, fought, drank and whored, and Jackfield catered for them. There were cheap boarding houses, brothels, and numerous pubs – the Severn Trow, the Black Swan, the Tumbling Sailors. It was almost accepted that they lived on the meat of sheep stolen from riverside pastures. By 1800 there was a towpath from Coalbrookdale to Bewdley and by 1811 this had been extended to Gloucester. Only then could horses be used. Between 1767 and 1771 George Matthews built 2 furnaces for smelting iron ore at **Calcutts** in the Broseley-Jackfield valley. The bellows were operated by a water-wheel and the water returned to the upper pool by a steam engine. By 1786 there had been added 2 forges, coke and tar kilns and a brick kiln. Under the management of Alexander Brodie the works became internationally famous for the cannon that were cast and bored here. The coking ovens were installed by Lord Dundonald. He lived at the gaunt 7-bay brick house called The Tuckies (from Tuck Eyes) at SJ691.024, behind the present Maws factory. (William Reynolds, the great ironmaster, lived here from 1800 to 1803, when he died.) Lord Dundonald constructed 12 coking ovens at The Calcutts between 1784 and 1786 and 8 more by 1800. Coal was burnt very slowly and the smoke was condensed on cold plates. The tar so obtained was heated again. This drove off gases which were condensed as oils (used in varnishes) and left a residue of pitch (used to caulk wooden ships). The Calcutts Ironworks prospered as both a furnace and a foundry, and later as an engineering works producing steam

Blist's Hill, the old iron furnaces.

engines until the 1830's. In 1836 the foundry was demolished and the railway was used for transporting coal only. The pottery industry all but ceased at Jackfield when the master potter John Rose, who had briefly taken over the Thursfield works, moved across the river to Coalport. He was probably enticed by offers of capital from William Reynolds, who had single-handedly planned and developed the new town. However, the economy of the area was sustained by the brick and tile industries which flourished well into this century. One of the largest of the manufacturers was Prestage & Co., whose Milburgh Tileries produced over 4 million tiles a year. Another, smaller concern occupied part of the old Calcutts Ironworks. The economy was further boosted by the development of the encaustic and tasselated trade in the second half of the 19th Century. The leading producers were Maw and Co. and Messrs. Craven Dunnill. Both were very large concerns and both exported their highly decorated tiles all over the world. The factories we see today date from 1883 (Maw's) and 1872 (Dunnill's). Until recently the Craven Dunnill works were used as a brass-foundry and today the Maw's complex is partly used as a tile museum. Both factories produced a variety of wares but it was for their encaustic tiles that they were famous. These had been made by hand in medieval times, quite possibly here, in Jackfield. Encaustic tiles have patterns created by inlaying different coloured clays, a bit like the pottery equivalent of marquetry in wood. When production ceased the craft died. The machines and plaster moulds remained but the skills had gone with the operators. A demand developed for replacement tiles to repair worn areas in floors (such as from the House of Congress in the U.S.A.), but no-one could fathom out how to work the machines. At the last moment 2 very old men, ex-operators, were found and they passed on their 'tricks of the trade'. The Maws factory is now going back into production in a small way. Beware of buying property in the Jackfield area. It is notorious for land slip and is well known in geological circles. All along the banks of the Severn are heaps of pottery spoil, slowly being washed into the muddy waters. Some good examples of broken encaustic tiles can be found, but be careful. This is a gorge and the river can flow very fast and deep. At the time of writing Jackfield is still somewhat run down and gipsy-like, but most pleasantly so. As to the settlement itself, there are a variety of small cottages and a few larger houses such as The Calcutts of about 1695, a 3-bay dwelling that is now a guest house. The church of St. Mary was built in 1863 to a design by Sir Arthur Blomfield, an early work in red, yellow and blue brick with stone dressings. Blomfield re-used some of the tiles from the abandoned Georgian church of 'Old' St. Mary which stood on top of the hill. The graveyard of the old church is said to be haunted by the spirits of the departed.

KEMBERTON *3m. SSW of Shifnal.*
It lies on a hill on a minor road off the A4168 and has long views over the gently rolling cornfields, meadows and copses of this red sandstone country. There are cottages of stucco and red brick, a pub called the Mason's Arms and hidden on a side road a delightful timber-framed cottage of 1675 called 'The Timbers'. There was once a colliery nearby with notable Pithead Baths. It was owned by the Madeley Wood Company who at its peak employed 769 men. It closed in 1967. The mine lay 1¼m. NW of the village and old photographs show it surrounded by cornfields. The church of St. John the Baptist was rebuilt, for the 6th time, in 1882 and the tower was added in 1908. It stands on the highest point in the village and the site is obviously ancient. In the main street there is a large Georgian brick house of 1722 with a columned porch and good views. There is a noticeable gap in the residential development on the other side of the road. All the old tradesmen are now gone and the school of 1857 closed in 1964. It is mostly commuters who now live here. Evelith Mill and cottage lie 1m. NE in the delightful valley of the River Worfe. Also on the River Worfe, ¾m. E of Kemberton, is an old mill site at which the Slaney family of Hatton had an iron furnace. This is known to have been producing in the 1690's and was believed to have been one of the furnaces to use coke instead of charcoal. However, by the 1720's it had reverted to charcoal and by 1790 had ceased to make iron and was grinding corn. (The mill was used to work the bellows of the iron furnace.)

KEMPTON *4m. ENE of Clun.*
A tiny place of stone-built cottages and farms with the occasional dwelling in half-timber. One such is the charming village shop made impossible to miss by the multiplicity of multi-coloured signs. The River Kemp lies in the valley below and beyond that, to the W, is the bulk of Clunton Hall, a part of Walcot Park, in which is one of Clive of India's several houses. One mile N are the impressive entrance gates to Walcot Farm. **Purslow** lies 1½m. S on the B4368. Here is little more than a farm and the substantial Hundred House pub. The tree-clad hill to the NE is Burrow Hill on which is a large, double-ramparted Iron Age hill fort.

KENLEY *2m. SE of Acton Burnell, which is 7m. SSE of Shrewsbury.*
An unspoiled, stone-built hamlet on high ground with superb views in every direction. The church of St. John is long and low and stands on a circular mound embowered in ancient yew trees. The Old Schoolhouse has attractive cast-iron, chapel-like windows. At the end of the 18th Century the rector was Archibald Alison, author of Essays on the Nature and Principles of Taste (1790). He was often visited at Kenley by Thomas Telford and together they spent the time conversing playing bowls. His son, who became Sir Archibald Alison and was an eminent 19th Century historian, was born here in 1792. The name Kenley is from the Old English and means '*Cenna*'s wood, or clearing in a wood'. There are early references to squatters at Kenley and by the end of the 16th Century there were 12 cottages on the Common. Some small stone cottages and irregular enclosures of squatter origin still exist.

KENSTONE *1m. SSW of Marchamley, near Hodnet.*
The first element of the name could be either from the Old English *cempa*, a warrior, or *Cena*, a personal name. Kenstone Hall is one of the Hawkstone Hills. This is an area of outstanding natural beauty famous for its golf club and motorbike trials course. Kenstone lies on the lower slopes of Kenstone Hill, a hamlet of less than half-a-dozen dwellings. Bank Farm is a very untidy and unsightly place littered with piles of rubble, cow slurry, and mounds of black plastic bags. The farm buildings are of stone, the courtyard is cobbled, and the house is of red brick. Along the lane to the S is a new house with a weather-boarded barn; a mellow brick house with a white painted extension; and the modest Manor House which has 3 bays, 2 storeys and brickwork laid using the Flemish bond. By the murky pond a track leads E and then N behind these dwellings to a disussed quarry in which is a brand new house of some size most handsomely positioned. The main lane leads S. and off it an unsigned track leads W. to Crin Cottage (SJ.586.277) which appears to be incorrectly named Daneswell on the Ordnance Survey map. Crin Cottage is delightfully situated at the head of a small, hidden valley that has in the past been quarried. There are some old caravans on the lawns amongst the trees and a tiny swimming pool in front of the cottage. This is a seasonal holiday camp used by 'naturists'. At the bottom of the hill is a very much restored and modernized cottage called Delta Lodge. On the hillside opposite is the Dane Well. It lies behind a wooden door in a brick chamber; one dwelling still uses it but most now have their own boreholes. South of Delta Lodge is the stone-built Daneswell Farm, which has 3 bays with a 1-bay extension to the left. There are geese and horses here beneath the wooded hillsides and black-faced, black-legged Shropshire Sheep graze the meadow opposite. Further S again, just before the cross-roads is the West Midlands Shooting Ground, an area of scrub woodland stocked with game birds which is patrolled by a grumpy Welsh gamekeeper.

KETLEY *1½m. E of Wellington.*
The name is Anglo-Saxon and means 'the wood of the wild cat'. The old, rural village was drawn into the industrial revolution in 1757 when Thomas Goldney, Abraham Darby and Richard Reynolds consorted to build an ironworks on 14 acres of land, S of the Holyhead Rd. and E of Ketley Brook, that they had leased from Earl Gower. They built at least 3 pools, and installed an atmospheric engine that was used to return the water back to the pools after it had fallen and driven the bellows that pumped air into the blast furnaces. By 1806 the ironworks had

Ketley Hall.

The guillotine lock,
Hadley Park, Leegomery.

The castellated windmill tower,
Hadley Park, Leegomery.

expanded and was the second largest in Shropshire, with a worldwide reputation both for its size and the innovations made here. There were 6 furnaces, a forge, a rolling mill making boiler plate, a slitting mill making merchant rods and 2 large pairs of shears. The foundry could make castings of almost any size and could turn and bore cylinders for steam engines. From 1789 there were kilns for coking coal and distilleries for processing the major by-product, coal tar, to produce pitch and oils for varnishes. The following year, 1780, saw Reynolds producing the Heslop Steam engine. Three of these engines were still in use 100 years later at the Madeley Wood Company's coal pits. The railway network of the Coalbrookdale Company reached Ketley in 1757 and by 1788 there was a line to the iron stone mines at Donnington Wood. Until 1767 the rails were of wood and the waggons were horse-drawn. In 1787 William Reynolds (son of Richard) constructed a 1½m. long canal to Oakengates. Between the canal head and the Ketley ironworks was a drop of 73 ft. This was overcome by the use of an Inclined Plane, the first in Great Britain to be used successfully on a canal system. The boats were placed on cradles and moved up and down along 2 sets of rails. The downward, loaded boat, hauled up the empty boat. By 1802–3 the system was so efficient that 24 hauls an hour were possible. It is not known whether the bottom canal stretched all the way to the works or simply acted as a basin. For many years it was thought that the position of the Incline was lost. It is now believed, with some certainty, that it was on the S side of Ketley Hall, falling Westwards. The Incline was out of use by 1818, probably because of the recent closure of the ironworks, and was thereafter abandoned. (The old ironworks were on the site presently occupied by the main Glynwed factory.) In 1835 the canal wharf at Wappenshall opened and by 1837 white limestone for the iron smelting furnaces was brought to Ketley from the Llanymynech and Trevor Rocks quarries on the Welsh border. Evidence of the existence of numerous coal mines around Ketley is plain to see. From the hill at the top of Woodside Road the landscape has the nature of hill country with the grassed-over humps of old spoil mounds making a distincive pattern. After the boom years of the Napoleonic Wars the iron trade in Shropshire went into a decline from which it was never to recover. By the middle of the 19th Century the glory days were over. As the sun set on Shropshire so it arose over the Black Country. But the industry is still at Ketley in the presence of Glynwed and their enormous works. Their most famous product is the Aga-Rayburn stove but all manner of other cast-iron products are made; at the Sinclair works, for example, they manufacture cast iron pipes and fittings. To describe Ketley today we will journey down the Holyhead Road from The Unicorn Inn, which stands at the junction of Shepherd's Lane, to the Ketley Brook Roundabout. Just W of The Unicorn is the Church of God, an old, yellow-painted chapel with a pediment and round-headed windows. The walls to the left of the church are built of glassy green and blue iron furnace slag. Behind the church is a large depression in the process of being back-filled. This is one of the furnace pools of the old ironworks. A little further down is a tall, 2½-storey, rendered dwelling called West Brook House, opposite which are 2 red brick schools. Both schools are embellished with cupolas and narrow round-headed windows, some arranged in a Venetian window pattern. The larger school is dated at 1897, the smaller at 1904. The track between the schools leads through a development of bungalows, across Red Lees Lane and through a short avenue of trees to Ketley Hall. For something that rarely gets a mention in books of this kind it is surprisingly handsome. Parts of the fabric are 400 years old but most of what we see today is in Georgian style. The old 13-bay centre is flanked by 2 newer, slightly projecting wings of 2 bays each. The house is constructed of red brick and stands on the wooded slopes of Red Lake Hill. Just S of the Hall is a modern housing estate. One of the new roads is called The Incline and is, indeed, in the vicinity of the original slope between the 2 old canals. Returning to the main road our eyes are beleaguered by the ugly blue panels and black bricks of the Ketley Recreation Centre. Beyond are playing fields, a swimming pool and squash courts. The junction of the Holyhead Road and Waterloo Road is a busy little place that is, to all intents and purposes, the centre of Ketley. Here is a pub called Twenty's, a handful of shops, a modern brick Methodist church, a tall block of white, concrete-clad flats and a lower block with an unusual and not unattractive scalloped roof. Just S of the crossroads is a business park, at the entrance to which is the uninspired, flat-roofed office building that is home to the Shropshire Star and all those who work behind the blue and green panels that fill the spaces between the windows. Opposite, a housing state lurks behind a bland wooden fence. (Developers really should be made to build proper brick walls when boundaries face public highways.) The main throughfare of the estate is Woodside Road. It leads past a row of shops, the Wren's Nest pub, and the Ketley Town County School and up the hill from the top of which are splendid views to the SW, over the coal spoil heaps already mentioned. Back on the Holyhead Road we pass the modest brick front of the Glynwed Rayburn-Aga factory. Behind the facade are acres of tall, steel-framed sheds that stretch along the line of the dismantled railway (now a public footpath). On the other side of the road are more shops, a service station and a tyre centre and behind them is the huge Glynwed works. Further along, opposite the Horseshoes pub, is the entrance to Glynwed Foundaries Sinclair Works. The pub stands on the old Holyhead Road that descends into the valley of the Ketley Brook. In the early 19th Century Thomas Telford by-passed this section of the road. He made an embankment of slag from the furnaces over the old forges and that is the line taken by the modern road as it approaches Ketley Brook Roundabout. Also on this now abandoned section is the Church of the Seventh Day Adventists housed in an old yellow brick chapel. Behind are the dour houses of a council estate complete with recent barracks-like blocks of flats. What is most striking about Ketley is the almost complete absence of older houses. The 18th and 19th Century ironworks were large and many men worked there, but of their dwellings virtually nothing remains. Ketley, in fact, was notorious for the number of squatter cottages built on waste ground by miners and other workers. In the early 19th Century there were dark and dirty company 'barracks', but by the 1840's these had been replaced by pleasant abodes, many of which had between a 1/6th and a ¼ acre plot of ground for the growing of vegetables and potatoes; most households kept a pig. Most households also kept a bulldog, by all accounts most ferocious beasts. Bull-baiting was an immensely popular 'sport'. It was outlawed in 1835 but 2 years earlier the Duke of Sutherland's agents had taken the law in to their own hands and killed some 500 bulldogs. Cockfighting, though, continued for many years, and some will tell you has not yet ceased. (*See Red Lake.*)

KETLEY BANK *2m. ESE if Wellington.* The 18th Century name was Coalpit Bank. Today Ketley bank is neatly confined within the Queensway to the E, the M54 to the S, Mossey Green Way to the W and the Holyhead Road to the N. Ketley itself lies 1m. to the W and is a separate settlement. Perhaps the building of greatest historical interest at Ketley Bank is Bank House, the home of Richard Hartshorne (d.1733). He was the leading coalmaster of his day in Shropshire with mines in all parts of the coalfield. To contemporaries he was much better known and of a far higher business and social standing than his contemporary Abraham Darby I, to whom he supplied coal, coke and iron ore. The reason for Hartshorne's lack of renown today is that so little is known about him; he is an obscure figure. (Note: He had another house in Wellington, probably the Old Hall.) Another leading industrialist who lived at Ketley Bank was the ironmaster Richard Reynolds, a Quaker born in Bristol, who married Hannah, the daughter of Abraham Darby II in 1757. They set up house here and in 1758 their first son, William, was born. Today much of the old village still stands on top of the hill, and most attractive it is too. If the traveller leaves the Greyhound Roundabout (named after the old pub that faces it) and turns into Marquis Terrace he finds himself heading up Greyhound Hill – a few old brick cottages, modern bungalows, brick and tile-clad houses, scrub covered spoil mounds, a short Victorian terrace and a pub, the Stafford Arms. (The sign depicts a castle, 2 Staffordshire Knots and a lion.) At the top of the hill he should bear right into Main Road, past a Duke of Sutherland Cottage, to a pleasant road junction with trees all about and a grey-rendered Wesleyan Methodist church with round-headed windows and a pediment. Adjoining the red brick Victorian Grange is the red brick Victorian school (now a community centre) and a large playing field. A little further along is a handsome group of red-painted terraced houses with a cluster of old work buildings behind and a striking yellow and brick house beside them. Beyond the high sandstone wall, out of sight, is the tall Methodist church of St. Paul. We are now in the centre of the old village. Here is the Lord Hill pub (spoiled by a tatty front verandah) and many old cottages with some modern infill. Dominating the group is the tall, 2½-storey, cream-stuccoed house, No. 40. Again, the presence of trees gives maturity and character to the area. When West Road appears on the right, turn left and just past a few modern bungalows is the once elegant Bank House, dated at 1721, where lived Richard Hartshorne. It stands in a wooded garden, a substantial red and orange brick house of 2 storeys and 5 bays, with Georgian sash windows and a clumsy right-hand extension of 2 bays. The side is 4 bays wide and the roof was originally hipped – it still is to the left. The back of the house is a disgrace. Various bits have been added but worst of all are the old caravans, boarded fences, and blue-painted sheds of a garage-workshop. Adjacent to all this is a small steel-stockholder's yard with yet more caravans. At the junction turn left into Bank Road, Mossey Green. This new road leads past new houses, coal spoil heaps and a barracks-like modern block of council flats, back to Greyhound Hill and down the road whence we came. On the turnings to the left are older houses; to the right is a council estate. And that is Ketley Bank. We have described it in some detail because old residential areas like this are few and far between in Telford these days. Many were cleared away lock stock and barrel for either redevelopment, like Dark Lane, or by open-cast mining operations, like New Dale.

KING STREET *1m. W of Cross Houses which is 5m. SSE of Shrewsbury on the A458.*
Don't waste too much time trying to find King Street. It is a crossroads on a rise in undulating country. The field hedges have been chamfered at the corners to improve visibility, an unusual courtesy for such a remote little place. By little we mean one cottage. The fields are mostly arable.

KINLET *8m. S of Bridgnorth on the B4363.*
Travelling from the N the first glimpse of Kinlet is its church, a tower protruding from a small wood, the best part of a mile from the road. It looks like a forgotten ruin, but is still the parish church and in regular use. The approach to it is along the track to the Hall, down a slope and then up again. The entrance to the church is through an ivy-clad tunnel. Inside there is an enormous brick wall forming the churchyard boundary to one side, with lesser walls to the other sides. A strange place. The church has a Norman nave with 14th Century transepts and chancel and the whole was sympathetically restored in 1892. Monuments abound to the dead of local families – the Childes, Baldwins and Blounts. Amongst them is the imposing tomb of Sir George Blount, "the Terror of Scotland", (d.1584). Close behind the church is Kinlet Hall, hidden by the trees. The present house was built of red brick by Francis Smith of Warwick in 1727–9. It is now a private educational establishment. In the village there is a Post Office, a village hall, a stone-built school, and a pub. The Eagle and Serpent got its name from the inn sign which depicts the Childe family coat of arms. Why are the Hall and church so far from the village? The answer is landscaping. In the early 18th Century a huge Park was created by the owners of the Hall, the Childe family. The village spoiled the view and so was destroyed. The road was also moved and at a later date a new, small village was built, mainly to house workers on the estate. Not being able to bring themselves to destroy the ancient parish church they hid it from view by planting a small wood. The curious tale is told that in years gone by a Squire Blount (some say it was the great Sir George himself) of Kinlet Hall disapproved of his daughter's marriage. When he died his spirit came back to haunt her by dashing across the table on a war charger whilst she and her guests were seated for dinner. The lady was distraught and called in clergy to reduce his spirit and 'bottle it'. The bottle was then placed in his tomb in the church and she was troubled no more. Such 'bottlings' were not uncommon. The name Kinlet is thought to be derived from Old English *cyne-hliet*, 'royal share', and indeed at the time of the Norman Conquest the manor was owned by the Anglo-Saxon queen, Edith.

KINNERLEY *1m. S of Knockin, which is 3m. W of Ruyton XI Towns.*
Kinnerley has council houses, a mini-middle class suburbia, a Post Office and general store, the Cross Keys pub, and a church built on a mound. St. Mary's has a Perpendicular tower with a Georgian nave, chancel, and apse, all of 1774, by Thomas Farnolls Pritchard. (Pritchard is thought by some authorities to have been the designer of the famous Iron Bridge at Ironbridge.) John Bridgeman, a Bishop of Chester and one of King James' chaplains, is buried in the churchyard. For many years the old font was used as a piece of garden furniture at a house in West Felton. It now stands by the S porch; Lady Ida's Well, a disused Chalybeate Spring, lies hidden close by. At Belan Bank, ½m. S of the village, is a motte and bailey castle. In 1086 the manor of Kinnerley was held by Ernucion and 'one Welshman pays one hawk in revenue'. It was later given by Henry II to his Welsh interpreter, Iorwerth Goch. The name Kinnerley is from the Old English *Cyneheard's-leah*, the first element being a personal name. To the E Kinnerley adjoins **Dovaston**, a pretty little place with a pretty little cottage opposite the gaunt, sandstone-built Royal Oak pub. There is a red brick United Reform church and a riding establishment, which accounts for the plump young ladies on plump little ponies that are a feature of the landscape hereabouts. At **Tir-y-coed**, 1¼m. SSW of Kinnerley on the road to Crosslanes, a 15th or early 16th Century 'long-house' has been identified. These buildings, wherein the farmer's house and his cattle shed are under the same roof, were common in the Middle Ages but very few remain standing.

KINTON *1m. W of Nesscliffe, which is on the A5, 9m. NW of Shrewsbury.*
The lane from Knockin Heath to Kinton passes through some lovely, lush country with the Welsh hills dark in the distance. Kinton is a hamlet with a large farm and a small black and white Manor House of about 1580. Attached to the Manor House is the Tithe Barn, part brick and part weather-boarded and newly converted into dwellings. The name Kinton is Old English; the first element is from a personal name such *Cyneheard*, *Cynehild*, or *Cynehelme*.

KNOCKIN *6m. SE of Oswestry.*
The name is probably from the Welsh *Cyncyn* meaning 'a low rise or hillock'. Knockin is a village with some character that lies along the B4396 in flatish country. There is a Post Office and general store; a pub, the Bradford Arms; a shop that sells venison and trout; a variety of houses, mainly of red brick with some timber-framed cottages; a church; and a castle. The strange little church of St. Mary was founded in the late 12th Century. It has a Norman chancel, nave and N aisle, but was heavily restored in 1847. With its bright red sandstone porch and chancel and yellow brick bellcote it is a remarkable sight. At the back of the church is a small, entrenched stream and beyond this is a wooded mound, the motte of the castle which even in Leland's time was 'Knockin Castel . . . a ruinous thing'. The keepers of the castle hastened the end of the church by stealing the stones to make the churchyard boundary wall. The most striking of the timber-framed houses is Top Farm at the W end of the village. It has a gable to the street with close uprights on the ground floor, cusped lozenges in 3 tiers on the first floor, and more lozenges in 3 tiers in the gable. The windows are clumsy though – iron, with diamond shaped lights. **Knockin Heath** lies 1½m. W of knockin. This is a scattered hamlet and in places most attractive. It is heavily wooded with mature broadleaf trees, flowering horsechestnuts and bluebells in the dells. There is a flourishing Methodist church and a few old cottages. It was at Knockin Heath that Thomas Elkes was hanged for the murder of his ward. He drowned her in a tub of water and then ran away to Hertfordshire. His pursuers are said to have found him by following several ravens who led them to Elkes' hiding place, a haycock. Just S of Knockin, on the road to Kinnerley, is a radio telescope, white, round and gleaming in the sun. There are also many turfed

brick and concrete buildings scattered in the fields about, presumably of 20th Century military origin.

KNOWBURY *4m. E of Ludlow.*
A fairly large, scattered village lying on the slopes of Clee Hill to the S of the A4117, the Ludlow to Cleobury Mortimer road. Limestone quarrying and coalmining with some brick making were the most important industries here. Quarrying still continues on the summit of Clee Hill, 1½m. away, primarily of the hard, dark, basaltic dhu-stone used for making roads. Iron-working has a long history in the area also. Originally the ore was mined here and taken to the Bringewood furnaces (4m. W of Ludlow); later, in the early 19th Century, an ironworks was established at Knowbury. The church of St. Paul was built in 1839 and altered in 1885; it has a German stained glass E window of 1886 by Mayer of Munich and a reredos in the style of the German Primitives. The present vicar has to manage 5 parishes. There is a Post Office and general store and a Memorial Hall of 1960. In recent times the house of the school master of Knowbury School, Mr. Barber, was haunted by the ghost of a child of about 11 years of age. The spirit was clad in grey and cried and laughed and played mischievous tricks. An Express and Star reporter visited the school with Mr. Barber and at midnight sparks flew off the masonry walls. As they walked back to the house they heard sobbing and then laughter. The reporter intended to stay the night but by 2.30 am, having had lights dancing before his eyes and felt a ghostly hand gripping his ankle tightly, he became so ill he had to be taken away and spent the rest of the night in an hotel. Knowbury House is the home of Major Adrian Coles. In 1982 he founded the British Hedgehog Preservation Society. The major work of the society is to encourage landowners to install escape ramps in the pits below cattle grids. Thousands of small animals fall into these traps every year and die of starvation. The name Knowbury is probably from the Old English *cnoll-burh*, 'a fortified settlement on a small hill' (a knoll). Just to the E of Knowbury is Colleybrook Green where there are the ruins of a brick works. At Angelbank, on the A4117, is a petrol station run by a most likeable Polish gentleman.

KYNNERSLEY *6m. WSW of Newport.*
Most easily approached off the B5062 Newport to Crudgington and Shrewsbury road. There are several lanes leading to the village but one of the most interesting is that which leaves the B5062 ½m. E of Crudgington crossroads. This crosses the northern Weald Moors. Its foundations are supported on large wooden slats, but even so the surface pitches up and down. This road was built when the lakes and marshes of the moors were drained for the Duke of Sutherland by Thomas Telford in the late 18th Century and early 19th Century. Previous attempts had been made to drain the area but it had remained a wild morass. Even today much of the soil is almost pure peat. There is a complicated system of ditches, or strines, connected to the straightened and deepened rivers and there are large, fertile fields with belts of trees planted to give protection to the crops and the cattle from the wind. About ½m. down the road from its junction with the B062 is a crossroads. A lane cuts across the main track and in this lane, known locally as Crudgington Green, are about 15 estate workers' cottages, which used to be occupied by the men who kept the ditches and rivers clear. They also had a small piece of land, a smallholding, to cultivate and to keep a cow and a few chickens on, which helped them survive the winter when work on the land was impossible. In one of these cottages today lives Mr. Frank Teece. He has turned the front garden of his little cottage into a summer wonderland of colour, with a display of flowers that attracts visitors from all over the world. The village of Kynnersley lies on a slight rise. It is an old village and still has some timber-framed buildings. However, with the new prosperity that came with the extra productivity of the farms, made possible by the drainage schemes, it was largely rebuilt in red brick. This happened to most of the villages on the Weald Moors. In the village today there are some 'model' Victorian houses, some black and white cottages, a farm or two and the church of St. Chad, standing on the raised platform of its prehistoric site. The present church is largely of the 13th and 14th Centuries, with a tower of 1722 and some later rebuilding and restoration. It has an unusual double bellcote. Local tradition has it that the orchard of Whym Cottage (1559) is a burial ground wherein lie those unfortunates who were hanged from the tree on the Whym, often for comparatively petty crimes such as sheep stealing. (The Whym is a raised triangular piece of ground near the church.) Alleged offenders were first tried in the Court Room, a room in the Manor House. The shop, the Post Office and the school, like the sheep stealers, are no more. Nevertheless, Kynnersley is still an attractive, working village, surrounded in all directions by open farmland mainly laid to pasture. The name Kynnersley probably means '*Cyneheard*'s island', a raised area of dry land surrounded by swamps. (*See Lubstree.*)

KYNNERSLEY WALL *1m. NE of Kynnersley.*
There is only one farm and a pair of cottages here. Kynnersley Wall Farm (SJ.681.178) lies beside a low sandstone hill which is surrounded by ancient earthworks that form an irregular circle about 330 yds. in diameter. Except at the NE, which adjoins slightly higher and drier ground, the fortress settlement was surrounded by marshes. These have now been drained and, indeed, a deep drainage ditch cuts through the NE earthworks. (This is actually a natural stream which has been dug out and lowered artificially, probaly in the 18th Century when the Weald Moors were drained by the Duke of Sutherland.) Another ditch, or strine, abounds the site to the SE. The land beyond the enclosure is dark and peaty. The main defence of the fort consists of an earthen bank, 45 feet wide at the base, which today is at its highest in the SE corner, where it measures 6 ft. from inside the enclosure and 8 ft. outside, into the ditch. This was almost certainly originally higher and would have been topped by a wooden palisade to form a most substantial barrier. At the SE corner there are the remains of at least one other bank and at the NE corner a 'platform' juts out for about 65 ft. to connect with another separate bank and ditch. The modern road cuts through the site and the earthworks can be seen quite clearly on the E side from the road itself. On the W side the land has been used for growing crops and the banks and ditches have almost been ploughed out, though undulations in the fields reveal their position. Similarly, the far eastern side is in the process of disappearing. The fort was almost certainly occupied by Iron Age Celts and may even pre-date them. There is higher land to both N and S where crops could be grown and cattle could be grazed. Within the enclosure, or stockade, there would have been plenty of room for the stock in times of trouble. The fort itself was well protected, first by the marshes and then the earthworks. The main entrance was probably in the NE but there might well have been another entrance in the SW. There must have been many such lowland fortress-settlements in the Midlands but very few have survived as well as this. Most have either been completely ploughed out or built upon. At Kynnersley Wall (the 'wall' is the high SE embankment) there have been several 'finds'. A flint arrowhead was discovered just N of the ditches that lie over the road N of the farm; a Celtic bead was found in the peat about 200 yds. S of the farm; a number of large rocks, in an area where they would not have naturally occurred, were found just beyond the embankment S of the farm; a mysterious stone in the shape of an elongated pear drop with a hole at the broad end that could be either a Celtic ornament or a medieval knife sharpener; and 2 areas where fires were made, denoted by dark earth and charcoal fragments, together with numerous pebbles located just beyond the embankment ESE of the farm. (Early man heated water by first heating stones in a fire and then placing them in an earthenware pot containing water. If the pots of water had been placed directly on the fire they would have cracked.) A trial trench was cut through the bank in the SE corner near the road. It was shown to have been made and re-made at least 4 times at 4 different periods. Most interestingly, in the process of making a cess-pit lagoon measuring 100 ft. by 30 ft. at the back of the modern cowsheds of the farm in 1984, post holes were discovered. These marked the position of circular wooden huts together with post holes running in straight lines that remain unexplained. In the stream that runs by Wall Cottages, 200 yds. S of the farm, Roman coins have been found by local workmen. Kynnersley Wall Farm is constructed of red brick. The garage is a

Longford Hall, Newport.

Longford Mill Farm, Newport.

The Bull Ring, Loppington.

The aqueduct, Longdon-upon-Tern.

converted sandstone building which almost certainly formed part of the previous farm. There are also lengths of stone wall about the homestead. Many of the farms on the Weald Moors were enlarged and rebuilt in brick when the marshes were drained and a new affluence came to prevail. Wall Farmhouse is said to be haunted by a cloaked woman, and the wife of the farmer, Mr. Dobson, has heard unexplained footsteps late at night. Behind the farmhouse is a small quarry, 'the rock hole', from which, presumably, the stone for the earlier buildings were taken. In the 18th Century it is believed that this was worked again to provide stone for the Duke's Drive bridge, ¾m. SSE, over the Newport Canal (now disused and dry). A human skeleton was found in the quarry and re-interred by the previous owner of the farm. The horses and dogs of the present farmer refuse to enter this compound, although my dog had no qualms. Note: The highly ornate Dukes Drive bridge has now been dismantled. (*See Lubstree.*)

LACON *1½m. NE of Wem on the B5065.*
There is no centre to Lacon; it straggles along the main road and has outlying farms 1½m. to the N – Lacon Farm and Upper Lacon Farm. Lacon Hall is an irregular and somewhat ungainly house clad in white stucco and distinguished by a having a very large external chimney breast at the front, left of the entrance door: a most peculiar arrangement. The land is very flat, the soil dark and peaty and the fields are both laid to pasture and used for growing arable crops. There are a surprising number of small ponies hereabouts. The cottages and farms are spaced out along the road; the best house is Aston Grange, white with a hipped roof, tall windows, a columned porch and yellow chimney pots. The area is rather spoiled by the Caravan Park and the rows of wartime Nissen huts which are now the nucleus of the Wem Industrial Estate. There is a general untidiness barely compensated for by the goats and patches of natural scrubland. Soulton Hall lies ½m. E. (*See Soulton.*)

LANGLEY *1½m. S of Acton Burnell, which is 7m. SSW of Shrewsbury.*
Here is a small, rectangular chapel of 1564 with a stone-tiled roof and a weatherboarded belfry now in the care of English Heritage. It is noted for its early 17th Century Puritan layout in which the reading pew is placed so that the priest must face the congregation all the time, whilst those taking communion kneel all around the altar. The furnishings are early 17th Century, complete with a musicians' pew at the rear. It is a simple place but one with great character. Close by is a farm into which are incorporated fragments of the old Langley Hall, the home of the Lee family. The gatehouse still stands with stone to the front and timber-framing to the rear. There was a castellated wall running from the Gatehouse to the road but this has been dismantled. The Gatehouse is in disrepair and covered in scaffolding. Repairs are planned but work is proceeding extremely slowly. Behind the Gatehouse are old Nissen huts and beyond them is the now dry basin of a medieval fish pool. The old, timber-framed Hall was pulled down in the 1790's. It stood to the left of the present house, in what is now the kitchen garden. Behind the house are depressions in the ground, earth banks and mounds. The countryside hereabouts is both beautiful and peaceful. The name Langley means either 'long wood' or 'long clearing in a wood'. The hamlet of **Ruckley** lies ¼m. W of Langley. In her Shropshire Folklore Charlotte Burne writes: 'By the side of the Roman road between Ruckley and Acton Burnell yclept(?) the Devil's Causeway, and half-way down the Causeway Bank, there rises out of a ferny, flowery bank a most beautiful spring which drips into a rocky basin, partly natural, partly formed of great grey slabs of stone placed there by the hand of man'. Here there are said to live 3 frogs, always seen together. These are imps transformed. Sitting alone and seldom seen is a large frog, Satan himself.

LAWLEY *1½m. SE of Wellington.*
A small industrial village with a shop, a school, a Victorian parish church of 1865 and a large common on high ground. There was formerly much mining of coal here and recently great opencast pits have ravaged the land. In 1987 the most recent of these gobbled up the historically important hamlet of New Dale, a model village of back to back houses built by the Darbys of Coalbrookdale along with a furnace, a foundry, a forge and a Quaker Meeting House. It is said to have been the earliest housing development of its kind in England. The name Lawley is from the old English *Lafa's-leah. Lafa* was a personal name. The last Anglo-Saxon to hold the manor was Erngeat. By 1088 it had passed to William Pandolf, one of Roger de Montgomery's 'leading men'. However, the lands here were waste.

LAWTON *1m. SSE of Diddlebury.*
It lies at the end of a track by the River Corve. Several footpaths meet at the bridge over the river. Lawton was once much larger than the hamlet we see today. A number of house platforms (caused by the gradual collapse of houses) line the old main street, now a hollow way, that runs through an orchard to the old village pond. By the bridge are what are believed to be the remains of a 14th Century mill. The name Lawton is from *hlaw*, a 'hill' and *tun*, a 'settlement or homestead'.

LEA *2m. NNE of Pontesbury.*
A tiny hamlet off the A488 Shrewsbury to Bishop's Castle road. It used to belong to the manor of Ford and in 1086 was part of the personal estate of Roger de Montgomery. The access lane from the main road crosses the Rea Brook (pronounced 'Ray') by a steel girder bridge with lattice walls which was constructed in about 1901. Before that there was a ford. There are a few modern red brick houses; the old Manor House, with a modern facade but original black and white work to the left gable and the whole of the rear; Lea House Farm, a substantial house with substantial outbuildings; and the most attractive South House, a new brick dwelling in Georgian style of 3 bays and 2 storeys with a one-bay extension to the left and above the doorway a segmental window within a broken pediment supported by pilasters. There are wide views from here over attractive country to Pontesford Hill. The name Lea could be from either the Old English *leah*, 'a forest, or a clearing in a forest'; or from the Celtic, *Lleu*, a Welsh god whose name is in turn derived from the root *lug*, meaning 'light'.

LEA (CASTLE) *2m. E of Bishop's Castle.*
A tiny hamlet on the lower slopes of the 'toe of the Long Mynd', the SW part of the great mountain that was cut off by the gorge of the River West Onny at Plowden during the last Ice Age. The interest at Lea is in the ruined Tower-House sometimes called Lea Castle. A modern farmhouse adjoins it and set above the porch of this is a wooden shield dated 1560 taken from a previous house. The tower is older than that, though, and was probably erected in the early 13th Century. It measures some 30ft. by 50ft. with walls about 6ft. thick. There are a few surviving details in the S wall: a late window in the basement, and in the hall a small window, a row of 4 corbels, and a doorway with a portcullis groove and drawbar slot. The original tower must have been at least 3 storeys high for the portcullis to have been operated. In 1645 the tower was held by the Royalists but fell to the Parliamentary forces. The garrison was taken prisoner and removed to the Red Castle at Weston-under-Redcastle near Hodnet.

LEA CROSS *6m. SW of Shrewsbury.*
The cross is the junction of the A488 and the minor road from Arscott to Lea. It is a tiny place and yet has a Post Office and general store, a pub and a church. The church was built in 1888 and is dedicated to St. Anne in memory of the daughter of the Rev. Hawkes. He had quarrelled with the vicar at Pontesbury and the church at Lea Cross was built as a gesture of defiance. It has never been consecrated and weddings and internments are not allowed. **Cruckton** lies 1⅔m. NE of Lea Cross. It is well known for the annual ploughing competition held here. Close by is **Cruckmeole**, on the Meole Brook, where there is a school (1969) and a fine old black and white Hall (1588). There also used to be a brickworks, the kilns being fired with coal from the mine at Hanwood, ½m. W. The 'Cruck' element of the names Cruckton and Cruckmeole is from the ancient British *cruc* meaning 'a hill'. Meole is from *moel*, Welsh for 'bare'. (*See Lea.*)

LEATON *4m. NNW of Shrewsbury.*
A tiny hamlet on the B5067 with a striking church by S. Poutney Smith. Holy Trinity was built in 1859 and the tower and N aisle were added in 1872. The gabled Old Vicarage stands adjacent. It is built of red brick with sandstone dressings to a Gothic design and has 7 bedrooms, a coach house, stables, dog pens, a wood, an

orchard and several acres of paddocks. North of the church is a brick house of 1683 with diagonally placed chimney stacks. To the E, by a ¼m., is the main line railway from Shrewsbury to Gobowen; to the W is the River Severn. The name Leaton could be from either the Old English *hleo-tun*, a settlement that grew around 'a shelter', or *laet-tun*, a settlement by 'a water-course'.

LEEBOTWOOD *3½m. NNE of Church Stretton on the A49 Shrewsbury to Ludlow road.*
A small village in 2 parts. The N part adjoins and faces the main road pub, the thatched Pound Inn (1650); and the S part lies off the main road where there are some pleasant houses in a mixture of styles. The church of St. Mary lies ½m. away on the road to Woolstaston, the turning for which is opposite the Pound Inn. It is medieval with nave and chancel in one and a Georgian tower. Inside there is a good hanging monument to Sir Uvedale Corbett (d.1701) and there are iron hat pegs in some of the box pews. The churchyard is essentially round and the wall is built on an encircling bank, both signs of a pre-Christian site. Why the church lies so far from the village is a mystery. There are no signs or records of the village having once been there and then moved, but this is nevertheless the most likely explanation. It has been half-heartedly sugested that the ground on the hill was firmer and better suited to take the weight of a substantial building than the alluvial soils where the village stands. As late as the 12th Century the area around the present village was described as 'waste', and in 1199 the Knights Templar of Lydley assarted (i.e. cleared) 40 acres of Botwood Forest. Haughmond Abbey had a large manor called Boveria, which contained Leebotwood. The Abbey established granges – farms on the border of cultivation – and by the end of the Middle Ages most of the area had been put to agricultural use. In later times the Pound Inn was well known because it was on one of the most important Welsh cattle drovers' routes. This route started at Montgomery, went through Bishop's Castle and E to Plowden, over the Long Mynd, along the prehistoric Portway, down hill to Leebotwood and thence to Shrewsbury.

LEE BROCKHURST *3m. SE of Wem.*
In the past the village was simply called Brockhurst, and occasionally Brockhurst Castle, although its early, Anglo-Saxon name was Lee (Lege in Doomesday Book). After the Norman Conquest Roger de Montgomery gave the manor to one of his huntsmen, Norman Venator, who in turn gave a part of it to Shrewsbury Abbey. The village lies just off the busy A49, Shrewsbury to Whitchurch road, at the point where it crosses the River Roden. By the bridge (of 1800 to a design by Telford) there is a large red brick Georgian house, a telephone kiosk, the kennels of the North Shropshire Hunt and a War Memorial, The rest of the village lies 400yds. down the road to Wem. Here are a few cottages and farms and the church of St. Peter. The church has a Norman nave and a chancel and bellcote of 1884. Across the road from the church, and to the right of a cottage called Vernon House, is a low mound upon which stands a large, modern, steel and plastic barn. This is the site of the much reduced motte of a castle though local people have always thought it to be a prehistoric burial mound. The authors of the Victorian County History are in no doubt, however, that it is a castle mound. Opposite the church is Lee Hall Farm. Under the pebble-dash rendering is a timber-framed house of about 1600. There are several other timber-framed buildings, including Manor Farm. The road westwards crosses the River Roden at Thistleford Bridge, a pleasant, wooded spot with cows grazing the waterside meadows. It then makes a sharp turn and climbs up Hilcop Bank to Aston, which is to all intents and purposes now a suburb of Wem. But to return to Lee Brockhurst; at the development near the A49 bridge there is a narrow unmade lane running from beside the War Memorial to the N, parallel to the main road. It runs for ¾m. before rejoining the A49. This is a section of the old Roman road that was by-passed when the new highway was constructed. The track is a little rough in places but can still be negotiated by a modern car. It runs uphill through a sandstone cutting, past 2 cottages and through a delightful oak-wood to a timber yard where there is a striking cedar-boarded house. From here the track is metalled and after passing a few modern houses joins the A49. There is is fascination in travelling such old, forgotten roads. Along this narrow lane the Romans marched followed centuries later by the Anglo-Saxons and the Normans. Stage coaches thundered by here. For centuries it remained virtually unchanged and then was fossilized when it was by-passed by the new road on lower ground to the E. The name Lee Brockhurst is from *Leye under Brockhurst*, probably meaning 'the clearing, or forest glade, by Brockhurst'. Brockhurst can be interpreted in several ways: 'badger-wood', 'badger-hill', 'stream by a wood', 'stream by a hill' etc. *Broc* is Old English for badger and *brocc* is 'a stream'. *Hyrst* can be either 'a wood', or 'a wooded hill'. The Brockhurst of the name could therefore be a local physical description, or might possibly be nearby Preston Brockhurst 'a clearing in the forest near (Preston) Brockhurst', for example.

LEEGOMERY *1m. ENE of Wellington.*
In Domesday Book Leegomery is called Lega. It was held both before and after the Conquest by Thored, though under the Normans the overlordship passed to Reginald the Sheriff. In 1086 there were 5 slaves, 2 villagers and 4 riders. The name Leegomery is derived from the old English *leah*, 'a clearing in the forest', and Cambrai. Alfred de Cambrai was a Norman Lord of the manor in about 1200. Today Leegomery is a modern, urban village on the northern outskirts of Telford New Town. Several small estates of red brick houses have now merged but not unpleasantly. There are open spaces, a pool, a children's recreation area, a County Junior School and a Community Centre with a Spar store, a video hire shop, a coffee bar, a doner kebab take-away and a pub, the Thomas Telford. The social centre is housed in an old farmhouse and there are a welcome number of trees in the area. The eastern boundary of Leegomery is Hadley Park Road. Here are older houses, including some Victorian dwellings. About half-way along, on the W side of the road, is a footpath (part of the Silkin Way) that leads past a pony paddock, over a footbridge (that crosses the Harley Brook now flowing through a 12ft. deep drainage ditch) to a lightly wooded area and the Leegomery Mill and mill cottages. The old, dark brick mill was grinding corn until 1945. At the rear is a willow-fringed pool. This is now dry but once fed a waterwheel, though the mill could also be powered by steam. Much of the internal machinery is still in place. In 1978 the building was damaged by a fire, but is now in the process of being restored and converted into a craft shop with living accommodation. To the E of Leegomery are the huge Hadley Castle Works (where automotive body parts are made) and to the W is the huge new Telford Hospital. At the time of writing this has been constructed but is not yet open. It is a rather grim looking place of yellow brick beneath a red pantile roof that is locally already known as 'the prison'. Near the main entrance is the new Mytton Priory estate of attractive yellow brick houses and bungalows. The estate sign is very well done indeed. West of the hospital is **Apley** Park. Here are delightful woods with all manner of trees and bushes including a surprising number of yews. An inpressive avenue of limes strides across open fields from the Park to the Leegomery Centre. The fields will not be open for long though; a prestige development of superior houses is planned to take advantage of this noble avenue. In fact the Park was once the garden of Apley Castle. The first house on this site was a stone mansion fortified with battlements by Alan de Charlton in the early 14th Century. This was replaced in the 17th Century and in 1643 it was garrisoned for the king. It fell to Cromwell's men and was dismantled. This in turn was replaced in the late 18th Century by a large, red brick house designed by J. H. Haycock, which, in turn, was demolished in 1956. It was for long the home of the Charlton family. They owned lands at Wytheford, where they had an iron furnace–forge; in Wombridge where they had coalmines; in Wellington where they had a brickworks; in Wrockwardine Wood where they had ironstone mines; and in Preston-on-the Weald Moors where they had a saltworks. This was in addition to their agricultural estates and extensive woodlands. The platform on which Apley Castle stood has been landscaped and sets of steps lead through a charming fleur-de-lis path and lawn design set amongst yew trees. Parts of the foundations of the house are exposed and with the richly coloured

woods all around this is a delightful spot. The extensive collection of parkland flora is extended by the wide variety of hedgerow plants that line the old Wappenshall to Wellington bridle path, which cuts across the lime tree avenue. To the W are the stable, extensive brick-walled gardens, workshops, and the laundry. These are sadly neglected but still standing. Pride of place must be taken by the stable block, or rather the substantial 17th Century stone-built stable cum lodging house that was incorporated into it. This has mullioned and transomed windows and and re-set arched doorways taken from the original castle. Roofless and derelict it may be but the walls, some 2½ft. thick, still stand straight and strong. To the right there is a long depression that could have been either a pond or part of a moat. Apley Castle is not way-marked but there is a garden centre in the large walled garden called Watkin's Nurseries, and this is signposted off Whitchurch Road N of Apley Castle roundabout. It can also be approached off the A442 between the Leegomery Roundabout and the Shawbirch Roundabout – look out for the gap in the low sandstone wall; the rough dirt track will take you straight to the stables. There are 3 pools in the park, the largest being Apley Pool which adjoins the A442 near the Maxell UK Ltd. factory. Apley Park is owned by the Telford Development Corporation and is a public area; the Silkin Way footpath skirts the edge of the woods. The eastern boundary of Leegomery is Hadley Park Road; at the S end of this street is the Guru Nanak Sikh Temple (housed in a redundant school) and at the N end is the Leegomery Methodist church. Near the church is the Malt Shovel pub and almost opposite the pub is the well signposted entrance to Hadley Park Farm, in the area now known as Hadley Castle. There are a number of interesting things to see here. (*See Hadley Castle*.)

LEIGH *3m. WSW of Minsterley on the B4499.*
A tiny, wooded, place on the edge of the plain of the Rea Brook. It lies at the bottom of a steep valley cut into Whitsburn Hill by a tributary stream. Leigh was almost certainly one of the 13 outliers of Worthen at the time of Domesday Book and this belonged to Roger Corbet. In 1526 Sir John Corbet was described as 'of Lee' and in 1585 his family built a new fortified house here. In 1644 this was garrisoned by Sir Pelham Corbet and held for the King. In 1645 it was captured and burned by Cromwell's men, though it was subsequently repaired and re-occupied. The moat survives, except for the S corner, and still holds water. Inside, a 3ft. thick wall enclosed an area some 53 yds. by 26 yds. All that survives of this is a 2ft. high fragment in the SW corner and some buried foundations in the NW. The gateway was also in the NW and was approached by a bridge, later replaced by a causeway. The name Leigh is from the Old English *leage*, the datative form of *leah*, meaning either 'a forest' or 'a clearing in a forest'.

LEIGHTON *4m. WNW of Ironbridge on the B4380.*
It lies within 300 yds. of the River Severn, a small, pretty vilage with trees seemingly everywhere. The Kynnersley Arms was formerly a mill, and near to it once stood an iron furnace. Iron was smelted and coal mined here from at least the 16th Century. The Royalist armies at Shrewsury and Oxford used canon-balls and musket shot made at Leighton. Today, it is a peaceful, country place. There are clipped yews and box trees, and small 19th Century cottages with leaded lights. The red brick Hall, of 1778, was the childhood home of novelist, Mary Webb. Virtually in the garden of the Hall is the church of St. Mary which was rebuilt in 1714 using, in part, the old medieval masonry. In the church are monuments to the Leighton and Kynnersley families, and in the churchyard the cast-iron tomb of Cornelius Reynolds, d.1828, can be seen. The name Leighton could be from the Old English for either *leac-tun*, meaning 'the settlement where leeks are grown', or *lyhtedon*, 'the bright, light hill'.

LIGHTMOOR *1¼m. N of Ironbridge.*
This is one of the few remaining hilly landscapes of pit mounds and waste tips that have been left under their naturally regenerated cover of scrub woodland. Lightmoor lies just N of the Woodside (Madeley) housing estate and stands around the lanes of Cherry Tree Hill, Lightmoor Road and Brick Kiln Bank. Coalbrookdale, with which Lightmoor was closely associated, lies ½m. SW. The top soils at Lightmoor are glacial sands and gravels but below lie Carboniferous rocks. However, the Lightmoor Fault has resulted in the coal seams to the E of the fault-line being much deeper than those to the W. Coal had been mined in the area from at least the early 18th Century and by 1755 a forge is known to have been operating hereabouts. In 1758 the Lightmoor Furnaces (SJ.681.053) were erected, a 3-acre pool was excavated and a pumping engine (to return the water to the pool) was installed. This pool used to be where the Pioneer Cement Works and Q.A. Kitchens stand. By 1790 the works was capable of manufacturing engines, which means engineering skills had been developed. By 1796 there were 3 blast furnaces at Lightmoor and the works were run by the Homfray family. By 1839 2 rows of cottages for the workers had been built, Coker's Row (5 houses) and Pool Row (6 houses). Adjacent to the furnace works there were coal, iron-ore and clay pits. From 1839 the works were operated by the Coalbrookdale Company. They closed in 1878 (not 1883). The other major industry at Lightmoor was, and still is, brick making. The local clay was suitable for fire bricks, a valuable commodity. There were 3 brick and tile works close to each other. The earliest was the Cherry Tree Hill works of about 1767 which closed in 1905. Shutfield Tileries opened about 1825 and closed in 1950. Lightmoor Brickworks opened before 1779 and was taken over by the Coalbrookdale Company who, in 1862, were making a wide range of products: 'Bricks, Tiles and Ornamental Works in terracotta, consisting of Garden Vases, Flower Pots etc. etc.' The Coalbrookdale Company operated the brickworks until 1935. In 1950 the site was taken over by Coalmoor Refractories who rebuilt the works to produce refractory brick linings. In 1984 it was taken over by Ibstock Ltd. and modernised once more. They make only one product: a hard, blue facing brick. Anyone exploring Lightwood today is faced with the problem of old maps not showing new roads. During the last few years a new road has been built that slices through the heart of the area, and a most attractive road it is with hills, woods and meadows on both sides for its entire length. It runs from Jiggers Roundabout in the W (on the A4169) to Castlefields Roundabout in the E (on the B4373). A journey along it will let us see most of what Lightwood has to offer. We will start at Jigger's Bank. (This is sometimes caled Gighouse Bank, and was the old packhorse way out of Coalbrookdale. It was later used by the railway incline that was dismantled in 1802. The road, which runs from Coalbrookdale to Wellington, was constructed in 1818.) But we digress. The new road leads off the roundabout and up a slight rise. To the left is a council rubbish dump and to the right is the entrance to Crackshall Lane (the name is said to be from cracked-shell; the moulding casts from the foundry in Coalbrookdale which were used as road metals). This delightful, wooded track runs parallel with the new road for ¾m. or so. There are several cottages, some of stone and some of brick. One or two are derelict but are about to be renovated. This is typical squatter country. Coalminers and other workers built themselves simple little houses on waste land. The landowners usually turned a blind eye; after all, their labourers had to live somewhere and even when the ironmasters and coalmasters did build dwellings they never built anything like enough. The lane continues on and crosses the course of an old canal. At the point where the lane rejoins the new road there is another track to the right. This leads to the old incline winding engine house and an old uncapped mine shaft. Many deep shafts have still to be either filled or capped with concrete. This is one such and lethal it looks with only a timber cover, part of which is missing. The new road now descends, winding through woods and meadows. Some people will find it hard to believe that almost all this landscape is artificial, that most of the hills are waste spoil mounds. To the right Cherry Tree Hill joins the new road. This leads downhill to Coalbrookdale. There is a parking place half way down the hill and walks in the woods of Oilhouse Coppice. Back on the new road the next junction on the right is Brick Kiln Lane. There is no need to go down this road to see the brickworks; they can be seen quite clearly a little further down from the new road. Opposite the entrance to Brick Kiln Lane, over the new road and over the old Lightmoor road that runs parallel with it, is the site of the old Shutfield brickworks

A squatter's cottage on A518 near Lilleshall.

The limestone cliffs of Llanymynech.

St. Michael's, Llanyblodwell.

Longdon-upon-Tern, the Hall.

and tileries – a large shallow basin surrounded by trees with foundations of some brick buildings and some concrete pads. The new road now passes the brickworks of Ibstock Brick, Telford Ltd. The chimneys stand all of a row. Opposite the main entrance to the brick storage yard, on the other side of the road, are the ruins of part of the long abandoned Lightwood Furnaces. Much of the brickwork is camouflaged by undergrowth but a 5ft. high 'in situ' section stands proud and is easily visible from the road – it is only a few yards from the highway, against the bank (this piece of brickwork is probably part of a retaining wall, not the furnace itself). In the yard of Ibstock's works there are several areas where blue-green-black furnace slag can be seen both in heaps and used as walling by the Brick Kiln Lane entrance to the offices. If we now backtrack a short distance we can turn off the new road, down a short access link and into Lightmoor Road. Facing us is the sign of Q.A. Kitchens. This company and Pioneer, a ready mixed concrete works, share a large depression. This was the main furnace pool for the Lightmoor Furnaces. The lane to the left at the Q.A. Kitchens sign is Lightmoor Road. Turn into this and a short distance further on is the track of Burroughs Lane. This is a meandering path that leads into an area well known for its old squatter communities, all of which have now been detroyed. One of the cottages was dismantled and re-erected at the Blists Hill Open Air Museum. It dates from about 1830, is built of rough sandstone, measures 20ft. by 12ft., is single storey with 2 rooms and had an earth floor. In the 1860's a man and his wife and 7 children aged between 5 and 24 lived there! It was inhabited until the 1930's. Burrough's Bank connects with **Holywell Lane** where there was a squatter settlement of some size. In 1772 there were 6 cottages built on the roadside verges. By 1825 there were 26 dwellings and by 1882 they numbered 32. The ground then belonged to Lord Craven, an absentee landlord, who turned a blind eye. Not so the Telford Development Corporation who destroyed the village in the 1970's. These squatter settlements were in fact socially more stable than the company built 'rows'. When a squatter changed jobs he kept the same house; when a company house dweller (usually a more skilled man) changed company he changed house also. But to return to Lightmoor Road at Q.A.Kitchens. Up the hill, past the entrance to Pioneer, is a modern 'squatters village'. That is not true, of course; the people here own their land, but it has a most engaging gipsy-like air. It is called Leasowe Green, Lightmoor Village. At the entrance is a derelict brick cottage. Beyond, past the cedarwood shed in a woodland clearing, is a self-build scheme. Modern, timber-framed houses of some size and individuality encircle an oval shaped green. The owners take absolutely ages to complete their houses and in the meantime live in old caravans. It is a very higgledy piggledy place but with much honest charm. Up the hill again, over the railway now long gone, and the large scrapyard of Telford Motor Spares is to be seen lurking amongst the trees to the left. The new houses at the road junction ahead belong to Little Dawley. To complete the journey the traveller should return down Lightmoor Road, turn left in the new road and continue to the Castlefields Roundabout. The mine head scaffold is not genuine; it is if you like, a modern folly.

LILLESHALL *2½m. SW of Newport.* The village is by-passed to the W by the A518, Newport to Wellington road. It is an ancient place and it is thought that there was a Saxon church on *Lill*'s Hill from at least 670, though the present church of St. Michael is Norman with later medieval alterations. These days the name is familiar to most people because the Hall was chosen to become the National Sports Centre. Lilleshall lies on the NE boundary of the Shropshire coalfield. A charcoal blast furnace for smelting iron was built within the parish in 1562, one of the earliest in the county. The Lilleshall Company, formed in 1802, was centred at Oakengates where the last blast furnace to operate on the Shropshire coalfield, 'blew out', as they say, in 1959. The first canal in the county ran from Donnington Wood to Pave Lane, 2m. SE of Newport on the A41. It was constructed in 1765–68 to move materials and goods around the Lilleshall estate. Though long abandoned its course can still be traced as far as the Duke's Drive to Lilleshall Hall. (This entrance drive was made in the 19th Century and cuts through the head of the old canal.) On the northern fringe of the village is a wooded area containing the old limestone quarries and ruined kilns which belonged to Earl Gower. (Note: Earl Gower became Lord Stafford and his son became the Duke of Sutherland.) He built a series of canals to connect with the Donnington Wood Canal for the purpose of bringing coal to his lime works and for carrying away the lime and limestone. The levels between the canals were different, so an Inclined Plane was built at Hugh's Bridge. It stretches for about 100 yards and has a fairly easy gradient. At first the goods in the boats in the lower canal were unloaded, packed into panniers and carried up the slope to the top canal by horses. Then a tunnel was dug at the level of the bottom canal to a point below the upper canal and the boats were floated along the tunnel. The goods were already on pallets and these were hauled up a vertical shaft to the top canal. The third and final stage of development was the Inclined Plane proper of 1796. The tunnel and shaft were blocked. A track was laid on the slope between the 2 canals and the boats themselves were hauled from lower to upper canal. There are 3 buildings at the site of the Lilleshall Inclined Plane. Two are houses. They were originally the cottages of the men who operated the system, but have now been modernised and extended. The third building is a stable in its original state, home to the horses that worked on the Incline. The Incline itself is quite evident. The lower canal has water in it still, but the upper, the Donnington Wood – Pave Lane canal, is dry and totally overgrown with tall trees and dense undergrowth. The Inclined Plane is situated at Ordnance Survey map reference SJ739.152. Somewhat surprisingly it is not shown on the one inch Ordnance Survey map. The canals are shown but, confusingly, look as if they are one. On a rocky hill close by the village is a 70 ft. obelisk, raised to the memory of George Granville Leveson-Gower Duke of Sutherland (d.1833). Lilleshall Abbey lies ½m. SE of the village. It stands by the disused Donnington Wood – Pave Lane Canal, already mentioned, and old monastic fishponds. This 12th Century Norman abbey was one of the finest in the county. The Abbey church was 207 ft. long and has a magnificent W door. The Abbot owned large and widespread estates; at Atcham, for example, he kept 2 ferry boats and later built a bridge over the River Severn and charged tolls for crossing it. After the Dissolution of the Monasteries Lilleshall Abbey was bought by the Leveson family but it was badly damaged during the Civil War. Much of the masonry was used to build Lilleshall Grange and also to construct canal bridges; what is left is now in the care of English Heritage. Lilleshall Abbey has a locally well-known ghost, a monk who was seen by a custodian of the ruin as he, the monk, kneeled in the area once occupied by the altar and who spoke of 'the secret of the Abbey' without disclosing just what this might be. Strange noises have frequently been heard in the early evening. Lilleshall Hall lies about 1½m. from the village. There is a direct road from village to Hall, but visitors are encouraged to use the main entrance on the A41, 1½m. S of Newport. (It is somewhat annoying to find that to get from the Abbey to the Hall by the signposted route one has to travel the best part of 7 miles, when, as the crow flies, it is less than 1 mile.) The Hall was built in 1829 by Sir Jeffry Wyatville for the Duke of Sutherland. The Duke had been driven out of his palatial Trentham Hall (the grounds of which are now the well-known Trentham Gardens) by the unbearable smell of sewage in the Trent. The Duke's new house at Lilleshall replaced an older Leveson home. Wyatville did him proud with a beautiful Tudor style mansion in white stone with mullioned and transomed windows. The Hall is approached along the Dukes Drive. It is tree lined all the way from the A41 entrance to the Hall, a distance of 2m., long enough to impress anyone. The park extended to 600 acres and the family lived in some style. A census of 1811 lists the occupants: Anne, Duchess of Sutherland, 2 of her daughters, 26 resident house servants, 9 postilions, grooms and helpers in the stables, and 4 visitors with 5 visitors' servants. This degree of staffing was unusual, however, and was by no means representative of other big country houses. On his farms and in his fields the Duke employed all the latest modern methods and the cottages he built for his workers are locally held in great esteem to this day. In 1920 the Hall and the estate were sold. For a time the Hall became a pleasure park. 'See Lilleshall and know the thrill of living', was the slogan used to attract customers. If the Duke was to return to Lilleshall today he would be horrified. The National Sports Centre has seen fit to construct the most tasteless additions to the Hall. The new buildings are acceptable in themselves, but in conjunction with the Hall are an abomination. The front entrance to the house is completely surrounded and cars are parked almost everywhere. It is a mess. Only the garden aspect has been left untouched, giving a glimpse of what things were like once upon a time. From the high to the humble: on the A518 within the shadow of the rocky hill is a tiny stone squatter's cottage (SJ.724.154). It lies beside the road, crammed in on the narrow verge. It measures 16 ft. by 13 ft. and has a brick chimney and a tiled roof and is still lived in. Opposite is a road junction guarded by 2 large and battered sandstone lions. Note: George Granville (surname Leveson-Gower) is quite properly called the Duke of Sutherland. However, it is not widely appreciated that he was only a duke for 8 months. He had married the Duchess of Sutherland, but her title did not pass to him. The authorities only granted Granville his dukedom very grudgingly and only after influence was brought to bear. So, for almost all his life, the Duke was known simply as George Granville Leveson Gower.

LINLEY *4½m. NW of Bridgnorth on the B4373.*
Do not confuse this Linley with that near Bishop's Castle. (*See More.*) The small Norman church of St. Leonard is much in its original state, and has several unusual details. The tower has notable bell openings of twin arches set in recessed fields and above the blocked N doorway is a tympanum decoration consisting of a carving of the Green Man. The Green Man was a pagan fertility figure, a naked man standing with legs apart and foliage issuing from his mouth. The church is set in a wooded drive that leads to Linley Hall. This is an Elizabethan stone building with a Georgian brick facade. It is now divided into flats. Down the hill is the little village of Linely Brook. The lands of Linley are part of the Willey Estate, administered by its owner, Lord Forester. Linley station still stands, more than 1m. E at Apley Forge, on a now abandoned stretch of the Severn Valley Railway that is now used as a lane. It regularly won the prize for best kept station but now the platform is derelict and the station building has been converted to a house. The name Linley is from the Old English meaning 'the clearing where flax is grown'.

LITTLE DRAYTON *1m. SW of Market Drayton.*
Although Little Drayton is now a part of Market Drayton it was for a long time an independent community. Under Anglo-Saxon rule the land here was owned by the Countess Godiva of Coventry, but after the Norman Conquest it passed to Torold de Verlio as part of the Manor of Betton. He gave 'Drayton Parva' to Shrewsbury Abbey. (Market Drayton belonged to the Abbot of Combermere in Cheshire.) Today Little Drayton is essentially a residential suburb of the larger settlement. It is bounded to the N by a modern by-pass, which has stolen the designation A53 from the old Shrewsbury Road, and to the S by the marshes of the Tern Valley. There is, perhaps, a surfeit of council houses, but there are some cosy little lanes with mellow Georgian and Victorian cottages and houses. There is also a centre, of sorts. This lies at the junction of the Shrewsbury Road, Buntingsdale Road and Christ Church Road. Christ Church was built in 1847 to a design by S. Pountney Smith. It is constructed of red sandstone and has a tower with a little turret at one corner. The Old Vicarage stands beside it, a substantial stone house embowered in trees. It is a very mature looking building and in the past has seen service as a school. Opposite are playing fields and the Village Hall. The King's Head pub lies on the main road, a little to the N, and to the S, hidden down a lane lined with allotments, is Quarry House. This is now an old person's home run by the County Council, but is, in fact, the Old Workhouse, a somewhat grim Victorian edifice in red brick with iron-framed windows and an octagonal corner tower. We were told that several of the older people of Little Drayton refuse to enter the place. When visiting friends or relatives resident here they will go no further than the front door, such is the distaste they feel; so painful are their memories of the human indignities suffered within these walls. The old quarry itself lay to the S and has been back-filled and levelled. Building plots are, from time to time, advertised for sale here on this made-up ground; the wise are wary. It was from this quarry that stone for the church was taken. In Christ Church Road there are 2 buildings that are poles apart: No. 21 is a lovely, old, stuccoed cottage with dormer windows and a tall, hand-operated water pump in the garden; and just up the road, opposite the church, are the offices of an ugly, noisy modern scrapyard which pollutes the place from behind tall, corrugated iron fences.

LITTLE NESS *See Nesscliffe.*

LITTLE WENLOCK *3½m. S of Wellington.*
It seems likely that Little Wenlock was a daughter settlement of Much Wenlock, planted by the Abbey. In the early 16th Century coal was being mined here and transported to the Abbey's iron foundaries in Shirlett Forest. At the beginning of the 18th Century Abraham Darby laid wooden railway tracks to transport coal down to his works in Coalbrookdale. Mining continued as a major industry well into this century but has now ceased and the numerous spoil heaps are in the process of being landscaped. Huge machines rumble through the countryside creating hills, valleys and lakes where there were none before. Little Wenlock itself is an attractive, well-heeled little place. The stone church of St. Lawrence stands in the centre of the village on a raised mound. The tower dates from 1667; a new nave and chancel were added in 1865, at which time the old nave and chancel were converted into an aisle and side chapel. The new nave is built of brick. The Old Hall stands close by. It is an Elizabethan house with gables and mullioned and transomed windows, but now much reduced in size. The interior was re-worked in early Georgian times. The only social facility in the settlement is the Village Hall.

LLANBROOK *1m. WNW of Hopton Castle and 1¼m. ENE of Obley.*
Llanbrook is one of several farming hamlets clustered in a steep-sided valley NW of Hopton Castle. As well as Llanbrook there are Llan Farm, Llanadevy and Llanhowell Farm. *Llan* is Welsh for 'church' but there is no record of a church here. A church, of course, need not be a building. It is, technically, a body of people who meet in communal worship. In early Christian times such meetings were often held in the open, by conspicuous trees, stones or hollows in the hills. Perhaps, The Llan, as the valley is called, is a folk memory of such a long-forgotten place. Near the stone-built Llan Farm, high on the valley side at SJ.352.794, is a rough rock, believed to be a tombstone. Carved on it are a 'W', a heart and a star. At Llanbrook, as late as the 1940's, the belief was strongly held that to cut saplings when the sap was rising was to court ill-luck, even disaster. One farmer cut small wood to mend his fences at the wrong time, and what was even worse burned the clippings in the house. Straightaway his mother fell ill and took to her bed for a year and many lambs were lost that spring. This talk of wood brings to mind a bit of lore that has been scientifically proved. That is that good luck comes if timber used in construction – of houses, fences and anything else – is set so that the wood stands as it did naturally, 'top as is to top as was'. Fences erected in Elizabethan times obeying this code have stood without rotting but fences using equally good oak can rot in a few years if placed in the ground upside-down. A possible explanation is that moisture at the bottom can rise up through the capillaries and be evaporated higher up. Placed upside-down this cannot happen and the water so trapped rots the wood.

LLANFAIR WATERDINE *4m. NW of Knighton.*
The name means: 'the church of Mary in the valley of the river'. The church of St. Mary is of 1854 to a design by T. Nicholson, a simple stone building of nave and chancel with pointed-arched windows which stands within a tree-fringed churchyard. It replaces the previous church and to this day local people are bitter that old St. Mary's was done away with. Even at the time there was resentment towards the vicar who commissioned the new church. Inside, some of the old woodwork is preserved. Of particular note are the altar rails on which are carved a woman in a

long dress and a man with a beard together with pigs, rabbits, birds, dogs, a deer, a lion and a dragon. In the churchyard is a row of lime trees and the grave of a Romany gipsy. Opposite the church is the long, low Red Lion pub the older part of which seems to be to the left. The car park sees more than its fair share of Mercedes and BMWs. There are but the merest handful of cottages in the village but there is a good, stone bridge over the shallow and stoney-bedded River Teme and an even better 'big house' called Nant Iago. This is a long, stuccoed dwelling with 6 gables, 2 bay windows, and an off-centre porch. It stands amidst lawns behind a lake and a thin screen of trees. Unusually for this part of the world there is a minimum of social facilities at Llanfair Waterdine. As well as the pub and the church there is a Post Office, a grocer's shop and a meeting place called Everest Hall, which is housed in what appears to be the old school. It is called Everest Hall because Lord Hunt, one of the 2-man team which first climbed Mount Everest, has a weekend cottage here. This is called **Cwm Cole** (which could mean 'beloved valley') and lies 1m. N of the village at SO.243.778. It is a white painted stone building with 3 dormer windows a central porch and brick chimneys. Alongside the lane is a small stream which runs down the valley. Amongst the hill pastures there are broadleafed woods (some of which have been coppiced) and ½m. S is Black Hall, an old gabled house in the process of being renovated. It is beautifully situated at the junction of 2 valleys, but is spoilt by its collection of unsightly farm buildings. This is sheep and cattle country. **Cwm Collo**, 'the lost valley', is the name of a traditional hill-farm ½m. NW of Cwm Cole at SO.238.783. The old stone farmhouse and its stone and weather-boarded outbuildings form an open ended quadrangle. They huddle together on a sheltered slope above a stream in their little, wooded valley. It is a pity that the Council has inflicted a concrete and steel tube bridge on them. **Cwm-brain** lies ½m. down the valley, a stone built house with corrugated-iron sheds, a field full of goats and some open, broadleafed woodland. From here one can head SE to **Tregodva**, 2 houses and Runnis Chapel (1837), where there are steep wooded hillsides to the N and splendid views over the River Teme to the S. The river can be crossed at Dutlas and the B4355 will take the traveller SE to Knighton. This road is actually in Hereford and Worcester but from it one can look northwards into Shropshire at some very fine scenery indeed. The view of Nether Skyborry from just SE at Knucklas is especially beautiful. (There is an opposite view from Graig, a farm ½m. S of Llanfair Waterdine. An old road leads to a wood of tall trees which stands on a high point, at SO.240.759, from which there are absolutely splendid views over the winding Teme into Herefordshire. The earthworks of the pre-historic Knucklas Castle stand clear above the woods on the hillside below in which green conifers spell the initials RP within a background of red-brown trees. This was the work of a Mr. Price.)

LLANYBLODWELL *7m. SW of Oswestry.*

It lies in the valley of the River Tanat. The name is Celtic and means 'the church at Blodwell'; Blodwell is the name of a tributary stream of the Tanat. The village is quite charming and has several notable attractions. The approach from the E, along the A495, takes one through wooded country with very active quarries in the hillsides. Shortly after taking the B4395 to Llanyblodwell one comes upon the striking, stone-built, village school and schoolhouse. The school is now a house and part of the schoolhouse is a Post Office – the tiniest we have seen. The lane continues on to the church with its remarkable tower. St. Michael's was designed by the Reverend John Parker, a vicar here, and built about 1850. Pevsner, in his 'The Buildings of England – Shropshire', describes the church as 'absurd'. This is more than cruel; many people find the building quite charming. It is certainly different and in many ways unique. Local belief is that the church stands on an ancient Druidic ring. Next to the church is a good red brick house of some substance. A fine red sandstone bridge of 1710, with 3 arches, spans the shallow, wide, and stony-bedded river. The central arch is unusual in being ogee-shaped and the cut waters continue up to parapet level to provide pedestrian refuges. Upstream, the valley is heavily wooded and fishermen frequent its banks. By the bridge is a group of black and white buildings. The Horseshoe Inn is dated at 1445 and has a weather-boarded barn attached. Opposite is a timber-framed cottage with flowers in the front garden. All in all, this is a tranquil and delightful place. Blodwell Hall was once a place of some significance but all that remains of Sir John Bridgeman's house is the stone-built servants' wing, 2 pillars with lions and a handsome summerhouse dated 1718 and bearing the Bridgeman arms (they were Earls of Bradford). There had long been a Welsh princes' house here and Joan, the last of the line married into the Mathews family. Ursula Mathews became the wife of Sir John. In 1989 the Member of Parliament for the North Shropshire constituency is W. J. Biffen; his home address is Tanat House, Llanyblodwell. From the hills above the village there are views of the Welsh mountains that Bagshaw describes as being 'of sublimity perhaps unsurpassed in any part of Wales'. These are best seen on a summer's evening with a slight haze to add mystery to their grandeur. We took the road from Llanyblodwell to Llanerchemrys, part of which is an unmetalled track, and were not disappointed. A short diversion into Wales that can be recommended is to take the B4396 from Llanyblodwell to the Green Inn, a distance of about 3½m., and from there turn N on the road to Llansilin. At the top of the hill is a splendid view over the valley of the River Cynllaith and a Norman motte hidden in trees behind a farm. The farmhouse is bedecked, in a style common to this area, with white-washed walls and a black-tarred roof. At the bottom of the hill one can turn right on to a lane that rejoins the B4396. A little to the SW, off the B4393, is the delightful Welsh village of Llanfechain. Many of the cottages are decorated as just described and there is a beautiful church in a raised churchyard.

LLANYMYNECH *7m. S of Oswestry.*

The name is Welsh for 'the church of the monks'. Today the settlement is half in Wales and half in England. Above the somewhat gaunt village looms the reason for its being: the sheer rocky slopes of Llanymynech Hill. The hill has probably been mined from prehistoric times. It is riddled with shafts and tunnels. The Romans worked it extensively and built forts to defend it. They extracted lead, copper and zinc, but also mined the limestone to use as mortar when they built Uriconeum (Wroxeter). The town is one of several sites that are possible locations of the lost Roman city of Mediolanum. The cavern of Ogof's Hole, which is approached via the golfcourse at the top of the hill, has many legends connected with it: that it was the prison home of British slaves; that is the entrance to Fairyland and the long lost castle of Carreghofa; and that a blind fiddler and a harpist wander around the uncharted workings playing mysterious, wistful music. In 1965 some small boys found treasure in the Hole. The 33 silver Roman coins they found were declared treasure trove and are now in Oswestry. In the 17th and 18th Centuries limestone was extensively quarried on the hill. Some of the stone was used as a flux in iron smelting furnaces and some was burned in kilns to produce lime for which there was a large agricultural demand. The great landowners were enclosing commons and open fields and improving the soil. Lime was also a prime ingredient of the mortar used for the construction of the new industrial towns. The Shopshire Union Canal passed close by the kiln and was used to transport the lime. The red brick kiln chimney still stands and the canal, though no longer used, is still water filled in places. Both are visible from the main road. Limestone is still quarried on the other side of the Hill at Blodwell. Llanymynech is built about a crossroads, S of which is the grey and yellow stone church of St. Agatha. This is neo-Norman, of 1845, to a design by Thomas Penson. The Lion Hotel is divided by the border; the bars are on the English side. South of the town is a fine bridge that carries the A483 over the broad River Vrynwy. One pier is built on a mid-stream bayou (island). Offa's Dyke makes a long detour to the E, between Llanymynech and Whitehaven, 2½m. N, to enable it to keep its westerly aspect. (The Dyke was built so that it always faced Wales on a western slope in order that patrols could always look into Wales. If an eastern slope had been used the hill above would obscure the view and give the high ground to the enemy.) The old limestone workings, that created the great cliffs that hang above the town, can be approached off the Oswestry road. The

Llanyblodwell, the River Vrynwy.

road climbs the hill and at the end of the row of council houses is a stand of tall coniferous trees. Turn left here. At the end of this short road is a narrow footpath that leads through scrub woodland, past a partially filled, stone-lined shaft, and joins a wide track that leads up the hill at the foot of the cliffs. There is a ruined stone engine house that powered a tramway and just past this is the entrance to a dramatic, short tunnel. This leads through the hill but is fenced off. Continue up the track, which becomes a narrow path and, on rounding a corner, the great bowl of the hollowed-out mountain reveals itself. It is a splendid place. Sheer cliffs surround spoil mounds, a few stone buildings, and the large hole into which the tunnel leads. There are no pools of water as there often are in old quarries because limestone is pervious. After careful consideration the author had decided to relate a strange incident that occured on our visit to these workings. As we passed the tunnel we heard men's voices and the clanking of heavy machinery. On reaching the old workings a few minutes later we were surprised to find only one other person there, a bird-watcher at the far end of the great bowl. We had fully expected to see men working the quarry from the noises previously heard. This must be another example of the strange acoustics associated with the hill – the phantom fiddlers and pipers, etc. There are workings on the far side of the mountain but they are a mile away. What we heard seemed to be very close and was heard at no other time during our visit.

LLYNCLYS *4m. S of Oswestry.*
A crossroads hamlet at the meeting of the A483 and the B4396 (which to the W becomes the A495 for a couple of miles). The settlement consists of little more than a few red brick cottages, some council houses and the stone-built White Lion pub, at the back of which runs the track of a disused mineral railway. This connected the quarries of Nantmawr, Whitehaven and Blodwell with Oswestry, 4 miles to the N. Llynclys Hill lies to the W and here are several winding narrow lanes along which are not a few cottages and smallholdings. These may well have been built by the miners who were employed in the lead mines and limestone quarries. Much of the hill is common land – Llynclys Common – and some of the cottages were no doubt originally built by squatters. In the scrubby grassland here are 8 varieties of orchid and many insects, including the brown Argus butterfly. Llyn is Welsh for lake, and the lake of Llynclys lies just to the NW of the crossroads. It is said to be very deep and there is a legend that below the waters there is a city whose towers can be seen on a clear day. This city was ruled by King Alaric who had married a woodland spirit, the Maid of the Green Forest. But the king was disturbed, for every seventh night his wife left him and forbade him to follow. A wise man by the name of Willin offered to help the king. He followed the Maid to Ogo's Hole on nearby Llanymynech Hill (which is known locally as the entrance to Fairyland) and saw her emerge dressed as a radiant Fairy Queen. Willin was struck by a desire to possess her for himself. He used spells, first to give peace of mind to the king and then to make the Maid of the Forest marry him, Willin. He arranged to meet her at the cross by the White Minster but when she arrived it was as an aged crone. His spell had prevented her from returning to Fairyland to renew her youth and now he was compelled to marry a hag with rolling rheumy eyes, skin that crackled like parchment when she moved, a voice that cackled and croaked and who was all but bald on the head but well whiskered on the chin. King Alaric and his city were punished by being banished to the bottom of the lake and there they remain. On still, quiet days beautiful but sad music has been heard drifting across the waters.

LOCKLEYWOOD *5½m. SSE of Market Drayton.*
It lies on the Hinstock to Market Drayton road, the A529. We came to the crossroads and stopped to look at the map and were serenaded by a donkey. The settlement lies on a rise in flattish country and consists for the most part of small stone cottages. There are several small, disused quarries, presumably the source of the sandstone for the houses. The name Lockley probably means 'the enclosed field in the forest'. By the late 17th Century the forest was little more than scrubby woodland. In 1676 the manor passed to the Corbets of Adderley and in 1704 the commoners gave up their rights on the whole wood in return for ownership of a half. This was divided between them and enclosed, though the many small fields then made have since been enlarged by amalgamation.

LONGDEN *5m. SW of Shrewsbury.*
It lies on the old road from Shrewsbury to Bishop's Castle that, further S, passes through the hills of Stiperstones and Long Mynd. The village of Longden is long, but the name actually means 'long hill'. At the time of Domesday Book it was a sizeable settlement held by the Corbet family. It later passed to the Botterells and became the *caput* (the head) of their barony until about 1282 when it passed to Robert Burnell (the Bishop of Bath and Wells between 1275 and 1292). Today Longden is a substantial, though somewhat characterless, village. There is much new housing; a garage, the forecourt of which is littered with old cars in various states of decline; a Post Office; a pub, the Tankerville Arms; a farm which, with its corrugated iron barns and giant silos, looks more like a factory; the 19th Century church of St. Ruthin, which has a roof and S door of the 17th Century and a brick polygonal apse but which lies hidden behind a row of terraced houses; a Victorian octagonal Gazebo of red brick with blue brick dressings and a scale-patterned tiled roof built into a wall at the S of the village; a good brick and half-timbered dwelling called Cross Houses, situated at the top of the hill; and a Primitive Methodist Chapel of 1870. **Annscroft** lies on the main road 1m. N, a long line of houses and cottages in a variety of styles and ages but which for the most part are early 19th Century. There is little to remark on here. The Victorian church is of red stone with cream stone dressings, and has lancet windows and a dormer with a rose window. A little to the N, down a side lane, is Moat House which stands beside a pool. Most of the population of Annscroft worked at the Moat House Colliery; indeed the settlement developed largely because of the pit. To the W is the estate of **Longden Manor**. The lane from Longden to Plealey passes through this place which is situated on a rise and has enough estate cottages to constitute a village, by Shropshire standards anyway. The grounds are well wooded and well tended and it is a great shame that the Manor House is no longer with us.

LONGDON-UPON-TERN *4m. NNW of Wellington.*
A fairly compact, small village to the W of the Weald Moors. The Hall is Tudor, of red brick and sandstone. What remains is only a part of a much larger mansion. Today, it is a farmhouse. Alongside it is the small church of St. Bartholomew, rebuilt in red brick in 1742, with 19th Century alterations. The site is obviously old, possibly occupied previously by either an Iron Age or Anglo-Saxon fort or a small Norman castle. Near the church there was a wharf on the now disused and waterless Shrewsbury Canal. To the W of the village the canal crossed the River Tern by means of an aqueduct, 62 yds. in length, designed by Thomas Telford. It is believed to be the first aqueduct to be made of cast iron. The castings were made at Ketley. It is as long as it is to allow for flooding, which in these flatish lands can extend the width of a river considerably. The aqueduct performed perfectly adequately, despite fears that it might crack in periods of severe cold. It still stands and is visible from the road. However, it must be just about the most ugly structure Telford ever designed. Near the large Victorian warehouse by the river there used to be a watermill and a windmill. The name Longdon-upon-Tern means 'the long hill by the River Tern'. In recent years the village has gained some new houses but lost the school, the shop, Post Office, the garage and the railway halt.

LONGFORD *1½m. W of Market Drayton.*
A charming hamlet on a rise in undulating country. The land to the N was once marshy and the lane to Moreton Say is raised and characteristically undulating. The fields are mostly laid to hay, corn, and pasture grazed by dairy cattle. Longford Old Hall is a handsome black and white house with a symmetrical front. Two gables stand proud of the recessed centre in which there is a dormer window. There are tall chimney stacks to left and right, and that to the left has some elaborate brick diaper work. Opposite the Hall is a house built in brick using the Flemish bond, with red stretchers and white leaders; it is now used as an antique shop. Moreton Say lies 1m. NW. In the

Loppington, Shelton House, now a 'special school'.

churchyard there lie the bones of Robert Clive whose grave was only recently discovered. (*See Moreton Say.*)

LONGFORD *1m. WSW of Newport.*
One leaves Newport on the Longford Road. As the suburbs end so the long sandstone wall that protects the wooded park of Longford Hall begins. The house cannot be seen from the road which is a pity because it has a most handsome facade. The architect was J. Bonomi and he built this mansion for the Leeke family in 1794–7. A central block of 7 bays with giant pilasters is fronted by an imposing Tuscan-columned porte-cochère and flanked by 2-bay recessed wings. The house sits on a rise in flat country and has good views towards Lilleshall. Today it is used as accommodation for the junior boarders of Adams' Grammar School in Newport. By the road are the old stables and coach-house built around a courtyard, in the centre of which is a tower with round windows that looks like a dovecote. All the buildings are sandstone in the lower part and brick above. The lower parts are probably from an older structure. A hundred yards NW of the stable block is the now redundant church of St. Mary (1802–6). It is constructed of red sandstone and has a tower, nave and polygonal apse. The interior has been gutted and the building awaits redevelopment. The new church was erected alongside the old one. Part of this, the stone-built chancel, has been preserved and is believed to date from 1155. Inside is a monument to Thomas Talbot, d.1686, and wife, d.1706. The churchyard is raised and walled about to the NW. Beyond is a cottage, the core of which is a small sandstone building of the early 15th Century in which there is a Gothic pointed arch doorway. Alongside the track to the cottage is a long, narrow medieval fish pond, fed by its own spring and still stocked with carp. It lies above the level of the old flood plain of the Strine Brook and is separated from the lowland by a bank. The river now runs in a 10 ft. deep trench and the dark brown peaty fields of the broad valley are drained by smaller, straight ditches that feed into it. This is all part of the extensive works carried out by Thomas Telford for the Duke of Sutherland when he drained the Weald Moors, the greater part of which lie to the W. The road to Edgmond crosses the Strine Brook by a bridge guarded by an old concrete machine gun bunker. The depth of the watercourse below ground level can be seen here quite clearly. Before being canalised the stream, by nature of the broad valley it lies in, would have been wide and the ford across it therefore long, hence Longford. Many midland waterways have been entrenched and the result is a major loss to the landscape. What were once lazy, shallow, meandering and wood-fringed rivers are now deep, narrow ditches that are a danger to livestock, rarely visible but unsightly when they are seen. The irony is that the purpose of these drainage works was to reclaim land for permanent agriculture, land of which there is now an excess. Just S of the Strine Brook bridge is Longford Farm Mill (SJ.718.181). This is a strange structure. The old brick tower has been almost encircled in its lower parts by a range of 5 arches some 12 ft. tall and 2 or 3 ft. deep. These were added after the tower had been built but their purpose is a mystery. One is bigger than the others and has what might be a fireplace within it. The mill has not worked for at least 50 years and is now used as a store place. Behind it is a depression in the ground that at one time may have been a pond. To the side is a white-painted brick cottage. The farmhouse is of red brick, 2 storeys high and 3 bays wide. The map shows a broadleaf wood here but that is long gone.

LONGNER *See Atcham.*

LONGNOR *8m. S of Shrewsbury.*
The village lies just off the busy A49 and has a population of about 400. It is in the shape of an oval, and consists of old black and white cottages, a shop, farms, a new school and new bungalows. The black and white Moat House, with half of its substantial moat still water-filled, is the most noteworthy dwelling. Inside there are the remains of the original open roof of the 14th Century 'great hall'. There is a house called The Malthouse and another called The Maltings, a row of red brick 'Vineyard Cottages' and a barn full of hay in the main street. By the bridge over the Cound Brook is a watermill, newly converted to a dwelling; there has been a mill here since the time of Domesday Book. The bridge itself was built in 1925 of reinforced concrete to a design type known as 'multiple box culvert'. The small church is considered to be a perfect example of the Early English style (about 1260). It has box pews and clear glass windows. Close to the church is an area of disturbed ground. This is probably where part of the old village stood before the Corbets moved it when they landscaped Longnor Park. Longnor Hall was built for Sir Richard Corbett in 1670 in red brick with an oblong hipped roof. It is a good, symmetrical building which in recent years has been used as a Country Club but is now being restored and modernised as a house. Between the Hall and the road are some good specimen trees and a herd of black-horned sheep. At the rear (the sunny, S facing side) is a deer park with 3 streams, though 2 are little more than drainage ditches. By the flowing stream are 2 famous trees, the Black Poplars of Longnor. One of them has the greatest girth (24 ft.) of any Black Poplar in Great Britain, and the other is one of the tallest (124 ft.). The Black Poplar is native to Britain but is very uncommon. A Scots Pine, here in Longnor Park, has the county girth record of 16 ft. 3 ins. for the species. Richard Lee was born in Longnor. His parents were humble peasant folk and he became a carpenter's apprentice in Shrewsbury. Here he found an interest in Classical and Middle Eastern languages. By the age of 25 he had mastered Chaldee, Syriac, Samaritan, Persian and Hindustani. He earned his living first as a carpenter – it was he who made the pews in Longnor church (they are initialled R.L. and dated 1723) – and then as a language teacher at local schools. His talent was soon recognized and he was able to attend Cambridge University on a Missionary Society grant. He took his degree and went on to become an internationally known linguistic scholar, having mastered 18 languages. He became Professor of Arabic and later of Hebrew at Cambridge, and then retired to become a humble vicar at Banwell in Somerset, where he died. Longnor has a ghost, a young girl who has been seen at local dances and by the bridge where she was attired in a wedding dress. The last sighting was in 1975. The old name for Longnor was Longenolre from the Old English *long-alor*, *alor* meaning 'alder'. The name might therefore mean 'the long alderwood' or 'the tall alder trees'. There are, in fact, alder trees on the banks of the Cound Brook.

LONGVILLE *See Rushbury.*

LOPPINGTON *3½m. W of Wem.*
A very pleasant village. Set in the middle of the road, in front of the Dickin Arms, is the only known in-situ bullring in Shropshire. Bulls were tethered to it whilst baited with dogs, a 'sport' which was practised up until about 1835. Bears were also baited here, the last baiting being the highlight of revelries that followed the marriage of the vicar's daughter in 1825. There is a doggerel ballad entitled 'The Loppington Bear', for which see Charlotte Burne's 'Shropshire Folklore'. The very weathered yellow stone church of St. Michael is medieval with Jacobean furniture and 18th Century pews. It is built on a raised mound. During the Civil War the church was garrisoned by Cromwell's men and was damaged by fire when stormed and captured by the Royalists. The nave was rebuilt in 1656. Loppington Hall is a tall, red brick house of the 18th Century, 5 bays wide and 2½ storeys high with a Tuscan pilastered doorway. It was the home of the Dickin family, local landowners. Close to the church and the Hall is The Nook, a timber-framed house, and at the S end of the village is a cottage constructed on crucks (curved timbers forming an inverted 'V'). Opposite an attractive black and white house called Spenford is a red brick school with separate entrances for Girls and Boys. The restored village pond was formerly a tan pit. Out on the road to Wem is Shelton House, a small hall, set in an agricultural estate, now an Adult Training Centre. There are several large granite glacial boulders in and around the village. The name Loppington means 'the settlement of *Loppa*'s people'.

LONGSLOW *1¼m. NW of Market Drayton.*
A tiny hamlet on a rise in undulating country, the name of which is locally well known because of the Longslow Dairy, purveyors of milk to the public. Longslow Farm still dominates the settlement especially now that it has acquired large, unsightly modern barns and storage tanks.

Ludlow, the castle from Whitcliffe.

At the entrance to the farm is a mature oak tree with a huge aerial root which is almost as thick as the trunk; a most striking sight. The Old Smithy is half-timbered but most of the buildings in the hamlet are of red brick. The fields around are laid mainly to corn and to pasture for dairy cattle. There are views over the valley of the River Duckow to the white-painted Styche Hall, built by Clive of India. (*See Moreton Say.*)

LOUGHTON *1½m. S of Burwarton, which is 8m. SW of Bridgnorth on the B4369.*
A hamlet of 2 farms, a pair of semi-detached brick houses, a stone and brick house and a tiny church, all set on a hill with wide views over the attractive arable – pastural country around. The settlement lies within the boundaries of the Burwarton estate. The church is built of random stone and stands on an ancient pre-Christian site. There has been a chapel here since at least 1291 though what we see today dates largely from 1622. It has a nave, chancel, bellcote, two-light windows with round arches, Norman fabric in the Chancel arch and 14th Century fabric in the doorway. It is a most charming little church, simple but full of character. The interior is perfect for a small country place and the furnishings include an old pedal operated harmonium. However, the chief attraction here is in the churchyard. The Loughton Yew is completely hollow but is nevertheless a picture of perfect health. At about 1,000 years of age it is one of the oldest living things in the county. What is more, this is not supposition for the Loughton Yew has been radio-carbon dated, the first living yew in the world to be tested in this way. The tree was here long before the chapel and was quite probably of pre-Christian religious significance. Long after the country became nominally Christian by the conversion of tribal leaders (sometimes only for political reasons) the common people clung to their ancient beliefs. The name Loughton is thought to be from an Anglo-Saxon personal name such a *Lulha* or *Luka*.

LOWE *1¼m. NW of Wem.*
The country hereabout is on the wet side and the land is laid mainly to pasture. At the crossroads is a substantial red brick farmhouse of 3 broad bays, 2 storeys tall and dated at 1666. Bed and breakfast can be obtained here. To the S is the complex of Lowe Hill Villa, and to the E are the scattered cottages of Creamore Bank. The land is flat and there are mature trees in the hedgerows. There is also a moated site. It has been suggested that Lowe is the as yet unidentified *Lai* of Domesday Book (4.8.2).

LOWER FRANKTON *See Welsh Frankton.*

LUBSTREE *¾m. E of Preston-upon-the-Weald Moors which is 3m. NE of Wellington.*
It lies just N of The Humbers which is today largely a place of army houses. The name Lubstree is probably Old English meaning '*Hubba*'s tree'; 'tree' or 'try' often means 'a cross'. There is only one large farm here today but in the mid-19th Century the name was familiar to many as a canal terminus. In 1844 an arm of the Birmingham and Liverpool Junction Canal (later incorporated into the Shropshire Union Canal) was opened. It left the Norbury-Wappenshall line near the handsome Dukes Drive aqueduct and ran about 1m. to a wharf on the Preston-upon-the-Weald-Moors to Donnington road. The wharf was on land owned by the Duke of Sutherland. Iron was exported from here to the Black Country and coal was sold at wharves along the canal to the W. Limestone was also handled at the wharf. Ironically both Lubstree wharf and Wappenshall wharf (2m. W near Eyton) received the sleepers used to construct 2 railway lines (the Shrewsbury-Wellington-Stafford of 1849 and the Shrewsbury-Wolverhampton-Birmingham of 1850) which were to steal their trade and seal their fates. The Lubstree wharf (sometimes called the Humbers wharf) was on Humbers Lane, near the entrance to Lubstree Park, close to where the road crosses the Humber Brook (SJ.692.152). Lubstree Park is signposted off the road as The Strawberry Farm. The wide, flat acres of the Weald Moor are very peaty here and the light soil is suitable for soft fruit. The red brick house has 3 bays and 2 storeys. Behind it are a half-timbered barn and a red brick range with 5 arches, 2 dormers, and a gable. The courses of the main canal and the Lubstree spur are both best seen at SJ.686.164 where the Dukes Drive aqueduct used to stand. The aqueduct carried the canal over the long, straight track now called Kynnersley Drive on the map but still known locally as the Duke's Drive. The high, earth embankments upon which the canal lay still exist but the aqueduct was removed over 16 years ago. Parts of 2 round columns are now gate posts outside the Kynnersley Working Men's Club. The canals and embankments are covered in scrub woodland and dense undergrowth. It is very quiet and lifeless here and the deep drainage ditches are unwholesome, murky, looking things. The track is very rough indeed and is not really suited to a modern car, as mine will vehemently tell you. At the Kynnersley end of the lane is a row of archetypal Duke of Sutherland cottages. (*See Kynnersley, The Hincks, and Horton.*)

LUDFORD *¼m. S of Ludlow.*
Ludford is today really a part of Ludlow. It lies in a well-wooded area just over the Ludford Bridge which crosses the River Teme to the S of Ludlow. The narrow bridge is medieval and very fine. It is well worth the delays it causes to the flow of traffic. The church of St. Giles has a Norman nave and a chancel of about 1300. The stone-built St. Giles Hospital, to the NE of the church, was founded about 1216. To the NE of the Hospital is the Old Bell, a partly timber-framed former inn of 1614. At the bottom of its garden is the stone and timber Mill which faces the Ludlow Mill on the other side of the Teme. The most handsome and interesting Ludford House is just S of the church, and facing the church is the range that once contained the main entrance. It is either of late Elizabethan or early Jacobean date, of mixed stone and half-timber construction. Facing the main road is a cliff-like stone wall topped with 4 large brick chimneybreasts. Up river is Whitcliffe from where is obtained the best view of Ludlow castle. Here can be found the Ludlow Arms which has its own Fives Court. A narrow road leads from Whitcliffe, passing firstly through a farm and then through Oakly Park. The name Ludford is possibly from the Old English *hlude*, meaning the 'loud one', a reference to the noise from a rapid on the River Tame, and *ford*, giving 'the river crossing by the rapid'. For an alternative entymology for the Lud element of the name see Ludlow.

LUDLOW *26m. S of Shrewsbury.*
There can be little doubt that Ludlow is the finest town in Shropshire. Not too big and not too small, busy but not frantic, it has good houses, good shops, a castle and lovely country all around. Little wonder that the poet A. E. Housman, who knew the place so well, chose to have his ashes buried here. The town originated as a crossing place of the River Teme, either by a ford or by a bridge. Old Street, leading to Corve Street, at the top of the hill, was the road that connected with this crossing. (*See Ludford.*) It is quite probable that there was an Anglian hamlet called Dinham here before the Normans came but Ludlow was, to all intents and purposes, a 'plantation' of the French invaders, built to serve their castle. Ludlow Castle was started about 1085, either by Roger de Montgomery or by Roger de Lacy, and became one of the most important border forts in the Welsh Marches. In the late 15th Century the Lords President of Wales met and held their Court of the Marches here. The castle has formidable natural defences, with high, sheer cliffs falling to the Teme in the W and the Corve in the N. To the S and E was a deep ditch, now filled-in and built upon. The castle is entered from Castle Square, near the centre of the town. The visitor pays his toll at the booth and finds himself in the huge outer bailey, the 'killing ground', where invaders who broke through the outer defences could be trapped and fired on from the 12th Century Gatehouse Keep which stands beyond a now dry moat. To the right of the outer bailey are the barracks and stables. The Elizabethan gateway is of 1581. The chapel of about 1140 stands within the inner bailey and is almost unique in having a completly circular nave; the chancel has been pulled down. Along the N walls of the inner bailey are later buildings of some sophistication, one of which is the Council Hall where Milton's 'Comus' was first performed. Beside it are state apartments, the armoury and the Pendower Tower. Perhaps the most impressive view of the exterior of the castle is from river level at Whitcliffe. (*See the note at the end of this*

A. E. Housman.

Misericord carving, St. Laurence's, Ludlow.

Ludford Bridge, Ludlow.

article.) The town was built shortly after the early Norman castle was completed, almost certainly by Roger de Lacy, in whose manor the land lay. The first settlement seems to have been destroyed in the 12th Century. Rebuilding began with the High Street. This was wide and extended eastwards from the castle gates. It was used as a market place. In time temporary stalls became permanent buildings. From the High Street a regular grid of roads was laid out downhill towards the Teme. There have, of course, been alterations and additions to the grid but it is still quite plainly to be seen. The High Street was infilled, creating the narrow lanes that run from the Buttercross to the Castle Square. The parish church of St. Lawrence was given a whole square to itself, but this too has been built on, so preventing the impressive church (the largest parish church in Shropshire) from being seen properly. It is only when the town is viewed from a distance that the size of the church is appreciated. St. Lawrence's was built by the burgesses, not the lord of the manor, from the profits of the lucrative wool trade. Construction began in the 12th Century when an ancient prehistoric burial mound was removed to make way for it. The mound was called Ludan Hlaw (*Luda*'s Grave), hence Ludlow. The church we see today is mostly Perpendicular, restored by the eminent Victorians Sir Gilbert Scott and Sir Arthur Blomfield. There are many good things in the church but nothing better than the 28 misericords in the choir stalls. These are mostly from 1447; amongst the subjects carved in oak are a fox in a bishop's robes preaching to geese. The other major attraction is the area, marked by a wall plaque near the N door, where the ashes of the poet A. E. Housman (1859–1936) are buried. The town walls were begun about 1233 but took nearly a century to complete. Ludlow continued to flourish throughout the Middle Ages and had several royal connections. The 2 young sons of Edward IV, Edward and Richard, were sent here. The elder son was proclaimed King in Ludlow before he and his brother went to London to meet their deaths. (Edward IV was a Mortimer, one of the great Marcher Lord families, whose principal seat from the mid-14th Century was at Ludlow Castle.) Henry VII's eldest son, Arthur, lived here with his wife, Katherine of Aragon. He died and she was immediately betrothed to the king's second son, who became Henry VIII. His daughter, Mary, lived at Ludlow for a time, before she became 'Bloody Mary'. Handsome timber-framed houses were built during the Tudor period. Some of these can still be seen but many more have newer brick facades. The impression one gets today is of a fashionable Georgian town, especially in Broad Street, which is nationally known and highly praised. It is a shame that it has become one enormous car park. There are many attractive buildings in the town. Outstanding is The Feathers Hotel, a magnificent house of 1603. The timbers of the frame are elaborately carved and beautifully preserved. The Butter cross of 1743 was actually built as the Town Hall. It stands virtually in the centre of the town and faces down the hill along Broad Street. The actual Town Hall is of red brick and rarely receives praise. There are many public houses and inns because this was an important market town. Most of the local gentry had town houses here and there has always been a flourishing cultural and social life. Because Ludlow is affluent and thriving there is little need to worry about out-of-character development taking place. Here it is good business to have good taste. Not so in many less fortunate country towns in South Shropshire. Towns like Clun and Bishop's Castle are more than peaceful and quiet; they are dying. When property becomes empty and times are hard the developers watch with predatory eyes. It is often much cheaper to knock down an old building and rebuild anew than to repair the old. Shropshire is full of little Ludlows, but whilst she needs little help her country cousins need all they can get. There is one cause for concern, however, and that is the rapid rate at which barns of every size and description are becoming converted to dwellings in the Ludlow area. A barn sits quietly in the country; a modern detached, double glazed and centrally heated house does not. What is more, and the planners must know this, the farmer is more than likely to replace the lost storage space with yet another cheap and ugly modern construction. Note: Ludlow castle has several literary associations. Samuel Butler (1612-1691), the author of the satirical Hudibras (1663-1678) was Steward at the castle in 1661. John Milton (1608-1674) wrote his masque Comus in 1634 for the Earl of Bridgewater and it was performed for the first time at the Council Chamber of the castle. Philip Sydney (1554-1586) often visited the castle as a child. (His father was Lord President of the Marches.) Richard Baxter (1615-1691), the puritan divine, was tutored by Richard Wicksted, chaplain to the Council of the Welsh Marches, but was scathing of the education he received here. The castle also has a ghost, that of Marion de Bruyere who, during the reign of Henry II (1154-1189), was in the habit of allowing her lover secret access to the castle at night. On one such visit he came armed with 100 men who seized the castle. Ashamed, bitter and angry Marion killed her lover with his own sword and then lept to her death from the battlements of the Hanging Tower. Her ghost has often been seen since wandering around the ruins.

LYDBURY NORTH *3m. SE of Bishop's Castle on the B4385.*

The name Lydbury means 'the fort on the slope'. At Lydbury North there are traces of earthwork defences, probably erected in the 9th or 10th Centuries when Danes were striking inland by sailing up the River Severn. (In Worcester some Danes were captured and skinned alive, such was the hatred they had engendered.) Lydbury North was already a flourishing market town when Bishop's Castle was first founded in the 12th Century. It is still a fair sized village but is no longer a busy administrative centre. In recent years some new houses have been built in the village and the school of 1840 has been enlarged. Many of the residents work in Bishop's Castle. The large Norman church with its massive tower stands in the middle of the village on a raised mound. The transepts are named after the two most prominent local landed families, the Plowdens (N chapel) and the Walcots (S chapel). There is a room above the 17th Century Walcot chapel and John Shipton began a school there in 1663. The Plowden chapel was founded by Roger Plowden in thanks for his escape from prison at Acre whilst on a Holy Crusade. The church has clear glass and box pews. There are bullet marks on the entrance door. The 17th Century Red House stands by the church, a tall brick building with two projecting wings and Georgian windows. The house has a Cock Pit with a pyramid roof, a relic of the days when the cruel sport of cock fighting was commonplace. Half-a-mile away, across the River Kemp, on the opposite hillside is **Walcot**, a large red brick mansion with a Tuscan columned portico built for Clive of India about 1763 by Sir William Chambers. The ballroom was added in the 19th Century and the house has been altered in recent years. From outside it looks a bit ramshackle and very tired with flaking paint and an uncared for appearance. Young people with battered old cars were in occupation when we visited. The Park has many splendid specimen trees and a long, narrow lake made by damming the river. There is abundant bird life here, especially Canada geese. The woods on the hill behind the house were laid out to spell the work 'Plassey', the victory won by Clive in 1759. However, it was removed during the Second World War because it would have aided German aircraft navigators. To airmen it was a well-known landmark. The Clive family left in 1933 and moved to Powys Castle. Just over 1m. E of Lydbury North is Plowden Hall, a fine Elizabethan black and white house with a private chapel, set on a hill, remote, with views over the River Onny to the Long Mynd. The house was probably built by Edmund Plowden, a leading Elizabethan lawyer and it is still in the same family. The interior of the house is very fine, with much panelling. The name Plowden probably means 'the valley where deer play' but could mean 'the valley where sports or games are played'. Totterton Hall and its small park lie 1m. NE of Lydbury North. It is a late Georgian brick-built house of 5 bays and 2 storeys with giant pilasters and a porch with 2 pairs of tuscan columns. Across the valley, on the tree-clad slopes above the hamlet of Eyton, is the prehistoric earthwork of Billings Ring.

LYDHAM *2m. NNE of Bishop's Castle on the A488.*

It lies in a broad valley ½m. W of the River West Onny in the South Shropshire hill country and is in origin an Anglo-Saxon town, the centre of a substantial manor. The name Lydham is from the Old

Castle Hill House, Marshbrook.

Marche Manor.

Richard Reynolds (see Madeley).

Boulder by pulpit, Mainstone church.

English *hlid* meaning 'a slope', and *ham* meaning 'a low lying meadow by a stream'. The earthworks of the Norman motte and bailey castle lie to the W of the squat Norman church of Holy Trinity, a grouping suggestive of a planned village. The original 13th Century, church was restored in 1642 and largely rebuilt in 1885. There is a Jacobean pulpit and a good timber roof. Lydham Manor adjoins the A488 ½m. S of the village. The house lies on rising ground in a small, well tended park in which is the county's largest oak tree. Indeed, only the Bowthorpe Oak in Lincolnshire exceeds it in the whole country. The Lydham oak is heavily berried and covered with mosses, lichens, holly and elder. It measures about 30 ft. round its girth. Also in the park are 2 very large walnut trees and the largest fern leafed beech in the county. The Sykes family acquired the Lydham estate in 1897 and reside at the Manor House. (The old Manor House was pulled down in 1968.) Before then the Oakeleys were the major local landowners. The other large house in the village is Roveries Hall of 1810. As to the rest there are cottages, council houses, bungalows, farms and smallholdings but no shop, pub or village hall. Local employment is in farming and forestry but as everywhere these days there are many commuters. Lydham lies in an officially designated Area of Outstanding Natural Beauty. The **Roveries** are a handful of dwellings on the lower slopes of a wooded hill that lies between the A489 and the A488 about 1½m. NW of Lydham. On the hill are 2 prehistoric fort-settlements, the Upper and Lower Camps. The Lower Camp is protected by banks and ditches and the summit of the hill rises high within the defences. The Upper Camp lies at a level 50 ft. above the Lower but looks to be higher because of a dip between them. The SE defences, facing the Lower Camp, is a wall of rocks, and in the NE there are 2 impressive ditches cut into the shale. The enclosed area, for all this labour, is quite small: an oval 400 ft. in length. Lydham Heath lies 1m. SW of Lydham. There is a pool in a wood and the course of a dismantled railway. Rather more interestingly a hoard of Bronze Age weapons was found in Bloody Romans' Field during drainage operations. Included in the hoard were 'a lunate spearhead, other spearheads, and fragments of three more as well as part of three swords, one of them being leaf-shaped'. It is thought that these articles, and probably many more like them, were thrown into the water as propitiatory offerings to water gods.

LYTH HILL *See Bayston Hill.*

MADELEY *1½m. ENE of Ironbridge.*
It is now, for its sins, a part of Telford New Town. The N boundary is effectively a modern road, the Queensway; likewise, to the E the boundary dividing it from the Halesfield Industrial Estate is the Brockton Way (A422). Madeley now consists of essentially 4 parts: Woodside, a new housing estate to the W; the old town in the centre; Sutton Hill, another modern housing estate to the E; the Blists Hill museum complex to the S. The old town of Madeley is probably of Anglo-Saxon origin. The name means 'the wood (or clearing in the wood) by the Mad Brook'. In the 13th Century the manor belonged to Wenlock Priory which developed and extended the town by clearing more woodland. By 1300 it was flourishing and had a large number of burgage tenants with their plots running at right angles off the main street. Coal was likely to have been mined at a very early date. In 1711 deep mines were operating and later in the century there was an underground network of tunnels with wooden rails, along which were pushed tubs of coal to the River Severn where they were loaded on to boats. Between 1782 and 1801 the population of Madeley doubled and it doubled again between 1801 and 1871. The town can be divided into 3 parts: The area around the church, the site of the earliest settlement but which was rebuilt in the 18th Century; the High Street, which reflects the plan of the medieval ecclesiastical development of the late 13th Century; and the unplanned, irregular industrial area lying to the N. There are a substantial number of yellow brick buildings and several Nonconformist chapels. The octagonal parish church of St. Michael was rebuilt in 1796 by Thomas Telford but it still stands on the pre-Christian earthwork platform occupied by its predecessor. Lying within 100 yds, downhill from the church, are a pair of semi-detached, three-storeyed, Georgian style houses. These have brick facings but are mostly made of well dressed sandstone. This stone came from the old church. Charles II hid in the barn of Upper House, on 5th September 1651, during his flight after the Battle of Worcester. North-west of the church is the Old hall, a good brick house of about 1700 with a hipped roof. This has now been developed and used as the nucleus for a first-class complex of old person's flats. The coach house and lodge and other buildings have all been incorporated and most sympathetically restored and rebuilt. The red brick Italianate Anstice Memorial Club and Institute of 1869 is said to be the first Working Mens Club in England. North of the town by half a mile, in open ground, are the remains of the once magnificent grey limestone Madeley Court, now in the process of being restored to its former early Elizabethan glory. It was built by Sir Robert Brooke, the leading lawyer of his day and speaker of the House of Commons. Before the Dissolution of the Monasteries there had been a grange, an outlying farm, of the priors of Wenlock on the site. Brooke's mansion must have been an imposing building. The gatehouse, with its twin polygonal towers has survived virtually complete, probably because it had been converted into cottages. The 'L' shaped hall has gables and dormers and the windows are mullioned and transomed. Behind the house is a lake and in the left-hand garden is an 'astrological toy', a massive lump of carved stone that used to sit on a 15 ft. high pillar. In front of the Gatehouse is a tree covered mound of industrial spoil that is believed to be on the site of the medieval fish ponds. It was a later Brooke – Sir Basil Brooke – who built the Old Furnace at Coalbrookdale, in 1638, in which Darby made his experiments in the use of coal to smelt iron. In 1709 Darby took up residence at Madeley Court and lived there until his death. Three other illustrious residents of Madeley were Thomas Randall, William Fletcher and Lord Moulton. Thomas Randall had a small factory producing soft paste porcelain. His wares were never marked with his name so authentication is difficult, but his work was of the very highest quality and is therefore very valuable. The Reverend William Fletcher was vicar at Madeley from 1760 to 1785. He was a friend of John Wesley and wrote his biography. The old Rectory, where he lived, still stands beside the church, opposite the National School. Lord Moulton was a lawyer and an expert on high explosives. He left the Court of Appeal to join the War Office where he was responsible for the production of more than one million tons of explosives; and yet a man with so much blood on his hands has been described as 'one of the most delightful... men of our time'. The old ironworks at **Blists Hill** are now the centre of an international award winning museum which includes the Coalport Canal, the Coalport Inclined Plane, and the old Tramway System. All of these are in their original setting, together with buildings, such as printing works and toll houses that have been taken from elsewhere in the area and rebuilt on the museum site, and collections of machinery and castings. The Blists Hill site is now linked to these in Ironbridge and Coalbrookdale and together they form the Ironbridge Gorge Museum which attracts visitors from all over the world. (*See Ironbridge.*) Leading away from the SW corner of Sutton Hill is Great Hay Drive. This leads to Hay Farm which has a history dating back to 1238 when the monks of nearby Wenlock Priory granted permission for a hay, an enclosed deer park, to be made here. It remained in church hands until the Dissolution when it was bought by the Brooke family. It was Basil Brooke (of Madeley Court) who sold a derelict furnace to Abraham Darby in 1708. In 1771 the Darby family themselves purchased Hay Farm to provide hay for the horses that hauled tubs of minerals along the railways that linked their ironworks. In the 1780's Abraham Darby III sold part of the farmlands to Richard Reynolds whose son, William, established the new town of Coalport. After several changes of ownership the old Hay Farm is now the nucleus of the Telford Hotel and Country Club. The Telford golf course lies around the hotel. Hay Farm is a substantial building of brick with Georgian sash windows, hipped roofs and tall chimneys decorated with round-headed blank arches. The foundation plinth is stone and there is some stone in one of the large, handsome barns. An adjacent barn is dated at 1775. Behind the modern entrance block there are some tasteless

Maesbury, the church of St. John.

motel-like extensions, stuccoed with aluminium-framed windows. The views are marvellous. The Severn Gorge, the Iron Bridge, Jackfield, Buildwas power station, Benthall Edge, etc., are all laid out before one.

MAESBROOK *6m. S of Oswestry.*
Maesbrook lies on the B4398 between Llanymynech and Knockin. There are 2 parts to the village, the dry upper and the often wet lower. The River Morda passes close to the settlement near its confluence with the River Vrynwy. Both waterways are prone to flooding and periodically overcome the earthen banks raised to protect the fields. There were 2 mills, one ground corn and the other made paper, but both are now dwellings. There is a Methodist church of 1844 and an Anglican church, St. John the Baptist of 1878. The pub has survived but the village shop and Post Office is now closed and the old railway line ¼m. S has been dismantled. Judging by the farm names Maesbrook is more a part of Wales than England.

MAESBURY *4m. SSE of Oswestry.*
A hamlet consisting of a few dwellings along the main road and the quaint church of St. John in white-painted corrugated iron with a black, tarred roof. Next door is a chapel-like building with a bellcote and a candy-twist chimney, now a house. Maesbury Council School is ashlar and has a blue enamelled clock. The countryside around is flat but the foothills of Wales lie a few miles to the W. Cattle and sheep graze the pastures between fields of wheat. Just to the N of Maesbury is *Ball*, so named after the pub, The Original Ball, the sign of which depicts a globe of the earth. There is a stone chapel close by. To the SE of Maesbury is **Maesbury Marsh**, the largest settlement of the three. It has a school, a shop and 2 pubs. The houses are mainly of red brick with some stucco and there are a few stone cottages. For the most part this is an early 19th Century canal settlement that developed as a port on the Shropshire Union Canal. The Navigation Inn, looking quite derelict but actually a going concern, has stables around the rear courtyard for the horses that towed the narrow-boats. (An old man told us that at Pant, 2½m. SW, between 30 and 50 horses were used to haul an ice-breaker boat when the canal froze over.) The warehouse and wharf-crane stand in good order and the canal still holds water though it is now neglected and is rapidly becoming overgrown with weeds. There was also a bone-glue factory, a grain store and a flour mill at the settlement. The name Maesbury means 'the fortified place near the boundary'. The boundary is almost certainly Offa's Dyke, 3m. W. One mile SE of Maesbury at Woolstan is the beautiful St. Winifred's Well. (*See West Felton.*)

MAINSTONE *3m. WSW of Bishop's Castle.*
The village lies deep in the South Shropshire hills, in the heavily wooded valley of the River Unk, an old name if ever there was one. 'Mean' is Welsh for stone, so the name of the village literally means 'stone-stone'. In the church, beside the pulpit, is a small, grey granite boulder weighing something over 200 lbs. Exactly why it is there no-one knows. There are several jumbled legends involving this stone and its possible original use – as a measuring weight for bags of corn, as a test-of-strength stone etc. It could even be a pagan relic. The Celts were in the hills around here long after the Romans had gone and they worshipped stones, amongst other natural objects and elements such as trees and water. Early Christian churches often had 2 altars, one of which included pagan idols to lessen the culture shock of a change in religion. The early missionaries were no fools. They concentrated their efforts on Kings and Chiefs. Once a Chief was converted he would decree that henceforward all his people would adopt the new faith, like it or not. Most of his people, naturally, did not like it and the church had to make counciliatory gestures. The Mainstone Stone could well be one such gesture that has lingered. Local people, having long forgotten its original function, would then speculate on its meaning and purpose, and all manner of both fanciful and practical stories would develop over the years. Another curiosity is the position of the church itself. It lies 1m. W of the village in the Cwm Frith valley next to Offa's Dyke. Its sole companions are 2 cottages. Together they constitute the hamlet of **Churchtown**. So why should the parish church of Mainstone be at the hamlet called Churchtown? No-one really knows; not the vicar and not the Diocesan Archivist (of Hereford). The most likely sequence of events is that the original Anglian Church was at Mainstone. Then, the centre of population moved up the valley, a movement which we know did occur. Maps of as late as 1750 show Churchtown as a larger village than Mainstone and local farmers can point out the sites of now dismantled cottages at Churchtown. When the old church in Mainstone needed to be rebuilt it was decided to move it to the now more populous settlement up the valley, which was then called Churchtown. Then, Churchtown suffered a decline in population and Mainstone once more became the larger settlement. In Victorian times, when the church was again rebuilt, it was not moved but left at Churchtown. Churchtown is a delightful little place with a clear stream and a field of pasture surrounded by high hills. It is in fact a prettier and more friendly spot than Mainstone. In Mainstone today there is an air of decay and several stone-built cottages are in a ruinous condition. There are in all only some half dozen houses in occupation: a farm, a Primitive Methodist chapel of 1891, which is still well attended, and a telephone box. The road from Mainstone to Churchtown continues up Churchtown Hill into the centre of the Forest of Clun. There are spectacular views all around. In these hills are numerous prehistoric settlements and burial mounds. A wild, beautiful place. Note: It is quite possible that the name Mainstone is a corruption of an earlier name such as Manleystone which in turn may well be a corruption of the Old English Maegenstan, which means 'the great stone (or rock)'.

MALINSLEE *It adjoins the W fringe of Telford New Town Centre.*
The name is Anglo-Saxon. Lee is from *leah* which means either 'a forest', or 'a clearing in a forest' and Malin was an early lady owner of the land. Coal was mined here in the 18th and 19th Centuries and in 1797 Thomas Botfield was transporting his coal along the newly constructed railway that ran from Hollinshead (adjacent to the fringe of Telford Town Centre) to Sutton Wharf on the River Severn. The Botfields were, from all accounts, paternal masters. In 1828 Berriah Botfield III was married and some 2,000 colliers and ironworkers were invited to the wedding reception where they consumed strong ale and 4 fat oxen. There is still something left of the old village: houses of red brick, both humble and relatively grand; the church of St. Leonard, built in 1805 of yellow stone to a stretched octagon pattern with a sturdy tower and round leaded windows, all set in a lawned churchyard; a handful of scattered shops; and 2 pubs, the Church Wicketts and the Ring O'Bells. Early Council houses sheathed in grey pebble-dash infringe on the old centre adding little to the townscape. The ruined Norman chapel that once stood S of Malinslee Hall, in what is now Sainsbury's car park, was removed in 1971 and rebuilt beside the Withy Pool in the Town Park, a short distance to the E of Malinslee. It has a nave and chancel divided by a stone screen and was probably used as a shelter, a lodge, by medieval travellers traversing the woods and moors of the Mount Gilbert Forest. (Mount Gilbert was the Norman name for the Wrekin.) To the NE of Malinslee there used to be an industrial village called **Dark Lane**. This was cleared away to make room for the new Telford Town Centre. The village had some 60 cottages, arranged in 3 terraces, in which lived the operatives of the Old Park Ironworks, part of the Botfield empire. Between the Norman chapel and 'Wonderland' the Silkin Way footpath follows a length of the now quiet and wooded Dark Lane, which N of here was once the High Street of the village to which it gave its name. The new centre of Malinslee is amongst the new houses by the pagoda style Shropshire Lad pub. A Forbuoys supermarket, a newsagent, 2 boarded up shops and the Bethesda Surgery make a desultory little group. Hidden away in Wythewood Drive are 'youth houses' and a 'family centre'. Behind them rises the landscaped spoil heap called Spout Mound from where there are good views over that permanent building site called Telford New Town Centre.

MARCHAMLEY *1¼m. NW of Hodnet, on the A442 Whitchurch to Wellington road.*
A most charming village on a low hill with a mature mixture of old black and white

Lutwyche Hall, Wenlock Edge.

view south-east from Churchtown Hill, Mainstone.

Marton Pool, near Worthen.

cottages and houses of red brick and stucco standing behind hedges of holly and copper beech with sandstone boundary walls. There are new houses of some quality and on the main road is a thatched cottage with diamond patterns between pilasters on the brick chimneys. The Old Manor House, in School Lane, is timber-framed with brick infill and is dated 1658, with the initials W.B. When houses come up for sale at Marchamley they don't stay on the market for long. The village green was sold in 1808 to help pay the costs of the Enclosure Act that redistributed the land around the settlement. The entrance drive to Hawkstone Park Hall is in the village. Hawkstone Park itself is best approached through the golf club, 2m. W at Weston-under-Redcastle. At **Marchamley Wood**, 1¼m. NNW, is Vale Farm, a handsome mid-17th Century black and white farmhouse with extensive ranges of traditional brick farm buildings. The land is divided in use between permanent and temporary pasture and arable – winter wheat and winter barley. The name Marchamley is from the Anglo-Saxon *Merchelm's-leah*. *Merchelm* is a personal name derived from *Merce* meaning 'Mercian'.

MARCHE *¾m. SW of Half Way House which is 11m. W of Shrewsbury on the A458.*

Here is the enchanting Marche Manor, a black and white house of medium size beautifully situated on a rise amongst trees and well kept gardens complete with a babbling brook. Two sculptured birds of prey stand guard at the entrance. The house is dated at 1604 and has 2 gables; that to the left front is decorated with concave-sided lozenges and that to the right end has lozenges within lozenges. Marche Hall turns its back to the road but can be seen from high ground on the lane to Vennington. It is an ivy-clad red brick house of 2 storeys with a parapet and 7 front facing windows. However, what catches the eye is the magnificent Cedar of Lebanon in the field in front of the house. Several limbs are broken but it remains a splendid sight. There is little more to Marche other than a large farm called Partonwood and a handful of scattered cottages. The fields are both arable and pastoral. Just S of the railway about ⅓m. E of the manor house is an ancient ringwork (SJ.243.107). It lies on a gentle rise and measures 36 yds. by 27 yds. across with a 3ft. deep ditch still remaining on the W side. In 1066 Marche was 3 manors, held by Leofgeat, Daeging and Wyngeat, all freemen. By 1086 it was held partly by Roger son of Corbet and partly by his brother, Robert son of Corbet. The Corbet *caput* was Caus Castle, 2m. due S. The name Marche (*Messe* in Domesday Book) could be either from the Old English *Merce*, meaning 'Mercian people' or *mearc*, meaning 'boundary', or *merece*, meaning 'smallage' (wild celery).

MARDU *See Cefn Einion.*

MARKET DRAYTON *18m. NE of Shrewsbury.*

A most attractive and largely unspoiled market town with a full range of shops and services. The settlement probably originated as a prehistoric fort which guarded the crossing of the River Tern now utilized by the Whitchurch to Newport road. St. Mary's church stands on the likely site, a headland with steep slopes facing the river. The Anglian town of the mid-13th Century was part of the estates of the Cistercian Abbey of Combermere near Whitchurch in Cheshire. The Abbey operated a deliberate plan of expansion which can be recognized in the street layout of today. In the country around Market Drayton the old medieval open-field system, whereby each farmer had a strip, or strips, in each of several large fields (so that any variety in the quality of the earth would be shared equally) was maintained very late and only came to an end when the land was enclosed in the 19th Century. This is quite posibly the latest date in the whole country. Building development in the later 19th Century reflects the framework of the old field system. The town is mainly red brick with some good black and white – the Crown Hotel, the Star Hotel (1669), Sandbrooke's Vaults (1653), the adjacent Tudor House Hotel, and Freddie's Chinese Restaurant. Many of the town buildings are Georgian, the foremost being the ivy-clad Corbet Arms, the now dowdy Red House and perhaps the most handsome of all, No. 41 Shropshire Street. In Cheshire Street is the Buttercross, a small open-sided market hall of 1834 with Tuscan columns and a pediment. On top is a small bellcote with two fire bells. On market day (Wednesday) the stalls fill Cheshire Street from end to end. There used to be a quaint custom here, a survival from medieval times, whereby persons who quarrelled on market day could have their disagreement settled at a 'court of pied poidre' – you stood before the judge in your 'dusty feet' to settle the matter quickly. The livestock market, called Smithfield, is on the fringe of the town adjacent to the Raven Hotel. By the church are the Clive Steps which lead to the old Grammar School, founded in 1558 by Sir Rowland Hill, a Lord Mayor of London and a local landowner. This was the school attended by Robert Clive (1725–74), later to achieve fame and great wealth in India. His parents lived at Styche Hall near Moreton Say, some 2m. NW of Market Drayton. The church of St. Mary is mostly of the 14th Century but includes an earlier Norman doorway and capital. It is a large church much restored by the Victorians; inside there is some good stained glass. As a boy Clive is said to have climbed the tower – no easy task, but less harmful than the protection racket run by him and some of his young friends, whereby they guaranteed the safety of local shopkeepers' windows for a small weekly consideration. The most handsome road in Market Drayton is Great Hales Street which runs from the Clive Steps to the large, red brick house called the Grove which is now a school. The Shropshire Union Canal runs along one side of the school playing fields and close to the road by the bridge is a Second World War machine gun bunker. Adjoining the bridge is the Market Drayton canal basin. This was once a busy inland port and several of the ancilliary buildings still stand. Today it is a marina for pleasure craft. There used to be a substantial water mill by the river, below the church, but this is now gone. To the N of the town, in Cheshire Street, is a Victorian cornmill, part of which is still used for its original purpose. The tall, 6-storey, 3-bay block was erected in 1899. Across the road is the modern Palethorpes factory. On the A529, the road to Newport, on a small hill overlooking the River Tern, is Tyrley Castle. The original castle was probably built by Ralph de Botiller, Baron of Wem in 1247. By 1406 it had passed to Margaret Corbet of Moreton and by 1619 to Gilbert Lord Gerard of Bromley. At the time of Dr. Plot's visit in 1689 it was owned by R. Church Esquire. Today the castle mound is quite evident but the castle has given way to a handsome farmhouse. Just down the road, towards market Drayton, is the open air swimming pool, a popular summer attraction. There are a number of large houses and halls in the country around Market Drayton. Tunstall Hall lies 1½m. NE, just off the A53. It is a large, 9-bay, red brick house of about 1723 and stands above the River Tern. It was once a school and is now a home for the elderly. Buntingsdale Hall, 2m. SW of the town centre, was built in about 1730 of brick with stone dressings. It has 9 bays to which was added an extension of 1860. The Hall had a large park and a lake but the house and the estate were vandalised by the R.A.F. before being sold off in bits and pieces. The airforce is still in the area, though. (*See Tern Hill.*) Market Drayton is essentially a country market town and there is little industry here – a pork products factory, a cornmili, a milk bottling plant, a timber treatment centre and sand and gravel pits at Almington and Pipe Gate, both operated by A.R.C. Western; all are country-based enterprises. The old names for Market Drayton were *Draitune* and Drayton in Hales. The first element of the name is from the Old English *draeg*, a word which is never found alone but always in compound words. It can mean several things but the general sense seems to be 'a place where something is dragged', e.g. a boat taken out of a river and dragged over land around an obstacle and re-launched beyond it. Market Drayton stands above and beside the River Tern and this porterage meaning is the most likely. (*See Little Drayton.*)

MARSHBROOK *3m. SSW of Church Stretton.*

It lies at the point where the A49 leaves the route followed by the previous Roman road and makes a diversion to the E. It rejoins the Roman road just S of Wistanstow 3½m. to the S. Marshbrook is a busy little hamlet with a pub, a garage and a level crossing on the main line railway which runs from Shrewsbury to Ludlow. The signal box is of interest to enthusiasts, we understand. A little to the S, overlooking the main road, is a pretty black and

Countryside between Morville and Much Wenlock.

View over Ironbridge from Hay's Farmhouse, Madeley.

Horses in morning mist, at Mickley.

Meeson Hall.

white house of the late 16th Century with upright studding, 2 small sandstone extensions, leaded lights, carved bargeboards, 2 dormer windows, brick chimneys and a yew tree in the garden. This was formerly known as Post Office Holding (the owner had some fields across the road alongside the river) but in later years was re-named Castle House. Local tradition has is that there was a stone-built castle on the hill (Castle Hill) just SE of the house at SO.445.897 and that there are some remains in situ. A lane leads from the busy A49 to Acton Scott, well-known for its Farm Museum. A track leads N of this to **Swiss Cottage**. This is a hamlet of 3 stone dwellings and a half-timbered cottage with an overhanging roof giving it the superficial appearance of a Swiss mountain chalet. In the lower part of the valley, by the stream, is a plantation of young conifers.

MARTON *1½m. NE of Baschurch on the B4397.*
A scattered hamlet. If it has a centre at all it is the road junction between Marton Farm and Marton Hall. The Hall is screened from the road by an earthen bank and a small wood. It is a substantial house built of red sandstone with varying roof levels, mullioned and transomed windows, gables and dormers. To the left is an unbelievable accretion, a red corrugated-iron car-port. Such a handsome house does not deserve to be so treated. An Irish wolfhound prowls the grounds eying one suspiciously. To the left of the Hall are derelict mews cottages and the Hall Farm. The premises are now used by a corn merchant, W. L. R. Gwilt. Just to the E, on the main road, is another junction at which is situated an old toll house and a telephone box. To the SW is the most attractive Tan House a timber-framed building of about 1620 which was originally a tannery but is now a private dwelling. To the front of the house, by the stream which is the parish boundary, are some large stones. These are all that is left of an old mill. Tan House was thatched by a local craftsman and on the roof-ridge are the traditional straw birds. Just down the road towards Baschurch is the derelict Birch Grove Farm which overlooks the dark waters of the tree-fringed Birchgrove Pool. Marton Pool lies ½m. SE of the Hall and just S of that is another pool called Fenemere. This is a watery place, laced with small drainage ditches.

MARTON *15m. SW of Shrewsbury on the B4386.*
There are 3 other Martons in Shropshire: one near Baschurch (*see above*); and 2 W of Ellesmere, namely New Marton and Old Marton (*see New Marton*). The name means 'the settlement by the mere'. Marton Pool used to cover some 30 acres but has been reduced in size by drainage. It lies ¼m. E of the settlement. A dugout canoe was found in the mud of the lake and is now at Shrewsbury Museum (but is not the one currently on display). Between the pool and the village is a mound, believed to be a tumulus, a prehistoric burial place. Today the wood and reed-fringed pool is used for sailing and fishing. Alongside the northern shore is a residential caravan site, one of the best kept and most attractive parks we have seen. We didn't see it but a 'Heated Penguin Swimming Pool' is advertised. On a hill called Marton Crest is Bray's Tenement, the birth place of Dr. Thomas Bray (1656-1730). He began the Parish Libraries scheme which developed into the influential SPCK (the Society for the Propagation of Christian Knowledge). He was later sent to Maryland, U.S.A., where he founded the SPG (the Society for Propagating the Gospel). In the village there are some black and white houses; the stone-built church of St. Mark (1855); a Post office and general store; a Rural Crafts Centre, housed in an old stone chapel on a bad bend in the road; some modern bungalows; a pub, the Sun Inn, in which minor court cases were once conducted; an old school of 1864 which closed in 1948, reopened in 1951, and closed again in 1984; and the attractive Marton Pool Hotel (Lowerfield Brook) which has a thatched roof and used to have its own brewery. To the SE, across the valley of the Aylesford Brook, is Wilmington where there is a Norman motte and bailey castle, and 1¼m. S is Wortherton where there is a Norman motte. Binweston lies 1¼m. NE of Marton. This is really a single large farm. The house is of 2 levels, presumably built at different times. The lower parts are stuccoed and the upper parts are timber-framed with stucco infill. In the hills 1m. W is the strangely-named Flying Dingle. (*See The Beeches.*)

MEESON *1½m. NE of Waters Upton, which is 5m. N of Wellington.*
A spread-about little place with trim cottages of red brick and stucco, and sandstone boundary walls. It stands on a rise close by the River Meese from which it takes its name. Meeson Hall lies away from the hamlet embowered in mature trees. It is a handsome mansion of 1640 constructed of large ashlar blocks of red sandstone. It has 3 straight gables, tall red brick chimneys and small extensions to both sides. Inside there is panelled dining room with a particularly good carved fireplace. Meeson (Meiston) was originally a part of the manor of Isombridge (Asnebruge) which was held by the powerful Mortimer family. Shortly after 1086, however, it went to the holder of the office of, King's Forester of Shropshire. Isombridge declined in importance and became a part of the scattered manor of Great Bolas (Bowlas).

MELVERLEY *10m. WNW of Shrewsbury.*
Most easily approached off the B4393. The road from Crew Green to Melverley crosses the River Vrynwy by means of a single-track metal and wood-boarded bridge. The Vrynwy marks the border here between England and Wales. We meet her again in Melverley where she flows within a few yards of the 16th Century church of St. Peter. This is one of only 2 old timber-framed churches in Shropshire, the other being at Halston Hall. A farm stands next to the church. The village could be most attractive but is spoiled by the untidiness of the public house. There are good views to the Breidden Hills and Rodney's Pillar (an obelisk). Admiral George Brydges Rodney (1718-92) invented the naval warfare tactic of 'breaking the line' which was later used by Nelson at the Battle of Trafalgar. Melverley was one of only 7 parishes in the county to have their old open fields Enclosed by Act of Parliament. Most 18th Century enclosures and redistributions of common lands were made by local people simply meeting and handling the matter between themselves. (Acts of Parliament and the fees of Commissioners had to be paid for.)

MEOLE BRACE *1½m. SW of Shrewsbury.*
It is now a suburb of Shrewsbury. The large, red sandstone church of Holy Trinity (1867) is the third church on this site. It is known for the very fine stained glass by William Morris and Burne-Jones; that around the altar is considered to be the best they ever designed. These 3 windows depict, in the middle, the Crucifixion with the Virgin Mary and Angels, and to the left and right Holy Kings, Martyrs and Apostles. In the left apse window are scenes from the Old Testament, and in the right apse window scenes from the New Testament. Mary Webb, the novelist, was married in the church and is buried in the churchyard. The old part of the village is most attractive; Church Row, Church Lane and Upper Road encircle a wooded island on which are only a few house. This might well have been the old village green. Most of the houses and cottages are Georgian and Victorian and are now nicely mellowed. Meole Brace Hall is of brick, 2½ storeys high and 3 bays wide, and lies more or less secluded in nicely kept gardens wherein is the site of the castle, burned down in 1669. Lucy Bather, the children's author lived at the Hall from her birth, in 1836, to her death, in 1864. The village is bounded to the N by the busy Longden road. Here is a large, modern school looking like a red brick barracks. There is an Industrial Estate and amongst the tenants are a Veterinary Cash and Carry (there can't be many of them) and the succinctly named "Shropshire County Council Direct Labour Organization Central Depot Road Safety West Mercia Supplies". As to the name Meole Brace; after the Norman Conquest the Anglo-Saxon manor of Meole (the settlement stood on the Meole Brook) passed to a branch of the Mortimer family and from them to Andulf de Braci (about 1206) whose name became attached, Meole Braci becoming Meole Brace. Meole is quite possibly from the Welsh *moel* meaning 'bare', probably of a hill, or low rise.

MERRINGTON *6m. NNW of Shrewsbury.*
Percy Thrower, probably the best known gardener in Great Britain from the 1960's to the 1980's, lived at The Magnolias, Merrington. He is survived by his wife,

View over Minton.

A Maiden's Garland, Minsterley church.

Carved bench head, Middleton-in-Chirbury church.

Mitchell's Fold, Bronze Age stone circle, on Stapeley Hill, near Middleton-in-Chirbury.

Connie, and their personal garden is opened to the public occassionally – though, in fact, much of it can be seen from the road. Their house is a modern dormer bungalow. The settlement stands on a low rise, the name means 'pleasant hill', and there are views eastwards to the wooded ridge of Pim Hill. The hamlet is well-wooded and is very neat and trim with well clipped copper beech hedges and well manicured lawns. Merrington Hall is built of red brick under a hipped roof with dormers and has a flat-roofed curved bay. Merrington Grange is stuccoed and is presently run as a residential home. The Victorian village pump stands over a deep well paid for by Robert Aglionsby Slaney and his wife Elizabeth (née Mucklestone) heiress of Merrington and Walford, as we are informed by the commemorative plaque. At the back of the Hall is a part-timber-framed cottage, a reminder that Merrington is an ancient place. In 1066 the manor was held by Hunning. By 1088 it had passed to the Norman, Picot, (the nickname of Robert de Say from Sai in the Department of Orne, France). He had 9 slaves to work his land and 3 villagers, 4 smallholders and one rider also lived here. In Domesday Book it is called Gellidone and in the early Middle Ages was given to Haughmond Abbey. On the road to Walford is **Merrington Green** which seems to consist of little more than 2 cottages; a corrugated-iron bungalow, now rusty red and mossy green; a farm called The Hayes; and a scrubby heathland of bracken and birch in which a Nature Trail is being constructed. In the past clay was mined here and used at the Leaton Brick and Pipe Works and the Old Wood Brick and Pipe Works; both have now closed. During the war there was an American Army Camp on the common land. This was later used to accommodate German prisoners of war. Further along, on the way to **Old Woods**, the road runs through the conifers, broadleaved trees and laurel bushes of Old Wood Coppice, and then past the mostly deciduous Old Wood. The main line railway (Shrewsbury to Gobowen) is crossed by a red sandstone bridge and the hamlet of Old Woods appears. Beyond the pub, the Romping Cat, are cottages of stucco, brick and stone; a few new houses; a small pool; and a tangle of overhead electricity cables and telephone wires.

MICKLEY *4m. WSW of Market Drayton.*
The name means either 'the large clearing in the forest', or possibly 'the large forest'. The settlement was part of the Anglo-Saxon manor of Prees and as such belonged to the Eccleshall estate of the Bishop of Lichfield. At the time of Domesday Book the Lichfield diocese was controlled from Chester. Today Mickley is as it probably has always been, a scattered farming community. There are a few red brick cottages, a pair of stuccoed semi-detached houses and 2 substantial farms. Mickley House Farm is now a large bungalow and hides behind a screen of fast growing conifers. Upper Mickley Farm is a more traditional 3-bay, 2½-storey house. Both farms are of red brick. Mickley lies along the road between the A53, near Tern Hill Crossroads, and Fauls. This follows the line of a low ridge and for most of the way is an 'enclosure road' with deep ditches on both sides. Northwood Grange lies ¼m. W, a modest 3-bay farmhouse set back from the lane behind a field of pasture that contains an orchard and small flock of goats, one of which was busily ring-barking a tree when we were there. There are sheep and cattle in the fields. **Lostford** Hall and Lostford Manor lie ¾m. SE of Mickley at the end of the ridge and overlooking the wide, flat valley of the River Tern. The river has been deepened and is now little more than a drainage channel, little wonder the ford is lost – or was it originally 'last'. The 2-storey, 3-bay red brick Hall is of 1540 and with its farm and a screen of trees makes a pleasing and compact group. On the left of the concrete entrance drive, adjacent to a very tall barn, is an earthen mound now clad with trees. It looks like a Norman motte but according to local tradition it is a Civil War military work, a gun position perhaps. The present occupants of the house have some 30 black and white cats. At the beginning of the entrance drive is the site of the now filled-in and barely discernable moat of Lostford Manor. The present manor house stands on the opposite side of the lane. Along the main road, which runs parallel with the river are a handful of cottages and houses. The embankments and bridges to the E used to carry the now dismantled railway that ran from Market Drayton to Wellington.

MIDDLETON *1½m. NE of Ludlow.*
A hamlet with a pink, thick-walled, sandstone Norman chapel much restored in 1851. Middleton Court stands close by. The Hopton Brook runs southwards and ½m. down-river is Henley Hall, a large red brick house on the A4117 Ludlow to Kidderminster road. It is of Elizabethan origin with later additions and alterations. The stables are Georgian, as are the excellent wrought iron gates (which were made at Wirksworth in Derbyshire). On the other side of the road, a little further N, is Henley Farm, partly stone with mullioned windows and partly half-timbered.

MIDDLETON SCRIVEN *5m. SW of Bridgnorth.*
Most easily approached down a mile-long lane off the B4363 Bridgnorth to Cleobury Mortimer road. A most pleasant village in downland-like country. The church of St. John the Baptist was rebuilt in Early English style in 1843–8. In a field on the opposite side of the road to the church are 2 old yew trees. They suggest that the original church was adjacent to them and that it was moved when rebuilt. In the vicinity there are some medieval fish ponds which are unusual in that they were established by a lay lord, not a religious body. The name Middleton Scriven is from the Old English *Middel-tun*, 'the middle settlement' and 'scrivener', a public notary, a drafter of documents and sometimes a money-lender.

MIDDLETON IN CHIRBURY *4m. ENE of Montgomery.*
Not really a village at all but a scatter of cottages high on the western slopes of Stapeley Hill. The church of Holy Trinity (1843) is well known for the many fine carvings it contains. They are the work of Waldegrave Brewster, a former vicar of Fitz, and were completed in the early part of this century. On a transept capital he depicts a legend connected with Mitchell's Fold. Mitchell's Fold (SO.303.984) is an important prehistoric stone circle set on the ridgeway on top of Stapeley Hill. Fifteen stones remain out of a probable 30. The tallest is 6 ft. high and the circle is 42 ft. at its widest. There are several legends connected with the ring. One tells of how, during a famine, a good fairy brought the villagers a wonderful cow. Each day the cow would appear in the circle and fill any container placed beneath it with milk. However, a witch milked the cow into a sieve. The cow collapsed exhausted and disappeared into the ground, never to reappear. Nearby are other prehistoric stone monuments such as the Whetstones and Marsh Pool Circle, believed to be Bronze Age. (*See Black Marsh.*) Due S is the conical Corndon Hill, and 4m. to the W is Offa's Dyke, where for a stretch of 2m. it forms the boundary with Wales, as it did in Anglian times. When exploring these hills it is as well to have the largest scale Ordnance Survey map you can obtain. Mitchell's Fold is easily accessible from the road that leads from Priest Weston to The Marsh. It is even possible to drive right up to the Circle, though this should be discouraged for all but the physically disabled. Before the Norman Conquest the manor of Middleton in Chirbury was held by the Saxon warrior Lord Edric the Wild whose exploits have moved into the realms of legend. After the Conquest the Normans came and have left their mark. Between Weston House Farm and Middleton Hall, on the other side of the road, is a medieval moated site and adjacent to this are fish ponds. Less than 1m. E of these are what are described as 'pillow mounds' on the Ordnance Survey map. They are, in fact, a warren, artificial burrows where rabbits were kept and bred. The Normans introduced the rabbit to England and it was an important foodstuff. The population of the area was much greater in times past than it is now. When the nearby lead mines closed the people left. The school and the vicarage are now houses.

MILSON *1½m. WNW of Neen Sollars, which is 3m. SSW of Cleobury Mortimer.*
A small village in pleasant, rolling country. The Norman church of St. George has a squat 13th Century tower with a pyramid roof. Most unusually, the chancel arch has been replaced with a large lintel. The pulpit is Elizabethan. Opposite the parish church is the substantial part brick and part black and white Church House set amongst attractive gardens. At Lea Fields, ¼m. SE on the road to Neen Sollars, there is a moated site. In 1086 Milson was an outlier of the manor of Neen Sollars and was held by

Osbern son of Richard. There were '3 riders, 3 villagers and 3 ploughs'. The name Milson is probably derived from the Anglo-Saxon *Myndel's-tun*.

MINSTERLEY *10m. SW of Shrewsbury on the A488.*
The 17th Century red brick church of Holy Trinity is a collection of 18th Century Maidens' Garlands hung high on a wall, with one in a glass case for visitors to inspect more closely. The garlands are essentially wooden-framed crowns about 1 ft. high and adorned with many-coloured paper ribbons and flowers, though the colours have long faded. Each garland represents a young girl who was engaged to be married but who died before her wedding day. They also sometimes commemorate a woman whose betrothed died before their marriage and who remained true to his memory by never marrying. The W front of the church is most unusual; the doorway incorporates carvings of cherubs and skull and crossbones. Holy Trinity was built by the Thynne family when they moved to Minsterley Hall after their old home, Caus Castle near Westbury, was dismantled. (One of the Thynnes worked as an editor for Chaucer.) The village is large and most attractive with several black and white cottages, numerous trees, a stream and a feeling of being well-lived-in. The restored timber-framed Hall can be seen through the wooded gardens from the main road. There is a Post Office, a general store, a newsagents, a craft shop, 2 chapels and 3 pubs. The Crown and Sceptre is timber-framed and has parts dating back to 1240. It was formerly a house and became a pub in 1840. The Angel changed its name to the Bath Arms, a courtesy to the Marquis of Bath who, until 1920, was the major landowner hereabouts; and the Bridge Hotel was formerly the Miners' Arms, a reminder that coal was mined just to the N, and lead just to the S. Probably the major local employer today is the large Express Dairies factory, situated on the edge of the village and not altogether a pretty sight. The name Minsterley is Anglo-Saxon and means 'the forest, (or clearing in the forest) belonging to the monastery'. It was a Royal Manor held by King Edward which passed to Roger de Montgomery after the Norman Conquest.

MINTON *2½m. SSW of Church Stretton.*
Minton is a very ordinary little place in a magnificent position on the southern slopes of Long Mynd. The name is from the Old English and means 'the settlement on the hill'. The village is most interesting because it has maintained its Anglo-Saxon layout. Farms, cottages and the old stone manor house lie haphazardly around the substantial village green. Behind the manor house is an earthen mound; some think this to be of Anglo-Saxon origin, others that it is the motte of a Norman castle. Perhaps it was both, an Anglian mound re-used by the Normans. King William's men had a 'castle and a sergeantry' here to defend the forest of the Long Mynd and the hays of Bushmoor (*Bissemar*) and Hawkhurst (Hauechurche) in Cwm Head. Minton had been a Royal Manor held by King Edward before the Norman Conquest. By 1088 it had passed to Earl Roger (Roger de Montgomery) who held all the land of the county in the place of King William but who held Minton as one of his personal estates. Most of the manors of the county were let to 9 tenants-in-chief who in turn sub-let their estates to some 140, mostly Norman, families. From the hill above the village there are superb views of the Stretton Hills. The pastures hereabouts are rough and grazed mainly by sheep.

MONKHOPTON *6m. W of Bridgnorth.*
It lies just off the Bridgnorth to Craven Arms road (A458 to Morville and then the B4368). The rendered, orange-coloured church of St. Peter has a Norman nave and chancel and a tower of 1835. Opposite the church is the 17th Century red brick Monkhopton House, built for the 3rd son of the Earl of Wenlock. The village school of 1849 closed in 1983 and is now a dwelling. There used to be a pub called the Wenlock Arms but this is now a farmhouse. Monkhall Grange, 1m. NNW and approached off the road to Acton Round (when the road forks, go left), was founded by Wenlock Priory in the early Middle Ages. Spoonhill Wood covers the hills here, a part of the once great Shirlett Forest. The village failed to develop and survives only as a single farm. There were 3 other similar attempted settlements within 1m. to the W: Masons Monkhall, Harpers Monkhall and Woodhouse Field. Today, they too are only single farms.

MONTFORD BRIDGE *5m. NW of Shrewsbury, on the A5.*
There has been a bridge over the River Severn at Montford from early medieval times. The river-crossing here was a traditional meeting place for English and Welsh negotiators in times of discontent. In 1283 Daffyd ap Gruffydd (son of the great Llewellyn ap Gruffydd), the last true Prince of Wales, was brought in chains to the bridge by his own countrymen and handed over to the English. In 1284 Parliament was summoned to Shrewsbury and adjourned to Acton Burnell. The main business was the trial of the Welsh prince who, as a baron of the English realm, was accused of treason. He was found guilty and his sentence illustrates the bitterness felt by the English to the Welsh. Daffyd ap Gruffydd was tied to a horse's tail and dragged through the streets of Shrewsbury. He was then hanged (at High Cross) and his heart and intestines removed and burnt. Finally, his head was cut off and his body quartered, each quarter then being displayed in a different part of the kingdom. His head and that of his brother Llewellyn, who had been killed in battle in 1282, were displayed on lances at the Tower of London. Note: High Cross, in Shrewsbury, was the scene of several such barbarous executions. The old cross is long gone but it has been replaced by a new, white stone cross on the same site. It stands at the top of Pride Hill and there is a plaque on the wall of Barclays Bank (on the other side of the road) commemorating the death of Daffyd ap Gruffydd, and also the similar fate that befell Harry Hotspur after the battle of Shrewsbury. But to return to Montford. The present 3-arched sandstone bridge was built in 1792 to a design by Thomas Telford, slightly W of the old bridge. Prison labour was used in its construction. Telford's bridge was built at right angles to the river which created awkward bends on the approach roads. Later, he slightly re-aligned the bridge and built an embankment on the S side. His road by-passes the old town of Montford Bridge which now lies E of a small toll house (1793) in a cul-de-sac. During the 19th Century a cross-road settlement developed. Today there is a pub, The Wingfield Arms; a garage; a caravan dealer; a Post Office and general store; and a variety of houses and cottages. The quiet, unpretentious, village of **Montford** lies 1m. SW of the bridge in undulating country close by the Severn. It stands on a hill and the entrance to the churchyard is unexpectedly impressive with 5 wide stone steps and white gates. The church of St. Chad was built in 1737 and restored in 1884. Buried near the tower are the mortal remains of Robert Waring Darwin and his wife, Sussanah Wedgwood, daughter of Josiah Wedgwood, the potter. They were the parents of Charles Darwin. Half-a-mile NE of Montford Bridge is **Forton** Aerodrome used only by civilian aircraft these days. Forton itself is a tiny hamlet of red brick houses and cottages with roses by the roadside. In the summer of 1988 a 155 metre water bore-hole was being drilled next to No.8. Sheep graze in the adjacent field. **Mytton** lies 1m. NE of Forton. It is a pleasant hamlet in the valley of the Perry. The red sandstone bridge over the river is of about 1800. In the gently rolling fields around are crops of corn and potatoes and pastures grazed by cattle and sheep. Mytton Hall, stuccoed with an Ionic porch, incorporates parts of an earlier 18th Century farmhouse and Mytton Mill is now home to several light industrial companies engaged in wood-turning and glass decoration etc. **Bromley Forge**, just S of Mytton, lies close to the confluence of the River Perry and the River Severn. There was a small wharf on the Severn to service the works. Today a grey rendered house stands amongst galvanized iron barns and a great litter of black plastic bags containing hay. **Preston Montford** lies 1m. S of Montford Bridge. It stands on the banks of the Severn at the end of a track off the B4473. At the time of Domesday Book it was a tiny place with 2 villagers and was held by the church. It is now owned by the Council but is still tiny and is really the estate of Preston Montford Hall. This is a good, 18th Century red brick house with 5 bays, 2 storeys, a hipped roof and a pedimented doorway. It is now a Field Study Centre.

MONTGOMERY *18m. W of Shrewsbury.*
Lovely country all around and a castle on a crag make Montgomery a town not to miss. It guards the most important route into mid-Wales and has been occupied by

Hendomen, near Montgomery: the motte & bailey castle of Roger de Montgomery.

Montgomery Castle.

Montgomery, an English town in Wales.

man since Neolithic times. Montgomery is included here because it is an English town and until the late Middle Ages was actually in Shropshire. What is more, Roger Montgomery, who gave his name to the town, held all the land of the county of Shropshire in the King's name and played an important part in its history. The earliest settlement was on **Ffridd Faldwyn**, a hill 1½m. NW of the present town. There are in fact two forts here. The small, inner settlement is Neolithic and covers about 2 acres but later, in the Iron Age, this was incorporated into a much larger fort covering 11 acres. It has a huge rampart and an elaborate SW entrance. The next fort was built by the Romans near the River Severn, ½m. S of the farm called **The Caer** (which means 'fort'), and about 2m. NW of Montgomery. This fort is one of the most important Roman sites in Wales. It was built in AD75 and was occupied and abandoned several times before being held until the Romans left britain in the middle of the 4th Century. It is still being excavated but there is little to see except the outline of the much-ploughed ramparts. After the Romans had left the Anglo-Saxons invaded and settled in England. In 779 the Anglian King, Offa, ordered the construction of the defensive boundary bank and ditch between England and Wales called Offa's Dyke. A stretch of this immense work can be seen in the delightful grounds of Lymore Park just E of Montgomery. To get there take the B4385 towards Bishop's Castle. Just outside Montgomery is a gate on the left. Continue eastwards passing between 2 lakes and into a copse which actually lies on the Dyke. At this point it still forms the boundary between England and Wales. The ditch is on the Welsh side. It can also be seen where it is crossed by the B4386 to Chirbury. After the Norman Conquest large parts of Shropshire (Montgomery was in Shropshire then) and Sussex were given to one of the King's best friends. He was Roger, Lord of Montgomery in the Pays d'Auge in Normandy. During the Conquest he had remained in Normandy as Joint Regent with William's wife, the Duchess. In 1067 he came to England, was created an Earl and built a castle at **Hen Domen** (the old mound). This was a typical Norman motte and bailey castle and the earthworks are still in good order. It lies about 1m. NW of the town. To get there leave Montgomery on the B4385 and after about ¾m. turn right at the sign for Hen Domen. At the village park near the telephone kiosk and walk back some fifty yards. On the right-hand side of the road is a stile. Over the stile, to the left, is the castle amongst some trees. It is being extensively excavated so do not cause any disturbance to the ground works. Earl Roger was one of the most powerful and influential of the Marcher Lords. He would have often stayed at this castle but he also had several others. At the crossroads at Hen Domen turn left, i.e. go NW, past an eyesore of a garage, and over a level crossing with gates that you have to open youself. (If the red light shows check with the railway on the phone provided because the light is often faulty and sticks on red.) The lane comes out at a T junction. Just across the road is an area where you can park amongst the trees, next to the river. The river is the Severn and at this point it can be forded. In the summer it is only about 1 ft. deep. This is the famous **Rhydwhyman**, the spot where for centuries the English, on this side of the river, met the Welsh on the other side. Here the Celtic chiefs met the Marcher Lords to make bargains and treaties. It was also, of course, an important trade crossing, one reason why through the ages this area has always been fortified. The castle at Hen Domen passed to Earl Roger's son, Hugh, and it was during his time that it was sacked by Cadwgan ap Bleddyn, Prince of Powys in 1095. The castle was rebuilt, and in the reign of Henry I it passed to Baldwin de Boulers. It was from him that the town of Montgomery got its Welsh name, Tre Faldwyn (Baldwin's Town). Through the 11th and 12th Centuries the castle was involved in the constant border warfare between the Welsh and the Marcher Lords. In 1223 Henry III ordered a stone castle to be built on the high crag overlooking the present town. By 1225 the castle was complete and the town started to develop almost immediately. Within only 2 years, in 1227, the burgesses obtained a charter from the King allowing them to enclose the town with a defensive earthen wall and to hold fairs on St. Bartholomew and All Saints days and a weekly Thursday market. All these fairs are still held and the Town Walls can still be traced for almost their whole length. (They can be clearly seen on the left of the path up the Town Hill, which crosses them.) The castle was beseiged several times by the Welsh but never fell. It is built on a ridge of rock some 500 yds. long by 60 yds. wide, of which the castle occupies about 300 yds. There are several wards (enclosures) separated by ravine-like ditches. Around the whole castle is a natural rock platform from which there is a precipitous fall of 250 ft. to the road below. Hubert de Burgh (died 1243), Justiciar of England, was a Constable of the castle, as were Prince Edward (afterwards Edward I) and Edward the Black Prince. It then passed to the powerful Mortimer family. In the early 15th Century Owain Glyndwr sacked the town but not the castle. By the end of the 15th Century the Welsh Tudor family were on the throne of England and in Montgomery the Herbert family came to the fore. Lord Herbert of Chirbury, the poet, is of this family. Richard Herbert, his brother, was M.P. for the county in 1601 and Ambassador to Paris in 1619. During his lifetime the castle met its demise. It held out for the King, but surrendered to the Parliamentary forces. Then, Herbert changed his mind and the King's army made an attempt to re-take it. A battle was fought and 3,000 Parliamentary troops defeated 5,000 of the King's men. In 1649 the castle was partly demolished, since when much of what was left has been plundered by the townspeople to build houses. However, enough remains to make a striking ruin and a romantic sight when seen from a distance. The church of St. Nicholas is a handsome building with a rich interior. It was originally a chapel to the priory of nearby Chirbury and can be dated to about 1225. At the end of the 13th Century the transepts were added to the nave and chancel and the whole church was lengthened. At a later date a tower was added but this was pulled down in 1816 and rebuilt. In 1878 the church was restored by Street who, to his eternal discredit, removed the old 2-decker pulpit and box pews. There is a good timber roof with hammer beams. It is believed that the eastern portion of the Montgomery screen, together with the stall-work and the misere seats, were taken from Chirbury Priory at the Dissolution of the Monasteries by Henry III. The S, or Lymore, transept contains monuments to the Herbert family. Most noteworthy is the Elizabethan tomb of Richard Herbert (d.1596), probably the work of Walter Hancock of Much Wenlock. On the floor of the transept are 2 medieval effigies which have given rise to much speculation and conjecture. In the churchyard is the Robber's Grave. It consists of a plain cross cut in the earth on the W side of the path leading from the tower to the N gate. Legend has it that John Davies, hanged for highway robbery in 1821, protested his innocence and to prove it promised that no grass would grow over his grave for at least one generation. No grass grew. The best half-timbered house in the district was Lymore Hall but, regrettably, this was pulled down in 1930. There are several 16th Century black and white cottages in the town but most of the houses were either rebuilt or had facades added in red brick between 1750 and 1850. Today, Montgomery is essentially a Georgian town. Broad Street, which looks like a market square, is most attractive. At its W end is the late-Georgian Town Hall, built in 1748 with the upper storey rebuilt in 1828. The whole character of the town is English, a colonial outpost in Celtic territory. In Gaol Street are the substantial remains of the old County Gaol. We stumbled upon it by accident (the handbooks make no mention). Outside the handsome Gatehouse gaily-coloured laundry fluttered on washing lines. A most strange sight. The front doors are original cell doors. Henry III made Montgomery the capital of the new county of Montgomery and so it remained until the local government changes of 1974. Today, it is a district with a Town Council, equivalent to a parish council in England – a sad decline in status. The town has a population of about 1,000. Except on market days and Saturdays Montgomery is very quiet. Several times we have visited and seen barely a single soul on the streets. Despite this it seems a healthy place in no danger of dying, as are some similar towns like Clun and Bishop's Castle in Shropshire. There is no empty or derelict property worth mentioning and there is a convivial atmosphere. Westwards lies the big, open country, Mid-Wales for which Montgomery makes a good touring centre. About 2½m. SSW of the town is **Cefn-y-coed** and here the film actress

Julie Christie has her home at an old farmhouse called the White House (White Hall Farm on the Ordnance Survey map).

MOORTOWN *1m. WNW of Crudgington which is 5m. N of Wellington.*
A town it most certainly is not. A sunken lane leads north off the B5062 and comes to a dead end at the settlement which consists of 2 farms, a pair of semi-detached cottages, and a stand of broad-leafed trees. The Chestnuts is an irregular brick house with some good and varied traditional farm buildings most of which, like the house, have hipped roofs. The Firs is also an irregular brick house, somewhat spoiled by having had aluminium framed double glazed windows inflicted upon it. The farm buildings are very substantial indeed. The semi-detached cottages have a central block with lower recessed wings and yellow brick decoration. Moortown is a trim and well kept little place. The country is gently undulating and the mainly arable fields are very large. The Weald Moors lie to the E but Moortown is not really a part of them. The land here is much drier. To the W there was an old wartime aerodrome, now put to other uses. (*See Osbaston.*)

MORDA *1m. S of Oswestry.*
The settlement takes its name from the Morda Brook. Morda means 'the great dark river' from the Old Welsh *mor-taf.* The old part of the village lies around the stone-built bridge. Here there is the substantial 4-bay, stone-built, Weston Mill warehouse of 3 storeys with a 4th in the pediment-gable; it is now partly occupied by a craft centre. To the right is a row of brick cottages and behind them a graveyard for old cars. Opposite the warehouse is a red brick church adjacent to which is the site of the old workhouse; the flat-roofed building was the workhouse office block. Morda Hospital is a regular red brick building with a central pediment and pedimented projecting wings of 1791. South of the river the high land is now a place of Victorian terraces and bland modern houses amongst which are the stone-built pub called The Drill, a garage, a Primitive Methodist chapel of 1871, and a Post Office-come-corner shop. North of the river are more old brick cottages. These are adjoined by the mature southern suburbs of Oswestry; large Victorian houses, the cricket pitch and the ambulance station, lie alongside the main road amongst trees and hedges. The road to the W, from The Drill pub, leads past modern bungalows and the Hen & Chickens into a landscape of gentle hills dotted with sheep and colour-washed farmhouses. At **Coed-y-go** (*coed* means 'wood') is a 'holiday centre', a farm with tourist accommodation behind the corrugated-iron silos.

MORE *2m. NNE of Bishop's Castle.*
The name is from the Old English *mor* meaning 'a moor, or a marsh or a fen'. The village stands around the church and together they make a fine group. There are farms, cottages and black and white houses. A narrow road completely encircles the high-mounded churchyard. The church of St. Peter has a squat Norman border tower capped with a double pyramid; the rest, including the More chapel, was rebuilt in 1845. In fields to the W of the church are the remains of More Castle, a low mound about 25 yds. in diameter with a ditch and 2 large rectangular baileys. The area is known locally as the Moatlands. The More family live at Linley Hall, 1m. NE of the village. **Linley** consists of little more than half a dozen cottages. (Note: Do not confuse this Linley with that near Bridgnorth.) The most striking feature of the area is the mile-long approach avenue to Linley Hall. This is, of course, a totally artificial creation with no other purpose than to impress a visitor. The occupants of the Hall rarely use it because there are more convenient accesses. The Hall itself was built by Henry Joynes in 1742. It is a square Palladian-style stone house of moderate size, the first of its kind in Shropshire. It replaced and partly incorporates an older house. There is a good stable-block, a pretty lodge and a classical temple on an island in the lake. The More family were descended from a follower of William, Duke of Normandy and were given these lands after the Conquest. They have lived here ever since, in an unbroken line. (Norman lords often adopted the name of their new English manor, in this case More.) Robert More was an 18th Century botanist and friend of the great Linnaeus. In 1783 he introduced the first larch trees to England, planting some in the Park of Linley. Roman lead workings have been identified in the hills behind the Hall and ingots of lead stamped with Hadrian's name (117-38) have been found hereabouts. There are remains of a Roman building at the NW corner of the avenue leading to Linley Hall and beneath the road from Lydham to Norbury that separates the avenue from Linley Park. A small building with three rooms was found in 1856 and an aqueduct that connected to it was traced for 880 ft. to the NW and probably carried water from springs N of the Hall. Close by were black earth and ashes from a furnace which probably processed lead from the nearby mines. Many other remains were found, mostly walls some of which are 12 ft. thick, in fields covering some 12 acres to the SW of the three-roomed building. These are probably the remains of a Romano-British village.

MORETON CORBET *10m. NNE of Shrewsbury and 1¼m. N of Shawbury.*
One of the most spectacular ruins in the county. It is a surprise to come upon such well-cut masonry standing half fallen in a meadow near a busy RAF airfield. It must have been a fine house, but possibly makes an even better ruin. The mansion is Elizabethan with strong Classical influences: ogee gables and Tuscan columns, mullioned and transomed windows, and flat walls with slightly projecting bays, all of which combine perfectly. The house is dated at 1579, the same year that Sir Andrew Corbet died. It was he who had ordered its construction and the work was continued by his sons. To the N of the house is the castle, a Norman keep of about 1200 with a good fireplace. This was besieged by Cromwell's troops in 1644 and it was they who dismantled the castle and burned down the great house. The church of St. Bartholomew stands close by the castle. It has a Norman chancel; a S aisle of 1330-40; an unusual W window; a tower begun in the 1530's and finished in 1769; and a S chapel of about 1778. The monuments include memorials to the Corbets, the earliest being to Sir Robert, died 1513. Around the castle can be seen earthwork remnants of the medieval village defences. There were at least 13 houses around the castle before the Corbets built their mansion. These were demolished when the park was created and the road diverted. In those days parks were more agricultural estates than the landscaped grounds of the 18th Century. In 1635 the 'House Controller' of Moreton Corbet Castle was John Dutton, a giant of a man who was 8 ft. 4 ins. tall and an ancestor of the Shropshire Giant. (*See Stoke upon Tern.*) In recent years an unusual spirit has been encountered in the lanes between Moreton Corbet and Shawbury: a multi-coloured witch who attaches herself to motor-cars and hitches a ride for a mile or so before vanishing into thin air. The name Moreton is Anglo-Saxon and means 'the settlement (or homestead) in (or by) the moor (or marshy place)'.

MORETON MILL *1½m. NE of Shawbury on the A53.*
Sometimes spelled Morton. The name is from Moreton (Corbet), 1m. WNW, and means 'the settlement by the marsh'. The mill still stands, a red brick building but with several areas of grey stonework in its lower parts. It is not large but has 2 long sheds attached to it, one running to the right, the other to the rear right. They are both only one storey high and we have seen similar buildings used as fulling sheds. Today the mill and its outbuildings are used as a farm store. There is a stand of tall trees in wet ground and a leat (a water channel) leading from the River Roden to the mill. Just to the E is a block of 3 brick-built Mill Cottages and on higher ground close by them is a black and white timber-framed house. The wood that lies alongside the river ¼m. N is called Forge Coppice. There were several furnace-forges on the River Roden hereabouts. (A furnace-forge was not a blacksmith's forge; it was a place where crude 'pig' iron was re-heated and beaten with great hammers to remove impurities. In this way brittle cast-iron was turned into malleable, and more useful, wrought iron.) The furnace-forges used charcoal as a fuel until the early 19th Century and forge owners were responsible for the planting of many of the small woods we see in the country today. A small, main road development is also classed as being a part of Moreton Mill. There are 2 chapels here, a small Wesleyan and a larger Methodist, both with pointed-arched windows. Adjacent to them, to the left, is a 4-bay house. This used to be a

More, the beech grove on Linley Hill.

pub called The Dog. The date 1900 over the door is of the re-building. To the left of this house is a long, low shed built of large, red sandstone blocks. This was originally a malt-house where the inn-keeper malted his barley. The country around is very flat.

MORETON SAY *2½m. W of Market Drayton.*
Here lies the body of Clive of India. After an eventful, and by all standards a successful, life he committed suicide by cutting his throat in a lavatory. His grave is unmarked because in the 18th Century people who deliberately killed themselves were not allowed to be buried in holy ground. He should have been interred outside the cemetery walls. In fact he was buried, local people say, beneath the small side entrance door of the church which lies 30 ft. to the right of the main porch. Clive was born at Styche Hall, 1¼m. from the village, and went to school in Market Drayton. The Hall was then an Elizabethan timber-framed house. The present brick mansion was built in 1760-4 by Clive for his parents. It is a modest place, quite spoiled by having been covered in white paint which is now dirty and flaking. Even worse, it has been somewhat haphazardly converted into several flats. Multiple occupation is rarely good for a building and this is no exception. The park is separated from the house and is used for agricultural purposes. The land was originally quite marshy and is drained by deep dykes to the River Duckow. Seen from a distance the Hall looks quite well, set on a low hill with woods to its rear. In the village the church of St. Margaret is equally unimpressive in its outward appearance. However, the Georgian red brick is a casing for a stone building of about 1200. The tower was added in 1769. The village itself is a pleasant mix of old cottages and new houses. Somewhat surprisingly there is a school, in orange brick of 1871, and a substantial, shiny, red brick village hall of 1911 called the Clive Memorial Church House. The red brick, Victorian-looking Old Hall lies on a rise just to the NE of the church. The white stuccoed house opposite the church is Church Farm. Beneath the rendering is the fabric of 3 timber-framed 14th Century cottages. As to the name: Moreton is Old English and means 'the settlement by (or in) the fen, and Hugh de Sai held the manor in 1199. Note: Robert Clive joined the East India company in 1743 as a clerk. Opportunities arose of which he made the most. In 1751 he led the troops that captured Arcot and in 1757 he took Calcutta and defeated the Nawab of Bengal at Plassey. Clive virtually ruled Bengal until 1760 when he returned to England. He went back to India as Commander in Chief of Bengal in 1764. He made many reforms but also added greatly to his already considerable fortune. In 1767 he returned to England and was attacked in Parliament for improper behaviour in India but successfully defended himself. Nevertheless, he committed suicide shortly after.

MORVILLE *3m. WNW of Bridgnorth on the A458.*
An old settlement, flourishing before nearby Bridgnorth was even built. It takes its name from the Mor Brook. At the time of Domesday Book the town had 17 dependent berewicks (villages or hamlets) and was held by Roger de Montgomery. The church of St. Gregory was the only church mentioned in the whole of the Saxon Hundred of Alnothstree in the 1086 census. During the rule of Edward the Confessor it had 8 canons, but by 1086 this had dropped to 3 priests. In the 12th Century the church of St. Gregory established many daughter chapels in the surrounding countryside, including those at Tugford and Aston Eyre. The village green at Morville was created in the 16th Century, when the Hall was built (1546) and the village re-modelled to improve the view from the big house. The house stands on the site of Morville Priory which had been abandoned after the Reformation. Indeed, the old stones were re-used in the new Hall. Sir William Acton, the Lord of the Manor, had flourishing iron mills at Morville (Morveld as it was then called), and in 1561 was granted a licence to fell trees in Shirlett Forest for use as fuel. In the 18th Century the Hall was virtually rebuilt in grey stone by, amongst others, William Baker in 1748-9. The present church is almost entirely Norman, the nave and chancel being dated at 1118. In a chancel window is an early 14th Century stained-glass picture of Christ being crucified. There is a Whipping Post at the road junction and an Animal Pound in the car park of the Acton Arms. The pub was originally the Abbot's Lodgings and is reputedly haunted by Richard Marshall, 28th Abbot of Shrewsbury. A quarter of a mile NW of Morville, on the A458, is **Aldenham Park**. Travelling northwards out of Morville the splendid wrought-iron entrance gates, with the Hall standing at the end of a long avenue, can be seen on the right-hand side of the road. The house of 1383 was rebuilt in 1691 by Sir Edward Acton, and there are later Victorian alterations. There is a chapel with a white marble Madonna, in memory of Gabrielle Acton who died in her teens on returning from a pilgrimage to Lourdes. The first Lord Acton was a respected historian and Roman Catholic. His extensive library was bought by the multi-millionaire, Andrew Carnegie, and, via Lord Morley, was donated to Cambridge University. The park around the house was probably laid out at the time of rebuilding in 1691. Interestingly, a map of 1725 shows that an iron furnace ('furnis') was situated close to the house and there were small woods, or coppices, planted in the park to provide fuel. Today, there is little left of the park save the ½m. drive – an illusion of grandeur. (By coincidence, the present owner of the house made his fortune in metal working.) Father Ronald Knox lived at the Hall during the Second World War whilst making his translation of the Bible. Aldenham Hall is almost certainly the original of P. G. Wodehouse's Matchingham Hall which appeared in some of his Blandings stories. (*See Acton Scott.*)

MUCH WENLOCK *8m. NW of Bridgnorth.*
An ancient town, famous for its ruined Priory. The name Wenlock is from *gwenloc*, Celtic for 'the white church', suggesting a pre-Roman origin. Throughout the Middle Ages Wenlock had an alternative Welsh name, Llan Meilen, meaning 'sacred enclosure', which implies that there was a large Welsh contingent amongst the population. The Priory of St. Milburga was founded about A.D.680, shortly after the establishment of the Anglo-Saxon Kingdom of Magonsaete, which encompassed Shropshire, S of the River Severn, and the Herefordshire plain. The site of the Priory was purchased from Merewalh, the first King of Magonsaete. (He was the son of Penda, King of Mercia.) Merewalh's daughter, Milburge, became the Abbess. At the time of her death the Priory was the wealthiest in the whole of West Mercia, with extensive, but very scattered, estates. By the 9th Century the monastery was out of use. Some believe it was destroyed by the Danes. In the middle of the 11th Century it was re-founded as a minster church by Leofric, and re-endowed with most of its original estates. After the Norman Conquest Earl Roger de Montgomery raised the status of the church to a Priory once again and brought over Cluniac monks from France. But the buildings we see today are not those of St. Milburga or Earl Roger. They are almost wholly Early English with only a few Norman parts remaining, though these are some of the most interesting, especially the arcading in the Chapter House and the fine carvings (now replaced by glass fibre replicas) in the Lavatorium. The Lavatorium was where the monks washed. Wenlock Priory flourished until the Dissolution (1539). It had wide economic and commercial interests, including a toll bridge over the Severn, copper and silver mines, coal mines and iron works, not to mention its vast agricultural and forestry holdings. The Prior's Lodge, built about 1500, is a very fine building, one of the best pieces of domestic architecture in the country. Today it is, fittingly, a private house. It has a stone slated roof, as do many of the older buildings in the town. This stone was quarried at Hoar Edge, 6m. W of Much Wenlock. The arches and mullions are made from Alveley sandstone. The town possesses many other good buildings. The quaint black and white Guild Hall of 1577 is supported on wooden posts, one of which doubles as a whipping post with handcuffs attached. Like most timber-framed buildings it was prefabricated in the carpenter's yard and only took 2 days to erect. The upper storey has a panelled court room and council chamber. The handsome Ashfield House, half stone and half black and white, was once an inn and Charles I lunched there on his way to Bridgnorth in 1642. At Tickwood Hall, 2½m. away, he met local landowners in the Audience Meadow to ask them for financial assistance. The large sandstone church of Holy Trinity is Norman with a Transitional tower and has 13th Century alterations. The houses, inns and shops of the village

Moreton Say, the church of St. Margaret. Clive of India is buried beneath the small door just off the picture to the right.

Much Wenlock, the Guild Hall.

Much Wenlock Priory.

make an attractive collection, with the local limestone well represented. The Georgian and Victorian buildings blend nicely and modern development has been rightly banished to the suburbs. It has been suggested that Much Wenlock owes its very existence to St. Owen, a French monk who, in the 6th Century, joined the monastery on Bardsey Island off the coast of Caernarvon. He is said to have visited Much Wenlock, and the holy well bearing his name became a place for pilgrims, one of whom was Milburge, the King's daughter. The well can still be seen, beside a classic example of a cruck-built cottage, in the street opposite the lane that leads to the Priory. At Westwood Common, near Much Wenlock, there lived a 19th Century witch called Nanny Morgan. She kept live toads in her cottage and it was said she possessed 'the evil eye', that is, she could make misfortune befall upon anyone she wished. The Olympic Games were first revived at Much Wenlock. The modern movement was initiated by William Penny Brookes who was born in the town and later became a doctor here. The first sporting meetings that he organized were of a light-hearted nature – blind wheelbarrow races and the like – but true athletic events increased in number and by 1870 it was a leading track and field meeting. Prizes included laurel leaves and medals and classical poetry was recited in praise of achievements. In 1890 Baron de Coubertin visited the Olympic Games at Much Wenlock and in 1896 he organized the first international modern Olympics at Athens.

MUCKLETON *2m. E of Shawbury, which is 6m. NNE of Shrewsbury.*
The hamlet lies in flat land off the A53. It must once have been a charming place. There is an old 3-bay, red brick Hall with a hipped roof, a black and white cottage and a good stand of mixed broad leaf trees, but the old brick barns are now engulfed in charmless, galvanized structures; the Hall has been disfigured by the insertion of cheap, skinny-framed modern windows; and the black and white cottage is suffering haphazard 'improvement'. There are cows and sheep in the fields around and 1m. NW of Butlersbank are the battery hen houses of Heal's poultry farm. There are more of their timber 'egg sheds' on the nearby main road, the A53.

MUNSLOW *14m. WSW of Bridgnorth on the B4368.*
It lies in Corvedale with views across the valley to Brown Clee Hill. Most of the village lies off the main road and much of it is constructed of Wenlock limestone. The church of St. Michael has a Norman tower with later work atop it; the chancel and the S side of the nave are Early English (13th Century); and the N nave has interesting early 14th Century windows. Inside there is a large brick from the Great Wall of China, brought here in 1884. The Vicarage is late Georgian with 2 pairs of Ionic columns. Half a mile up the road towards Bridgnorth is Beambridge Smithy, a 19th Century roadside folly in Gothic style and castellated. The Crown Inn was once the Hundred House at which medieval manorial 'Courts Leet' were held. The gabled stone house close to the War Memorial is the birthplace of Edward Littleton, a prominent 16th Century judge, who became Chief Justice for North Wales and later Solicitor General. Just NE of the village, and with an entrance from the main road, is Millichope Park. This is noted for its beautiful lake and superb collection of specimen trees. The entrance drive passes through a sandstone cutting and leads to the attractive Hall which was designed by Edward Haycock and built in 1840 for the Reverend Robert Norgrave Pemberton. It is a square, grey stone house with a portico of fluted Ionic columns and has many original features. In the Park is a rotunda, dated at 1770. North of the Hall, on the road to Rushbury is Upper Millichope Farm, a very rare early 14th Century stone 'tower house'. The upper half-timbered part is of a later date. There was a Norman motte and bailey castle at Munslow but it was obliterated during the 18th Century. The village green was sold to pay for the cost of the local Enclosure Act of 1838; houses were built on it but the road still goes around the site. The name Munslow probably means '*Mundel*'s burial mound', or perhaps '*Mundel*'s homestead by the burial mound'. (*See Aston Munslow.*)

MUXTON *4m. SW of Newport.*
The first element of the name is possibly from the Welsh *mochros*, 'a swine moor'. Muxton was once a separate village but has become a part of the suburban sprawl of Telford. It adjoins Donnington and the huge Army Central Ordnance Supply Depot. There are still several black and white cottages along the Wellington Road, which was the A518 before that designation was transferred to the new road that now runs parallel to it. Timber-framed houses lie to both sides of the stuccoed White Horse Hotel and on the other side of the road is No. 57, a quaint little thatched cottage in good but not over-restored order. Very small, old cottages like this are much rarer than larger houses. A modern pub, the Sutherland Arms, is a reminder that the land hereabouts was part of the estate of George Granville Leveson Gower who became the Duke of Sutherland 8 months before he died. There were once 2 corrugated iron mission churches in the village but St. Chad's was dismantled and re-erected at the Blists Hill Museum, Madeley. St. John's in Muxton Lane is still here, though, and still in use. Also in Muxton Lane is No. 40, a good timber-framed cottage with a jettied first floor. Muxton Lane is destined to become well known because it is one of the access routes to the new Granville Country Park. Some 350 acres of the wild, forgotton wastes of Donnington Wood have been made safe by the capping of old mine shafts and laced with paths that connect areas of archaelogical and natural interest. The star feature is the Lodge Furnace complex which stands next to the old Granville Colliery. (*See Donnington Wood.*)

MYDDLE *8m. SE of Ellesmere.*
It lies on the A528 Ellesmere to Shrewsbury road. Up until the 16th Century the land around Myddle and its 3 large open fields was covered in dense forest. Between 1550 and 1650 this was cleared and sold for fuel to local glass furnace owners and ironworks. The castle was built in 1307 by Lord Lestrange of Knockin. In the 16th Century it was owned by the outlawed nobleman Humphrey Kynaston (*see Nesscliffe*) and whilst in his charge it fell into disrepair. Today it is little more than a much reduced mound with what is virtually a folly standing on top. It lies at the back of a farmyard with chickens and a horse for company. There is a great deal of fully dressed stone around the farm, presumably from an older, now dismantled, building, possibly the castle and its outbuildings. The church has a tower of 1634 but the rest of it was rebuilt in 1744 and Gothicized between 1837 and 1877. Myddle is famous, not for itself, but for the book 'The Antiquities and Memoirs of the Parish of Myddle' by Richard Gough, 1635–1723. This gives an almost unique glimpse into the everyday lives, gossip and general goings-on of a Shropshire county village and the neighbouring market towns of Shrewsbury, Ellesmere and Oswestry. As to the village today, there is a handsome old inn and too many new houses. Just W of Myddle, on the road to Baschurch, is a sign to **Fenemere**. A short way down the lane is a good black and white thatched cottage with a matching garage called The Oaks, and a striking blue brick house with yellow brick ornamentation and tall chimneys. Fenemere Pool is in fields ⅔m. SW.

MYNDTOWN *4m. E of Bishops Castle.*
Set at the foot of Long Mynd, you approach by narrow lanes to find a tiny hamlet consisting only of a roughcast church, a few cottages and a farm, but the views are reward enough. A mile and a half to the NE, on top of Long Mynd, are the ancient Portway and the modern Gliding Club. They can be reached from Asterton, 1m. N of Myndtown. The road from Asterton is very, very steep. South of the Gliding Club are several burial mounds.

NANTMAWR *4m. SW of Oswestry.*
Here, near a disused quarry, is a Shropshire Trust for Nature Conservation reserve covering about 7 acres of limestone, woodland, cliff and scree. It is called Jones' Rough and is at Ordnance Survey map reference SJ.247.247. Part of the wood is thought to be ancient and it contains many yew trees and some wild cherry. On the ground are spurge, laurel and stinking hellebore. The cliff and scree above the woodland have a wealth of wild thyme, common rock-rose and, hairy rock-cress. The name Nantmawr is Celtic and means 'the great brook', *nant* being 'a brook'. Offa's Dyke runs ¼m. to the E. Today, now that the quarries, lime kilns and brick works have ceased production, the population has fallen to a level that has forced the closure of the shop, the

Nesscliffe, Kynaston's Cave.

school, the United Reform Church and the pub, the Carver's Arms.

NASH *2¼m. NNE of Burford, which lies over the river from Tenbury Wells.*
It lies just off the B4218 amongst small hills and valleys and narrow lanes with views towards Titterstone Clee Hill. Milk and cereal production are the mainstays of agriculture here. Nash is a tiny red brick place with some houses painted white. The much renewed church of St. John the Baptist is in parts of the early 14th Century with a N aisle of about 1865. Nash became a parish in 1849. The large brick house of 5 bays and 2½ storeys with a parapet that can be seen from the churchyard is Nash Court, built in 1760 by the Arbuthnot family. It has a 4-columned porch to the front and a stable block with cupola to the rear. The school of 1846 closed in 1958. Down the steep hill, ¼m. E of Nash, is Whatmore Farm. Opposite the farm is the 450 year old Nash Oak, a huge, knarled tree with some dead, white wood but which has recently been pollarded and has much good, new growth. The name Nash is from the Middle English *atten ashe* meaning 'the ash tree(s)'. North of Nash, on the B4214 road between Cleehill and Burford, is *Court of Hill* a handsome brick mansion of 1683 which stands in a small park. It was built by Andrew Hill and incorporates parts of the previous 13th Century house called Hulla. The Hill family left in 1927. The present house has 7 bays, 2 storeys, a hipped roof, stone dressings and a particularly good eaves-coving. Inside is some good woodwork and an early 19th Century plaster ceiling with intertwined foliage. Outside there is a dovecote with a lead lantern and a Georgian pavilion with a facade of Tuscan columns topped by a pediment.

NEEN SAVAGE *1m. N of Cleobury Mortimer.*
The River Rea is crossed by a ford here and just to the E is the track of a disused railway. There were paper mills and a brickworks in the country around but now it is quiet amongst the trees and the hills. The church of St. Mary has a Norman tower which was damaged by fire caused by lightning in 1825. The wooden spire was replaced with battlements. The Screen is 15th Century, but much restored. As to the name of the village: Neen is the old name of the River Rea and in the 13th Century the land here was owned by the Norman, Adam le Savage, (from Adam de Sauvage). Today there is a parish hall but no shop or pub and the school closed in 1964. There are only some 100 houses in the whole parish of 4,400 acres. The oldest is Cherry Orchard, a timber-framed cottage said to be of the 11th Century. One mile NE of the village, adjacent to and on either side of the B4363 to Bridgnorth, are the earthwork remains of Wall Town, a Roman fort. Detton Hall lies 1m. NNW of Neen Savage. The house is probably of the 17th Century and has 2 ranges. The E range is the older and the stonework pre-dates the timber-framing. There are star-shaped chimneys and inside is a good staircase.

NEEN SOLLARS *2¼m. SSW of Cleobury Mortimer.*
The village is situated between the Mill Brook and the River Rea and consists of little more than a pub, several black and white cottages and a 13th Century red sandstone church with a shingled spire. All Saints is a cruciform church and inside is a monument to Humphrey Conyngsby, 'a perfect scholar . . . and a great traveller . . . '. He was Lord of the Manor here and died in 1624. Neen (or Nene) is the old name of the River Rea and about 1195 Roger de Salariis was the Norman lord – hence Neen Sollars. At Tetshill, ¼m. S, is the track of a dismantled railway. There is also a weir on the river here.

NEENTON *6½m. SW of Bridgnorth.*
A small, main road settlement. In 1066 Azor held the manor but by 1086 it had passed to Ralph Mortimer. Neen is the old name for the Rea Brook which is here crossed by the B4364. Just W of the bridge the road makes a right angle turn by the Pheasant pub (formerly the New Inn) and the church. There are stone built cottages and red brick houses of recent origin plus a row of ubiquitous Council houses. The church of All Saints stands on an ancient, mounded site but the pink sandstone building we see today is of 1871 to a neo-13th Century design by Sir Arthur Blomfield. Inside there is a Norman font, a hightly decorated 15th Century chest and an E window, probably by Morris & Co. of about 1920. Outside there is a bellcote that seems out of proportion to the rest of the church, and 3 trees that stand somewhat stiffly. One mile S of Neenton is the ruin of the Charlcotte iron-furnace. (*See Charlcotte.*)

NESSCLIFFE *8m. NW of Shrewsbury.*
The village straggles along the busy A5 below the quarried sandstone cliffs of Nesscliffe Hill (*ness* is Old English for 'a ridge, a projecting headland'). The houses are mostly either sandstone or red brick; there is a modern school but the church was pulled down in the early 18th Century. Probably the oldest building here is the Old Three Pigeons Inn. The left central section is dated at 1405. Shruggs Common, on Nesscliffe Hill, is the smallest common in the county, possibly in the country, at only a quarter of an acre. It might have escaped enclosure because of tenurial problems. There is a prehistoric fort on the hill, now covered in trees, and here, also, is the cave of Humphrey Kynaston, the nobleman who became a highwayman. At the close of the 15th Century Kynaston roamed the countryside in the manner of his more illustrious predecessor, Robin Hood. He was outlawed in 1491 for his part in a murder at Church Stretton but became a popular hero, defying authority and stealing from the rich and giving to the poor. He had a remarkable horse called Beelzebub that once jumped the River Severn at Montford Bridge when the wooden slats of the roadway had been removed to prevent his escape, so legend has it. Kynaston's ancestral home was Myddle Castle and it was whilst he was on the run that it fell into decay. In 1518 Henry VIII gave the outlawed nobleman a free pardon and he passed his last years in peace on a small estate near Welshpool where he died in 1534. Kynaston's Cave on Nesscliffe Hill is reputed to have been his home for several years whilst he plied the trade of highwayman. It is a handsome place in a handsome setting and well worth visiting. The public footpath to the cave commences at the 5-bar gate on the A5 directly opposite the Old Three Pigeons Inn. The track is wide and deeply sunken. Mature trees line the way up the slopes of the wooded hill. A fairly steep set of earth and timber steps lead to the base of the 60 foot high cliff. A further set of stone steps, hacked out of the rock, lead to the cave. This has a door opening, a window and chimney vent. There are 2 compartments: one measures 11ft. by 11ft. and the other 7ft. by 5ft. There are half-a-dozen storage niches cut into the walls and the roof has been primitively carved to give the effect of vaulting. The trees of the woods are mixed broadleaf and coniferous. The whole area is most attractive and has great atmosphere. One mile SE of Nesscliffe, off the main road, is the village of **Great Ness**. It has a sandstone church with a sturdy Early English tower and some good Georgian brick houses. One mile NE of Great Ness is **Little Ness** where there is a small Norman castle motte situated next to the red sandstone Norman church. The church has been much restored but there is a Norman S doorway, a Norman arch with zig-zag decoration and the remains of a Norman font. The triptych paintings are early 16th Century German in early 19th Century English frames. The present (1989) Member of Parliament for the Shrewsbury and Atcham constituency is D. L. Conway of Queen's Court, Little Ness. Half-a-mile E of Little Ness is **Adcote**, a fine, grey stone neo-Elizabethan house of 1876–9 by Norman Shaw. The best exterior view is from the S. Inside is a splendid great hall, with a large fireplace which has a hood 15 ft. high. The Hall was built for Rebecca Darby, widow of Alfred Darby of Coalbrookdale. She had a fortune in her own right coming from a family of wealthy hatmakers. Today, it is used as a school for girls. The name Adcote means '*Adda*'s cottage'. Just to the E of the Hall on the River Rea is Adcote Mill. The area is well wooded. A mile to the SW of Nesscliffe is Wilcott where there is a military training camp. During the Second World War the Central Ammunition Storage Depot was located in some 2,000 acres of the low lying and badly drained land that lies between Wilcott and Shrawardine, 3m. SSE. (*See Wilcott.*)

NEWCASTLE *3m. WNW of Clun.*
A stone-built village situated on the B4368 at the confluence of the Folly Brook and the River Clun. The Clun valley is one of the most scenic in South Shropshire. It cuts through high hills from the Forest of

Nash, the Nash Oak opposite Whatmore Farm.

Clun in the W to Craven Arms in the E. On the hill immediately above Newcastle is the Iron Age fort of Fron Camp. The church, St. John the Evangelist, of 1848 lies ¼m. E of the village and has a unusual revolving lych gate. From here can be seen a good section of Offa's Dyke climbing the hill to the E, keeping to a W facing slope as it does for its whole length. To the SE of Newcastle village, on the river bank, is a Norman motte. Small castles such as this could be built in as little as 2 weeks, yet with their defensive ditches and pallisades they provided a secure base camp for cavalry patrols. Sometimes they were abandoned after only a few months as the force moved on to subjugate another area or because a more substantial castle was built in a better position nearby. It is remarkable to think that something thrown up in a few days has lasted for 800 years and is still a major feature of the landscape. On the main road, ½m. E of Newcastle, is Lower Spoad Farm. The house lies directly in the line of Offa's Dyke, but is it for the carved lintel over the fireplace that the farm is famous. This is probably Elizabethan and most handsomely depicts a hunting scene: a stag, and a doe pierced through by a spear, being set upon by 10 hounds. At **Whitcott Keysett**, 2m. E of Newcastle, there is said to be an 81 ft. menhir, an undressed stone of religious significance to pre-historic man. (*See Cefn Einion.*)

NEWDALE *2m. SE of Wellington.*
Newdale today is a greenfield site rapidly dissapearing into the great hole of British Coal's open cast coal mine. A huge spoil mound has arisen and the noise is tremendous. The ground literally shakes as an army of great machines thunders about. The amount of coal one can see appears so little as to hardly warrant such frantic activity. The Ordnance Survey map still shows the village of Newdale but it has now quite obviously gone. It was built about 1760 by Abraham Darby and his partner Thomas Goldney – cottages, a Quaker meeting house and a foundry – and there is a story that the partners intended to move their Coalbrookdale Company here. They did'nt, of course, and it is doubtful that the foundry furnace was ever fired; Newdale became a coal mining community. By 1851 there were 90 adults and 39 houses.

NEW INVENTION *3m. S of Clun.*
A tiny place on the road between Knighton and Clun with a Methodist Chapel of 1864. The 'invention' made here is said to be the ruse of reversing the shoes of a horse to confuse pursuers, reputedly those of Dick Turpin's mount. This is splendid, rolling, South Shropshire hill country. It is noticeable that the upper slopes of the hills in this area are often divided into large fields by long, dead straight, well trimmed hedges which contain only one species of plant. These are signs of comparitively late enclosure by big landowners. In the valleys the fields are usually smaller, more irregular in shape and often contain several species of plants which are frequently allowed to grow to their full, natural height. Indeed, hedges in olden times were often treated as a crop, with fruit trees and winter fodder plants, such as holly, playing a not unimportant part in the rural economy. Some plants were grown for purposes now forgotten, such as damson which was used for making a cloth dye.

NEW MARTON *5m. W of Ellesmere.*
A place of corrugated iron barns, some red, some black and some shining silver. New Marton Farm is an attractive, ivy-clad brick house with dormer windows and traditional farm buildings around a cobbled courtyard. To the right is a remnant of a water-filled moat. To the right of that is New Marton Hall, a modest 2-storey house clad in stucco beneath a hipped roof. In the garden is a good Scots pine and some alien, fast growing conifers. Opposite Sandhole Farm is the most organized duck pond that we have seen with shelters and compounds separating the ducks and the swans, and more fast growing conifers. Nearby, down the hill, is a black and white cottage hiding behind a screen of, yes, fast growing conifers. **Old Marton** lies on high ground ½m. SE of New Marton. It consists of little more than 3 widely separated farms. That on the main lane has an unsightly concrete bunker on which are piled even more unsightly motor car tyres. These languish beneath cables and pylons belonging to the Central Electricity Generating Board. The land is mainly laid to pasture and there are hollys, ivys and oaks in the hedgerows. The name Marton can mean either 'the settlement by the mere', or 'the settlement by the boundary'. There is no sign of a lake today so perhaps 'boundary' is the more likely explanation, possibly referring to an Anglo-Welsh dividing line. There is, however, an area of low land about 1m. W that is drained by a strine. This could have been marshy, or even a mere, in days gone by.

NEWPORT *8m. NNE of Telford New Town Centre.*
The town is not old. It was planned and laid out as a commercial venture in the 12th Century when it was granted a Borough Charter by Henry I. The site, which was in the Anglo-Saxon manor of Edgmond, lay close to many ponds and meres, noted fisheries. Some of these still exist but most have been drained. The town's crest incorporates 3 fishes and the burgesses had an obligation to supply the Royal Household with fresh fish whenever it was in the area. There was a Vivary Pool stretching NE from the main road (A41) where it crosses the now disused canal. The road itself acted as a dam. This pool was later drained but a small part of it was kept and formed a wharf basin for the Shropshire Union Canal in Water Lane. This was in turn mostly drained and filled in by the Council and it is now a car park for the adjacent recreational field. The small pool that now remains is used for fishing once again. To the W are the Weald Moors. Before Telford drained these 'wild moors' there were shallow meres and pools amongst the marshes and the whole area, thousands of acres, was frequently under water, sometimes for months on end. The church at Newport was named after the patron saint of fishermen, St. Nicholas. It stands isolated on an island in the centre of the town with the busy main road to one side and St. Mary's Lane, a quaint cobbled street with attractive shops, to the other. The tower is 14th Century but most of the rest was rebuilt between 1866 and 1891. It is constructed of red sandstone and has some good stained-glass windows by Burne-Jones and William Morris. The roads around it formed the old market place, but this central area was later built upon – the Middle Row – and the market is no longer held here. St. Mary's Lane is named after the chapel and chantry in the church that was founded by Thomas Draper. As well as praying for souls it provided a school. The Chantry College was dissolved at the Dissolution but the school continued until 1879. The site of the College is now occupied by Royce. The old almshouses of 1446, built by William Glover in the churchyard, were dismantled in 1836 and rebuilt in Vineyard Road where they still stand. The cross in the road by the church is the Puleston Cross erected in memory of a 13th Century knight, Sir Roger de Pulestone, who was killed in a battle in Anglesey. For many years it stood by the Butter Market and came to be know locally as the Butter Cross. The Rectory is downhill from the church. It was built in about 1840 as a private house and is called Hurlstone. Thomas Jukes Collier lived here and later it was used as the Grammar School boarding house. It is a handsome, colonnaded building with good iron railings without and good plasterwork within. The long 'S' shaped main street is variously called Upper Bar, High Street and Lower Bar. Running of at right angles are the medieval burgage plots, fossilized in modern streets and buildings. When the fisheries ceased and the pools had been drained Newport continued to flourish as a market town for the newly made pastoral and arable farmlands that replaced them. In 1764 Newport Marsh was enclosed by Act of Parliament, the first common land in Shropshire not to be enclosed by private agreement. In the 18th Century the market ceased to be general and specialized in cattle and livestock. Only in recent years has a general market revival been attempted, without that much success. In 1665 Newport had a great fire and much of the town was destroyed. The fire started at a site now occupied by Barclays Bank. Only a handful of timber-framed balck and white houses escaped the inferno, most notably the Old Guildhall of 1615 and the charming Smallwood Lodge which is now a bookshop. The Georgian period is well represented and there are a large number of 3-storeyed, red brick town houses, many of which are now used as offices and shops. One of the best of these is Beaumaris House (1724), on the main road N of the church. It has steps and railings leading to the front door, 5 bays, giant angle pilasters and a parapet.

above: Myddle Castle.

top: carved panel in the lavatorium (washroom) at Much Wenlock Priory.

below: Brindrinog, near Newcastle, near Clun. Offa's Dyke, middle left.

Smallwood Lodge, Newport.

Charles Dickens (1812–1870) stayed here when it was the Bear Inn. Further N is Chetwynd House, the old home of the recluse Elizabeth Parker on whom Dickens based the character of Miss Havisham in Great Expectations (1861). The Royal Victoria Hotel, of about 1830, and the slightly subsiding Adams' Grammar School, founded in 1656 but much rebuilt in 1821, are 2 of the town's most notable buildings. The Town Hall, in St. Mary Street, was built in 1859 to an Italianate design by J. Cobb and was not unhandsome before it was painted blue and shops were inserted on the ground floor. In Wellington Road there is a Classical style Congregational Church with the old British School adjacent. Lord Shrewsbury paid for the Roman Catholic Church of St. Peter and St. Paul (1832) in Salter Lane. There are 2 old inns, the Barley Mow and the Shakespeare, one of the oldest buildings in the town. Modern houses litter the suburbs and have regrettably just started to encroach on the northern fringe of the town centre near the Canal. On the bridge that crosses the canal the council has placed a self-congratulatory plaque proclaiming their pride at having reduced a magnificent and well-known flight of canal locks to a concrete terrace and a drainage conduit. Newport is essentially a 'service town' these days but there is some local industry. Until recently the leading company here was Serck-Audco Valves which started as an agricultural machinery repair business and became a part of BTR Indusries. In 1988 the 4 acre factory site was sold to a housing developer. Companies in the town include Zip Up Scaffolding, Newport Silos, Newgill Foundry, Whitecroft Dairy, Taylor's Haulage, Classic Garden Furniture and the Advertiser Printing Works. The largest industry of the area is, of course, agriculture. As late as the beginning of this century the lands around Newport were almost entirely owned by 3 great landlords: the Leveson Gowers of Lilleshall (Dukes of Sutherland), the Bougheys of Aqualate (in Staffordshire) and the Boroughs of Chetwynd. The farms were tenanted and relatively small. Today they are mostly owner-occupied and considerably larger. The soils are light and barley, wheat, sugar beet, and potatoes are important crops locally. Generally speaking, there is very little wrong with Newport. It is very ordinary but very nice, with a good selection of shops, a ful range of professional services and a brand new ring road.

NEWTOWN *3m. NW of Wem.*
A scattered hamlet in flat, pastoral country with a moated site at Northwood Hall and another ½m. NW of that. The Church of King Charles the Martyr is of 1869 by E. Haycock junior and not to everyone's taste. The first church was a converted private house, the work of High Churchmen during the Commonwealth. Inside the present building there is a stone reredos depicting the Last Supper, a wrought-iron screen and pictures of the execution and burial of King Charles. The name Newtown is almost certainly the most common of all English place names. Many 'new-towns' are now very old, some dating back to the Anglo-Saxon period when they would have been called *Neowatun*, meaning 'new settlement or homestead', and not, by any means, necessarily 'a town'.

NEWTOWN ON THE HILL *1m. ESE of Myddle, which is 8m. NNW of Shrewsbury.*
Half-a-mile E of Myddle on the road to Clive is Balderton Hall, a large, ivy-clad, timber-framed farm house with red brick infill, tall chimneys in clusters, 2 projecting gables and specimen trees in the garden. Alongside the road are walls made of large, dressed, sandstone blocks. Newton lies on a rise ½m. SE, a tiny place with a few brick cottages, a sandstone farm and an old chapel with a shaped gable. Most of the hamlet still lies on the Balderton estate. Richard Gough was born at Newton in 1635. He wrote most of his celebrated History of Myddle in 1701 and died in 1723 aged 88. A little S on the main road is Lea Hall where the Mayall family practice organic farming. (*See Preston Gubbals.*)

NOBOLD *2m. SW of Shrewsbury on the road to Longden and Pulverbatch.*
A tiny place on the outskirts of Shrewsbury. The name was formerly Newbold, a contraction of New Buildings. There used to be clay and gravel pits and a brick-kiln and in the field opposite Sweet Lake Cottage (on the Longden road) there was a small coal mine with 2 shafts. At Conduit Head there is a visitor centre on the site of Broadwell, a spring from which water was chanelled to the town of Shrewsbury from as early as 1556.

NONELEY *2½m. WSW of Wem.*
The name is Old English and could mean either 'the settlement of *Nunna*', or 'the settlement belonging to the nuns'. At the time of Domesday Book Noneley was a part of the manor of Baschurch and as such belonged to St. Peter's church (Shrewsbury Abbey). It lies on a slight rise surrounded by very flat ground; ½m. S is Sleap aerodrome, just beyond the muddy drainage ditch called Sleap Brook. The nucleated village is very much a place of farms: the grey rendered Forrester's Farm; the handsome Noneley Hall, red brick with 2 projecting gables, irregular quoins and a round-leaded window above the entrance; Grafton's Farm, a timber-framed house now encased in brick with a central gable-pediment; and the Manor Farm, a modest red brick house on the fringe of the village. There is also a timber-framed dwelling called Primrose Cottage at present being restored with concrete blocks and fibre board. Opposite is a brick cottage that was once the Smithy. Noneley is not picturesque but it is very pleasant here beneath the big, open skies. In the field in front of the Hall horses, chickens, turkeys and sheep all live contentedly together. In the distance are many small copses and when we were here there was persistent gunfire, each shot signalling the bloody end of some small creature. The hamlet of **Ruewood** lies 1m. E of Noneley. Here is a charming small farmhouse with an orchard to the front and farm buildings to the rear. The house is clad in grey render with red quoins and green and white paintwork under a slate roof. The boundary walls are of sandstone cut into large blocks topped by green painted iron railings which escaped the clutches of the Second World War metal collectors. Small aeroplanes pass by overhead preparing to land at the aerodrome. (*See Sleap.*)

NORBURY *3½m. NE of Bishop's Castle.*
Norbury is a small village lying on the lower slopes of Norbury Hill. It is known for the ancient yew tree which stands by the churchyard path. This tree is not only one of the oldest (between 800 and 1,600 years of age) and largest (with a girth of 35 ft.) in Shropshire, it is also in very good health. The church of All Saints stands in an Anglo-Saxon churchyard and has a 14th Century tower with a 19th Century spire. Most of the rest was rebuilt in Victorian times. Next door is a noisy turkey farm. There is supposed to be something variously described as 'a rough boulder by the sanctuary steps', and 'a lowering of the altar step by the S side of the altar rail'. We could see nothing. The name Norbury means 'the north fort', or the 'northern fortified place'.

NORTHWOOD *5½m. ESE of Ellesmere.*
An unremarkable main road village of red brick and stuccoed houses that straggles along the B5063. There is a pub, the Horse and Jockey, and an unusual Italianate dwelling clad in cream stucco with a hipped roof, tall windows and iron window railings. The country is flat and mostly laid to pasture.

NORTON *5m. N of Bridgnorth.*
The busy A442 and mundane new houses spoil this otherwise most attractive village. There is a pub called the Hundred House, several black and white houses, a thatched barn and a grey-painted redundant school, which looks for all the world like a church. On the small village green there is an oak tree, 2 granite boulders and the old village whipping post and stocks. The name Norton means 'the homestead or village to the north'.

NORTON FORGE *½m. S of Norton-in-Hales which is 3½m. NE of Market Drayton.*
There are no buildings shown on the 2½ inch to 1 mile Ordnance Survey map (sheet SJ 63/73) but buildings there most certainly are. The access lane leaves the main road at Norton-in-Hales between the bungalow called Norway and the red brick cottage called Oakhurst. This is actually called Forge Lane but is not signposted. The metalled surface soon gives way to a rough track which leads through fields and a rocky cutting, past a small, disused quarry and down into the delightful valley of the River Tern. Here it is little more than a stream, wandering lazily through the meadows which run up to the water's edge. This is how many of our Midlands

New Invention, view from the A488.

rivers used to be until they were deepened to act as drainage ditches, a fate which befalls the Tern a few miles downstream. On the W bank, against a low sandstone cliff, there was a pair of old cottages. These have been demolished and in their place is a cedar-boarded bungalow. Just N of this the river flows over a small weir and is crossed here by a wooden foot-bridge. Some 50 yards downstream is another, red brick, vehicular bridge. In between the river widens to about 8ft and flows through a sandy-floored and sandstone-walled channel. Just downstream and now to the left of the bungalow is a triangular shaped pond about 25 yards long and at a slightly higher level than the river. At the S end is a concrete dam wall (a pond bay) with a pipe that empties into an overflow leat (water-course). Opposite the bungalow on the other side of the river are a collection of wooden-framed farm buildings in which are vast numbers of pigs, chickens, cows and geese. To the S of these ranges is a small red brick building with a stable-door and larger stone-built structure which measures 22ft. by 18ft. Between these buildings and the river are areas of black earth, a sign of iron-working. This area was probably the site of one of the several 17th-18th Century iron forges that were located on the Tern. Other known sites are at Atcham, Tong Forge, Upton Magna, Withington, Tern, Great Wytheford and Moreton Corbet. Most of these have very scant remains, far less than here at Norton Forge. They were not blacksmith's forges. There were ironworks where crude pig iron was re-heated in a furnace and then beaten with great hammers to remove inpurities. The resulting wrought iron was much useful and therefore valuable. The furnaces were fuelled with charcoal and the hammers driven by waterwheel. The field boundaries behind the farm are stone walls, most unusual for this area. Perhaps the stone was taken from the old forge buildings; this 'quarrying' was a common practice. Whilst photographing and taking notes here at Norton Forge we became the centre of attraction to a large black horse and its tawny companion, a little pony. They frisked and frolicked quite delightfully but a little to close to us for comfort. Whenever I went to take a photograph the horse nudged my arm and the pony did a little dance in merriment. South of the pond the river becomes natural once again and winds its way through a tangled wood before being dammed to form the lake of Oakley Park, just over the border in Staffordshire. (The River Tern forms the County boundary for several miles hereabouts.) In the park, only some 200 yards or so to the E of Norton Forge, are the Devil's Ring and Finger. These stones probably formed a part of a pre-historic burial chamber and were dragged to their present position when the field was cleared for ploughing. One stone is flat with a circular hole in it and the other is like a cross-shaft. They now form part of a wall by a tree. (Less than 2m. S near Broomhall Grange a Bronze Age axe was found in 1913.)

NORTON IN HALES *3½m. NE of Market Drayton.*
The red brick settlement lies around its small village green. Village greens are very rare in Shropshire, especially old ones like that at Norton. (There have been some deliberate 19th Century creations by romantic landlords). A green implies a degree of planning when the village was first created. Unplanned settlements tend to be very irregular and scattered. It is supposed that a group of settlers would make a clearing in virgin forest and then build their houses in a rough circle around a central communal 'green'. They could then provide each other with mutual help and protection whilst continuing the slow process of clearing the forest to make fields. One must remember that in Anglo-Saxon times, and earlier, bears, wolves and huge wild boar were in the forests. Wolves were a real peril as late as the 13th Century in many areas. On the village green at Norton is a large granite rock, probably deposited in here by a glacier. It is called The Bradling Stone. It seems likely that it had some pagan religious significance but all that remains of this is a quaint custom: anyone found working after midday on Shrove Tuesday will be bumped against the stone. However, our enquiries suggest that the custom has been discontinued. The church of St. Chad has a Norman tower but the rest was mostly rebuilt in 1864–72. The apparently Norman door in the tower is not original but a Victorian masquerade. Inside there is a striking alabaster monument to Sir Rowland Cotton and his wife (she died in childbirth in 1606), and in the churchyard there is a large, ancient yew tree. The Market Drayton – Stoke on Trent railway (1860) used to pass through the village; the station and station master's house are now dwellings. There is a Jubilee Hall; a cricket pitch; a pub, the Hind's Head (formerly the Griffin Arms); and a Post Office and general store. As to the name Norton-in-Hales: Norton is a common name meaning the 'settlement to the north'; to avoid confusion with other Nortons the qualification of the old parish-manor, Hales, was added. Just over 1m. NE of Norton-in-Hales, on the B5062, is **Bearstone** Mill. The present water-mill is some 300 years old but there was a mill here on the River Tern at the time of Domesday Book. The hamlet, with an old moated site, lies ½m. to the N. The foremost building here is Bearstone Farm, an Elizabethan house that has a handsome black and white gable with mainly diagonal strutting and 2 overhangs. There is a charming stable-block and nearby there are 2 good stone barns. Half-a-mile SW of Norton-in-Hales is Brand Hall. This is a handsome early 18th Century 7-bay red brick house with stone dressings. It is 2½ storeys high and has a 3-bay pediment, giant pilasters, and lower wings to both sides. To the rear is a detached cottage and in front is a lawn protected by a deep ha-ha. To the front also, at right angles to the facade, are 2 rather odd free standing archways with pediments. The house stands in a pleasant, small park with oak trees and cows grazing in buttercup strewn meadows. A quarter-of-a-mile NE of Norton, on the road to Bearstone is a cemetery with a tiny random-stone chapel consecrated in 1865. Four old, carved stone heads are incorporated into the facade.

NOX *6m. WSW of Shrewsbury.*
It lies in undulating country on the B4386 Shrewsbury to Westbury road, which here follows the line of a Roman road. There are sheep in the meadows and by the stream is a row of pollarded trees some of which have been allowed to grow to an unusual height – a strange sight. The big red brick house with sandstone corner pillars, blocked windows and a hipped roof is called The Lynches. Nox House is a tall, red brick house of 3 bays and 3 storeys with a stone barn (now a dwelling) to the left and a timber-framed outbuilding to the right. Add to those a modern bungalow, some rendered cottages and a farm and Nox has been fairly described.

OAKENGATES *1¼m. NNW of Telford Town Centre.*
It adjoins and lies to the N of the M54 in a hilly, industrial landscape. There were coal mines all around and their spoil heaps are their memorial. These mounds are now being landscaped and are not by any means unattractive. The settlement spreads up a hill from which are wide views to the NE, over the town centre, to the great plain of Shropshire beyond. The last steel works in Shropshire was the Lilleshall Company, the works of which were just out of the town on the Holyhead Road at Snedshill. They closed in 1959. (*See Snedshill.*) During the 19th Century the company virtually dug up the whole of Donnington Wood in its search for iron-ore. The last iron works, as distinct from steel works, in Oakengates were those of John Maddocks. They ceased production in August 1987. Today, the area still has some character. The town centre is definitely a town centre, despite what others may have written, and has a wide range of social facilities. Strangely, for a settlement surrounded by high ground, there is a feeling of light, of open skies. Coal had been mined in Oakengates in a small way for many years when in 1738 the great coal-master Richard Hartshorne installed an engine at the Greenfield mine. Greenfield was the area now occupied by the playing field at Hadley Park Road. On the death of Hartshorne the mines passed to the Charlton family who lived at Apley Castle, Leegomery. In 1747 new pits were sunk at Horsepastures, high ground to the NE of the town centre. A wooden railway was laid to transport the coal, probably to Watling Street. Most of the coal was used for domestic heating and cooking. Charcoal was still used to fuel the iron furnaces. From 1761 Richard Reynolds, the ironmaster, was working the mines and in the 1770's iron smelting furnaces were established in the area. In 1787 a 1½m. canal connecting Ketley ironworks with the Oakengates coal mines was begun, and at the Ketley end of this was

The old Vivary Pool, latterly a canal basin, at Newport.

The River Tern at Norton Forge, near Market Drayton.

The pool at Nib Heath.

the famous Ketley Incline. (*See Ketley.*) The main line railway arrived at the town in 1850 when the Shrewsbury to Wolverhampton line was opened by the Shropshire Union Railway and Canal Company. This line still exists and Oakengates still has a station, albeit an unmanned halt – you get your ticket from a blue-painted machine. The Oakengates tunnel is visible from the platform. Ominously there are 'for sale' signs on the yellow brick station office building. In the 19th Century Oakengates was a flourishing market and shopping town, a prosperity it largely owed to the presence of the railway. It was also a centre of blood sports, such as cock-fighting and bull-baiting, and, incongruously of Nonconformity. The Primitive Methodists and the Wesleyans were both active in the town in the 1820's but their converts still practiced their cruel pastines. Oakengates Wakes were far-famed and attended by many thousands from far and wide. Drunken orgies were the order of the day and the priests despaired. The last bull-bait in Shropshire is said to have taken place at Oakengates in 1833. The town's most famous son is probably Sir Gordon Richards, the jockey and trainer of race horses. His father was a coal miner here. Sir Gordon had his first win at Leicester in 1921 and retired 33 years later having ridden 4,870 winners. He was Champion Jockey for twenty years. The name Oakengates probably means, quite simply, 'the gates made of oak'. Like Snedshill, St. Georges is also considered to be a part of Oakengates. (*See St. Georges.*)

OLDBURY *½m. S of Bridgnorth.*
It lies on the B4363 and is now, to all intents and purposes, a southern residential suburb of Bridgnorth. The name Oldbury is not uncommon and means 'old fort', the fort usually being pre-Anglo-Saxon, that is Romano-British or earlier. Domesday Book records that in 1086 the manor was held from Earl Roger de Montgomery by Reginald the Sheriff, that 3 Frenchmen, 2 cottagers, 1 smallholder, and 7 slaves and their families lived there and that there were 2 mills and woodland for fattening 100 pigs. The village lies on high ground and there is a neat Victorian church. In the Middle Ages Oldbury was in the parish of Morville. The parish church of St. Gregory had set up a 'daughter' chapel at Oldbury by 1140. At a daughter chapel it was usual for religious services only to be held; Christenings, weddings and funerals were still the province of the mother church. Oldbury is now a parish church in its own right. There is a good Victorian house in the village called Eversley, 'a gentleman's residence', set in 2 acres of gardens. It has a 3-storey projecting porch demi-tower and pedimented doorway.

OLD PARK *½m. NW of Telford Town Centre.*
Old Park was an estate owned by Isaac Hawkins Browne. There were deposits of coal and iron ore here and in 1790 Thomas Botfield (1736–1801), who was employed as the manager of the estate, built an ironworks. This was the beginning of the famous Old Park Company which by 1806 was the largest ironworks in Shropshire and the 2nd largest in Britain. There were houses for the workmen, well equipped mines, at least 4 blast furnaces, and a rolling mill with a Boulton and Watt engine in the forge (which was using the new puddling process). When Thomas Botfield died he was a wealthy man. He had been born the son of a Dawley collier who had a small interest in leased land at Lightmoor; he died owning estates in Hopton Wafers (where he had paper mills), Ystrad-Fawr in Breconshire, and Norton in Northamptonshire, and of course had considerable interests in several industrial concerns. William Botfield, one of his 3 sons, took over the management of the Old Park Works and saw the company to its pre-eminence. The company went bankrupt and closed in the 1880's. Today Old Park consists mostly of coal and iron ore-mine spoil mounds, grassed over and windswept. The only habitations are along Park Lane, which leads off the new Old Park Way (B4373) near Mossey Green Roundabout. Here are a few old cottages, a few new houses and a red brick Wesleyan Chapel of 1853 with pointed-arched windows containing good cast-iron frames. At the time of writing it is used to store boxes of core samples from a drilling rig. The southern part of Old Park, around Vinyard Road, is already being developed as a residential estate. Hundreds of houses in Telford are built on old spoil heaps. Such ground is known to be unstable and one wonders what the future might hold for these buildings, especially those situated on slopes.

OLLERTON *3m. SE of Hodnet.*
It lies in the flattish northern lands of the county. Ollerton is a sleepy, unremarkable agricultural hamlet of red brick farms and cottages with some new houses and black, corrugated iron barns. A house with a good lattice-worked iron porch stands adjacent to a timber-framed barn. The early medieval name for ollerton was Alverton. The name may be derived from the Old English *alor-tun*, meaning 'the homestead (or village) among the alder trees'.

ONIBURY *5½m. NW of Ludlow.*
It lies just off the busy A49. The River Onny, the road and the railway intertwine here. Where the road crosses the railway there is a level crossing and a pretty blue and white painted station. The pleasing church of St. Michael has a Norman chancel arch, some Early English work, a roughcast nave and chancel, a W gallery, and ironwork by Detmar Blow who restored the church in 1902. The churchyard is beautifully kept and is the regular haunt of crowing cockerels from the large stone-built farm on the opposite side of the road. The village buildings are a mixture of red brick, limestone and half-timber. There is a school, a village hall, some council houses, a Post Office and general store, and a pub, the Hollybrush Inn. The white painted Onibury House is now a home for the elderly. Half-a-mile to the SW is Stokesay Court, a large but stern neo-Elizabethan mansion of 1889 designed by Thomas Harris. It is constructed of rusticated yellow stone with smooth, grey, Grinshill stone dressings and has mullioned and transomed windows and shaped gables. It sits in a well-wooded park and is approached down a long drive which crosses a tiny stream over a large bridge. The substantial and most handsome stone stable-block can be glimpsed from the drive. The front of the house faces a courtyard beyond which is a huge bank of rhododendrons, quite spectacular when in full bloom. The property is currently owned and occupied by Magnus Allcroft and the gardens are occasionally open to the public. Just over 1m. W of Onibury along the lane that leads past Stokesay Court, is the 'Wernlas Living Museum of Rare Poultry'. The owners live in a house that overlooks a sloping field in which are countless numbers of poultry houses. Two donkey stand guard near the car park.

OSBASTON *1m. NNE of High Ercall which is 7m. ENE of Shrewsbury.*
From afar it can be placed by the line of tall, elegant poplar trees that stand like sentinels amongst the red fields and green pastures of this wide, flat country. The name is Old English and means '*Osbeorn*'s homestead'. Today there are still one or two cottages in the lanes but the area is dominated by the large, black hangars of the wartime aerodrome. These are now used as commercial warehouses. The main administration buildings, built of brick with hipped roofs, now house the Road Transport Industry Training Board's 'Multi Occupational Training and Education Centre', M.O.T.E.C. for short. With its white-painted guard-room pillars, yellow-painted curbstones and neat green lawns it is probably as trim as it ever was in its military days. The complex is quite substantial, with accommodation blocks, offices, a conference hall, a garage, training areas, a 'skip room' and storage hangars. Outside the security fence is a group of modern houses.

OSWESTRY *17m. NW of Shrewsbury.*
The story begins in the Bronze Age. The place we now call Oswestry lay on the Ffordd Saeson, a trackway that linked Anglesey with the River Severn and was used by Irish axe traders. Later, during the Iron Age, the massive fort of Old Oswestry was constructed. It is the most outstanding work of its kind in the Welsh Marches and covers 40 acres. Work on it began about 250 B.C. It originally had only 2 earth banks, or ramparts. At a later date a third was added and finally the whole was enclosed in a huge double compound. It was abandoned shortly after the Roman Conquest, but was re-occupied when the legions left and England was flung into the Dark Ages. It lies to the N of the town and is open to the public. Shortly after the Romans had left the Angles and Saxons began their conquest of Britain, and by the 7th Century

Oswestry, Old Oswestry hill fort is in the distance; in the foreground a hedge is being ripped out.

they had taken control of the country. In A.D.641 Penda, the Anglian King of Mercia, defeated Oswald, a Christian King of Northumbria, at the battle of Maserfelt. The site of Maserfelt is now occupied by the playing fields of Oswestry Grammar School. Oswald was killed and his body nailed to a tree. Legend has it that an eagle carried off one of the dead Christian king's arms. However, the arm escaped from the bird's grasp and fell to earth whereupon a spring burst forth from the ground on which it landed. These are healing waters and have cured many sicknesses, according to local people. St. Oswald's well can be seen today, amongst the trees of a little park in Maserfield Road, on land adjoining the Grammar School grounds. A section of Wat's Dyke, a rampart and ditch similar to Offa's Dyke, runs by Oswestry. Constructed in the reign of Athelbold (A.D.716–57), it extends from the Dee estuary to Morda Brook S of Oswestry. For a few hundred yards, from Whittington to Old Oswestry, the modern road follows Wat's Dyke. The Normans built Oswestry Castle almost immediately after the Conquest in 1066. The motte (mound) is still very prominent and the bailey (enclosure) defences, though long gone, are reflected in the horseshoe shape made by Willow St., Cross St., and Leg St. Buildings now occupy the site of the bailey but one street is still called Bailey Street. The church of St. Oswald was Norman but all that remains of that time are the lower part of the tower and a lancet window. The town was created in the 12th Century by William Fitzalan and a Borough Charter was granted in 1228. The settlement lay in the Manor of Maesbury but it quickly prospered and soon took over not only the Manor but also the Hundred of Mersete. The name Oswestry is derived from 'St. Oswald's Tree' in honour of the Anglo-Saxon saint who had died here. The Manor of Oswestry became one of the major Marcher Lordships. The lord of a Marcher Lordship had almost regal rights in return for keeping the peace with the Welsh. They had their own courts, and what today might be called Emergency Powers which could override the Civil Law. The town prospered and today 6 major roads converge here, a mark of its economic and social status in the area. Despite being a centre of English government the town itself was more Welsh than English. It became an important market town and most of the farmers were Welsh. It helped to be Welsh, or at least Welsh speaking, to do business in the town. In the 15th Century Guto'r Glyn, a Welsh Bard, was made a freeman of the borough for a poem he wrote in praise of Oswestry. Not until the 18th Century did the English influence prevail amongst the townspeople. During the 19th Century the population of Oswestry increased by nearly 200 percent largely due to the railway trade; Cambrian Railways had their headquarters offices and workshops here. It was during this period that most of the many rows of terraced houses were built. Despite its prosperity through the ages the town has had a turbulent history. In 1215 King John burnt it to the ground; it was burnt again by Llewellyn, again by Owain Glyndwr, burnt accidentally several times in the 16th Century, badly damaged and the castle demolished by Cromwell, and finally burnt down once more in 1742. This accounts for the paucity of building dating from before the 18th Century. Llywd Mansion, 1604, is one of the few black and white buildings to have survived. It is in the town centre and the ground floor now houses a shop. Abounding the parish churchyard is the timber-framed Holbache House which dates from 1407 when it was established as a free grammar school. It was altered in Tudor times and until recently housed a Museum of Childhood. One of the most charming black and white buildings in Oswestry is the Black Gate Inn, on the road to Shrewsbury. The 'new' grammar school stands on the outskirts of the town on a hillside. There is a good Masters House of 1776 and some pleasing modern buildings. (There are also some awful modern buildings.) Canon Spooner of the Spoonerism was educated here. The Wynnstay Hotel is good Georgian; the Victoria Works (1870) once housed an iron furnace; the DHSS occupy a chapel of 1830; the handsome, stone-built Council Buildings are of 1893; and the steel-framed Secondary Modern School of 1956 by Richard Sheppard has its admirers. The town seems prosperous and alive and there is a good selection of shops and a cheerful atmosphere. Oswestry is a very pleasant place to visit and a good centre from which to explore North Wales. Sir Henry Walford Davies, the organist and composer, was born in Oswestry in 1869. One of the more unusual local industries is balloon manufacture. In 1987 Per Lindstrand and Richard Branson piloted the Virgin Atlantic Flyer across the Atlantic, a distance of over 3,000 miles and a world distance record. Their hot air balloon was manufactured in Oswestry at the works of Thunder and Colt. Wilfrid Owen, the First World War Poet, was also born in the town at Plas Wilmot, Weston Lane. The north-western part of Oswestry adjoins the park of **Brogyntyn** (formerly Parkington, and before that Constables Hall). The town entrance gates to the Hall are now chained up and entry is obtained off the B4579 to Sellatyn. The Hall is about a mile out of the town. It is the seat of Lord Harlech, though it looks somewhat unlived-in, and the park is part of a well wooded agricultural estate. There are good views both towards the hills of Wales and over the earthworks of Old Oswestry to the English plain. Castell Brogyntyn is situated in Brogyntyn Park in woods near the lake ¼m. NE of the village of Brogyntyn. It is a substantial and strong earthwork with a circular summit of 130 feet diameter, surrounded by a 6 ft. bank, an 18 ft. escarpment and a 6 ft. ditch at the base. The entrance was in the NE. This fort is believed to have been constructed by Brogyntyn, a son of Prince Owen Madre of Wales. Llanforda Hall, 1½m. SSW of Brogyntyn, was built in 1780 but only the handsome red brick stables survive. There is a central entrance arch; the end bays are raised like short towers, capped with pyramid roofs; and the ground floor windows have pediments. Today they serve as farm buildings. One-and-a-half miles NNE of Oswestry on the road to Gobowen is the famous Orthopaedic Hospital established here in 1921 by Dame Agnes Hunt and the surgeon Robert Jones (1870–1948) in the premises of the old military hospital at Park Hall. The hospital has a world-wide reputation for its innovative work. Park Hall itself was a superb black and white mansion of 1560. Many are of the opinion that it was the finest timber-framed house in the county, until it was destroyed by fire in 1918. South of Oswestry is a suburb called Morda. (*See Morda.*)

ORETON *See Farlow.*

OVERLEY *2m. W of Wellington on the A5.*

A hamlet on the A5 with a Hall, a garage, a handful of red brick cottages, a chemical works and an old Roman road. The Hall stands amongst trees, a large, irregular Victorian house of 1882. It has varied roof lines, tall chimneys, black and white gables and dormers, areas clad in hung tiles and a battlement tower over the entrance hall. It is now a private school for mentally handicapped children. Adjoining it is an Esso garage complete with shop and cafe. Almost opposite the garage is a dirt track. This leads up Overley Hill. To the left is an ancient but now disused roadstone quarry called Lea Rock. This is now delightfully overgrown with gorse and mature oak trees. At the top of the hill the track bears sharply right. The traveller is now on a section of the old Roman road called Watling Street. In the early 19th Century Thomas Telford diverted the Holyhead Road so that it skirted around the northern slopes of the hill; the old Roman road came straight up and over it. The hedge to the right has many different species of plant, a sign of great age. The hedge to the left is an enclosure boundary, all of hawthorne. At the end of the lane is Overley Cottage; the Lea Cottages shown on the Ordnance Survey map seem never to have existed. What does exist is a nasty bit of wartime Ministry of Defence building, a flat-roofed, brick and glass, iron window-framed piece of junk that can be seen for miles around. It was built in 1944 to house an optical research establishment. Today it is a toxic chemicals warehouse, the Overley Works of Cox Chemicals. If one returns down the lane back to the Holyhead Road (A5) and turns right and then first left you are on the road to **Leaton**. This is even smaller than Overley, which it adjoins. Leaton Grange is a good looking house with dormers and under the stucco there is a building of some years. The occupants of the Grange and the handful of cottages beside it must be fed-up with the noise from the roadstone quarry. The crushers and graders of Berwyn Granite chunter tirelessly and periodically there are explosions. One cottage is derelict, the former residents quite possibly having fled

Brogyntyn, Lord Harlech's seat, Oswestry.

Aston Hall, Oswestry.

Park Hall, Oswestry, (Destroyed by fire in 1918).

to some quieter corner of the county. The stone mined here is a dark grey basalt called Olivine Dolerite, not granite as the map declares. Tar will stick to this material; it will not stick to the pink coloured rock of the abandoned quarry of Lea Rock. Leaton was part of the manor of Wrockwardine, ½m. NE. The name Leaton is Anglo-Saxon and means either 'the settlement by the stream', (from *laet*) or 'the settlement by the shelter' (from *hleo*). There is another Leaton NW of Shrewsbury; this has its own entry.

OVERTON *2m. SSW of Ludlow.*
A main road hamlet on the B4361 to Richard's Castle. Overton House is of yellow stone, like its lodge, and is set in a small park. It is the home of Roger Salwey, a local landowner whose family used to live at Moor Park. (*See Batchcott.*) Behind the house, to the W, are the extensive woods and plantations of Haye Park in which is a pre-historic Enclosure. Overton Grange on the B4351 is a rendered, gabled, part ivy-clad Victorian-looking place now an hotel. The name Overton is probably from the Old English *ofera-tun* or *ofer-tun* both of which mean 'a settlement on a slope or a ridge'. (*Ofer* can also mean either 'river bank' or 'border'.) **Sunny Gutter** (now called Sunny Dingle) is a conifer-clad valley in Haye Park Wood close by Overton. It was in the old deciduous woods of this valley that the 3 children of the Earl of Bridgwater (the Lord President of the Council of Wales) lost their way and caused a panic in the President's household. They were found unharmed but this minor happening was the basis of John Milton's masque, Comus, first performed at Ludlow Castle in 1634. (Many people think that the incidental music by Henry Lawes is better than the drama.) Sunny Dingle is the lower part of the Mary Knoll Valley, which is marked on the 1¼ inch Ordnance Survey map. There is a parking area (SO.500.700) at the entrance to the dingle, opposite the entrance to Moor Park School. (*See Batchcott.*) Higher up is a wishing well. Drop a stone into the water and your wish will be granted immediately.

PANPUNTON *Adjacent to Knighton (Powys).*
The hamlet consists of a line of stone cottages and brick houses crammed in between the steep, wooded slopes of Panpunton Hill and the flood plain of the River Teme. Across the river is Knighton. The 4-storey, stone-built Kingsley House, dated at 1880, stands by the bridgehead. Opposite are the red painted sheds of Owen's Coaches garage, an assemblage that does little to enhance an already rather dismal townscape. The railway line to Shrewsbury sweeps by in a curve close to the backs of the houses and follows the Teme valley for several miles. The name Panpunton has 3 elements. The first might be from the Old English *pen* meaning 'an enclosure', or *panne* meaning 'a pan', perhaps a description of a landscape feature, or most probably it is derived from *pant*, the Welsh for 'a valley'. *Pun* might be an abbreviated personal name, and *ton* means 'a homestead, or settlement'. A suggested interpretation is 'Punna's homestead in the valley'. South of the River Teme is the most attractive market town of **Knighton**. It stretches up a steep hill and has a full range of facilities. However, it is, of course, in Wales and beyond the scope of this book; Knighton railway station, though, lies on the Shropshire bank of the river.

PANT *5½m. SSW of Oswestry.*
Pant is a substantial main road residential village with a population in excess of 1,000. It lies between Llynclys and Llanymynech on the A483. In the last century it was a busy port on the Montgomery Canal (closed in 1936) that handled the limestone quarried on Llanymynech Hill which lies immediately to the W. On the main road, above the old part of the village, is the Clifton Cafe. By the side of this is a lane that leads down to the wharf and the lime kilns. On the other side of the cafe is another, short, lane which leads to the top of the Inclined Plane. Here is a winding drum that was used to lower the stone down the slope to the canal. There are 3 general stores and a Post Office in the village. The name Pant is Welsh and simply means 'valley'.

PAVE LANE *1½m. SSE of Newport, and ½m. SSE of Chetwynd Aston.*
It has been suggested that the name is a folk memory of an unrecorded Roman road. (Roman roads were paved with stones.) However, it was in the 18th Century that this tiny hamlet found its footnote in history as a terminus of the Donnington Wood Canal. This was constructed in 1755-7 by Earl Gower (later the Marquess of Stafford) to carry coal from his Donnington Wood mines to a point of sale on the old Wolverhampton-Newport Road (the A41) at Pave Lane. There were also several branches that linked his Lilleshall limestone mines to the Hugh's Bridge Incline (see Lilleshall) which connected them to the higher main line canal. The coal was carried in tub boats of 3 different sizes and by 1798 there were 109 such craft on the canal. The waterway closed on Christmas Day 1882, though a short section that linked the Donnington (flour) Mill and the Trench Lock Incline remained open until 1929. Today Pave Lane is a nucleated hamlet in attractive country just off the new Newport by-pass (the A41T). It stands around the junction of the old A41 and Pitchcroft Lane. It was near this junction that the Donnington canal wharfs were situated. There are a few red brick houses, a couple of Duke of Sutherland cottages with their characteristic dormer windows, the stuccoed Norwood House Restaurante and a farm with ugly black and grey corrugated, iron barns. These lie at the northern foot of the conifer-clad Muster Hill, the old military meeting place for the men of the North Bradford Hundred. Pitchcroft Lane leads through the farm to an open area where the rectangular reservoir can be seen. To the left (W) of the reservoir some low undulations in the field can be detected if the season and the light are right. These are the almost ploughed out earthworks of a pre-historic fort-settlement; very few such lowland settlements have survived at all. This site was only discovered comparatively recently and does not appear on the 1:25,000 first series Ordnance Survey maps. A little further along, past the stable paddocks (there are numerous young ladies riding horses hereabouts) and small plantation of Christmas trees, the lane crosses an old red brick and stone canal bridge. To one side water emerges from a culvert under a field (the canal having been filled in) and on the other is a ditched stream that follows the course of the old canal through the garden of a cottage. The lane continues through flat country and crosses the main entrance drive to Lilleshall Hall. At the junction is a Victorian neo-Jacobean black and white lodge. The traveller can turn right here and will emerge on the old A41 at the Waverley Garage and an old persons' home called The Rubens that used to be a restaurant. South of here is the Fox and Duck, a modern brick pub with an amusing sign. On the opposite side of the road is the old, original pub (now a house) behind which is a coal merchant's yard. Is it more than coincidence that the Pave Lane coal wharfs were just up the road?

PELL WALL *1m. S of Market Drayton.*
Just S of Market Drayton and adjacent to the W flank of the A529 is an area of neglected woodland in which are hidden the old Hall and its ancilliary buildings. The name Pell Wall might be derived from the Anglo-Saxon '*Peola*'s well'. There is a large, hidden pool to the W of the Hall that appears to be spring fed. It lies in a hollow surrounded by mighty banks of rhododendrons and a few specimen trees. In the early summer this is an enchanting place, to our mind without equal in the county. A small stream runs out of the lake into a steep-sided little valley in which there are 2 smaller pools, now much overgrown. The Hall was built in 1822-8 to a design by the architect Sir John Soane for a friend of his, Purney Sillitoe. It was to be his last major work. Despite Sir John's considerable reputation it is difficult to be enthusiastic about Pell Wall Hall. The house is 'L' shaped with a porch in the indented external corner. The porch is, in fact, a later addition; the original entrance was a recessed loggia with columns and an arch. The building is constructed of red brick clad in ashlar stone with incised line patterns and pilaster strips. The dormer windows are somewhat unusual. Today the Hall stands in a ruinous state, a condition many find morbidly romantic. After many years of neglect, during which it was systematically vandalised, it was finally gutted by fire in the mid-1980's. The absentee landlord had for long been seeking permission to demolish the mansion and re-develop the site. Regardless of cost the County Council is determined that the Hall shall be restored. The setting is very fine, a gentle slope overlooking the valley of the River Tern and the town of Market

Drayton perched on a steep bluff to the N. The present entrance drive to the Hall is shared by the Court House, a cream painted brick building with round-leaded windows and hipped roofs. Like the Hall it is embowered in rhododendrons, laurels and mature trees. Higher up the hill is the drive that leads to Pell Wall House, stuccoed with dormers and 3 bays, and to the red brick Garden House, which stands alongside the large walled garden. Higher up the hill again is the track that leads to Pell Wall Court, a red brick house with mock Tudor black and white upper elevations. This stands beside an area of carefully coppiced woodland. Beyond is the Stable Block constructed of red brick with yellow brick ornament. Four ranges enclose a quadrangle. The rear range is occupied as a house and the entrance range has an archway, decorative buttresses and a clock cupola. Opposite is The Factory, a rectangular red brick building last used as a pottery, and The Cottage, a mid-20th Century dwelling with a hipped roof. On the main road N of the Hall is the Lodge, an architectural 'toy' if ever there was one. It is constructed of grey stone to an hexagonal plan and the windows have inverted 'V' shaped arches. On the flat roof there are parapets and a central lantern; in short it is a Classical-Gothic fantasy. Attached to the rear is a modern white-painted brick extension with small modern windows and a green felt roof. Sir John Soane would not have been pleased. Together with the Home Farm the buildings of the Pell Wall estate detailed above constitute a settlement more substantial than many Shropshire villages.

PENTRE *See Wilcott.*

PENTRE CEFN *2m. ESE of Llansilin and 4½m. WSW of Oswestry.*
An almost unspoiled little farming hamlet in a somewhat bleak landscape. The buildings are of stone under slate roofs and the fields of thin soil are laid to pasture. The lane is winding and muddy and flanked with well-trimmed hedges. Plas Pentre Cefn (*plas* means Hall or country house) is a handsome 2-storey, 4-bay mansion with 2 projecting gables and a porch offset to the right. It is constructed of stone, stuccoed and painted cream, with brown window frames and stands on the highest point in the settlement. Opposite is a modern chalet bungalow, a distinct blot on the landscape. North of the Hall is a small, disused quarry, one of many hereabouts.

PENTREHEYLING *3m. SSE of Montgomery.*
The name is probably from the Welsh *pentre*, 'a village'; and *haining*, Middle English for 'an enclosure', hence 'the enclosure by the village'. Pentreheyling is a main road settlement on the A489 near Church Stoke. There are 2 farms, a black and white house, and a bungalow with a hipped roof which is now a cafe called Meadowcroft. The River Caebitra runs a few yards to the S and beside it stands Bacheldre Water Mill. This is an 18th Century building of stone and brick, and the mill is in working order. There is a tearoom and a caravan park and the complex is open from Easter to the end of September. The village of Bacheldre lies a little to S. This, like the mill, is in Wales, the Caebitra being the boundary between Shropshire and Montgomeryshire hereabouts.

PEPLOW *3½m. SSE of Hodnet.*
It lies in flat country off the A442 Whitchurch to Wellington road. There is no village as such. The Hall was built in 1725 and stands behind a good pair of wrought-iron gates. It was enlarged in later years. The Chapel of the Epiphany (1879) is to the design of Norman Shaw. It is of timber and brick with a tall roof and lies in a meadow alongside the access lane. At the back of the Hall is a lake. This was created by damming the River Tern and is not altogether ornamental because it acts as a reservoir pool for a substantial mill situated half-a-mile downstream. This functioned initially as a corn mill but was later converted to operate a generator that supplied electric power to the estate. It has recently been sold and is now being converted to a house. The Hall was built for Sir Francis Stanier, d.1900. He was a multi-millionaire industrialist who operated in the Potteries, amongst other places. The story is told that he had a collection of rare Japanese and American geese which he tended himself. As he lay on his deathbed they all flew into the air and beat against the bedroom window before flying away never to be seen again. The name Peplow is probably from the Old English *phyppel-hlaw*, meaning 'pebbly hill', or 'stoney mound'.

PERTHY *3m. WSW of Ellesmere.*
The name could be from the Welsh *perth*, meaning a hedge or a bush, or *perthyn* meaning to belong or be related to. The village lies just off the A495 about ¼m. N of Welsh Frankton. It is a mature and attractive little place of scattered red brick cottages on an undulating hill. There are paddocks with donkeys and horses grazing, a wood, and holly and ivy in the hedges. At the top of the hill is a group of charmless semi-detached houses clad in stucco. A little further N the lane crosses a dismantled railway and by the bridge are 3 stuccoed and colour-washed cottages. This is **Crickett**. The land is mostly laid to pasture grazed by sheep.

PETTON *6m. SSE of Ellesmere.*
The Domesday Book name of Petton is Pectone and this is probably derived from the Old English *paec-tun*, meaning 'the settlement by the peak, or hill', though in fact the nearest hill is some distance away. Petton is a small hamlet and lies on a lane off the A528 Ellesmere to Shrewsbury road. Petton Park was built in 1892, a large brick house in Elizabethan style with straight gables and a turret. It is now a school and lies in a well-wooded park. Just to the E of the mansion is an oblong moat. The entrance to the protected mound within is from the NW side. There is also a tumulus in the grounds. The brick church has a nave and chancel in one and a bellcote. It was built in 1727 and restored and altered in 1870 and 1896. Inside are box pews and a good Jacobean pulpit. The W gallery is supported on 2 columns that were formerly in the Council House at Shrewsbury. In 1066 Petton was held by Leofnoth. By 1086 it had passed to Ralph who held it from Robert Butler who in turn held it from Earl Roger.

PICKLESCOTT *5m. NNW of Church Stretton.*
The country is somewhat bleak and the village nestles in a hollow. It lies on an easy route across the Long Mynd at a crossroads. The Gate House was formerly a pub, the Gate Hangs Well. The school closed in 1864 and became a farm; the field opposite the farmhouse is still called The Schoolyard. It is distressing to be reminded that these quiet little country places used to have a full compliment of tradesmen. Here there was a wheelwright, a carpenter, a grocer, a blacksmith and an undertaker. Now there are none. A stream passes through the village and to the S is a waterfall.

PICKSTOCK *3m. NNW of Newport.*
A hamlet in flat country N of Newport. In a garden, partly hidden by a wall, is the ruined stump of the Pickstock Cross, of great but unknown age. To the N of the settlement is Chetwynd Airfield, established before the Second World War. It covers some 260 acres and is mostly used for training helicopter pilots. The River Meese flows by to the W, and NW of Pickstock it enters a steep-sided, tree-clad, little gorge where it is joined by the fast-flowing Goldstone Brook (sometimes called the Ellerton Brook). Near Deepdales Farm willows arch over the slow moving Meese and a sturdy concrete footbridge crosses the river. Near the farm buildings is a huge glacial boulder, an 'erratic' which quite probably originated in the Lake District. In the Middle Ages Pickstock (Pickestoke) was part of the manor of Edgmond. The name might mean '*Pica*'s religious place'.

PITCHFORD *6m. SSE of Shrewsbury.*
The village is so named because at one time pitch, or bitumen, oozed out of the ground nearby. An area 1m. NE, on the line of the old Roman road to Wroxeter, is still called Blackpits. The well-known Hall lies ½m. NE of the village. It is a very large 'E' shaped timber-framed mansion of about 1560, built for Adam Otley, a wealthy Shrewsbury wool merchant. The main entrance used to face S, but was moved to the N in 1887 when the house was restored. The uprights are infilled with diagonal struts and it has star-shaped brick chimneys, but little other ornament. Pitchford is simply a good, well proportioned house and impressively large. After Park Hall near Oswestry was burnt down (in 1918) Pitchford Hall assumed the title 'Shropshire's finest timber-framed house'. In 1987 the roof on the W side was being repaired but the N side is so covered in grass and weeds as to be in places more like a field than a roof. The

Hall is approached down a long avenue of mature trees and lies in wooded grounds in which there are 2 lakes. Close to the W front of the house is a magnificent, and very rare, broad-leaved lime tree. It was a large tree in 1692 when it was marked on an estate map. This makes it well over 400 years old. It has a famous Tree House, also marked on the estate map. The house we see today, though, is not especially old. It was completely renovated in 1970 when the tree was also propped with metal supports. The Tree House is black and white, and has an ornate plaster ceiling. Queen Victoria, as a child, stayed at Pitchford and noted in her diary that she visited 'the little house in a tree'. The tree and its house are unique in Britain. The church of St. Michael stands near to the back of the Hall. It has a blocked Norman window in the nave with herringbone masonry below that and a Norman arch on the N side of the chancel, but is mostly Early English (13th Century). Inside there are 17th Century box pews, a Jacobean pulpit and monuments to the Pitchford family including effigies of more than 50 children. The Hall and the church probably stand on the site of the lost Domesday settlement of Eton, or Little Etton as it was sometimes called. There are some interesting remains of Roman engineering works on a stream crossing ¾m. SSW of Pitchford at Ordnance Survey map reference SJ.525.025. The Roman road from Wroxeter crosses a little wooded ravine and upstream of the modern footbridge there are the earthworks of 2 sets of Roman embankments, made at different times, and a stone bridge abutment on the S bank, though this is of coarse construction and may well have been rebuilt at a later date. The higher, larger embankments were probably joined by a timber bridge. The site can only be approached on foot along the public footpath. Southwards the line of the road is marked by a hawthorne hedge and a fence and then is followed by the modern road which continues to Frodesley, Church Stretton and beyond. The route of the road is clearly shown on the Ordnance Survey 1¼" map. From the N the Roman road cuts across the modern road ¼m. S of Pitchford. It veers away to the W (the right if travelling southwards) first as a rough track and then as a green, grassy lane lined with hedges.

PIXLEY *¾m. WSW of Hinstock which is 6m. S of Market Drayton.*
A lane leads off the main road on the bend near the church. It is marked as a dead end road, but this is not true. About ½m. along the track is a small wood. In this are the remains of 2 pools called the Pixley Locks. In 1912 one of these was cleared, the bottom gravelled and wooden platforms erected around the sides for the benefit of local swimmers. This swimming pool was made by the local Squire, Mr. P. V. Williams, but the pools are now once again overgrown and marshy. South of the pools, across the lane on the black, peaty lower ground, is a narrow hillock which according to local tradition is the site of a Norman castle. Indeed, it used to be marked as such on the Ordnance Survey map. The lane continues westwards as a rough track through large, red fields used for growing arable crops. Pixley is one farm and 2 blocks of cottages. They huddle together on a rise, remote and windswept surrounded by most attractive undulating country. The farm is partly of red brick and partly red sandstone, cut into large blocks. The farm buildings are only used for storage and this is now really a residential hamlet of no more than 6 dwellings. One pair of semi-detached cottages has been rendered, striking a discordant note. There are a few ducks, geese and dogs, a swampy little pool, a few oak trees and little else at Pixley. In essence it would have been little different in Anglo-Saxon times. The lane continues on and emerges near Hungry Hatton. The name Pixley probably means '*Pica*'s clearing in the forest'; *pic* can also mean a sharp point – a peak – but the hill on which the settlement stands could not be so described.

PLAISH *See Cardington.*

PLEALEY *1¼m. ENE of Pontesbury which is 7m. SW of Shrewsbury.*
A small residential place with a good timber-frame and brick house called Brookgate Farm which is in the process of being restored. There are some modern bungalows of brick and render and a few Georgian houses. On the edge of the village is a tiny one-roomed timber-framed building called The Den with its thatch encased in corrugated iron sheets but painted and well tended. Also tiny is the attractive, white painted Methodist chapel. It has a porch supported by wooden columns, 3 round-headed windows to the side and a hipped roof. Plealey House lies in the same lane, stuccoed, painted white with black trim and possessing a good selection of trees in the garden. Altogether, a quiet, attractive village in quiet, attractive country. The delightful explanation of the name Plealey is 'the glade in which the deer play'.

PLOWDEN *3m. ENE of Lydbury North which is 2½m. SE of Bishop's Castle.*
The name means either 'the valley where the deer play', or 'the valley where sports are played'. The A489 descends from the SE, past a little cottage with a large topiary-hedge entrance and the embankments of the disused railway (beside which stands a substantial red brick building with blue brick dressings and carved large boards) to join the River West Onny at the Plowden Gorge. The river carved this steep valley in recent geological time. The Long Mynd ends here, cut off sharply from the hills to the S. (*See Horderley.*) On the hillsides are pastures and mature mixed woods with a few scattered farms and cottages. By the road are the sheds of L. V. Harris, agricultural engineer, and behind his cottage is a house constructed of brick to the left, half-timber to the centre and stone to the right. Over the river, on the lane to Eyton is the red brick Catholic church and presbytery of St. Walburga, with its Gothic windows, gables, dormers and carved barge boards. Between the church and the bridge a lane leads S by woods stocked with game birds, a forestry yard and old oak trees to the superbly situated black and white Plowden Hall. It lies in a sheltered dip on the side of the hill with lovely specimen trees to the front and handsome 18th Century farm buildings of stone, brick and weather-board to the side. From the lane it looks magnificent. The house is Elizabethan and was probably built by the successful lawyer, Edmund Plowden, whose descendants still own the property. Inside there is much panelling, a good Jacobean mantelpiece, and a chapel with a brass of H. Plowden, d.1557, and wife. In medieval times Plowden was a part of the extensive manor of Lydbury North which belonged to the Bishop of Hereford both before and after the Norman Conquest. The lane continues on past the house and joins the Eyton-Lydbury North road. **Eyton** is a hamlet of white painted stone cottages and red brick houses dominated by large grey and black barns. There are sheep in the meadows and views over the broad valley to Plowden Hall. On the wooded hill behind the farm are the prehistoric earthworks of Billings Ring. **Totterton** lies ½m. W of Eyton. Like Eyton and Plowden it also belonged to the medieval manor of Lydbury North. The late Georgian Hall lies on a grassy slope with good views to the E and SE. Indeed, the name used to be Cot'dun which means 'a high place with a watch tower'. The house is built of red brick and has 2 storeys, 5 bays, a Tuscan columned stone porch, a central 1-bay pediment and brick pilasters. Along the roadside is a screen of evergreen shrubs, mature trees, and a small pool. To the right of the house is a walled garden.

PLOXGREEN *3m. SW of Pontesbury on the A488.*
Green in a place name usually denotes a medieval forest clearing. The first element of Ploxgreen might be derived from a personal name. It lies on the Shrewsbury to Bishop's Castle turnpike road, the A488. One of the more substantial buildings in the hamlet is Tollgate House, built of stone with red brick quoins. As to the rest there is a handful of brick cottages, a guest-house, a farm with stone outbuildings, a telephone box, a modern house, a bridge over a stream and a back road to Minsterley. A little to the S is the turning to Snailbeach and opposite that is the lane that leads to **Ladyoak**. This is the name given to a mile-long stretch of road along which stand widely spaced cottages and farms of red brick, stone and stucco. The narrow lane runs gently uphill and there are good views across the valley to Snailbeach and its glistening, white, lead mine spoil heaps. There is holly and ivy in the hedges and at the top of the hill are pastures and more good views, this time northwards over the broad valley of the Rea Brook towards Worthen. To the SW is the distinctive clump of pine trees on Bromlow Callow. The lane descends into a small valley, through a small wood of coppiced trees known as College Wood, past a black and white cottage and the hay

The beautiful 'hidden garden' of Pell Wall Hall, near Market Drayton.

Pontesford Hill.

New houses beside The Flash, Priorslee, Telford.

barns of Lady House Farm, to the small tree-clad mound of a Norman castle. Just past this motte, on the opposite (S) side of the road, is the entrance to Hope Hall. One is now in Hopesgate. (*See Hopesgate.*)

PONTESBURY *7m. SW of Shrewsbury.*
On Ponte*sford* Hill (the hill) is an Iron Age fort and S of the town at Ponte*sbury* Hill (the suburb of the town) is the Ring, an earthwork of similar age. Ponte*sbury* (the town) is of Saxon origin and was probably fortified in the late Dark Ages when the Vikings were sailing up the River Severn and raiding the country about. The 'pont' in the name is possibly from the Welsh *pont* meaning 'a hollow' or 'low lying'. Thus, Pontesbury means the 'fort in the hollow'. The church is placed at the highest point and lies roughly at the centre of the modern ring road system which follows the old earthwork defences. In the 7th Century a great battle was fought at Pontesbury between the Kings of Mercia and Wessex. That these were troubled lands is also indicated by the chain of 12th Century moated farmhouses in the Rea Valley: Woodhouse, Woodhall, Moat Hall and Hanwood. Woodhouse, meaning 'the house in the wood' is a name that usually describes an isolated farmhouse, deep in the forest with no neighbours to call on for help, and therefore usually moated. On Pontesbury Hill (1,047 ft.), to the SE of the town, lead has been mined for generations. The ore obtained is galena, a sulphide of lead. It was smelted at Pontesford and the chimneys had hundreds of yards of flues, used to recover by-products such as arsenic. The old industrial complex lay between Pontesford and Pontesbury on both sides of the A488 between the Nag's Head Inn to the W and the lane to Pontesford Hill on the E. The coal which fuelled the works was mined locally in numerous small pits – Poplar Pit, Heighway's Pit, Gibbet Pit, Corner Pit, etc. The ruins of a colliery engine house and an ore smelting works still stand. There was a mineral railway line to Pontesbury from the now disused lead mines at Snailbeach. The church has a 13th Century sandstone chancel. Most of the rest was rebuilt in 1829 by John Turner. The Old Rectory is of cruck construction. For 2 years Mary Webb lived at Pontesbury at the house called Roseville, where she wrote her first and best-known novel, 'The Golden Arrow'. She lived here in poverty with her husband, who was also an author. He grew fruit, flowers and vegetables in their garden and she sold them from a stall in Shrewsbury market. Unfortunately, Pontesbury is one of the towns that was chosen to carry the Shrewsbury population overflow. New houses abound. They are completely out of character with the area and are built with bricks and tiles made outside the locality, thus ensuring that they will not blend with either the old houses or the countryside around. The new Comprehensive School is named after Mary Webb. Pontesford Hill has several legends associated with it. One, used by Mary Webb in her novel, tells of a Golden Arrow, lost during the great battle between the Saxon Kings of Mercia and Wessex. Whoever finds it will come into a vast fortune. At the top of the Hill is an ancient yew tree, said to be haunted. On Palm Sunday it was the custom of young people to compete to be the first to pluck a twig from it. The winner was guaranteed good luck for the 12 months ahead. "The next proceeding," writes Charlotte Burne, "was to race down the hill to the Lyde Hole, where a little brooklet . . . turns and falls into a basin-like hollow at the foot of steep walls of rock, forming a deep circular pool which 'folk used to say there was no bottom to . . . .' Whoever could run at full speed from the top of the hill down the steep sides of the Hole . . . and dip the fourth finger of his right (left?) hand in the water, would be certain to marry the first person of the opposite sex whom he or she happened to meet. 'You could not choose but that one must be your fate'." The high summit of Pontesford Hill is technically another hill called Earl's Hill. Here is an Iron Age fort. The rock of the hill is more than 1,200 million years old. It is now a Shropshire Trust for Nature Conservation reserve. A varied flora and fauna exists from the stream at the foot of the hill through the ash, wych-elm and oak woods to the pastures, mosses and lichens of the bare upper slopes. One of the rarer plants found here is the rock stonecrop which covers the western screes and in summer provides a blaze of colour. There is a good bird population on the hill, from dippers and pied wagtails to tits, woodpeckers, sparrow hawks, buzzard, merlin and raven. The wide range of habitat attracts many butterflies; 29 species have been recorded. Badgers and bats also live here, and there is an excellent visitors' centre where a leaflet on the area is available. This reserve was first developed by the family of Mrs. Lily F. Chitty. This redoubtable lady was awarded the O.B.E and an Honoury M.A. for her services to archaeology. It was she who established the existence of the Bronze Age trading route now called the Clee-Clun Ridgeway.

PONT FAEN *1m. SW of Chirk.*
A single-arch stone bridge spans the stoney hedded River Ceiriog which here forms the boundary between England and Wales. It is an attractive river, about 14ft. wide and quite shallow with tree-fringed banks. On the N side of the valley are the broadleaf trees of Baddy's Wood; on the S side are hill pastures, a stone-built farm and a row of cottages in stone, brick and stucco. The valley is narrow here but broadens out to the W where there is a fish farm, its rectangular ponds hidden behind a hedge. There are patches of red bracken, dark green conifer plantations, and, near Castle Mill Bridge, a small wood of ancient oak trees. All there is at **Castle Mill** is one row of 3 cottages, a pair of semi-detached cottages, a tiny stone-built church, St. Catherine's and the stone-built bridge which carries the lane that leads up the hill to **Bronygarth**. Bronygarth is an attractive hamlet with cottages in a variety of styles, a green mission hut, some modern houses, and good views both down into the valley and across to Chirk Castle in Wales. On the road to Pentre Newydd, which it adjoins, are a battery of well-preserved lime kilns. On the hill above are 2 disused quarries. Offa's Dyke passes to the W of these and crosses the valley at Castle Mill. It continues up the northern, Welsh bank and passes through Chirk Castle Home Farm.

PORTH-Y-WAEN *5m. SSW of Oswestry.*
It lies just off the B4396 in the heart of mining and quarrying country. Copper, zinc, lead and even gold were mined by the Romans on Llanymynech Hill to the S. Limestone and granite roadstone rocks are still quarried in great quantities and the Steetley Lime Company is a major employer of local labour. Nevertheless, social facilities are scant. The shops and pubs are gone, the railway now only carries minerals and the school of 1839 closed in 1985. The Band Hall was built by the Porth-y-waen Silver Band. The Offa's Dyke Path passes just to the E of the village.

POYNTON GREEN *2m. SSE of Shawbury which is 7m. NNE of Shrewsbury.*
It lies in pleasant, rolling country with winding, high-hedged lanes. A red brick Methodist church with lancet windows stands next door to Loynton House which adjoins what appears to be a transport depot with tractors, trucks and trailers in substantial numbers. A large red brick house stands half derelict at the crossroads with more unsightly clutter, opposite – piles of broken concrete, mounds of earth and old machinery. **Poynton** lies ½m. S of Poynton Green. As places go it could hardly get much smaller. Thornhill Hall is white painted and part half-timbered. Amongst the numerous farm buildings are the remains of an old, grey-stone chapel. It consists of a gable end with one traceried 3-light window and stands adjacent to the road. It is now a part of a timber-framed barn with a tiled roof. Chickens scavenge in the farmyard and opposite is a cottage.

PREES *5m. S of Whitchurch.*
The name Prees is Celtic and means 'brushwood'. In Anglo-Saxon and early Norman times it belonged to the Bishop of Lichfield's Eccleshall estate. In Domesday Book it is shown as being held by the Bishop of Chester because the Lichfield diocese was for a few years administered from Chester. The nondescript buildings that line the main road of the village are a comparatively late development. The old, Anglo-Saxon centre lies up the hill to the E. Clustered about the War Memorial are some attractive cottages, the church and the Hall. General Hill, Wellington's right-hand-man at the Battle of Waterloo, was born at Prees Hall. The 9-bay Georgian red brick building has lain empty and neglected for many years, but is now being restored. The red sandstone church of St. Chad is mostly 14th Century with a tower of 1758 and 19th Century restorations. Inside there are monuments to members of the Hill family. It is believed that the

church stands on the site of an Anglo-Saxon College of Monks. Across the road from the church is a half-timbered cottage which has recently been totally renovated. At the rear of the Constable's House is the 18th Century ashlar sandstone 'lock-up'. It has a cornice and a stone roof with a ball a-top and is reputed to be one of only 5 such structures in England. The Grindley Motor Cycle Depot is long gone but the 'Grindley Special' designed and built here is still remembered by enthusiasts. At the bottom of the hill, on the other side of the main road, is the upstanding main drive-shaft of an old water mill, clearly displayed by the roadside. A mile further W is Prees railway station which is, somewhat surprisingly, still served by local passenger trains. The railway, the LNWR, came in 1853; the station ceased to be staffed in 1966. In the 18th Century Prees was an important coaching stop where horses were changed. At one time there were 12 inns one of which, the Hawk and Buckle, was commonly frequented by highwaymen and was forcibly closed in the late 1820's. It still stands as part of the Horse and Buckle Farm. Today there is only one pub, the New Inn.

PREES GREEN *6m. S of Whitchurch.*
A scattered village loosely centred on the junction of the B5065 and the A49 about 1m. S of Prees. Most of the buildings are of brick, including the Methodist church of 1933. The settlement lay within the estate of Hawkstone Park until well into this century and much of the population worked for Lord Hill. The estate timber yard was at the house called Oaklands just to the E of the village. West of the A49 and N of the lane to Quina Brook, was a large marsh part of which was common-land. Dogmore, or Bagmore as it was sometimes called, was one of the earliest of the Mercian 'mosslands' to be enclosed and drained and became the centre of a bitter dispute. Although covered in scrub woodland and often flooded it was nevertheless of some value to local people both as rough pasture and a source of small timber for use as poles and firewood. The land was owned by the Bishop of Lichfield and Coventry, who, in the person of Rowland Lee, was also the President of the Council of the Marches. (Common-land was not owned by the commoners, they merely had rights upon it.) In 1539 Rowland Lee sold 200 acres of Dogmore to a Cheshire landowner Sir Richard Brereton. Sir Richard enclosed this land by fencing and hedging and proceeded to improve it by 'stocking, dytching and mending of the moore'. Then began a long-running wrangle. The commoners were incensed at the loss of their land. They attacked one of Sir Richard's servants and even laid hands on the Bishop's horse nearly causing the cleric to fall. However, the courts decided for Sir Richard who proceeded to build 2 houses on the land and to grow hops and saffron most profitably. But the commoners were not done. In 1549 a hand of about 100 destroyed one of the houses and laid the crops waste. The Justice of the Peace, one John Dode, was taken diplomatically ill and did not attend to the matter. The dispute dragged through the courts and eventually Sir Richard won. (In the 13th Century the land had been leased and was technically still the property of a decayed manor.) Later, the whole of Dogmore was enclosed by Act of Parliament. The small fields were made large, the boundaries straightened and hedged and regulation width roads were constructed. These land-scape features remain to this day. The large, flat fields have earth of a quality suitable for arable crops, not just pasture. The main drainage stream that runs south-wards to the Soultan Brook is some 10 feet below the field surface level and even as we write new drainage works are in progress just to the N of Aldersey Farm. Drainage ditches are evident S of Prees Green too, in the area called The Nook where a rough track connects widely spaced cottages and houses. There is a feeling here of homesteading, of the frontier.

PREES HEATH *2½m. NNE of Prees.*
At Prees Heath the A41 and the A49 meet. This junction is about half way between the Birmingham-Black Country connurbation and the holiday resorts of North Wales. In the summer many thousands of West Midlanders pass through Prees Heath and to cater for their needs a small town of eating houses has developed. The old road along which they stand is now conveniently by-passed by the new A41(T). Here are the Happy Eater, the Raven Cafe, the Midway Truckstop, Brekland Cafe, the Raven Hotel, the Cherry Tree and the Little Chef. The numbers of holiday makers have declined in recent years but a more stable trade has developed catering for long distance lorry drivers and commercial travellers. As well as the cafés there are half-a-dozen houses, a garage and a large, white painted brick farm. To the S of the settlement on the A49 is a dead straight 2-mile stretch of Roman road and along this are the pale green warehouses of a government Grain Store. The land is very flat here and just S of Prees Heath, on the A41, is an old aerodrome from which a parachute school now operates. Nearby is a plant nursery and there are several battery chicken egg producers hereabouts. North of Prees Heath, opposite the Shropshire Car Auction premises on the A41, is a lane that leads to Ash and the remarkable Brown Moss, an area of pools, woods and marshes which is open to the public. (*See Ash.*)

PREES HIGHER HEATH *3½m. SSE of Whitchurch on the A41.*
A spread-about but substantial village by Shropshire standards. W of the main road is the main settlement; to the E are the remnants of the heath and the plantations of Twemlows Big Wood. The name Twemlows is from the Old English *twaen hlawn*, which means 'by the two hills'. Hills in this flat country are gravelly rises that stand proud of the surrounding glacial drift, clays and sands and old peat bogs. Twemlows Farm lies 1½m. E of the main settlement. In the early 19th Century the farmer who worked the land here suffered the loss of his hay harvest when the ricks were deliberately burned. In desperation he sought the aid of the local wizard, 'a big, handsome, black-looking man', who lived at Whitchurch. He led the farmer into a small back room which instantly took on the appearance of his farm. In the darkness a light flared and the farmer recognised 2 local youths as they put a torch to his ricks. He went to the authorities and laid a complaint against them. They were arrested and tried and found guilty on the evidence of the farmer who swore that he had seen them commit the crime with his own eyes. Before they were hanged they admitted their deed. A true story or not, or perhaps half true, it illustrates a belief in witchcraft and the supernatural which in earlier times was widely held. Twemlows Farm is a substantial brick house which has been rendered to the front and painted grey. It is a handsome building of 2 storeys with dormers, a slate roof, and wood mullioned windows. To the rear is a range of good traditional farm buildings, home to a dairy herd. Sheep and potatoes are also a part of the local rural economy. The farm stands amongst large, flat fields drained by deep ditches and belongs to the estates owned by Tim Heywood-Lonsdale of Shavington. He inherited the lands hereabouts when his elder brother was killed in a fox hunting accident; he fell from his horse which in turn fell on him. Today, the house stands empty and 2 men and a bonny black dog work a farm which before the war employed 30 hands. Twemlows Hall lies ¾m. NNW. This is another estate farm, also of red brick but much older looking. Two stone foxes face each other on the roof of the porch and in the cobbled yard there is a barn with 5 sets of stags antlers, trophies from Scottish hunting trips. Between Twemlows Hall and the A41 is an old airfield used by a local parachuting club called Sport. Amongst the trees of the Big Wood there are still many wartime huts. (The large, pale green, hangars are now government grain stores.) On the other side of the road is a surprisingly large development of 20th Century housing. Most of the dwellings are bungalows, though there are a few houses, and hidden away out of sight is a Council estate full of them. There are stands of pines, birch trees, copper beech hedges, pony paddocks, a poultry farm, the transport yard of Grocontinental, a main road garage, a village hall and several acres of market-nursery gardens. The Manor House lies at the S end of the settlement along with a handful of other older houses. It is a modest red brick building with several blocked windows and a corrugated asbestos roof. Some of the internal fabric is said to be medieval. All that seems to be known about it is that the Assizes were held here in 1247.

PRESTHOPE *3m. SW of Much Wenlock.*
A tiny hamlet on the B4371 Much Wenlock to Church Stretton road. Limestone quarries lie all about for this is Wenlock Edge. A mile SW, along the main road, is Easthope Wood. Opposite the pub at the

junction of the road to Easthope is Ippikin's Rock. Ippikin was a robber well known in Shropshire. He lived with his gang in a cave at the base of the crag and had many adventures and amassed a vast fortune. Legend has it that a boulder fell from the rocks above and blocked the cave entombing the robbers and their treasure. Ippikin's ghost still haunts the rock and it is said that anyone who stands on the cliff and cries "Ippikin, keep away with your long chin", will thus summon the phantom. But be warned, if Ippikin thinks you are taunting him he will pick you up and hurl you to a certain death on the rocks below. (*Ipa* is an Anglo-Saxon personal name.) On the main road, the B4368, near to Ippikin's Rock is a gap – a 'glat' in local dialect – in the fence. A path used to lead through this to a steep cliff that overlooks Apedale. This gap, legend has it, can never be blocked; mend it and the hole appears again overnight. Many years ago a man murdered his father and dragged the corpse in a wood. He made up the gap in the fence with some thorns but come the morning they had been removed and the body exposed. The work of the fairies they say. The gap is called the Clattering Glat. The name Presthope means either 'the valley where the priest lives' or 'the valley belonging to the priests'.

PRESTON BROCKHURST *8m. NNE of Shrewsbury on the A49.*
The handsome, late 17th Century grey stone Manor House stands on high ground overlooking the main road village. It is a substantial building 2 storeys high with dormer windows in the roof and 3 bays wide with 3 gables and a projecting porch. To the right is a good, stone stable block and to the left an arboretum of specimen trees which adjoin the detached and somewhat isolated red brick walled-garden. Along the main road are a few black and white cottages, some old and some of the 19th Century. There is a working blacksmith's shop in a red sandstone building that looks like an old coach house and some modern bungalows. The countryside around is most pleasant. To the NE are the wooded hills of Hawkstone Park and to the SE the quarried hills of Grinshill. To the S is Acton Reynald and one of the mansions of the great Corbet family. Preston means 'the settlement of the priest(s)'. For an explanation of Brockhurst *see Lee Brockhurst*.

PRESTON GUBBALS *4m. N of Shrewsbury.*
A small village that lies just off the A528 Shrewsbury to Ellesmere road in pleasant, undulating country. The manor was held by the church of St. Alkmund in Shrewsbury before the Norman Conquest. The man who held the land from the church in 1066 was called Godbold from whom Gubbals is derived. Today Preston Gubbals is a parish and the parish church is St. Martin's. This was rebuilt in 1866 by S. Pountney Smith and all that remains of the medieval church is the S wall of the aisle with its round-headed priest's doorway and a Perpendicular window. Inside is an interesting 14th Century memorial which contains the bust of a man with a foliated equal-limbed cross on his chest. The village stands on a rise and Scots pines are a feature. There are several black and white dwellings: the Old Schoolhouse, Hillfields Cottage and a sizeable farm that has sandstone barns. Gubbals House is of red brick with stone dressings and has tall chimneys. A pleasant, unspoiled little place. Lea Hall lies ¾m. N on the main road. It is a gabled Elizabethan brick house with slightly projecting wings on both sides. Inside is a noteworthy fireplace dated 1584. There is a good dovecote, also in brick. Adjacent to the Hall is Pin Hill Farm which has a farm shop, advertised on the main road.

PRESTON UPON THE WEALD MOORS *3m. NE of Wellington.*
It lies on the SW edge of the wide, flat, former marshlands of the Weald Moors. Most of the drainage work was carried out under the supervision of Thomas Telford and commenced in the late 18th Century. The good pasture land that was reclaimed revived the economy of the area and the moorland villages were largely rebuilt in red brick. (*See Kynnersley*.) Amongst the farms and cottages of Preston is a surprise. The fine hospital of 1725 founded under the will of Lady Catherine Herbert (daughter of the 1st Earl of Bradford), is approached through a short avenue of trees leading to good wrought-iron gates. The buildings lie around 3 sides of a quadrangle. Facing the gates is the impressive hall. To the left is a colonnaded wing (the oldest part of the building), which housed widows 'of the better class'. In the opposite wing were 12 girls who were being trained for domestic service. (It is only in comparatively recent times that the word 'hospital' has become exclusively associated with the provision of medical care. It used to mean a place of hospitality and accommodation, either for travellers or the elderly and homeless.) At a later date 2 lodges with huge pilasters were erected at the entrance to the drive. Lady Herbert established the foundation in thanksgiving for having been saved from death by St. Bernard dogs in the Alps. The old church of St. Lawrence was demolished in 1739 and replaced by the present red brick building. By its side is a small village green, a rarity in Shropshire. Just SW of Preston is Hoo Hall, part half-timbered and dated at 1612. There is a school in the village but there has not been a pub here for 200 years. Most of the residents are commuters. The course of the Newport branch of the Birmingham and Liverpool Junction Canal touches the N fringe of Preston. The canal keeper's cottage still stands, next to the school, but the bridge was destroyed in 1985 as part of an engineering research programme into the effects of heavy traffic on old arch bridges. The name Preston means 'the settlement or homestead of the priest'. In the 18th Century there were 2 saltworks at Kingsley Wyche. The Charity Saltworks was operated by the Honourable Thomas Newport from 1707 to 1736 and the other belonged to the Charlton family of Apley Castle (now in Telford). Their works extended over one acre. Brine was pumped up (by a horse turning a wheel) and boiled in large, iron pans to drive off the water. The fires burned coal from the Charlton family's mines at Wombridge (also now in Telford New Town). By 1799 the works were derelict. In 1833 a hoard of 5 Middle Bronze Age axe heads was found at Kinley Farm.

PRIORSLEE *¾m. NE of Telford New Town Centre.*
The headquarters of the Telford Development Corporation are in the early 18th Century brick and stone Priorslee Hall. However, a new centre has been purpose built for the Corporation and it will be moving there within a year or so. Priorslee Hall was built by the Jordan family but in the early 19th Century it became the headquarters of the famous Lilleshall Company (*see Oakengates*) and was the residence of the managing director up until 1963 when it was bought by the Telford Development Corporation. To one side of the Hall there are cheap, steel-framed factory units and to the other expensive, modern, red brick houses. There is an especially attractive development by the Flash, a large pool which lies amidst old coal mine spoil heaps that in places have generated a natural scrub cover and in others have been landscaped artificially. The water's edge to the side of the pool has been built up as a wharfage-promenade and small houses of varied styles, materials and colours face the pool. A lot more of the residentail development in Telford could have been like this. As to social facilities we could only find a small newsagent's shop but then Telford Centre is very near. All that is left to remind one of the extensive mining and iron working industries of the 18th and 19th Centuries are a terrain made hilly by old spoil heaps; occasional patches of black furnace earth (the Lilleshall Company had blast furnaces here from 1851); a row of handsome miners' cottages with ornate cast iron windows dated at 1839 (near the Hall); and a handful of Victorian houses by the old main road where also is the only pub for some distance, The Lion.

PULVERBATCH *See Church Pulverbatch.*

PURLOGUE *4m. N of Knighton (Powys).*
It lies in lanes W of the A488, a few stone-built farms with weather-boarded barns in a watery valley drained by the River Redlake. In the hedges are ferns and a variety of wild flowers; on the hills around are grazing sheep. The first element of the name Purlogue may be from the Old English *pur* meaning the bird 'the bittern', a large marsh bird the male of which has a booming call.

PURSLOW *4½m. W of Craven Arms.*
A crossroads hamlet with an old inn and a Hall in the valley of the River Clun. Purslow Woods lie ½m. S. It was here that in the 1880's an horrific crime was committed. An unmarried girl, who had a child of about 3 years of age, came home

Preston Brockhurst Manor.

opposite, right: Transport café at Prees Heath.

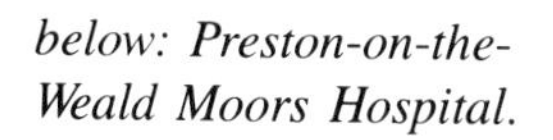

below: Preston-on-the-Weald Moors Hospital.

opposite, right: Mary Webb.

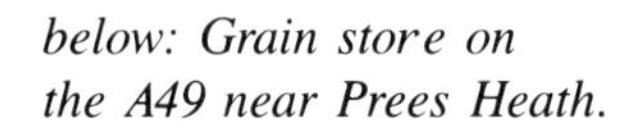

below: Grain store on the A49 near Prees Heath.

from Knighton market one day without her little daughter. She said that a woman had taken a fancy to the little girl at Hopton Heath station and she'd let her take the child to Manchester. A few months later a ploughman investigated the source of a terrible smell in Purslow Woods and found the rotting corpse of a little girl in a briar bush. The young mother admitted murdering her daughter and was sentenced to 10 years imprisonment. The reason she gave for the killing was that her boyfriend, the father of the infant, had said he would marry her only if she could find a home for the child, and that she had been unable to do. The name Purslow used to be spelled Pursloe and before that Pusselawe. It means '*Pussa*'s burial mound'. (*See also Kempton.*)

PYLLAU'R-MEIRCH *1¼m. SE of Llansilin (SJ.222.272).*
A tiny hamlet with some stone-built cottages that have brick quoins. A winding lane leaves the main road here and heads SW through a gentle valley with a stream, sheep on the rolling pastures, rocky outcrops, isolated farms, patches of scrub, stands of broadleaf trees and a few black cows with white faces. At **Wern ddu** are 2 newly renovated stone cottages. Just beyond them are the long, low cliffs of the Craig Sychtyn Quarries which stretch southwards for half a mile. Just off the road is what appears to be the remains of a lime kiln.

QUATFORD *2m. SSE of Bridgnorth.*
The village lies on the A422, crushed in between sandstone cliffs on one side and the banks of the Severn on the other. Some of the settlement lies along the steep, winding Chapel Lane that passes through chestnut woods in its higher reaches. There is a Norman motte and bailey castle by the river and a Watch Tower folly in the rock face overlooking the road. The church of St. Mary Magdalen stands on a hill above the village and is dramatically positioned on a sandstone crag amongst pine trees. Beside it is an ancient oak tree, a remnant of Morfe Forest. St. Mary's was built by Roger de Montgomery, who became the Earl of Shrewsbury and who held the whole of Shropshire on behalf of the King. His second wife, the Countess Adelisa, had been caught in a terrible storm at sea and vowed that should her life be saved she would build a church on the spot where she had met her husband. The chancel, chancel arch and tower arch are original Norman; the nave and tower were rebuilt in 1714 and the S aisle and S porch were added in 1857. There had been settlements at Quatford long before the Normans arrived in 1066. It is very likely that prehistoric tribes occupied the site, though no evidence remains. Certainly the Danes were here. They had begun their attacks in 793. In 865 a large army arrived and in 893 a force of Danes from London came to Quatford and built a camp by the river. In 912 the Mercian Queen, Ethelfleda, constructed 10 forts, or burgs, in the Midlands to defend the area from Danish attacks. One of these was at Quatford. At the time of Domesday Book only Quatford and Shrewsbury were recorded as being boroughs. Quatford was of importance to the Normans because it had a bridge over the River Severn. In 1101 it lost this advantage for in that year the resident lord, Earl Robert de Bellême, abandoned his castle at Quatford and moved 2m. N. There he built a new bridge. The new bridge was N of the old one so he called his new town Bridgnorth. 'Quatford Castle' lies near Daneford, ¾m. N of Quatford. This most handsome, battlemented house of 1830 was made to look like a castle and is set high above the road amongst dense woods. It can occasionally be glimpsed through the trees from the road. John Smallman, a builder, designed and constructed it for himself. As to social facilities, there are few. The vicarage, schoolhouse and 3 cottages were pulled down when the A442 was widened in 1960, and the ferry service over the River Severn ceased in 1940. There is a village Hall, though, that in its time has been a chapel and a stable. Recreational fishing on the Severn a major industry; here the Birmingham Angling Association lease the rights. The first element of the name Quatford could be from either the Old English personal name, *Cwatt*, or *cwead*, meaning 'a muddy place'.

QUATT *4m. SSE of Bridgnorth.*
'Model' 19th Century stone and timber cottages line the main road which here runs along a sandstone ridge parallel to the River Severn. Lovely rolling country and distant views are to be found all along this highway (the A442 from Bridgnorth to Kidderminster). The village is old – its very name sounds ancient – though much of it has been rebuilt. The church of St. Andrew was also much rebuilt in 1763 but the chancel, N chancel chapel and the N arcade are all medieval. Inside there is a monument to Mary Wolryche (d.1678), holding a lute, and a Jacobean pulpit and readers desk. Opposite the church is the early 18th Century red brick Dower House with 9 bays, 2 storeys, 2 front doorways, and a hipped roof. The name Quatt is either from the Old English personal name *Cwatt*, or *cwead*, meaning 'a muddy place'. **Dudmaston** Hall lies ½m. NW of Quatt. It is in Queen Anne style, built about 1730, in red brick with stone quoins but altered in the 19th Century with the addition of Grecian motifs. It has 9 bays, the middle 5 being recessed. The house stands in a well-wooded park with lakes and walks through the valley of the Severn. In the grounds there is some modern sculpture by the Claverley artist, Anthony Twentyman. The house was bought by the present owner as a gift for his wife but she did not like it and refused to live there. Inside there are collections of Dutch flower paintings, botanical art and watercolours, and to cater for summer visitors there is a shop and a tearoom.

QUEEN'S HEAD *3½m. SE of Oswestry.*
Queen's Head is a red brick hamlet just to the E of the junction of the newly improved A5 and the reed fringed Shropshire Union Canal. It takes its name from the Queen's Head Hotel. The corn mill received its supplies of wheat from Ellesmere by narrowboat. In the fields behind the green corrugated iron shed that stands beside the canal there were sandpits. The sand was transported in wagons drawn by donkeys who hauled it through a low tunnel that emerges in the middle of the shed. One mile WNW of Queen's Head, approached down a minor road off the motorway-like A5, is **Aston** Hall. The gateway could be mistaken for a whimsical folly: 2 large, Greek Doric columns support a simple entabulature next to the random stone lodge. They do no justice to the 7-bay Hall which is a fine, dignified building of grey ashlar stone built in 1780 to a design by Robert Mylne. It stands in an attractive, wooded park in which there is a lake, complete with swans, and a formal garden. To the left of the Hall is a complex of cottages and stables, and to the right is a roofless church. This was built in 1742 of red brick with stone dressings. The tower and walls still stand, slightly ruinous, but cemented over at the exposed edges to stop water penetrating and causing further damage. Specimen trees surround the church. The grounds in general are immaculately tended and there is a newly planted avenue of mixed broad-leafed trees along the entrance drive. **Hisland** Farm lies ¼m. W of Aston Hall. This stands on the site of an ancient township. It is approached down a dead straight avenue of mature trees at the end of which is a substantial tree-clad mound, probably the motte of a Norman castle. This is 18 ft. high and has a level summit is about 37 ft. in diameter. The farm has a pedigree herd of Fresian cattle.

QUINA BROOK *3m. NNE of Wem.*
A red brick hamlet that straggles along the B5476 to the N of Wem. There is a garage, a shop, a telephone box, a row of stuccoed council houses, overhead power cables, a stand of mature broadleafed trees, an abandoned school and a Methodist church built of hard, red brick. Here, too, can be seen the old lime kilns, once serviced by a branch of the Shropshire Union Canal which was filled-in during the 1970's.

RADMOOR *2½m. SSE of Hodnet.*
A crossroads hamlet on the A442 in the flat northlands of the county. Adjacent to the main signpost is a huge pile of old horse-shoes, rusty-brown and easily missed. It is some 10 ft. tall and 6 ft. square at the base. The old blacksmith's shop still stands, empty but in good order, under a chestnut tree. The blacksmith still owns the premises but now lives in America. The few dwellings are mostly brick and tile estate houses built by Francis Stanier in 1877. The Stanier family lived at nearby Peplow Hall and made their fortune as iron-masters and coal-owners in North Staffordshire. There is an avenue of mature trees along the main road, a wood with a timber yard beside the track of the old railway, and a garage. The name Radmoor is probably a corruption of Redmoor.

Rhyn, near Oswestry. The 'cottage in a wood' at Glynmorlais.

RAGDON *1½m. due S of Church Stretton.*
The name is from the Old English *ragudun* meaning 'the hill where lichens grow'. The sign off the main lane says 'Ragdon Farms' and indeed that is all there is. Amongst the farm buildings are 3 houses, the oldest being Georgian in appearance with a hipped roof. Oldest of all are a range of stone barns and stables. The Hough lies ¼m. SW. Here are donkeys, dogs, geese, sheep and old scrapped cars galore. To the NW is Ragleth Hill and from Ragdon to Church Stretton the road leads one through some lovely hill country. It joins the B4371 by the Caer Stone, a rocky crag in the shape of a dumpy pinnacle that stands clear against the slopes of Hope Bowdler Hill. The highway from here to the A49 is called Sandford Road, one of the most favoured residential areas of Church Stretton.

RATLINGHOPE *4m. WNW of Church Stretton.*
Ratlinghope, pronounced by the locals as Ratchup, is not a village in the accepted sense. The cottages, pub, church and manor are scattered about a lovely valley in the NW hills of the Long Mynd. Black Welsh cattle are to be found on the steep, green meadows. The Darnford Brook runs down the wooded valley and past the pub to join the River Onny half a mile down the road at Bridges. Almost every hill hereabouts has prehistoric remains – earthwork forts, burial grounds and trackways. These are all marked on the Ordnance Survey map, without which one would be unwise to venture on foot into country of this kind. Castle Ring, the earthwork of Ratlinghope Hill, ½m. N of the church, is the fort described by Mary Webb in her novel 'The Golden Arrow'. In the 12th Century there was a cell of Angustinian canons attached to Wigmore Abbey at Ratlinghope. The church we see today was rebuilt in about 1788, using in part at least, materials from the old church. It has grey stone walls, a slate roof and a weather-boarded belfry with a pyramid cap. Remote churches like this rarely had a resident vicar. In 1865 the Rector of Woolstaston had to conduct services both there and at Ratlinghope, nearly 4 miles across the wildest parts of the Long Mynd. On Sunday, 29th January he was returning from Ratlinghope to Woolstaston when he was caught in a terrible snowstorm. He was lost for 22 hours, snow blinded and almost dead. (One man did die in the snow that night whilst searching for his ponies.) The Vicar wrote an account of his ordeal called 'A Night in the Snow'. Ratlinghope Conglomerate is a rock consisting of quartz pebbles in a sandy matrix. There is a stretch of road at Ratlinghope along which a phantom funeral is said to travel in the failing light of early evening. The name Ratlinghope is from the Old English and has 3 elements: the first is possibly from a personal name such as *Rotel*; the second 'ing' meaning, 'the people of' and the third 'hope' meaning 'a valley'. The name therefore probably means 'the valley of Rotel's people'.

RED LAKE *It adjoins Ketley which is 1¼m. E of Wellington.*
It lay on the Lilleshall estate of the Leveson Gower family, later to become Dukes of Sutherland. In the 18th Century a squatter community developed here on the land made waste by coalmining. There were numerous small pits on the hill; in Cow Wood they were being worked within living memory. Many of the cottages were badly constructed; some were mere huts. The Leveson Gowers tolerated the situation but in 1812 James Lock took over the management of the estate. He made the cottagers official tenants, charged rents, destroyed unsafe dwellings, repaired others and generally had a paternalistic attitude. A man who behaved well was loaned money to buy a cow and given a piece of land upon which to graze it. Some of the dwellings were remodelled in the style of Duke of Sutherland cottages and have dormer windows, square chimney pots and herringbone lintels. One such is Shrubbery Cottage, dated at 1772. There are still, however, numbers of old stone cottages now mostly clad in stucco. The cottage on the right hand side of the lane that leads from the church to Cow Wood was originally a one up and one down stone cottage which was later extended in brick and then rendered. It still retains its piece of land, its smallholding. The handsome church of St. Mary the Virgin and the large, red brick vicarage stand amongst mature trees in the village centre and complete a picture which is totally rural. St. Mary's was built in 1838 and consecrated in 1839. It cost £1,300 and was paid for by the 2nd Duke of Sutherland who is reputed to have attended a service here incognito and to have been much affected by the singing. The fabric is a dark grey stone; the style is mixed, with lancet windows and a Norman doorway and a low pyramid roof on the tower. Amongst the old cottages are more substantial Victorian houses, some modern infill, and the sheds and vans of Upton's the removers, the only blot on the landscape. Many of the boundary walls are built of grey stone. The name Red Lake is said to have originated as a description of the lake (there were actually 3) at the bottom of the hill on the other side of the Holyhead Road. (*See Ketley.*) These pools were made as reservoirs for the Ketley ironworks and the flames from the furnaces reflected on the water.

REDNAL *5m. ESE of Oswestry.*
A tiny hamlet known for the railway – canal interchange, ⅔m. W of the settlement, where the Chester – Shrewsbury railway crosses the Shropshire Union Canal (SJ.351.276). Just N of the modern crossover is a large, ruined warehouse which was later used as a fertilizer factory. A channel from the canal passes before the warehouse and leads to a large overgrown basin which connected with the railway sidings. The interchange complex was abandoned in the 1850's. Close to the railway bridge there is a square brick and timber building which is believed to be the terminal of a short-lived high-speed canal passenger service which ran between Newtown and the Rednal and West Felton railway stations. Alongside this building is a rare 'roving bridge' which allows a horse towing a narrowboat to cross from one bank to the other without being unhitched. Note: This section of the Shropshire Union is sometimes called the Montgomershire Canal.

RHYN *3m. N of Gobowen which is 3m. N of Oswestry.*
The name might be from the Welsh *rhynnu*, 'to starve with cold'. Rhyn lies 1m. due W of St. Martin's and the parkland has given its name to the modern Rhyn Park School. Head W from the school and after about ¼m. turn N off the B5070, just past the sewage farm at Nefod. Nefod Lane runs along a wooded valley. Amongst the trees on the other side of the stream runs a section of the ancient and mysterious Wat's Dyke. Turn right at the 'T' junction and the traveller is now in Rhyn Lane. At a kink in the road there is a collection of wooden sheds one of which used to be a rope manufactory. The rope maker's cottage, No.3, Rhyn Lane (SJ.308.372), is stuccoed and painted cream with black 'timber framing'. The adjoining cottage is red brick. There is no centre to Rhyn; it consists of a scatter of cottages and farms. The Park lies to the W of Rhyn Lane. Opposite a stone-built farm there is a turning to the E that leads down into the valley and the delightful settlement of **Glynmorlas**. The descent is very steep. The mainly stone-built cottages lie about the fast flowing stream which froths and foams as it leaps over a weir by the stone bridge. There are sheep on the hill pastures, geese by the brook, a baby goat in a paddock, ivy-clad broadleaf trees and stands of conifers. Past Saw Mill Farm and an abandoned weather-boarded hut and downhill a-way is an enchanting spot. A narrow foot bridge leads across the stream just below a meeting of the waters and leads to a white stone cottage. A wisp of blue smoke from the chimney drifts through tall trees and one is reminded of the folk song that begins 'Oh, there's a cottage in the wood, all underneath hill it stood'. Just beyond the valley opens out into the flood plain of the stoney-bedded River Dee which is crossed by an elegant one-arch bridge at **Pont-y-blew**. The settlement consists of less than half-a-dozen cottages and farms. Above, hidden from view, is Brynkinalt, an impressive mansion which is the home of Lord Trevor. The road leads up, out of the valley and passes the sturdy stone walls and towers of Forge Farm, looking quite castle-like with its thin, slit window-vents. But this is in Wales; the Dee forms the boundary here.

RICHARDS CASTLE *4½m. SSW of Ludlow on the B4361.*
Richards Castle lies just over the border in Herefordshire but is included in this book because it has historical connections with Shropshire. The Richard after whom the settlement was named was Richard le Scrob. He was a Norman nobleman who had settled here and built himself a sub-

Bridge over abandoned railway line near Ridgwardine.

Rorrington Hall.

View from The Long Mynd near Ratlinghope.

stantial motte and bailey castle before the Conquest of England by Duke William. It has been suggested, but with no corroborative evidence, that Shrewsbury – originally called Scrobbesbyrig – was named after him. Not long after the Conquest the land on which now stands the town of Ludlow was given to Richard's son, Osbern Fitz Richard, and from him was held by Roger de Lacy. It was at this time that Ludlow castle was begun to be built in stone and the new town laid out. The remains of the castle at Richards Castle are on a wooded slope 1m. NW of the modern village. The road runs uphill and at the top, as it veers right, is the Old Village green on the left. There are 2 cottages, one in stone and one in half-timber. A path leads to the church and the great moat (which is up to 30 ft. deep) and the motte of the castle are revealed. There are far ranging southerly views from here and it is an evocative place. Tall pine trees now stand sentry with their feet in a tangle of briars, ground ivy, nettles and grass. Beneath this ground cover is a great deal of fallen masonry. The mound itself was for long thought to be just that – an earthen base for the original wooden fort. However excavation has revealed that in the early 12th Century a great octagonal tower of stone 50 ft. in diameter and with walls 20 ft. thick was built on top of the mound. As this tower collapsed the material from the upper parts surrounded the base to first floor level and the whole appeared to be a part of the motte. The original mound was 35 ft. high; as we see it today it is 55 ft. above the original ground level. In the early 13th Century the curtain wall and a residential tower to the E were added. There is evidence to show that in addition to the defences of the castle there was a town wall that encircled the whole of the top of the hill so that the church and the settlement were also afforded protection. The town, for it was a borough, was not insignificant. By 1340 there were 103 burgages (housing plots) and it had a weekly market and an annual fair. Richards Castle is of great interest because it is quite possibly the earliest true castle in Britain, and was certainly one of only 3 pre-Conquest castles in the country. The Celts, the Romans and the Anglo-Saxons fortified settlements, not private houses. Indeed the local Anglo-Saxon nobility complained bitterly to the King (Edward the Confessor) about Richard le Scrob's castle and tried to have it dismantled. But the King was pro-Norman and allowed the fortress to stand. By 1540, when Leland visited the area, the castle was no longer lived in and was beginning to become ruinous. The large and impressive stone-built church of St. Bartholomew stands 100 yds E of the castle. The Norman church, probably built by Richard Fitz Scrob in the 11th Century consisted of a nave and a chancel. The present N wall of the nave is original Norman and has 2 round headed windows; the rest of the building is 14th Century except the 15th Century W window and the later E window. Although now a redundant church St. Bartholomew's is well maintained and open to the public. Inside there are box pews and the canopied Salwey family pew. There is a tangible atmosphere here, almost monastic, and very medieval. The roof stands high supported by thick, bare walls some of which lie at odd angles. As one approaches the church a bricked up door and window can be seen below the chancel. These belong to St. Anthony's Bower, a crypt that was possibly once a hermit's cell. Beyond the chancel is the detached bell tower, square and strong like a castle keep and indeed it was designed as a part of the village defences. There are 3 bells beneath the pyramid roof. The church was abandoned in 1892 when the new church of All Saints at Batchcott was consecrated. (*See Batchcott.*) Richards Castle is known for its Bone (or Boney) Well. The name is misleading, for it is really a section of a stream rather than a spring or a well that is referred to. This is situated about 100 yds W of the castle at Ordnance Survey map reference SO.481.502. It is most easily approached along the firm dirt track that leads through the farm of the old 16th Century Manor House now called Green Farm (which is on the road between the modern Richard's Castle village and the church). At the bottom of the hill the track crosses a stream in a wooded dingle. This is the Boney Well. The ground is marshy and there are signs of springs in the slopes to the right. A water pump 'ram' lies in a pit covered by corrugated iron and a brick spring cap can be seen nearby. The well gets its name from the fact that at times of heavy rain numerous spiny bones accumulate here, washed out of rocks laid down in a Silurian sea. (In a strata of rocks called the upper Ludlow beds there is a thin layer – about 1½ ins. – of fish bones.) In former times they were believed to be frogs legs! The modern, main road, village of Richards Castle is, in fact, not so very modern. There are several black and white houses which must be at least of the 17th Century, and many cottages of the local yellow stone. There is a handsome pub, The Castle, a red brick Methodist church, a crescent of council houses, a modern estate, and a village hall; but the Post office, shop and school are with us no more. The settlement lies on a slope surrounded by fields laid mainly to pasture. Between the main road and the castle there are several good houses, notably the Court House. This is where parish matters were discussed and decided. The black and white part is the original building; the N wing of stone was added in 1620–30 but was subsequantly shortened and rebuilt at the W end. In the garden is a handsome Dovecote. This is at least 14th Century and has nearly 1,000 nesting holes in its circular, 3 – 4 ft. thick walls. At the NW corner of the house there is a 17th Century cider mill. To the W of the village are well-wooded hills and amongst these is Wigmore, 5m. W of Richards Castle. This was the early home of the powerful Mortimer family who also held lands in Shropshire. In the 14th Century Roger Mortimer (d.1360) made Ludlow castle their chief seat. King Edward IV was a Mortimer and his 2 sons were the ill-fated Princes in the Tower.

RIDGWARDINE *3m. NNE of Market Drayton.*
A hamlet of farms on a short ridge in undulating country. Most of the land is old pasture, the ground characteristically uneven and the grass hummocky, with oaks and brambles in the hedgerows. Ridgwardine Manor is a 17th Century timber-framed house with 2 gables that has been encased in brick and extended to the left. The framing was exposed at the rear of the building during renovation. The stone chimneystack is some 10 ft. in width and is approximately central to the house as it now stands. At Upper Farm there is as a small, sunken roadside pool and in the fields to the SW are many 'kettle-hole' pools. Adderley Wharf, on the Shropshire Union Canal, lies ½m. NW. Here there are locks, a turning space, moorings and a cottage; a pleasant spot. The road N from here to Adderley runs alongside the track of the disused railway line that ran to Market Drayton. The name Ridgwardine means 'the homestead (or enclosure) on the ridge'. The second element of the name is from the Old English *worpign*.

RINDLEFORD *2m. NE of Bridgnorth.*
It lies on the road that runs from Worfield to the A442. This is a curious little place. In the summer it is virtually an inland resort. There is a small sandy beach by a footbridge over the River Worfe which on sunny weekends is well attended. Nearby there is a gipsy-like shanty town of wooden chalets and make-shift huts. There are areas along the river that are marshy and overgrown and others where there are meadows and country walks. The dominant building is the old, now disused, stone and brick mill. It has 3 bays, is 3 storeys high and is somewhat gaunt. Close to it are an old cottage; a substantial Georgian style house; some solid red sandstone farmbuildings; the remains of at least one cave-house; and some black sheds. Overall this is charming, untouched, ramshackle place and it would be a shame for it to be 'developed'. The path to the beach is sometimes guarded by a large Anglo-Nubian goat. He is essentially a friendly creature but does'nt seem to know his own strength.

THE ROCK *1m. WNW of Telford Town Centre.*
No need to ask the name; rocks lie all a-jumble on the roundabouts of the wide new roads hereabouts. How much more enterprising it would have been if they had bothered to raise a proper megalith. The old settlement is tiny and has been left undisturbed on its windy little plateau. A few stuccoed cottages of the squatter type line the steep hill called The Rock on the higher slopes and Mannerley Lane lower down. At the top of the hill there is a Primitive Methodist Chapel of 1861 and adjacent is a second dated at 1877, a rare pair. Not so rare is the modern development called **Overdale** that lies just to the N of the village. This is a bleak, treeless place of modern houses and bland wooden

boundary fences. There is a Post Office and a handful of shops in the first development phase; the most recent houses have mock-Elizabethan facades. To the N Overdale is bounded by the M54, and to the S by the great opencast coalpit at Newdale. Not the nicest of neighbours.

ROCKHILL *1½m. SSW of Clun on the A488.*
It lies on the A488 between Rock Hill to the W and Clun Hill to the E, a scatter of stone-built cottages and farms around the junction of the main road and 2 country lanes. One of these leads to **Treverward** (*tre* is Welsh for 'hamlet'). The settlement lies in a valley in pleasant hill country. Treverward House is partly of stone and partly grey rendered with yellow brick chimneys and has an attached farm with buildings in red brick and concrete block. A large, modern barn has been adorned with a small dovecote complete with white doves. By the concrete bridge is a white stuccoed house with pale green shutters and a bungalow with a hipped roof. There are sheep and horses on the pastures and on the hill ⅛m. SW is an attractive old wood. Skirting the edge of this is an abandoned sunken trackway (SO.276.779) which used to be a continuation of the lane to Treverward. We find such old roads as intrigueing and evocative as old buildings, and this trackway is an especially fine example. In past ages the summits of bleak hills were often inhabited. The remains of one such long disused homestead lies ½m. WNW of the old trackway at SO.263.781.

RODEN *See High Ercall.*

RODINGTON *7m. E of Shrewsbury.*
A small, nucleated village of red brick houses and farms on the banks of the River Roden. The now abandoned Shrewsbury Canal was carried over the river here on a 3-arched brick built aqueduct. This has now been demolished along with several attractive 'lifting bridges' like those still found on the Shropshire Union in the N of the county near Whixall Moss. The canal wharf at Rodington was opposite the entrance to Rodington House. The pub, the Bull's Head, is said to be the second oldest in Shropshire. Bull baits and cockfights took place here until the middle of the last century. Attached to the old school is a perfectly awful modern flat-roofed annexe. There are 2 churches: the Methodist chapel a ¼m. W of the village centre, and the parish church of St. George. St. George's was rebuilt in 1851 but there has been a church here since at least 1086 when Domesday book also records that there was a priest, a mill, 2 villagers, 3 smallholders, and 3 riders. The manor was held by Thored both before and after the Norman Conquest, an uncommon feat. The overlordship belonged to Reginald the Sheriff but this later passed to the Fitz Alans, one of the Marcher Lord families. The name Rodington simply means 'the settlement by the River Roden'. **Rodington Heath** lies ½m. NW of Rodington. It is very much a place of new houses and bungalows but is pleasantly maturing. The land here is very flat indeed.

RORRINGTON *4m. SW of Worthen, which is on the B4368 Shrewsbury road.*
An attractive, small village in the hills SW of Worthen. Halliwell Wakes held on Ascension Day at Roddington Green were a joyful affair. Originally a ceremony to appease the water spirit of the Holy Well the occasion became a feast day. The well was adorned with a bower of green boughs, rushes and flowers and a Maypole was set up. The people walked and frolicked around the hill to the beat of the fife and drum, and in the evening they banqueted and danced to the music of fiddles. In 1854, just before the Crimean War, a miner's daughter saw the Anglo-Saxon lord Wild Edric and his ghostly army of horsemen in the hills near Rorrington. She described Edric's wife, Godda, as having long, golden hair and a green dress with a belt of gold. These phantoms are said to make an appearance on the eve of all great wars. The name Rorrington probably means 'the settlement of *Hror's* people'. *Hror* is from the Old English and means strong, vigorous, as often a tribal leader would be. The settlement is very small but quite delightful. The Hall is a friendly looking mansion with gables, part red brick and part black and white. It stands amongst mature trees and paddocks in which scamper sheep and baby pigs. There is a little more than a handful of cottages, a bit of 20th Century mock-Tudor, and a few scattered farms. A stream passes through the hamlet and ¼m. SE is the motte of a Norman castle, a mound with a flat top some 44 ft. across that slopes upwards from the E to the W with a minimum height of about 5 ft. On the road to Wotherton, hidden behind a wood, is Rorrington Lodge, a modest red brick house with stone dressings which looks Victorian. During the 19th Century lead was mined near the village. In 1853 the vein was being worked by the Rorrington Company but by 1856 the ore stock was depleted and the company ceased trading. A new company put out false geological reports to attract shareholders but was wound up 3 years later having found no lead. In this century the mine was worked for barytes but production came to a halt in 1920. Today all that remains is one small building and some open stopes.

ROSEHILL *2½m. S of Market Drayton.*
Rosehill is a main road hamlet on the busy A41 some 2m. SE the Tern Hill crossroads. The highway here follows the route of an old Roman road. Rosehill Manor is a Georgian looking house which is now a hotel; Rosehill Court is a dachshund breeding establishment; Rosehill Mill House stands in a delightful setting above lawned gardens and the old mill; and Rosehill Industrial Estate is housed in old military shacks on the road to Stoke Heath. There are also a few cottages and a shop. Rose in a place name can mean many things other than the flower. It can be derived from *Ralph*'s; from *hreysi*, Old Norse for 'a cairn'; from *hrossa*, Old Norse for 'horse'; or from *rus*, meaning a rush, or rushes.

ROWTON *6m. NNW of Wellington.*
A hamlet on a low hill in lanes to the W of the A442 Wellington to Hodnet road. Richard Baxter (1615-1691) was born here at the house of his mother's parents and baptised at nearby High Ercall. He spent the first 10 years of his life at Rowton and for 6 of these received an education of extreme low quality. Baxter was to say, later, that all 4 of his tutors were ignorant, 2 were immoral and one was a drunkard. He was to go on to be offered a bishopric, which he refused, and to write The Saint's Everlasting Rest (1650). The nave of the chapel at Rowton has medieval masonry but was rebuilt with a new chancel and a tiny bellcote in 1881. (Note. Do not confuse this Rowton with that which stands 1m. W of Rowton Castle which is W of Shrewsbury on the A458.) Rowton is Rugheton in Domesday Book and probably means 'the settlement in rough country'.

ROWTON CASTLE *3m. W of Ford and 9m. W of Shrewsbury.*
It is situated on the A458 Shrewsbury to Welshpool road, a romantic castle-mansion of the early 19th Century, designed by George Wyatt who modified and added to an existing Queen Anne House. However, the site was previously occupied by a real medieval castle which was razed to the ground by Llewellyn, Prince of Wales, in 1482. This was rebuilt only to be destroyed once again in the 16th Century by Commonwealth soldiers. In this century the interior of the Wyatt castle has been brutalised by alterations carried out when it became a school for the blind. The exterior has also been sadly neglected. In the grounds are several modern houses and a sports club; they do little for the ambience of a once romantic spot. The castle, and the magnificent Cedar of Lebanon that stands before it, can be seen quite clearly from the road. (Cedars of Lebanon were first brought to England in 1646.) The village of Rowton lies 1m. W and adjoins Wattlesborough Heath on the main road.

RUDGE *7m. ENE of Bridgnorth.*
A handful of houses around a road junction (SJ.811.975) at which there is a circular stone wall about 4 feet high. This encloses a compound in which lost and stray animals – cows, sheep, horses, goats, etc. – were placed. The owner had to pay a small fine before he could retrieve his property. Rudge Hall lies a little to the N, on higher ground, in a small park well stocked with mature trees. The house has a 3-bay centre with a hipped roof and projecting one-bay gables with stone pilasters to both sides. There is a segmental pediment above the door and the chimneys are tall. The undulating country around is most attractive. The name Rudge is probably from the Old English *hryg*, meaning 'a ridge'.

RUSHBURY *14m. S of Shrewsbury and 4m. ESE of Craven Arms.*
It lies in Apedale at the foot of the wooded Wenlock Edge. It seems isolated but the Victorians could travel here by train. The railway is now gone but it's course can still be traced along the bottom of the Edge. Rushbury is an old settlement. The suffix 'bury', 'a fortified place', and 'Rush', meaning 'rushes', implies a marshy terrain, which was often favoured by the Anglo-Saxon because it provided a natural defence. The church, the school complex and the Old Vicarage stand together. The church of St. Peter has a chancel of about 1200 and most of the rest is later Norman. It has a squat tower and there is some herringbone masonry in the N wall and hammer-beams in the chancel. The brick-built school, schoolhouse and 2 almshouses were constructed and endowed in 1821 by Benjamin Wainwright. The Old Rectory is mock-Gothic of 1840. On Shrove Tuesday the church is 'clipped'; the school children form a circle around the church with their backs to the building and walk sideways around it. Rushbury Manor is a handsome timber-framed house with a large stone chimney. Opposite the Manor is another black and white property, the substantial Church House presently occupied by an 'old colonial'. On a low ridge, just NE of the church in what is called Bury Field, is an almost circular mound about 10ft. high and 140ft. in diameter surrounded by a badly worn ditch about 30ft. wide. The Lakenhouse Brook flows by about 150yds to the S. Half-a-mile SW of Rushbury is Roman Bank where, tradition has it, there was a Roman Fort. There is a one-arched pack-horse bridge here across the Ecton Brook. Stanway Manor lies 1m. SE of Rushbury. It is an irregular brick house of 1868 with farm buildings designed by James Brooks. About 3m. WNW of Rushbury, near **Longville in the Dale** is the most handsome Wilderhope Manor. This is tucked away, out of sight, in a valley on the slopes of Wenlock Edge. Built of stone in 1586 it has no regular pattern. To the front it has 4 gables. At the back of the house is a semi-circular stair turret with a conical roof. Inside there are some good ceilings, probably by the same craftsmen who worked at Morville Hall and Upton Cressett. Wilderhope was bought by John Cadbury and given to the National Trust to be used as a Youth Hostel. Longville itself consists of little more than a pub and a few cottages but near the bridge, over the now dismantled railway, is the old yellow-brick railway station. It still stands in reasonable repair, complete with platform, at the end of a wooded drive. The name Longville does not mean 'long-town'; it was originally *Long-leah*, meaning 'a long clearing in woodland', but was corrupted by the Normans.

RUYTON XI TOWNS *9m. SE of Oswestry.*
The full name of the village is Ruyton of the Eleven Towns, meaning it was the Ruyton which belonged to one of the eleven towns of the Manor of Ruyton, to distinguish it from other towns with the same name. The town was 'planted', that is deliberately created, in the early 13th Century by the Fitzalans, one of the great Norman Marcher-Lord families. It was granted a borough charter in 1308 and remained a municipal borough until 1886. There was a Norman fort here from 1155 but the stone castle was built by Edmund Fitzalan in the early 14th Century and it was he who obtained the borough charter. (The Fitzalans acquired lands in the S of England and with them the title, Earl of Arundel.) All that remains of the fortress today are 3 sections of the rubble keep walls which lie on a low mound adjacent to the tower of the church. The handsome, red sandstone church has a Norman nave and chancel with Norman windows but has been much restored. The village has some charm. There is a circular toll-house by the Platt Bridge of 1701 (designed and constructed by a local builder, not Thomas Telford) which crosses the River Perry; several black and white houses; the Admiral Benbow opposite the church; and the Talbot Inn which has a sign depicting a white dog with black spots – a Talbot, a dog of French origin. The sight of the modern, silver coloured Express Foods factory comes as something of a shock. This began as a farmers' co-operative in 1915, but was taken over by Kraft's in 1936 and then by Express Foods in 1954. As a young man Arthur Conan Doyle (1859-1930), the creator of Sherlock Holmes, worked in the village as a medical assistant to a local doctor, Dr. Eliot. On the wooded slopes of Grug Hill, about 1m. W of the town on the B4397 is Ruyton Manor. The entrance is not signed but the startling orange-red sandstone 'castle' can be glimpsed from the road. It is a castellated 9-bay house built in 1860, a most attractive place with a machiolated tower, a recessed turred, terraced gardens and tennis courts. The River Perry flows past the town. This drains the Baggy Moors, a low-lying area that stretches for 8 miles or so northwards. The drainage of this wild, swampy place was the last major land improvement scheme to be carried out in the county. In 1752 it was the largest marsh in Shropshire, but after an Act of Enclosure of 1861 the river was straightened and dredged and all obstacles such as weirs, dams and fords were removed. Deep ditches were dug, connecting with the River Perry, and the water was drained off the land. Today it is a flourishing agricultural area, both pastoral and arable. The name Ruyton is from the Old English *ryge-tun*, meaning 'a rye farm'. Note. The 11 towns of Ruyton Manor were: Ruyton, Shotatton, Eardiston, Coton, Shelvock, West Felton, Wykey, Haughton, Sutton, Rednal and Tedsmore.

RYTON *3m. WSW of Albrighton (near Shifnal).*
A small village in pleasant country. The area by the bridge over the River Worfe is most charming, a charm which owes much to Stirling Cottage which has flowers in the fore-garden and climbing plants on the walls. The church stands on top of the hill. It has a tower of 1710 and a chancel of 1720 with Gothic restoration in Victorian times. Near the church are the worn down motte and barely distinguishable bailey of a Norman castle. By the river there was a slitting mill for cutting iron sheets into nailors' rods. This was owned by the Slaney family of Hatton Grange and was built before 1692. Today there are many new houses of little character though the yellow brick school is better than most of its kind. Like Ruyton the name Ryton is from the Old English and means 'a rye farm'. One mile E, on the road to Albrighton, is Caynton Hall, a good, rendered and white-painted Regency house with a central bow window surrounded by 6 Tuscan columns.

RYTON *(near Condover) See Great Ryton.*

SAINT GEORGES *3m. E of Wellington.*
The old crossroads settlement of Pains Lane in the parish of Lilleshall became a mining village in the early 19th Century. Most of the miner's had moved here from elsewhere and were not able to attend their parish churches. In 1861 Isaac Hawkins Browne (a local landowner with mining interests) and the Lilleshall Company provided funds to build a church on land donated by Granville Leveson-Gower, the 2nd Marquess of Stafford. The church was called St. George's and shortly after the settlement became known by this name. In old documents, of course, the area is still called Pains Lane. The architect of the new church was G. E. Street later to be both prolific and famous. This is one of his early works and is considered to be one of his most interesting buildings. The substantial tower is almost detached from the body of the church; the window groups and tracery are unusual; the arcade piers inside are of different shapes; and inside there are a variety of materials; grey stone, buff stone, red stone and red brick. It stands in a wooded churchyard and opposite is the Quarry House Inn (which has a most evocative sign) and a row of good brick cottages. Adjacent to the church are playing fields and the hideous modern Sports and Social Club. The substantial Old Vicarage is of red brick, has tall chimneys and stands well back from the road amongst lawns and trees. All this is in Church Street which continues E to become Limepit Bank. Here is a modern house built in a small limestone quarry. The centre of St. Georges is W of the church at the old crossroads where West Street – Church Street and Gower Street – Stafford Street intersect. Here are a few shops, the Post Office, a fish and chip shop, and the Methodist Church with its steeply pitched roof. There are many modern houses but there are still some Victorian 'semis' and the occasional larger Georgian dwelling such as Grove House. There are pubs, the Barley Mow, the Elephant and Castle, the Bush and Hamlett's Restaurant; an old Victorian school with lancet windows and dormers, now a youth club; St. George's Church of England School; and in Chapel Lane, off West Road, is an old Methodist Chapel of

Rushbury, the Manor House.

1862 with a yellow brick facade and tall, round-leaded windows. This is now the Wrekin District Action Centre. It is they who organized and now oversee the extensive Granville Country Park. (*See Donnington Wood.*) In Gower Street are the old Lilleshall Company's New Yard engineering works which were established here in the mid-19th Century. New Yard was an important works making blast furnace blowing engines, pumps, locomotives, winding engines and engines for mills of all kinds. The buildings now house a small trading estate where, amongst other things, security fencing is made. At the bottom of the hill, near Cappoquin Drive, there is rough ground now tree-clad. The wooded hill above is Cockshutt Piece an area of old coal, clay and iron ore mines. On the other side of the hill is Wrockwardine Wood where the Lilleshall Company had 2 blast furnaces.

ST. MARTINS *5m. NNE of Oswestry.*
A rather cheerless place with much institutional housing and very few visible amenities. The Rhyn Park School (1954-7) is a modern, steel-framed affair by Sir Basil Spence with flat roofs and timber-cladding and not an ounce of the character of the old school of 1810 (which occupied the Almshouses of 1638) in Church Terrace. This is little more than a rectangular box with a central pediment but feels as right as its modern counterpart feels wrong. The handsome parish church of St. Martin's of Tours stands on a raised, roughly circular mound. It is of mixed styles and very irregular, but pleasing because of it. The tower is dated at 1632. The tall box pews and double-decker pulpit were removed in 1980 but now that the worth of such old furnishings is being realized the pulpit and one pew are being installed at the back of the nave. To the W of the settlement is the Bank Top Industrial Estate which occupies an old Second World War prisoner of war camp. Further W again, close to the junction of the B5070 and the A5, is a striking gatehouse with battlements and looking very castle-like. This is one of the lodges of Lord Trevor's estate at Brynkinalt. The Hall lies 1m. N, over the border in Denbighshire. Just to the NE of St. Martin's is **Ifton** where there was a colliery which in 1928 provided employment for 1,357 men and on this count was the largest mine ever in Shropshire. It closed in 1968, but the Miner's Welfare Institute still stands and has become the village hall. At Oaklands is Oaklands Mount, an oval mound some 130 feet by 75 feet surrounded by a now partly destroyed ditch. This was probably the site of one of the many small Norman castles of which there were some 150 in Shropshire. **St. Martin's Moor**, ½m. SW of St. Martin's, is a canal settlement that did not exist before the Ellesmere Canal was constructed. It was originally intended that this canal should join the River Mersey and the town of Chester with Shrewsbury and Central Wales, but it was not completed for commercial reasons and never did get to Shrewsbury. Today, the canal is mostly used by pleasure craft, narrow-boats kitted out as water-borne caravans.

SAMBROOK *4½m. NNW of Newport.*
A well matured village that lies off the busy A41 near Stanford Bridge in agreeable, well watered country. There is a stud farm and a gun dog breeding establishment; yellow brick council houses, detached villas, and old sandstone cottages; the Three Horseshoes pub, and an unusual house of random sandstone with a simple timber-frame that is now a 'country residence for the retired.' The neat little church of St. Luke is of 1856 to a design by Benjamin Ferrey with stained glass by Charles Gibbs. The school, now the village hall, adjoins the churchyard and though of a similar construction to the church was actually built before it. The vicarage was built after the church and is embowered in mature trees. Sambrook Hall has gables and 3 broad bays and is built of hard yellow brick with black brick decoration. The boundary wall appears to be much older and is of weathered red brick with sandstone dressings. A farm is attached. Sambrook Manor is a modest red brick house with sandstone dressings of 5 bays spoiled by the skinny-framed modern windows on the first floor. Over the door is a plaque: O.A.F. 1702. The initials stand for Obadiah and Frances Adams. Their 4th son George inherited the Shugborough estate (near Stafford). In 1773 he changed his name to Anson and his grandson became 2nd Viscount Anson and later the 1st Earl of Lichfield. It is nicely sited above a tributary stream to the River Meese. On this stream, just to the N of the Manor House, is Sambrook Mill. This probably stands on the site of the Domesday Book mill and the iron-forge mill owned by Richard Ford II when he was made bankrupt in 1756. The building we see today is dated at 1853. It is built of red brick with stone dressings, 4 narrow bays wide, 3½ storeys high, and is presently occupied as a dwelling. It was working until 1975 and the wheel is still in place. The mill pool is dammed by the road bridge, ringed with reeds and trees and is an altogether pleasant spot. The old Mill House stands opposite the mill and has the same cast-iron windows. Today, the village is largely inhabited by commuters and all the old tradesmen have gone. The name Sambrook is thought to mean 'the sandy brook'. **Showell** Mill lies ¼m. SE of Sambrook. It was built in 1778 by Thomas Kynnersley and enlarged in 1851. The first tenant was Richard Hazeltine whose family were in the iron working trade and it is likely that iron was produced here though it has for long been a corn mill. Unusually it was supplied not by a stream but from a pond fed by a spring. The mill is disused but the pools are now operated as a trout hatchery and fishery.

SANDFORD *5m. SSE of Whitchurch.*
This pretty hamlet lies on both sides of the busy A41 Whitchurch to Newport road. The manor was granted to Thomas Sandford by William the Conqueror and close to the Hall, by the stream, is a small mound (on which is now a water tank), the site of the original Norman castle. The village consists of several red brick houses and farms and 3 black and white houses all of a row. In the front garden of the middle cottage is a circular sandstone trough about 10 feet in diameter with a stone mill wheel standing vertically. These are the remains of a cider press. The wheel moved around the trough and crushed the apples. To the left, on the other side of the drive of the adjacent cottage is a stone pillar with a wooden horizontal arm and old metalwork; this is another part of the cider press. The third cottage also has something of interest, namely a stone-stepped mounting block, used by older people to mount their horses. Opposite the cottages is Sandford Hall, the old home of the Sandford family who had lived here from Norman times until they were bankrupted by a local butcher about 1930 and had to sell the house. It is now owned and occupied by a recluse, Mr. Meakin, of the pottery manufacturing family. The brick built house has 3 bays, 2 storeys, a hipped roof with dormer windows and a large central attic window with arched top. There are several extensions and the stable-block was rebuilt by Thomas Sandford in 1863 after a fire. At the bottom of the garden, near the road and W of the Norman castle mound, is a lake. This was created in 1330 as a fish pool and the road was diverted around it. (The bend so created has been the cause of many accidents in recent years.) At a later date the pool was used as a reservoir pool for the brick-built mill that stands by the roadside hidden amongst vegetation. There is a small waterfall here. We were intrigued by signs hereabouts which direct the traveller to a 'Civic Amenity Site'. This is located on the road from Sandford to Prees and turned out to be a Council rubbish dump. The name Sandford means 'sandy ford'.

SELLATYN *3m. NW of Oswestry on the B4579.*
It is believed that Sellatyn was originally an Anglo-Saxon village called Sellaton, from the Old English *Sulh-acton*, meaning 'the settlement in the oak trees by a gully'. In the early Middle Ages the village was, it is thought, colonized by the Welsh who altered the name to give it a more familiar Celtic ring. The country around is hilly and wooded. The settlement does, indeed, lie on the side of a steep little valley and the road makes an acute angle turn to follow the contour of the land. The old stone cottages near the Cross Keys pub and the shop have been supplemented in recent times by modern houses of brick and stucco. The Rectory is a handsome mansion in yellow stone with stone dressings, gables and bay windows hidden from the road by a screen of evergreen bushes and specimen trees. The church of St. Mary has a medieval nave and chancel with a tower of 1703 and transepts of the 1820's. The whole was restored in 1892 which included the fitting of stained-glass windows by Kempe. An early 18th Century vicar was Dr. Sacheverell, who created a sensation by his attack on Dissenters and the Whig party, for which he was brought before the House of Lords on a charge of high treason and suspended from preaching for 3 years; his sermons

Sellatyn, Mr. Hughes, an English farmer.

were publicly burned by the hangman. After his trial in 1710 he travelled back to Sellatyn accompanied by supporters who, on occasion, are reported to have numbered as many as 4,000 on horseback and an equal number on foot. One mile SW, at a field boundary junction near Crown House, is a monolith (a standing stone). Offa's Dyke can be clearly seen at Carreg-y-big, where a farm has been built directly in its line. Wat's Dyke lies some 3m. to the E and the area between the 2 dykes was a kind of no-man's land between England and Wales. There are many overgrown tracks in the area, the purposes and origins of which are unknown though it has been suggested that they could have early military associations. **Pant-glas** lies 1½m. SSE of Sellatyn. This was an estate village that housed the workers at Lord Harlech's seat, Brogyntyn Hall. The school and laundry are now private houses. On the edge of the village there is an alder wood which used to be cropped to provide material for the soles of clogs. To the W the lane passes through a conifer plantation, a delightful little spot.

SHAKEFORD *4m. S of Market Drayton.*
The name is probably from the Old English *scahhari-ford*, which means 'the ford of the robbers'. Locally the name is pronounced Shakerford. The hamlet lies on the busy A41 which here passes over a causeway-embankment. This is the dam wall that held back the waters of a stream to make the mill pool that occupied the marshy depression to the W of the highway. The mill stood on the other side of the road. It had an overshot iron wheel of 12ft. in diameter and a malt house. The water was culverted under the road and at one time there was a second ¾ acre pool by the mill. The mill projected into the road and was a hazard to traffic. In 1931 the council demolished the old buildings and widened the road. Today there is a haulage yard attached to Shakeford Farm, a few red brick cottages, a rendered bungalow and where the mill stood are some black timber sheds and some greenhouses. Here one can purchase cabbages, onions, potatoes, lettuce, apples, farm butter, apples and mushrooms. The Four Crosses pub lies ½m. S at the crossroads of **Mill Green**. Here there are some 20th Century houses spaced out along the road with paddocks and tall hedges. At the junction with the lane to Lockleywood is Stafford House, an old stone and brick-built farm with yew trees in the garden. In a field just to the W is a spring of good water. This has been piped into a channel that runs parallel with the main stream which flows on to Shakeford. In this channel are the remains of watercress beds which were flooded by damming the channel with planks. They were abandoned soon after the end of the Second World War, mainly because the workers suffered from rheumatism caused, or inflamed, by the damp conditions.

SHAWBIRCH *1½m. NNW of Wellington.*
Before the new town of Telford was developed Shawbirch was an open field site adjacent to the Weald Moors. The new housing estates are a mixed bunch. There are some neo-Tudor designs, run of the mill red brick detached, and what looks like an army barracks but which is actually Council housing. As is often the case the later development is better than the early and pools complete with ducks, woods, and green areas are now being incorporated. There are no social facilities at all as yet. Facing Shawbirch Roundabout, where the A442 and the B5063 meet, is Shawbirch House, a substantial red brick farmhouse of 3 bays and 2 storeys with dormer windows and a red brick barn adjacent. As to the name: 'Shaw' is from the Old English *scaga* meaning 'a small wood', hence, 'the small birchwood'.

SHAWBURY *7m. NE of Shrewsbury on the A53.*
Royalty pass through this substantial village quite regularly on their way to the airfield at RAF Shawbury. Convoys of police and security vehicles speed along the lanes here with little care for other road users. We got caught up in one such convoy that was escorting Diana, the Princess of Wales, along the A53 and witnessed two near-miss head on collisions. Far from protecting the young lady they seemed to be attempting to kill her off. The RAF first came here in 1917 during the first World War. The station closed down in 1919 and opened again in 1935 as a Flying Training School and in 1944 the Central Air Traffic Control School took over. The Rotary Wing of RAF Tern Hill arrived in 1976. Maintenance on the helicopters and the fixed wing training aircraft is done by a private firm, Marshalls of Cambridge, who employ some 300 people; there are about 700 RAF personnel. The first British aircraft to fly around the world – the Aries – took off and landed at RAF Shawbury. The housing estates, offices and barracks of the airforce adjoin the village to the N and virtually consitute a separate settlement. The civilian village is almost certainly of Anglo-Saxon origin. It lies on a low rise above flat lands that were thus provided a natural defensive barrier. The church of St. Mary was built in 1140, of Grinshill Stone. It replaced an earlier timber Anglo-Saxon church, only the font of which has survived. The tower and the N chapel are Perpendicular and the porch is late 17th Century. Betwen the church and the River Roden is Moat Field, a public area where people picnic near old oak trees and the medieval moat, probably the site of the original manor house. A few miles N of Shawbury the river used to power papermills and iron-forges. In 1086 the Norman lord was Hamo Peverel but during the Middle Ages the manor passed to the Corbet family and became a part of their Acton Reynold estate. It was Sir Walter Corbet who built the Elephant and Castle inn that dominates the village centre. This was constructed of brick in 1734 and originally had 3 storeys, 18 letting rooms and stabling for 18 horses. It has since been rendered and colour-washed. Between the church and the inn is Church Street. No.122 is a most attractive black and white thatched cottage of the late 17th Century. On the main road is Ivy House, the timber-framed part is dated at 1640 and the brick block is late 19th Century. In the past it has been home (c.1700) to a branch of the Kilvert family (of Kilvert's Diary fame) one of whom ran a school in the house; a hunting lodge (c.1900) for the Mytton family; a private residence; and once more a school. The modern, flat-roofed shopping block is no ornament to the village; remarkably there are 3 hairdressing salons out of 11 shops in Shawbury. There is also a bank, a library and a village hall. Indeed, the place does verge upon a town. There are several modern housing estates and the population is about 2,500, of whom only a mere handful were actually born here. Similarly less than 30 of the 170 employees of R.A.P.R.A Technology Ltd. actually live in Shawbury. The Rubber and Plastics Research Association is said to be the only organization of its kind in the world. It does research work for many individual member companies and has a huge database that covers some 25,000 different plastics and rubbers. The company's modern premises stand on the site of the old Shawbury Workhouse. The name Shawbury means 'the fort in the wood'. In Anglo-Saxon names the 'bury' ending usually refers to a pre-Anglo-Saxon fort, i.e., Iron Age or earlier.

SHEINTON *3m. NNW of Much Wenlock.*
Under the Anglo-Saxons it was held as 3 manors by the freemen Azor, Algar and Saewulf but under the Normans it was consolidated and held by Ralph of Mortimer. The settlement stands on high ground and to the N overlooks the Leech Meadow, part of the broad flood plain of the meandering River Severn. The church of St. Peter and St. Paul was re-made by S. Poutney Smith in 1854. The timber-framed bellcote with its pyramid roof is probably 17th Century. Inside is a Jacobean pulpit and a 2ft. long stone effigy of a smiling woman, c.1300. The name Sheinton is from the Old English *Scena-tun* meaning 'the beautiful place', either of a single homestead or a settlement of several.

SHELTON *2m. WNW of Shrewsbury.*
At the time of Domesday Book Shelton was held by St. Chad's church, Shrewsbury, from the Bishop of Chester. It is now a red brick suburb of the county town and lies about the A458 to Welshpool. At the junction of Racecourse Lane and the main road is The Grapes pub and opposite is Christ Church of 1854 by E. Haycock in rustic sandstone. It has a bellcote and the roof is extremely sharply pitched. Adjacent is the school, made to a matching design. Behind the trees is the Royal Shrewsbury Hospital, large and symmetrical with a neo-Elizabethan facade of 1843 by Gilbert Scott. There is much new housing and there are several large estates in both Shelton and Bicton Heath which adjoins it to the W, notably Gain's Park. A little further W again is **Onslow** Park, a large, landscaped agricultural estate with a ruined Classical (Greek Doric) Lodge by

the roadside. There were other lodges but like the great house itself they have been pulled down. **Dinthall** Hall, ½m. WSW of the site of Onslow Hall, is red brick, 6 bays wide and 2 storeys high, and has a parapet, a gabled roof and a segmental pediment above the doorway. It is dated at 1734 and lies at the end of a ½m. long track S of the A458 and nearer Ford than Shelton. The name Shelton is from the Old English *scylf-tun*, meaning 'a settlement on a bank or ledge'.

SHELVE *13m. SW of Shrewsbury on the Shrewsbury to Bishop's Castle road, the A488.*
Shelve Hill runs parallel with, and adjacent to, the A488 between Black Marsh and The Marsh and is now partly clad in coniferous forest. The village of Shelve is on the E side of the hill. The area around Shelve is one of the oldest lead mining centres in Shropshire. The Romans were here, and quite probably the Celts before them. Roman coins, pottery, wooden shovels and candles have all been found in the workings hereabouts. The oldest 'Roman' workings are on the W side of Shelve Hill, near the top above the White Grit mines where 7 or 8 veins of lead outcrop on the surface, and further N up the hill beyond Gravel mine. These old workings were largely open cutting's but at the top of the hill large caverns were made with galleries running off in all directions. All the Shropshire lead mines lie within 3 miles of Shelve village and all lie on the Mytton Beds, a series of hard flagstones and gritstones some 1000 ft. thick. These outcrop in 2 oval shaped islands; one stretches from just W of Minsterley and runs SW to Rhadley Hill (2½m. S of Pennerley), and the other, smaller outcrop, runs from Hope SW to a point on the A488 due E of Corndon Hill. The lead was deposited as galena (a lead sulphide) in fissures in the rocks as hot liquids from the earth's magma rose and cooled. These mineral veins also contain a white calcite. Two of the most prominent monuments to the lead industry are the ruined engine house towers at Ladywell (SJ.326.993), built by Arthur Waters in 1875, and White Grit (SJ.319.980), erected in the late 1840's. Both can be seen clearly from the A488, the Shrewsbury-Bishop's Castle road. By 1180 lead was being exported to Gloucester, Builth Wells and parts of Wiltshire, probably carried by water from Shrewsbury. The mining here was never well organized and many miners had agricultural smallholding to supplement their incomes. Mineral mining industries have always had fluctuating fortunes and farming provided a degree of stability. The area was relatively densely populated during the peak production years of the 18th and 19th Centuries, but, with the collapse of the market at the begining of this century, mass emigration occurred. The wealth that had been created in the area was not spent here. The landowners and mine owners lived in Shrewsbury and London, and the roads, houses and churches they built for the miners were barely adequate for their purpose. Now the area is pastoral – small farms with large acreages of hill grazing. Some cottages are being bought up by townspeople as country retreats, and they make a not insignificant contribution to the economy of the area. The Forestry Commission has levelled many old workings and planted them with coniferous trees – a mixed blessing. Many people find the area very attractive. The old workings, white spoil heaps, ruined engine houses and cottages, and the bleak hillsides with sheep and distant views, certainly give the locality a strong character. The village of Shelve is little more than a handful of cottages and a church. All Saints is small, has a barrel roof and was refurbished in 1839. At **Blakemoorgate**, just below the summit of Stiperstones, is a small deserted mining hamlet. **Pennerley**, 1m. E of Shelve, is another of the once-busy mining communities. There are cottages on the hillside, a little chapel, some large mining ruins and the motte of an old castle. High above, on Stiperstones hill, is the Devil's Chair, a craggy outcrop of quartzite rocks. The early medieval name of Shelve was *Schalfe*, from the Old English *scylfe* meaning 'a ledge, a shelf, a bank, most commonly on a hillside. (*See Black Marsh, Snailbeach, Stiperstones, and also the Bog*.)

SHERIFFHALES *3m. NNE of Shifnal.*
Most of the more recent developments have been along the B4379 and the old village lies largely unspoilt just to the W. The access lane leads down a tree-lined hill, past a delightful black and white cottage with a traditional English garden, to the church of St. Mary. All that is left of the original church is the N aisle. The nave was added in 1661 and the piers and arches were looted from Lilleshall Abbey after the Dissolution of the Monasteries. The chancel is 18th Century. Below the church is a neo-Elizabethan lodge which guards the entrance to a 2m. drive lined with lime trees that leads to Lilleshall Hall. Sheriffhales Manor is a large black and white farmhouse which looks to have been heavily restored. It was in this house in the last quarter of the 17th Century that the Reverend John Woodhouse opened a university for Roman Catholics and Nonconformists who were unable to attend other colleges because of the Test Acts. Amongst the young men to study here were the statesmen Henry St. John and Robert Harley, better known as Lord Bolingbrook and Lord Oxford. Today the Manor is run by Slater Farms who are major landowners in the area. As to modern social facilities the village is blessed with a shop and a post office and that rarity a blacksmith; Mr. Eardley is also a farrier – not all blacksmiths shoe horses. The name Sheriffhales is explained thus: Hales is Old English and usually means a remote valley or a place hidden in the fold of a hill, and Sheriff is from the Domesday holder of the manor who was Rainald Bailgiole, the Sheriff of Shropshire.

SHIFNAL *8m. S of Newport.*
P. G. Wodehouse (1881-1975) knew the town well and it was probably the model for his fictional Market Blandings. The early history of Shifnal is not well understood. It has the look of an archetypal 'planted' town of the type laid out by the Normans, but there is no record of a plantation. There seem to have been 2 manorial estates here. In the 14th Century one was abandoned and the other developed into a town, Shifnal. In 1591 there was a great fire, during which most of the houses were burned down and the Norman church was badly damaged. It is possible that the regular layout of the town dates from its rebuilding after the fire. In 1562 The Earl of Shrewsbury built a charcoal blast furnace near Shifnal, one of the earliest in the country. The town received a boost when Telford brought his Holyhead road through the town. It came from Wolverhampton, through Shifnal and joined the A5 (Watling St.) just to the W of Oakengates. Today, the M54 bypasses the town, so bringing peace to the residents but a drop in trade for the shopkeepers. In 1794 an Act of Parliament was passed to enable the medieval open fields of Shifnal to be enclosed. This was one of the last enclosures in the country. It has been suggested that the open fields could only have survived as economic units if they had been intensively cultivated as market gardens, supplying the nearby industrial towns with comparatively high-priced fresh vegetables. Today, as a local regional centre Shifnal has a full range of shops and professional and social services. Until very recently the town looked very dowdy but a convivial regeneration is occurring. Even the ugly railway bridge (rebuilt in 1953) that cuts the town into 2 parts has been painted to make it less obtrusive. However, there is still much Georgian elegance gone to seed and some brash new development. The large red sandstone church of St. Andrew lies at the W end of the town. It is based on 13th Century foundations and still has some Norman elements, but its history is quite complex. In the 13th Century it had collegiate status with the financial advantages that bestowed. Inside there are several 16th and 17th Century monuments to the Brigges family. Buildings of some note include: Old Idsall House, a good timber-framed house with narrowly spaced upright strutting; the well restored Nell Gwyn, a large timber-framed public house; the 5-bay Georgian Vicarage; the old mid-19th Century station of white brick with tall, arched windows; Idsall House in Park Street, a William and Mary style dwelling of 5 bays with a hooded doorway; also in Park Street the old workhouse of 1817; and the Roman Catholic church of St. Mary in Shrewsbury Road of 1860 by J. C. Buckler. On the outskirts of the town is Haughton Hall, a once fine house of 1718. It is now a school and has been badly treated by the addition of clumsy extensions. It was whilst staying here that Bishop Percy found the old manuscript which became the foundation and the inspiration of his huge collection of Reliques of English Poetry.

SHIPTON *6m. SW of Much Wenlock.*
Shipton lies in Corvedale at the bottom of

Sherriffhales, Lilleshall Lodge and the church of St. Mary.

The Lodge to Brynkinalt, near Saint Martin's.

Shrawardine Castle.

View from Lordshill, near Snailbeach.

The holly tree wood on Lordshill, near Snailbeach.

the dip-slope of Wenlock Edge. The Hall is a splendid Elizabethan house of about 1587 built in local limestone for Richard Lutwyche of Lutwyche Hall. He gave it as a dowry to his daughter Elizabeth when she married Thomas Mytton. It takes the form of an 'H' and has a 4-storeyed porch tower, star-shaped chimneys, and mullioned and transomed windows. The Hall replaced a previous timber-framed house that had been badly damaged by fire. Whilst building his new house Lutwyche decided to improve the view and demolished most of the village. This was rebuilt some distance away around a newly created green. Several black and white cottages still remain. In the mid-18th Century the fine 9-bay stable block was constructed and the Hall extended at the rear. To the left of the stables is a restored 13th Century dovecote and the adjacent farm has a good weather-boarded barn. St. James' church stands beside the Hall. It has a Norman nave and chancel arch. The chancel was rebuilt in 1589 and the tower is weather-boarded with a pyramid roof. **Larden** Hall, an early 17th Century house part timber-framed with elaborate lozenge decoration and part stone with mullioned and transomed windows, which stood 1m. NNE of Shipton is no longer in England. In the 1960's it was sold, dismantled and shipped to America where it was carefully re-erected at Disneyland. There is a tunnel from Shipton that emerged in front of the fireplace of Larden Hall, though it is now only passable for a few hundred yards. The name Shipton probably means 'sheep farm'.

SHOOT HILL *½m. S of Ford which is on the A458 W of Shrewsbury.*
A handful of white-washed cottages; a splendid, small, brick built country house of 3 bays with stone dressings, stone bay windows, grey marble pillars to the porch and a hipped roof, all in a lovely garden; an old railway line along which is kept timber haulage equipment; some red brick houses; and flattish country all around.

SHRAWARDINE *7m. WNW of Shrewsbury.*
A pretty little place on high ground overlooking the valley of the River Severn. It is known locally as Shraden. To the N lies the 40 acre Shrawardine Pool, and in the distance, to the W, are the distinctive humps of the Breidden Hills. In the village there are some attractive black and white cottages, one called Whisperwood and another called Steps Cottage, which is an especially happy mixture of sandstone, brick and timber-framing. Opposite Yeoman's Cottage, standing forlorn in a field, is Shrawardine Castle; Castell Isabella it was called by the Normans. The mound is about 38 yds. in diameter and there are 2 upstanding sections of the sandstone keep-walls. There is much fallen masonry and on one side part of a wide ditch, or moat survives. The castle was probably built in the reign of Henry I (1100-1135), the third son of William the Conqueror, and was dismantled in 1645 by Cromwell's men. Later, most of the stone was taken away and used to repair Shrewsbury Castle. The church of St. Mary has a red sandstone nave and chancel, part of which is Norman. The nave was rebuilt in 1649, the chancel in 1722 and the whole was restored in 1893. There is no tower, just a weather-boarded belfry. Inside there is a Norman font and there are Jacobean panels in the screen. There is no shop, school or meeting room in the village but there is a small Post Office. Most of the residents are commuters, but there are some agricultural smallholders. Between Shrawardine and Pentre, 2½m. NE, is a flat once marshy area on which are now numerous brick buildings (constructed using the 'English Bond') and man-made earthen banks. These were erected during the Second World War to house the **Central Ammunition Storage Depot**. These 2,000 acres are now used as a military training area; the headquarters camp is at Wilcott, near Nesscliffe 3 miles N. If one follows the lane westwards from Shrawardine to the Grenade Range at building No.46 (SJ.382.158) and then takes the grassy track to the River Severn (following the Shropshire Civil Service Angling Club signs) one comes to a pleasant river bank spot where the waters can be reached with ease (SJ.380.154). The field is rough pasture, with tough, spikey grass, and there are often cows drinking in the river. Beyond lies a red brick farmhouse. This is White Abbey Farm (SJ.375.152) and in the right-hand end gable wall is an arched window set in stonework. This is a part of the remains of Alberbury Priory founded in about 1225 for Augustinian canons. (*See Alberbury.*) In 1066 Shrawardine was held by the Anglo-Saxon, Eli. After the Conquest it became part of the Palatinate of Earl Roger of Montgomery and was held from him by Reginald the Sheriff. In time it passed to the Fitzalans and became an important centre of their barony. The name Shrawardine may be from the Old English *scraef-warpign* which can mean several things but in this case is probably best translated as 'the enclosure or homestead by the hollow', the hollow being, perhaps, Shrawardine Pool. **Little Shrawardine** lies ¼m. WSW of Shrawardine on the other side of the loop of the River Severn. There is another Norman castle here, probably the first in this area. The mound has been partly eroded by the river but is still evident. The track of the old Shropshire and Montgomery Railway passes close by to the E.

SHREWSBURY *13m. W of Telford New Town Centre.*
The early medieval name was Scrobbesbyrig which is thought to be derived from the pre-Norman *Scrobb's-burg*, meaning '*Scrobb*'s fortified place. '*Scrobb* probably meant 'gruff' and may have been applied as a nickname to a dour, surly person. Alternatively Scrobb could be from *shrub* meaning 'brushwood or scrubwood' but this is less convincing. Shrewsbury is the county town of Shropshire. It lies in the centre of the county on high ground in a loop of the River Severn. The river protects the town on all sides except in the N where marshes and an old meander made only minimal extra defences necessary. In later times this gap was filled by the castle. The river also contributed greatly to the town's prosperity being a major trade route. During the Iron Age central Mercia was controlled by the Cornovii tribe from their camps and forts on the nearby Wrekin Hill. The administrative centre moved to Uriconeum (Wroxeter) after the Roman conquest. When the Romans left in AD410 the capital moved to Pengwern, and here is a mystery because no-one knows where this important place was. It could have been The Wrekin, it could possibly have been Shrewsbury and many think it was at The Berth, near Baschurch. It could quite easily have been almost anywhere else. All we really know is that Pengwern (the name means 'the end of the marsh') ceased to function as the capital in the 7th Century. It seems likely that Shrewsbury then took over, probably in the early 8th Century. Not until the Anglo-Saxon shire was created about AD1000 can we be sure that Shrewsbury was the regional centre. The problem, of course, is that Shrewsbury is such a densely populated and built-upon area that any Iron Age and Anglo-Saxon remains that may have existed will long ago have been destroyed or buried beneath existing buildings. It is most unlikely that such a strategic spot was not settled at a very early date. It has been suggested that the Anglo-Saxon burgh coincides with the ancient parish of St. Alkmund – that is the area bounded by High Street, Pride Hill and possibly St. Mary's Street. After the Norman Conquest Shrewsbury became a part of Roger de Montgomery's extensive estates and it was he who built the first castle here. This had an oval motte (mound), an inner bailey and an outer bailey, and was placed in the 300 yd. gap not defended by the river. In 1069 the castle was attacked by an unholy alliance of Welshmen, men from Cheshire and local Anglo-Saxons led by Wild Edric. The castle held out and relief was approaching when the rebels burnt down the town and left. In the reign of Henry II the castle was rebuilt in stone, and shortly after the Anglo-Saxon churches were rebuilt, also in stone. In 1083-6 Roger de Montgomery introduced a Benedictine monastery on the eastern side of the river. Shrewsbury Abbey started life in a small, wooden Anglo-Saxon chapel, but was later housed in substantial stone buildings. Today, all that remains of any substance is the church of Holy Cross in Abbey Foregate. In 1215 Llewellyn, Prince of Wales, attacked and captured Shrewsbury. The town was quickly re-taken by the Normans but its weakness to seige led to the construction of town walls. These were completed in 1242. The Normans made deliberate attempts to foster trade. Both Pride Hill and High Street were widened so that markets could be held there, and The Square and Mardol Head were probably laid out on reclaimed swamp land at the same time. Settlements were developing outside the town walls. The Abbey

Rowley House, Shrewsbury, now a museum.

attracted a trading community to the Abbey Foregate area around the English Bridge, and at Frankwell, by the Welsh Bridge, a self-contained town had been planted and flourished, largely on the strength of its leather industry. By 1200 Shrewsbury was the most important commercial centre in the area attracting business from Wales as well as the English Marches. In 1209 and 1266 the town obtained charters which gave it a virtual monopoly in Shropshire of the trade in hides and undressed wool. In 1326 Shrewsbury became a Staple town for wool and leather, with a monopoly in the whole of North, and most of Central, Wales. As early as the 13th Century more than 25% of the town's population was involved in the wool trade. Shrewsbury flourished and the great wool families dominated the business and social life of the town and the County. In the early 15th Century there was a decline in trade, but the business revived again in the mid-16th Century. It was during this revival that most of the timber-framed town houses that we see today were built: Ireland's Mansion (1575), Owen's Mansion (1592) and the Council House Gatehouse (1620). The earliest brick house in Shrewsbury is Rowley's Mansion (1618), which is an addition to the earlier timber-framed Rowley's House (about 1600). The wealthy merchants also financed the construction of public buildings such as the Draper's Hall (about 1580) in St. Mary's Place, the old Market House (1596) and The Old Shrewsbury School (1590's and 1630), now the town library in Castle Gates. In the 17th Century the settlement began to change from a boisterous market and trading centre to a more sedate, elegant and fashionable county town. The quality of life one led became as important as earning a good living. In the early 18th Century streets of new town houses were built: Belmont, Claremont Bank and St. John's Hill. The most ambitious house of this period is undoubtedly Swan Hill Court in Swan Hill, which was built for the Marquis of Bath by Thomas Pritchard (1723-77), who also built the Lion Hotel on Wyle Cop. (Wyle could be derived from Wil, Old English for tricky, hence Wiley: or a trap, meaning a gate, Cop being a corruption of top – 'the gate at the top of the hill'.) The Georgian period saw the development of squares and crescents (though these are so shallow as to not really justify the description) along Town Walls overlooking the Quarry. The Quarry is a 25 acre park which has miraculously survived development. The uninspiring new St. Chad's church (1790-2) was built by George Steuart who, on a better day, designed Attingham Hall at Atcham. Over the Quarry, facing St. Chad's, is the Shrewsbury School (1765) which was originally built as a hospital and was first occupied by the School in 1882. In 1790 Thomas Telford converted the Castle into a residence, for the millionaire M.P., Sir William Pulteney, and added Laura's Tower – octagonal, in red sandstone. At this time the bridges were rebuilt: The English Bridge in 1774, probably by John Gwynne, and the Welsh Bridge in 1795 by local architects, Carline and Tilley. During the 19th Century Shrewsbury was largely unaffected by the violent changes in fashion and values that occurred in many other cities. It became once again an ordinary country market town. Whilst the Victorians were busy pulling down and rebuilding the old hearts of the new industrial towns, and tarting up their town houses in fashionable spas, Shrewsbury trudged on quite happy with the gifts of yesteryear. There was some demolition of course. One of the town's worst mistakes was in the removal of some magnificent timber-framed buildings in Shoplatch and Claremont St., to make way for the new Market Hall in 1866. This in its turn was demolished and has now been replaced by a 20th Century structure that is best ignored. Nevertheless, Shrewsbury still has a good selection of quality buildings of many periods, including the Victorian, for they were not only vandals. Notable amongst their contribution is the Railway Station. It is the Shrewsbury on the hill, in the loop of the river, that is the real Shrewsbury. Outside lies another world. Even Abbey Foregate, with a good number of once fashionable houses and its Abbey church of Holy Cross does not partake of the atmosphere that is special to the town within the waters. Housing and industrial estates surround the centre. The better residential areas lie to the SW and the more modest housing and industrial developments lie to the NE. Shrewsbury is a major communications centre. Twelve main roads and 5 railway lines converge here. She has a wide array of light industry and is the seat of local government. As a business and social centre it is impossible to think of Shrewsbury doing anything but prosper in the future. The old town centre is probably safe nowadays, with the new appreciation of the need to conserve the best of our past, but town planners really must sort out the traffic. It is a nightmare both to drivers and pedestrians. The answer must ultimately be to keep the cars out, so why not sooner than later? There are more than 30 churches in Shrewsbury. The foremost are St. Mary's, St. Chad's (the new one at the Quarry; the old one in Belmont collapsed in 1792 and only a part of it remains), St. Alkmund's and, of course, the Abbey Church of Holy Cross in Foregate.The Roman Catholic Cathedral of Our Lady of Help on Town Walls was built by E. W. Pugin (not his famous father, A. W. N. Pugin), and is impressive largely because of its height. St. Mary's is in St. Mary's Place in the heart of the town, opposite the head Post Office. It has a very tall tower and some fine stained-glass windows which include the 14th Century Tree of Jesse. It stands on the site of an Anglo-Saxon church, which measured 27 ft. by 76 ft. and had one apse. St. Mary's is 185 ft. long. The present church is mostly Norman. The tower is Norman in the lower part and Perpendicular in the upper part with an octagonal stone spire, one of the three tallest in the country. It is now redundant and its future use is yet to be decided. Only the tower of St. Alkmund's remains from the medieval church. The rest of it was taken down and rebuilt in 1793-5 after St. Chad's had collapsed. It has a tall slender spire of nearly 200 ft. The E window was designed by Francis Eginton. It portrays 'Faith' and is controversial, which is a polite way of saying that many people find it not really very good. St. Alkmund's is, we understand, about to be made redundant. The new St. Chad's is also controversial. Nikolaus Pevsner thinks it 'distinguished' but many other authorities have been very scathing. Seen from almost any angle it seems to be out of proportion; from the river it quite frankly looks awful. The Abbey Church is probably the most attractive of Shrewsbury's churches. As already mentioned it was placed here by Earl Roger de Montgomery about 1080. In 1094 he entered the monastery as a monk himself. Three days later he died and was buried here but the place where he was buried is one of the areas that has not been preserved. The construction of the road that now runs beside the church necessitated the demolition of almost all of the monastic buildings. The nave of the church is Norman but the building as a whole was much restored in 1862 and the nave rebuilt in 1887. The Abbey features in the novel 'The Raven in the Foregate', a medieval who-dunnit by Ellis Peters (Published by Futura). West of the Abbey Church, at the junction of Whitehall Street (which runs parallel to Abbey Foregate) and Monkmoor Road, is Whitehall. This is an Elizabethan house of red sandstone built for the lawyer Richard Prince in 1578-82. It stands on ground formerly owned by the Abbey, and was constructed largely with stone 'quarried' from the monastery which had become derelict after the Dissolution. The house had a symmetrical gabled facade, an octagonal turret with ogee roof, and most of the windows are mullioned and transoned. To the front is a gabled Gatehouse and incorporated into outbuildings at the rear is an old Abbey barn and a Dovecote. At one time the house was white-washed and gained the name Whitehall. Today it is used as government offices and has attracted unsympathetic modern extensions.

Random Notes and Points of Interest: *Woolworth*'s stands on the site formerly occupied by the 'Raven', an inn that had a lively reputation. In 1705 George Farquhar stayed there whilst writing his famous play, 'The Recruiting Officer', which is set in the town. Incidentally, he married his wife on the understanding that she was the heiress to a fortune. In fact she was as poor as a church mouse. Farquhar died before his thirtieth birthday. Shrewsbury has many charming narrow passages and lanes called *Shuts*. The derivation of the word 'Shut' is believed to be a corruption of 'shoot' (to shoot or move quickly from one street to another) because they do provide very handy shortcuts. But no-one really knows the origin of the word. There were many more shuts, such as Sheep's Head Shut, Leopard Shut, Spoon Passage and Factory Shut, but building development and

The King's Head, Mardol, Shrewsbury.

Statue of Lord Clive in The Square, Shrewsbury.

Boats on the River Severn near Porthill Bridge, Shrewsbury.

shortage of space led to them being at first closed and then incorporated into adjoining properties. A place of some nostalgia is the ballroom of the *Lion Hotel* on Wyle Cop, with its original plaster ceiling and gallery reputedly by Robert Adam (about 1777). Here were held splendid county balls in the elegant Georgian period and here Jenny Lind sang. The *Music Hall*, in the Square behind the Market Hall, was built by Haycock in 1840; Paganini played here. The foyer and the bar are actually remnants of an earlier building, namely Vaughan's Mansion of the early 14th Century. *Shrewsbury Jail* was designed by Thomas Telford after consultations with Howard, the penal reformer. It lies behind the station. Public hangings were held on the top of the left-hand bay next to the main gate.

'They hang us now in Shrewsbury Gaol:
The whistles blow forlorn,
And trains all night groan on the rail
to men who die at morn.'

A. E. Housman

Some street names. The original name of the street called Shoplatch was Schetep-lachelode, the house of the Schutte family. Pride Hill is also named after an influential family who lived here – the Prides. Claremont and Murivance are essentially Norman names. Frankwell, over the Welsh Bridge, is derived from 'Franc' meaning free, a reference to the fact that this town beyond the walls did not have to pay borough taxes. A lane leading around the town walls called Beeches Lane was originally called Bispestaneslane. Grope Lane was a medieval red-light district. An alternative derivation of the name Wyle Cop is 'Hwylfa Coppa', which is Welsh for the steep hill (Hwylfa) and top (Coppa): the 'top of the steep hill'. Butcher Row, Fish St., and Milk St. need no explanation. (Saddler's Row and Cutler's Row have been demolished.) At the top of Pride Hill is *High Cross*; the present cross is white. This is where public executions took place in medieval times. (See Montford Bridge.) The *Shrewsbury Flower Show* is internationally famous. It is held annually at the Quarry public park. Over 4 million flowers are on display during the 2-day event. It was founded in 1875 and had its beginnings as a side show at the annual town pageant, which in turn had originated in the holy festival of Corpus Christi. The organizers are the Shropshire Horticultural Society who have made several benefactions to the town. These include the gifts of Shrewsbury Castle and the Park Hill Bridge. There are 10 *bridges* across the Severn at Shrewsbury: The English Bridge 1774 (entirely reconstructed in 1922), the Welsh Bridge 1795, Kingsland Bridge 1881 (a toll bridge), Telford Way Bridge 1964, Port Hill Bridge 1922, the Iron Railway Bridge 1848 (cast at Coalbrookdale), the Severn Railway Viaduct 1849, Greyfriars Bridge 1880, Castle Footbridge 1951, and Frankwell Footbridge 1979. At *Ditherington*, on the A49, in the NE suburbs of Shrewsbury, is the world's first multi-storey iron-framed building. It was built in 1796-7, to a design by Charles Bage, for John Marshal & Co., to manufacture linen thread and yarn (that is, it was a flax spinning mill). In 1886 it was converted into a maltings by Jones and Co., and it remained a maltings, except during the 2 world wars when it became a barracks, until it closed in 1987. At the time of writing it is up for sale. In recent years it was owned by an Allied Breweries subsidiary called Albrew Maltsters Ltd (Shropshire Maltings). Further along the A49 at the junction with Harlescott Lane is the *Cattle Market*. One of the largest, most modern and best equipped in the country, it covers 25 acres and has its own abattoir. During the reign of King Athelstone (AD.925-940) a *Mint* was established at Shrewsbury. Coins were produced until the early 13th Century. During the Civil War Parliament seized the Royal Mint in London. To provide coinage for the King local mints were established, including one at Shrewsbury. This was at Bennett's Hall in Pride Hill, the remains of which are incorporated in the old John Collier shop.

Some people Associated with Shrewsbury: *Admiral Benbow* (1653-1702) was born in the town. His father was a wealthy leather tanner. There is a memorial to the Admiral in St. Mary's church which bears the inscription: 'born at Coton Hill (Shrewsbury), in this parish, and died at Kingston, Jamaica . . . of wounds received in his memorable action with the French Squadron, off Carthagena, in the West Indies . . .' In the Square, with his back to the Market Hall, is the statue of *Robert Clive* (1725-1774), by Marochetti. Clive was M.P. for Shrewsbury from 1761 until he committed suicide in 1774. (He is buried at Moreton Say, near Market Drayton.) At the far end of Abbey Foregate, up the hill and opposite the modern County Council Buildings, is the statue of *Lord Hill* (1772-1842), the Duke of Wellington's right-hand man at the Battle of Waterloo. The huge Doric column on which is perched his effigy is the largest in the world. The total height of column and statue is 133½ft. *Sir Philip Sydney* (1554-1586), poet and soldier, attended Shrewsbury School. He died from wounds inflicted during the Battle of Zutphen. *Charles Darwin* (1809-1882) was born in Shrewsbury at the Mount in Frankwell, which is now occupied by the Inland Revenue. His statue stands in the gardens in front of the old Shrewsbury School, where he was educated and which is now the library. As a young man *Neville Cardus*, that most charming of writers on cricket and music, was the cricket professional at Shrewsbury School. He described the playing fields there as the most beautiful he had seen. Much of the landscaping in the Quarry was carried out by *Percy Thrower* who was perhaps the country's best-known gardener. He was Parks Superintendent in Shrewsbury for many years. He is survived by his wife, Connie, and their personal garden, at the Magnolias, Merrington, near Bomere Heath, is open to the public for a few days each year. *Wilfrid Owen*, one of the most respected of the First World War poets, lived at number 69, Monkmoor Road between 1910 and 1918. This leafy avenue is now part of a busy ring road that skirts the town centre to the E.

SIBDON CARWOOD *1¼m. W of Craven Arms.*

This is an estate rather a village, but then so were most villages at one time. It lies on a slope at the end of a long avenue off the B4368 but is not signposted – the only sign says 'Private Carriage Drive'. The lane leads past a stone-built lodge and well-tended gardens with rhododendrons, climbing plants, and carefully manicured hedges of yew with some topiary. Sibdon Castle is not a real castle but a medium sized stone country house of 7 bays and 2 storeys with round-headed windows which has been castellated. The house is believed to be mid-17th Century but was modernized in the 18th Century and the battlements added about 1800. There is a good range of stone stables and a cobbled courtyard. The church of St. Michael stands in front of the house embowered in trees. There has been a church here from about 1180, but the present building dates from the mid-18th Century and was made Gothic, complete with tower battlements, to match the house, in 1871. There is a cottage or two hidden in the trees, a fishing pool to the N of the house, meadows rich in buttercups, and this is altogether a most charming little place.

SIDBURY *6m. SSW of Bridgnorth.*

Sidbury is a small village in attractive country and lies just off the B4363, Bridgnorth to Cleobury Mortimer road. The church of Holy Trinity is early Norman and has some original herring-bone masonry. However, it was much restored in 1878-8 and again after a fire in 1912. Inside there are 2 early 18th Century monuments to the Cresswells and a Norman font. Close to the church is the site of the deserted village of Sidbury. The Vicarage is modern and the Hall was built in 1904, originally to serve as a hunting lodge. It stands amongst trees ½m. N of the settlement. There was a moated castle but this was pulled down by Cromwell during the Civil War and later replaced by a house. This was destroyed by fire – set alight by a jealous wife, they say – and only the stable-block remains. In 1830 the present Hall farm was built on the site; it has a very good weather-boarded barn. There are 'marker' yews in the hedgerows around the village planted to mark the way to the church. Opposite the church is the black and white Rectory Farm. The local landlords are the Reed family who have been squires here since 1469. Agricultural in the area is based on cattle, corn, rootcrops and sheep with some forestry and trout farming. There are no social facilities. The name Sidbury probably means 'the south fort'. When the Anglo-Saxons used *burgh* in a place name they were usually referring to an earlier defensive structure, not one of their own.

SILVINGTON *5m. NW of Cleobury Mortimer.*

As Burchard, son of the Anglo-Saxon Earl Algar of Mercia, lay dying in Rheims

Snailbeach, lead mine waste heaps.

he gave several of his English manors to the church of St. Rémy. Amongst these was 'a manor of one hide' at Silvington which the French church continued to hold after the Norman Conquest. The moated Manor House is built of stone to an 'L' shape. Of the building we see today the S end is medieval and has retained some lancet windows. The churchyard is mounded and a tiny stream passes by on one side. Sheep graze amongst the graves and chickens forage in the shade of the trees. The attractive church of Nicholas has a 13th Century Norman tower, nave and lower chancel; the windows are mostly of the 14th Century; the porch is of 1662; the panelling in the chancel is probably 17th Century; and there is a good wall monument to Edmond Mytton, a former lord of the manor. The tiny hamlet lies on the northern slopes of Titterstone Clee and indeed the name Silvington is from the Old English *scylfinga-tun*, meaning 'the settlement of the people who live on the slope'.

SKYBORRY GREEN *2m. NW of Knighton (Powys).*
Here is a clutter of farms with unsightly outbuildings in the valley of the River Teme where sheep and cattle graze the lush pastures beneath wooded hills. Hedges obscure the landscape but as the lane to Knighton approaches **Nether Skyborry** it climbs above the valley floor and there are splendid views over the dale to the hills of Wales. On the hillside there are a few scattered villas, modern looking and colour-washed but doing no harm. The river meanders, creating gentle, stoney beaches and small earthen cliffs, and cattle stand ankle deep in the waters as they drink – a charming scene in lovely country.

SLEAP *2½m. SW of Wem.*
A tiny hamlet in the flat lands of the north known for the old wartime air-training field which lies S of the Sleap Brook. The Shropshire Aero Club is now based here. The club has 8 instructors and about 200 students who pay in the region of £2,000 each to learn to fly. Each lesson costs about £30. Only the main, concrete runway is now properly maintained and except for some delightful delapidation the place is much as it was during the war years. The settlement almost certainly takes its name from the Sleap Brook. Sleap is from the Old English *sleow*, meaning 'a muddy stream'. The stream has been lowered as part of a land improvement scheme and is now little more than a drainage ditch. It is said locally that Sleap was once a place of some size but it burnt down and was never rebuilt. Note: There is another Sleap just S of Crudgington on the A442. This is described under the article headed Bratton and an alternative entymology of the name is given there.

SMETHCOTT *2m. W of Leebotwood which is on the A49 to the S of Shrewsbury.*
An isolated village with a castle, a church, 2 pools and a waterfall situated on the northern end of Long Mynd. The modern settlement was established in its present position about 1300. The old, deserted, village stood ¼m. SW around the church. St. Michael's was mostly rebuilt in 1850. It has a Norman window in the N wall of the chancel and the priest's doorway may also be Norman. The hammerbeam roof is the work of a local woodcarver, William Hill. Just to the W, on the other side of the lane are the earthwork remains of the Norman castle. The motte stands at the N of a heart-shaped bailey and streams flow to the S and W. Motte and bailey castles were surmounted by a tower of wood on the motte. Sometimes this was erected after the motte had been made but sometimes the tower was constructed first and the earth thrown around and inside it mainly as a defence against attempts to burn it down. This latter method was that used at Smethcott castle. It might have been that when a stay of only a few days was anticipated a tower alone was erected or simply a Roman style palisade. These would, of course, leave no permanent record. A forerunner of the 3-field system was probably practiced here. There were 2 fields – the Old Field and Lynch Field – but it is likely that only one was kept in permanent cultivation. Adjacent to the present village is Stocking Field – 'a field cleared of tree-stumps'. In 1066 Aldred held the manor; by 1088 it was held by Edmund from Roger de Montgomery whose 'Palatinate' included the whole of Shropshire. There was land for 3 ploughs, woodland for fattening 50 pigs and one small-holder and 2 riders lived here. An early medieval lord of the manor answered to the name of Eudo la Zouche. The name Smethcott probably means 'the cottage of the blacksmith'. At The Bank, on an E facing slope, is an ancient earthwork, an oblong enclosure protected by a double bank and ditch. Farming in the area is dairy cattle dominated, the Friesian cows producing milk and beef calves, but there is some corn grown and there is a flock of Suffolk Sheep. As to social facilities there are none: no shop, no pub, no Post Office no village hall, and the school closed in 1964.

SNAILBEACH *3m. SW of Pontesbury, which is on the A488 Shrewsbury to Bishop's Castle road.*
A delightful, run-down, lead mining area with white, shining spoil heaps of 'brightus' (barytes), stone cottages and ruined engine houses on the hillside. The tracks of the narrow-gauge mineral railway, which used to run down to Minsterley and Pontesbury, are still in place, embedded in the tarmac of the lane and the old locomotive shed still stands. Lead was being won from the ground at Snailbeach in Roman times. The output of the mine reached its peak in the late 19th Century when it was the foremost producer in Europe. The lead ores were deposited mostly in one rich vein, the Snailbeach Vein, which ran E-W along the eastern side of the Snailbeach Valley for about 1000 yds. After the Romans, who worked the outcrop ores, there is no record of mining until 1676 when Derbyshire miners came here. In 1782 the land was leased by the Marquis of Bath to Thomas Lovett of Chirk who formed the Snailbeach Company in 1783. Their mine was extremely profitable from the beginning. By 1797 it was 520ft. deep. The peak years of production were in the 1850's, with annual ore production running at 3,500 tons. In 1863 the old smelting mill at Pontesford was abandoned and a new reverberatory mill was built ½m. N of the mine. A long ground flue took the poisonous fumes up to a chimney on Lordshill. This in turn was abandoned in 1895. The 2ft. 5ins. gauge Snailbeach District Railway to Minsterley was constructed in 1877. It brought coal to the lead mine and took barytes and lead on the return journey. The cost of re-equipping the mine and a fall in the price of lead forced the company into liquidation in 1884. It was re-formed as a smaller operation in the same year but by 1905 only 200 tons of ore was produced. In 1911 the mine was allowed to flood and production ceased. Between the World Wars the upper parts of the mine were worked for barytes by the Gravels Trading Company and after 1945 by the Snailbeach Barytes Company who closed down in 1955. At its height there was a large complex of buildings – blacksmith's shop, winding engine sheds, offices, stables, dressing floors (where the ore was crushed and separated), ore house, weighbridge, carpenters' yard, and storehouses. Later, a crusher house, a compressor house and a spoil treatment plant were added. One may wonder why the old spoil heaps have not been colonised by plants over the last 60-70 years. The reason is that they still contain high levels of arsenic, lead and other such poisons. Up the track at the top of the hill, in a hidden valley, is **Lordshill** Baptist Church with a dwelling attached. It was built for the miners, now long gone, but is still well attended and was used as a setting in the film of Mary Webb's 'Gone to Earth'. The track that leads to the church carries on up to the summit of the hill where are found 'The Hollies'. This is an unusual small wood of scattered wild holly bushes. Between 150 and 200 still survive from a much larger wood. There are a few rowan trees also. Some of these have grown through the hollow centres of very old holly trees. The result is a complex tangle and it appears at first glance as if the holly trees are bearing rowan berries. Most of the shafts and adits (tunnels dug into the side of a hill, usually for drainage or ventilation) are now sealed but take great care when walking through undergrowth, some vertical shafts are only capped with corrugated iron. What is more, mining was so haphazard here that not all the workings were mapped. (*See also Stiperstones, Shelve, and Hope.*) The village of Snailbeach today consists largely of modern bungalows with some older colour-washed cottages. They stand on the slopes of the hill on both sides of the road. Social facilities consist of a village hall, a block of public toilets, a stone-built Methodist church with a slate roof and pointed-arched windows, St. Luke's, also of stone with a flèche, and a sports ground. The hills to the N are conifer-clad

The engine house of 1875 at Ladywell lead mine on Shelve Hill.

and those to the S are moorland with a little coppiced hardwood. **Crowsnest** lies ½m. SW of Snailbeach. This is a quaint little hamlet that straggles around the road as it makes a sharp 'Z' bend to follow the contour of the hill. There is a handful of red brick and white-washed stone cottages, a collection of ramshackle garden sheds, some caravans, a few beehives, a stand of pine trees and an old industrial building with a squat brick chimney which is being converted to a house. On the fringe are several bungalows and an unsightly collection of coaches belonging to Long Mynd Executive Travel. There are good views from the road between here and Stiperstones ¾m. to the S. Four hundred yards to the E, high on Oak Hill, is the pre-historic fort of Castle Ring.

SNEAD *3m. N of Bishop's Castle.*
Though now in Wales, Snead was for long in Shropshire. It lies just over the border on the A489 between Lydham and Church Stoke. The name is thought to be from the Old English '*snaed*', which means 'a piece of woodland'. The settlement was probably amongst the 50½ hides belonging to the Castle of Montgomery at the time of Domesday Book. It was later given to the Priory at Chirbury. Today it is a hamlet with a small church and 3 distinctive black and white cottages alongside the main road. The land is flat here, in the broad valley of the Camlad, but 2 miles to the W the road rises up the lower slopes of Todleth Hill and past the remains of Simon's Castle, a Norman motte and bailey, perched on a rocky crag.

SNEDSHILL *¾m. SE of Oakengates.*
The coal mines and ironworks at Snedshill were owned by John Wilkinson until 1793-4 when a dispute with his brother, William, forced their sale to the partnership of Bishton and Onions. In 1803 they passed to the newly formed Lilleshall Company who, in 1819, had 3 blast furnaces and a forge here. Wilkinson had installed his 'Topsy Turvy' engines in both the furnaces and the mines. The famous Snedshill Chimneys with their decorative tile panels were part of the Lilleshall Company brick and tile works. By 1918 they were also making Belfast glazed sinks which were installed in thousands of new Council houses. The works closed in 1977. Snedshill lies on both sides of the Holyhead Road between Priorslee Roundabout and the Greyhound Interchange. The name might be from Sneyd's Hill which in turn might be from the Old English *snaed* which means nothing more exciting than either 'a clearing' or 'a piece of land'. The settlement of Snedshill lies SE of the actual hill. South of the Holyhead Road there are no houses at all, not one. Here, on the site of the old steel works, (which closed in 1959) are: the Priorslee Trading Estate, where many of the businesses are connected with motor vehicles and there is a great litter of old cars but also a Snooker Centre (open 24 hours), a non-ferous metal casting foundry and the Steetley Concrete Plant hard by the red brick offices of the old ironworks; the Castle Trading Estate, where there are upholsterers, a tool company, and welders amongst others in new red brick and grey clad sheds; and a projected Business Park Enterprise Zone. Despite all this activity there are still large areas of rough ground and the foundations of old buildings. Across the shallow valley stands the building site called Telford New Town Centre. The area N of the Holyhead Road is almost entirely residential. The main throughfare is Priorslee Road. At the junction of this road and Snedshill Way are 3 irregular-shaped blocks of rendered cottages. They stand about a 'village green' and are quite probably of the 18th Century. Nearby is a stuccoed United Methodist Church, dated at 1850, with tall, round-headed windows. Most of the houses belong to the council but the oppression is lifted a little by the presence of trees; there is even an avenue of them. Continuing along Priorslee Road in a N-W direction the traveller is startled by the bright red, white and blue garage called Livian Continental but reassured by the short, neat rows of red brick terraced houses that lie beyond. These are complete in every original exterior detail – doors, windows, and door hoods – and must surely be preserved. There are 4 blocks, 2 on each side of the road. At the junction of Priorslee Road and Furnace Road there are 2 more blocks, one of which has been rendered and painted. The rest of the houses are mostly on new, privately developed estates. Priorslee Road continues on as Canongate. To the left, the S, is Snedshill hill, now covered in coalmine spoil. Below is the motorway called Queensway. This cuts through the spoil which in places has lost its fragile cover of grass and is now being washed away into the drains. Canongate crosses Queensway and the traveller is now in Oakengates. To the left stand the battered old ironworks of John Maddocks' Oakengates Foundry, now derelict, windows all smashed, and awaiting redevelopment. But to return to Snedshill. The Holyhead Road passes by the back of the red brick St. Peter's Church which stands in a wooded churchyard and to the front faces Church Road. A little further W, and downhill, the Holyhead Road runs adjacent to the old Lilleshall Company potteries with their famous Snedshill Chimneys mentioned at the beginning of this article. Today the complex is a trading estate occupied by several small firms. Above it looms the spoil-clad mountain of Snedshill and just to the W is the Greyhound Interchange.

SOUDLEY *6m. SE of Market Drayton.*
The name is pronounced Sowdley and probably means 'the south field'. It is a trim little country place with stuccoed and colour-washed cottages, some new red brick houses, a Wesleyan Chapel of 1837 with a projecting porch and pointed windows and a pub, the Wheatsheaf. Around the village are several brick built farms among which are The Nook, Robin Hood Farm and Soudley Manor which has Flemish Bond brickwork with stretchers in red brick and headers in yellow. Half-a-mile to the SW is **Little Soudley** where there is a handful of houses and a bridge over an attractive stretch of the Shropshire Union Canal. Half-a-mile to the NW is Cheswardine. In the reign of Henry II (1154-1189) 'Soudley Magna and Soudley Parva' belonged to the Cheswardine Manor of the Le Strange family and it was probably at about this time that the manor was moved from Staffordshire to Shropshire by adjustments of the county boundary.

SOULTAN *2¼m. NE of Wem.*
Today there is only the Hall and a modern cottage. Soultan Hall is an impressive Carolean House of 1668. It is built of red brick with stone dressings, is 3 storeys high and has 3 bays with mullioned and transomed windows. Inside there is some timber-work and plaster from an earlier Elizabethan 'E' shaped house which in its turn probably replaced an even earlier Hall. It stands in the midst of flat country with farm buildings to the right a walled garden to the left and trees to the fore. Two hundred yards to the E, on the other side of the road, is a circular mound (about 32 yds. in diameter at its base) surrounded by a ditch, or moat, which used to connect with the Soulton Brook that runs a few yards away. The ground around here used to be marshy and it is believed that there were 3 corn mills close by. (Two fields adjoining the brook are called Weir Field and Mill Field.) In Domesday Book the Saxon Brictric is recorded as being at Sulctune in 1066. The Anglo-Saxons favoured marshy places because they provided a natural defence. The Norman lord, Yvo de Saleton, took the name of the hamlet and it was probably he who constructed the mound and moat. It is quite possible that the Anglo-Saxon settlement was on the small hill now occupied by the Hall and that as times became more settled the Normans usurped the site and built the forerunner of the Hall we see today. (This information is from the farmer who worked the estate.) The mound and moat are clearly visible from the road. It has been suggested that the mound may cover a cattle plague pit, but local farmers are dismissive of this idea. Moated farmhouses are not uncommon in North Staffordshire but this one does not seem to have been recorded until now. The name Soulton is probably from the Anglo-Saxon *sulh-tun*, meaning 'gully-settlement'.

STABLEFORD *5m. NNE of Bridgnorth.*
P.G.Wodehouse (1881-1975) the writer of humorous novels and the creator of Jeeves, the butler, lived at Hay's House. Stableford, between the ages of 14 and 21. His father was a judge in Hong Kong who retired through ill health. Wodehouse had a great affection for Stableford and indeed for the whole area around. Many houses and villages in the district became fictional settings in his books. Hay's House is on Hay's Bank on the B4176. It lies beyond the modern detached Georgian style house that stands next to the large green shed of Telford Farm Machinery Ltd. One drives through the warehouse yard and the house is revealed. The right-hand side

Hay's House, Stableford. P. G. Woodhouse lived here.

above: Anglo-Saxon pilaster strip in St. Peter's, Stanton Lacy.

Stableford Hall.

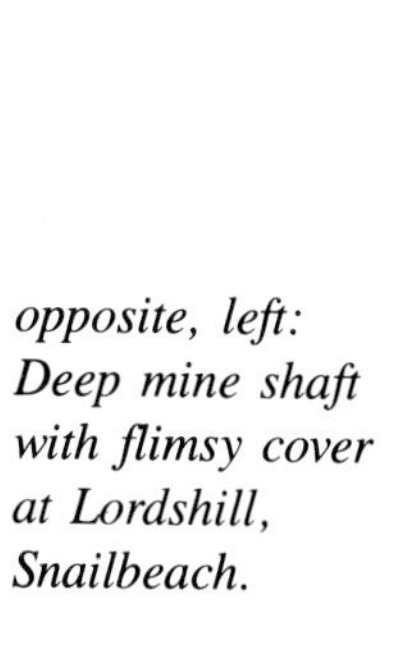

opposite, left: Deep mine shaft with flimsy cover at Lordshill, Snailbeach.

of Hay's House is built of red sandstone and carries the porch; the left-hand side is brick and render. Out-buildings lie behind, to the left. The house is some 400 years old and Grade II Listed. The present incumbent is 87 years old and of a most cheerful disposition. The croquet lawn in front of the dwelling was laid by Wodehouse's father. The Hay in the name of the house is from a former owner, Colonel Hay, who controlled Stableford, Rindleford and Grindleford during the Civil War. The village is small but most attractive with timber-framed houses, such as Gorsty Hayes, some modern bungalows, and the Hall. The latter has just been renovated and converted into 7 luxury apartments. It has 2½ storeys, 7 bays, a parapet, a stone porch with 3 arches and copper beech trees in the grounds, all surrounded by a high brick wall. One mile W of Stableford is **Ackleton** a mainly red brick residential village on a hill. A quiet and pleasant place with a pub, the Red Cow, and a small orangey-yellow concrete-block church. The undulating, wooded country along the main road, the B4176, is most attractive.

STANDFORD *4m. NW of Newport.*
The A41 is part of the old London-Chester Holyhead coaching road. At Standford the road crosses the River Meese, originally, as the name suggests by a 'stoney ford', but latterly by a bridge. The attractive old sandstone bridge of 1803 with its 2 cutwaters and 3 arches stands in a lay-by now that its unattractive modern replacement, constructed of pre-stressed concrete and iron, carries the newly widened and straightened road over the river. The dwellings are now in the lay-by also: 2 redbrick cottages painted white, the sandstone Willow Cottage, which has gardens down to the river, and the red brick Lodge Farm. Standford Hall, ⅔m. SSE of the bridge, is a substantial red brick house of 9 bays and 2 storeys with a hipped roof and dormer windows. It is a handsome building and can be glimpsed through the hedges that line the main road, the core of the house is Jacobean. Opposite the front of the Hall, on the other side of the main road, is **Deepdale** where there are 4 most attractive stone built cottages set well back from the road and framed by pine trees. Opposite the back of the Hall, and its brick and stone farm, is a lane that leads to **Calvington** which today consists of a 6-bay, 2-storey stuccoed country house with hipped roof and doorway offset to the left, a farm and a cottage all of which sit isolated amongst flat prosperous looking fields with well-wooded hedges. The name Calvington is from the Old English *calfatun*, 'the place where calves are kept'. The gable end of the great barn is a dovecote. Pigeon roosts of this kind are more common in this area than free standing buildings. Pigeons were kept primarily as a source of fresh meat in winter, though their eggs were also eaten. Until 1761 only the Lord of the Manor was allowed to keep a dovecote. His birds fed themselves on the fields of his tenant farmers which pleased them not at all.

STANMORE *1½m. ESE of Bridgnorth.*
Here is an industrial estate of some size but it is better known for the Midland Motor Museum that lies on the opposite side of the main road, the A548 to Stourbridge. Stanmore Hall houses a privately owned collection of over 100 vehicles, amongst which are classic sports cars and racing cars and a collecton of motorcycles. Amongst the historic cars to be found here are the Sunbeam Tiger in which Henry Seagrave broke the world landspeed record in 1926, the Delahaye which was Fastest Road Racing Car at Brooklands in 1939, and the Napier-Railton which lapped Brooklands at 143 m.p.h in 1933. The name Stanmore might mean either 'stoney-moor' or 'the stone on the moor', a reference to a megalith or perhaps a large glacial boulder.

STANTON LACY *3½m. NNW of Ludlow.*
An attractive and ancient village with cottages and farms of stone, brick, timber-framing and colour-washed stucco situated on the River Corve 3m. above its confluence with the River Teme at Ludlow. Stanton was an Anglo-Saxon settlement of some considerable size. Only when the Normans came was the suffix Lacy added to the name; the de Lacy's were the new lords of the manor. In 1086 there were 67 villagers, 28 smiths, 4 cottagers and 5 smallholders and their families, 28 slaves (in the service of Roger de Lacy) and 2 mills. It was probably Roger de Lacy who built Ludlow castle and 'planted' the town of Ludlow in the 11th Century. Up until the 19th Century Ludlow Castle was in the parish of Stanton Lacy. As Ludlow flourished so Stanton declined. The new town even usurped control of the old N-S trade route. The church of St. Peter at Stanton Lacy is a rare Anglian foundation. It is a cruciform church of good proportions that stands on a mound about 5 feet high. In many of the walls are typical Anglian pilaster strips, or leseves, of square section, some having short cross pieces. One begins above the typically Anglian N doorway with a bracket with 4 pellets. The main structure as we see it today is 11th Century Norman, with a 14th Century nave and tower, though it still retains the high and narrow proportions of an Anglo-Saxon church. It stands amongst yews and conifers. The story of the origin of the church is this: In 680 the daughter of King Penda of Mercia, the beautiful Milburga, was being pursued by a lustful young Welsh prince. Milburga crossed the River Corve and willed the waters to rise to make an unpassable torrent. They duly obliged and her maidenhead thus saved she founded a church in gratitude. St. Peter's is said to be haunted by the ghost of a man murdered by the troops of Oliver Cromwell during the Civil War. Today the village is a quiet little place with real character. There are tall, mature, broad-leaf trees; a large, mellow Vicarage, the only house in the village not owned by the Earl of Plymouth; a farm with weather-boarded barns; a Victorian school of 1872 (with an unfortunate modern appendage) that closed in 1982 and is now a craft workshop; a Queen Anne style house; and a bridge over the river that has the most horrible iron-girder safety rails; but there is no shop, pub, Post Office or village hall. The name Stanton probably means 'the settlement on stoney ground', but could also mean 'the settlement by the big stone', a megalith perhaps.

STANTON LONG *1m. SE of Shipton, which is 10m. W of Bridgnorth.*
An ancient, linear village in the lush pastures of Corvedale. It was an Anglo-Saxon settlement called Stanton and acquired the suffix Long in the 12th Century. The Normans often qualified towns which had common names to avoid confusion. (An adjoining manor was also called Stanton. It had its name changed completely to Castle Holdgate, now simply Holdgate.) Stanton was originally dependent on the village of Patton, the administrative centre of the Hundred of Patton. However, Patton declined and was finally incorporated into Stanton Long where a new parish church was built. The present church of St. Michael dates from the early 13th Century and the door and much of the fabric is original, though heavily restored in 1869. There are some good brick and timber barns in the village but the Methodist Chapel, Post Office and shop are now closed, and their is no school.

STANTON-UPON-HINE HEATH *5m. SE of Wem.*
An old but unremarkable red brick village with a most attractive stone-built church. St. Andrew's has a battlemented tower, 13th Century in the lower parts and Perpendicular above. The nave and chancel are Norman with some herringbone masonry and the whole sits on a raised, roughly circular, mound on the edge of the settlement. **Harcourt** Manor is a small, red brick country house tucked away out of sight off the road ¼m. NNE of the village. It has superb views towards Lee Brockhurst over the classic Shropshire farming landscape of the Roden Valley. The house used to be called 'The Woodlands' and Mary Webb, the novelist, lived here with her parents from 1896 to 1902 before moving to Meole Brace. It is now in the ownership of Major Hayes. A lane leads past the house, down the hill, and through the lovely park which provides a perfect setting for the village cricket ground. Scots pines and a cornfield flank the pavilion. In the woods opposite the cricket ground is an uncompleted sandstone cottage. Several people were killed during its construction and so it was abandoned, never to be lived in. The tree-clad slopes are called the Haunted Woods. Terrible screams have been heard here, though some of these are attributed to badgers fighting. The track continues past the estate farm and a large walled garden, in which is a derelict bungalow, and on to join-up with the public road near Moston. Between the farm and the public road there used to be a large house, now long gone. The name Stanton probably means 'the settlement on stoney ground'.

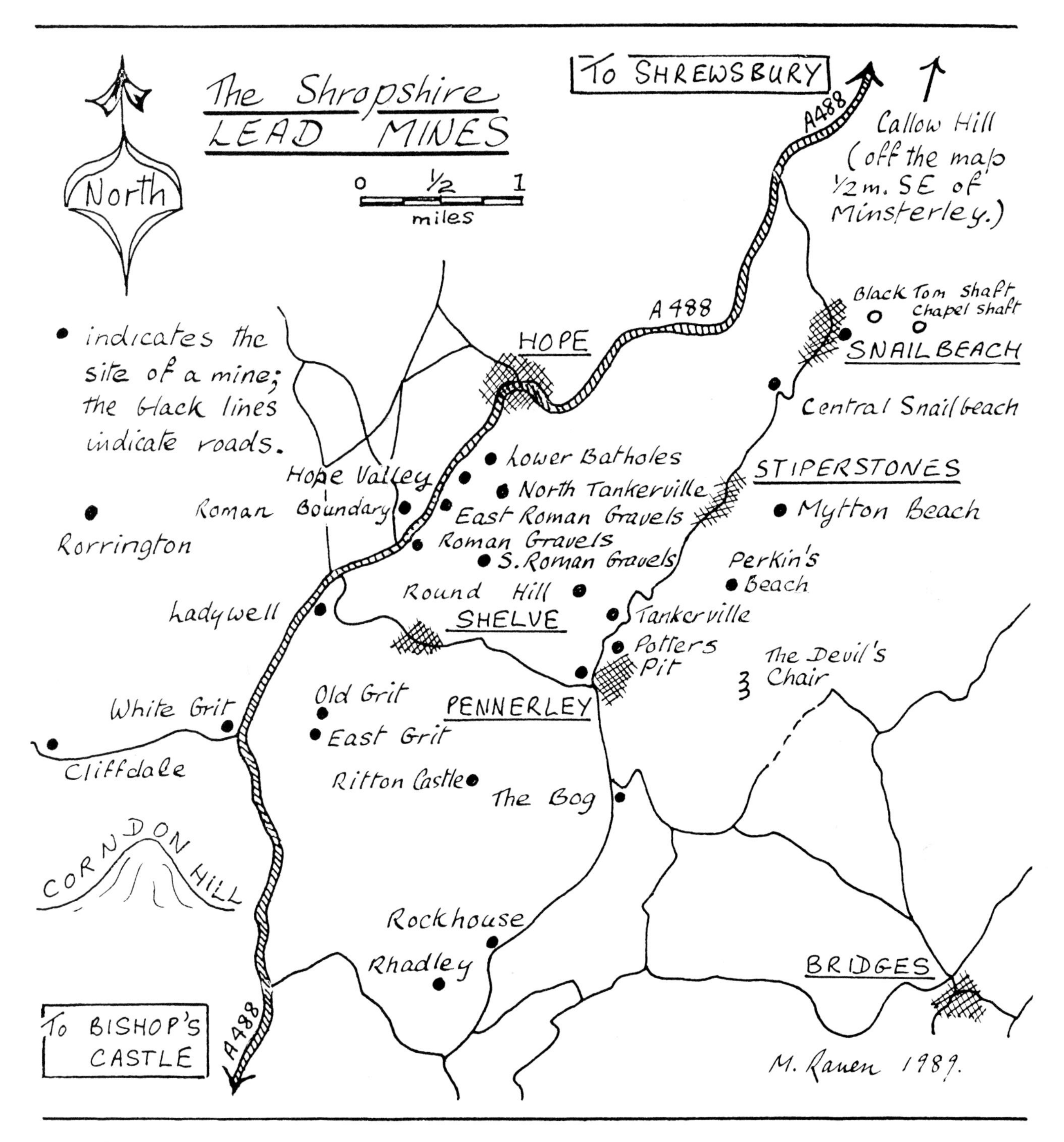

Map of the Shropshire Lead Mines.

STANWARDINE IN THE FIELDS *See Weston Lullingfields.*

STANWARDINE IN THE WOOD *See Weston Lullingfields.*

STAPLETON *6m. SSW of Shrewsbury.*
The name is not understood, but 'staple' might be from 'steeple' or *Steplewood*, a medieval forest. The village lies on a hill in undulating country cut through by narrow, high-banked lanes. There are red brick farms with weather-boarded barns, semi-detached Council houses, detached bungalows and an old water-pump which though neglected is nevertheless protected by a lich-gate-like shelter. This Pump House was erected in 1897 and re-newed in 1977. The interest at Stapleton, though, is in the church of St. John the Baptist and the area around it. The church is most unusual, if not unique, in the way an old undercroft has been incorporated by the removal of a floor. The original church, complete with undercroft was built about 1200. This 2 storey construction was unusual itself and is normally only found in palace chapels. In 1786 the church was remodelled and the floor between the undercroft and the main church was removed thus making the floor of the undercroft the floor of the church. The tower was added in 1840 and the whole building restored in 1867. The undercroft has slit windows with straight heads and a round-headed S doorway which is half-buried, making a strange sight. The upper windows, of the original church, are lancets. Inside there are 6ft. high German candlesticks made of wood and dated at about 1500. To the right of the church is the white painted Old Vicarage which has battlements to the upper storey at the rear. Behind the church is a meadow with shallow earthworks that resemble house platforms and a nearly semi-circular pond that may have been a moat. To the left of the church is the substantial and clearly evident motte of a Norman castle. On the other side of the road is the red brick Parish Hall. This is the only social facility in the village; the school of 1874 closed in the 1920's, the Post Office closed in 1958, and there is no pub and no shop. Moat Farm lies 1¼m. SW of Stapleton in flattish country. It presently consists of 3 red brick, timber-framed and white rendered ranges which stand on a stone plinth that rises some 18ft. above the level of the now filled-in moat. There are timbers in the house that date from the time of Henry VIII.

STEERAWAY *1¼m. S of Wellington.*
Steeraway is an isolated hamlet that was once famous for its limestone quarries and deep mines. The direct route to the settlement is along Limekiln Lane which joins the Holyhead Road at Old Hall School, Wellington. However, this has been blocked and so a diversion is necessary. Go to the crossroads, take the Dawley Road, turn right into John Broad Avenue, pass through the housing estate and turn left at the end. This pot-holed track will take you under the M54 and up into a lovely, broad valley. You pass through sheep-filled meadows fringed with woods to a cluster of 5 cottages (one of which is stone-built) and a red brick farm with dormers and tall chimneys. This is Steeraway. At the end of the lane there is a stile. Walk for 50 yds. and turn right into the bushes and trees and 2 of the Steeraway limekilns confront you. A substantial stone retaining wall supported by buttresses is built against a bank. At the bottom are two 24ft. long brick arched tunnels. The larger is 7ft. high and 6½ft. wide. At the top of the bank are the round, yellow-brick walled charging shafts. These are not properly filled in and could be very dangerous. There is at least one other kiln and the whole hill is riddled with old pits, quarries and shafts. Today nature has healed her ancient wounds and woods now cloak what was once a bare, rocky place. Limestone was being quarried at Steeraway (which was on the estate of the Forester family) from before 1700. By 1738 there was a railway from the hill to wharves near the Old Hall on the Holyhead Road where lime was sold to builders and farmers. The Horsehay ironworks were supplied with limestone from here during its first years of operation in the mid 18th-Century. By the 19th Century the workings were in the hands of the partnership of Rowley and Emery and after 1805 by Emery and Clayton. In 1803 a steam winding engine was brought in to work the mines and 1840 there were vertical shafts up to 120ft. deep. In 1842 the quarries and mines provided employment for some 100 men and 20 boys. A seam of stone 45ft. thick was worked by the pillar and stall method. Explosives were commonly used in limestone workings and the risk of a roof fall was high; limestone miners were therefore paid more than colliers. The workings at Steeraway closed down in 1850 and there are no signs of recent activity. One should take great care when walking the hill; an old shaft could be stumbled upon almost anywhere. Adjoining the woods to the W is the Wrekin golf course.

STEVENTON *1m. SSE of Ludlow.*
A little place on the River Teme just beyond the southern fringes of Ludlow. The railway crosses the main road here. There is a stone built Jacobean Manor House with a gabled 'E' shaped front and star-shaped chimney stacks. Inside, in the hall, is an original fireplace and wall panelling.

STIPERSTONES *4m. SSW of Minsterley which is on the A488 tothe SW of Shrewsbury.*
The Stiperstones is a range of hills after which is named the village of Stiperstones. The settlement lies along the minor road between Snailbeach to the N and Pennerley to the S. This is old lead-mining country and the hills to the E are riddled with old shafts. The village flourishes and has a Methodist church; a pub, the Stiperstones Inn; a Post Office-general store; a stone-built school (complete with children) and the clutter of coaches belonging to Minsterley Motors. There are numerous modern bungalows and some older houses with dormer windows. From the telephone box by the pub a track leads E, to Mytton Dingle, and the old shafts and spoil heaps of the mines known as Mytton's Beach (SJ.370.004). The workings here were comparatively small and their commercial life was short, from about 1850 to 1869, when the then owners, the Snailbeach Company, were liquidated. If one returns to the track near the telephone box another lane can be taken that leads S, past the Methodist church and past the tiny stone cottages of long gone lead miners, around the flank of Green Hill to a hidden valley (SJ.365.998). Here there are grassed over spoil heaps, derelict stone buildings, and fenced-off adits, that is, shafts driven horizontally into the side of the hill. This is Perkin's Beach. The lead ore lay in veins running E-W, the most important of which was the Big Spar Vein. It was being worked commercially in the 1850's by the Stiperstones Company who also operated the Pennerley mine. It changed hands several times but closed in the late 1890's. It is a pleasant if windswept little place. The hills around are moorland, covered with ferns, heather and gorse, and are now a part of the Stiperstones Nature Reserve. There is a great variety of moorland here: upland, lowland, and northern and southern types. Parts of it have been managed as a grouse moor for many years. In the winter the heather is cut back and burned to encourage vigorous new growth in the spring. Grouse are shot from 12 August to 10 December. A stoney track leads in a N-E direction up the hill to the summit of the Stiperstones. At the top one can bear either left for the deserted mining hamlet of Blakemoorgate and Lordshill (near Snailbeach) or right, and head S to the Devil's Chair. This is the largest of several quartzite rock outcrops that lie in a line and which seen from a distance look like the spine-crest of some gigantic primeval lizard. The name Stiperstones is from 'stripped-stone', an affect caused by frost action during the last Ice Age. It is a geological freak that is unique in Britain. There are many legends associated with the Devil's Chair: All the ghosts of the county meet here on 22nd December each year; if mist and rain hide it from view the Devil is sitting here; the boulders that lie around it were spilled from the Devil's apron as he stood up after resting on his way to dam the River Severn; cattle and sheep take fright if they graze too near it at night; Wild Edric, the Anglo-Saxon lord who led an uprising against the Normans in 1069-70, lies buried below in a great cave awaiting the day when an English king once again sits on the English throne, for then his lands will be returned to him. (The present Royal Family has too much German and Greek blood to qualify as English.) He is said to ride out with his ghostly army whenever war threatens the country, as a warning to the living to be properly prepared. Another, most charming folk tale concerns 'The Seven Whistlers'. Six birds are said to search the slopes of the hill looking for a seventh, a long lost companion. The day their friend is found,

Stirchley Chimney, a well-known Telford landmark.

however, will signify the end of the world. Half-a-mile SW of Stiperstones village the road makes a sharp right-angle turn. Here one can park the car and take in a splendid view to the N, down the valley over Stiperstones village and Snailbeach to the flatlands of the Severn Plain. A short distance further S the road makes yet another sharp bend to follow the contour of the hill. Here is **Tankerville** (SJ.355.995), a handful of cottages, a pottery and a guest-house. In the steep little valley below are the remains of the Tankerville Mine. (The Earl of Tankerville owned much of the Stiperstones hills.) Standing proud amongst dense undergrowth, are the tall brick chimney (pre 1870) and the stone-built water pumping engine house (1876). There had been a small workings here in the early 1800's but large scale production began in the 1860's under Heighway Jones. He passed the management on to Arthur Waters, an experienced engineer who had been involved with several other local mines. At 24 fathoms the mine was the deepest on the orefield and there were problems with flooding. Nevertheless it made money for its shareholders for 15 years until increasing costs and falling ore prices led to the company being liquidated in 1878. Three years later Arthur Waters died at the age of 53 having given most of his money and finally his health in the battle to work the mine. The pipe-shaped lode was the richest in the field, indeed it still is. The road continues S, passing the motte of a Norman castle to the W and the miners' chapel to the E, to the scattered cottages and farms of **Pennerley**. At the junction of the road to Shelve are the substantial ruins of the Pennerley Mine (SJ.353.989) which was operated jointly with the nearby Bog Mine. (*See The Bog*.) The land at Pennerley was owned by the Earl of Tankerville and lead had been extracted from small surface pits from at least the 1780's. In the 1840's shafts were driven to 120 fathoms (there are 9 feet in a fathom) and a steam engine was installed to pump out water. The ore lay in 2 major parallel veins that ran SW-NE: Big Ore Vein and Warm Water Vein. The mine passed through many hands and was profitable until 1876. Several later owners went bankrupt and it finally closed in 1901.

STIRCHLEY *1½m. S of Telford Town Centre.*

The name is Anglo-Saxon and means 'the pastures grazed by the horned beasts'. In the 18th and 19th Centuries the old farmlands of Stirchley were ripped apart for the coal beneath the ground. This was used to feed the iron furnaces on the surface and the landscape was made anew. In 1963 the Dawley New Town Development Corporation (re-named Telford in 1968) was formed and in the succeeding years the area was re-made a second time. Old spoil heaps were levelled, quarries filled in and shafts capped. It is now a place of new houses bordered by the Queensway (A422) to the E and the Town Park to the W. There are 7 parts to Stirchley: Stirchley Village, Stirchley Centre, Randlay, Halmer Lake, Brookside, Town Park and Hinkshay. **Stirchley Village** is the old, pre-industrial centre. The church of St. James stands on high ground. The stone built chancel is Norman and has kept its original windows in the N, S and E walls. Inside there is an ornate chancel arch with Norman zig-zag decoration. The rest was rebuilt in brick in 1741 and extended to the left in 1839. Today St. James stands locked and forlorn, a redundant church, but the fabric is still well cared for. To the right, in a wooded garden, is the large rendered and pink painted Old Rectory. To the left is the Old Post Office, one of several brick and stuccoed cottages. On the main road is the site of the Old Village School, now demolished; a tiny shop run by an Indian lady; the Rose & Crown, an old pub much extended recently; and Stirchley Grange Farm, a substantial red brick house which stands close to Grange Pool. The Grange once belonged to Buildwas Abbey and later to Isaac Hawkins Browne, M.P. for Bridgnorth, who bought the property in 1774. He was a major investor in iron and coal companies; his famous Old Hall Company was run by Thomas Botfield. (Just across the pool is Northwood Terrace a row of houses built for the clerks of the Old Hall Company in 1813.) Today the Grange houses the Stirchley Grange Environmental Interpretation Centre, which is open to the public. The most handsome building in Stirchley is Stirchley Hall. It turns its back to the road but with its outbuildings forms a most charming ensemble. Varied roof lines, gables, a blank arch, 3 tall chimneys, irregular windows and the stone-built barns and stables give the rear of the building great character. The gabled front is stone-built and dated at 1653. (The initials MF stand for M. Forster.) Whether this date refers to the building of the house or re-modelling we don't know but certainly parts of an earlier house are incorporated – the hall ceiling and the chimney stack in particular. The estate originally belonged to Buildwas Abbey and then to the Earl of Arundel (until 1560). The brick extensions are of the 17th and 18th Centuries and there have also been some modern contributions. The whole complex has been fully restored and converted most handsomely into a number of separate dwellings. From the churchyard of St. James one can look E over large playing fields to the new **Stirchley Centre**. Here are 3 flat-roofed schools, a centre for the physically handicapped, a library, a social club and a sports club all in pale yellow brick. The Centre lies on low ground and the open pyramid concrete spire that stands atop one of the buildings acts as a useful landmark. To the N and E are housing estates. Adjoining Stirchley Centre to the N is **Randlay**. 'Rand' is Old English and means 'the border, margin, edge' or even 'shore'; the name therefore means something like 'the clearing by the border'; of what we do not know. Today Randlay is a large housing estate with a small community centre based on the Lower Brands Farmhouse (part of the Grange Farm lands). There is a pub, called the Randlay Farmhouse, a fish and chip shop, a greengrocer's and a Spar superstore. A footpath leads westwards, down the hill, to **Town Park**. Stirchley Chimney lies in the valley bottom. Despite its low position its great height (62 metres) makes it a landmark for many miles around. The chimney is red brick with a yellow stone top and is dated at 1873. Beneath it is a fire pit. This was part of a Benjamin Huntsman crucible furnace that produced uniform, slag free steel but was never actually used for this purpose. The chimney stands close to the Great Wall, a massive ramp-like structure built against a hill. What was it built for? No one knows. It has massive, sloping red brick walls with arched tunnel entrances at the bottom and 2 areas of stonework. The ground level has been raised by debris and spoil. Opposite the Great Wall are much smaller structures. These consist of brick-arched cavities surmounted by a stone wall. These are said to be the remains of the very first iron furnaces on this site, built in 1725-7. All around are mounds of black earth, and green glassy slag now covered in scrub woodland. The whole area is today surrounded by security fencing and a close approach is not normally possible. The industrial history of the area is this: In 1824-5 the Botfield family, leading ironmasters who controlled the Old Park Company, bought land adjoining the Shropshire Canal at Stirchley and Hinksay and proceeded to build 2 blast furnaces at each place. There are remains of both sets of furnaces, though the visible remains bear no resemblance to each other, perhaps because different parts are exposed. When iron production ceased the Stirchley works were taken over by the Wellington Iron and Coal Company. They built the Stirchley Chimney and a railway and possibly had a hand in the construction of the Great Wall. (The Great Wall has become quite an enigma.) Before commencing production, however, the company went bankrupt and the works were taken over by the Wrekin Chemical Company in 1886. They produced charcoal, wood naptha, lime, salt, and sulphur. There was a little railway along which trundled 2 little locomotives, Arnold and Donald. The works finally closed in 1932. Just to the NE of the Great Wall is the Blue Pool now surrounded by woods. At the N end of this pool was situated the Randlay Brickworks which operated from 1856 to 1965. The pool is very deep and at the bottom is an engine house, now the home of a large pike with a fearsome local reputation. The pool occupies an old clay pit and the waters are blue from clay particles held in suspension. In its hey-day the works could make enough red bricks to build 43 average semi-detached houses. The chimney and the Great Wall are both built of Randlay brick. After the 1st World War several companies set up making concrete curb stones and paving slabs using the old furnace slag (valued for its hardness) which they crushed to use as the body of the mix. In recent years the old spoil heaps and slag heaps have been landscaped and now provide many green and grassy acres of playing fields and wild, dingly woods. At the N end of the Town Park there are formal attractions – an ampitheatre, a children's Wonderland overlooked by the red and green Wrekin giant, the Chelsea Gardens and the Withy (Willow) Pool by the side of which the old Norman chapel from Malinslee has been rebuilt. South from the chapel is a wooded lane that leads to the Wonderland car park. This is **Dark Lane**. It used to extend northwards and along it were several rows of miners' and iron workers' cottages and 2 blast furnaces (built in the 1830's). The settlement was called Dark Lane but was demolished to make way for the Telford Town Centre. Within ½m. to the W of the Stirchley chimney is **Hinksay**. Here there is a wooded Field Studies Area adjacent to the Stirchley Pools and the Ever Ready factory. The name Hinksay might be derived from the Old English *Hengest's-hay*, and could mean 'the field where the stallion is kept'. Behind the Ever Ready factory are the restored remains of a structure the purpose of which is as yet unknown. It consists of red brick arches surmounted by stone walling. There are 3 small arches and one larger arch with a security grid that is definately the entrance to a tunnel. These are *not* the Hinksay furnaces as the local walk leaflets and the park information boards say. These are situated to the SW on the bank above the Top Pool. They are hidden away in the woods enshrouded in ivy and undergrowth but are just visible from the top path that runs by the sports field. The once tall chimney-like structures have callapsed and now look like ruined towers. They are constructed of red brick and the interior surfaces have a metallic black coating. These are the iron smelting furnaces built by the Botfields in 1825-7. The Top Pool above which they stand was built as a canal reservoir originally, though it was later used as a furnace pool. It was filled with water pumped out of old mines. The Shropshire Union Canal (1793) flowed by in the valley bottom. Parts of it still exist, including a length that lies alongside the Bottom Pool from which is separated by a narrow causeway. This pool was originally not a pool at all. It was a railway depot and canal wharf interchange. The railway lines were called Jerry Rails and the old ironworks were nicknamed the Jerry Furnaces. (Note. In 1860, some years after the ironworks had closed, this arm of the Shropshire Union Canal was drained and replaced by a railway – not the Jerry Rails, but another 'main line' railway. This followed the route of the canal but had to cut a straight line when the canal meandered. The loops thus left are still water-filled and used as fishing pools. There are several in the Hinskay area.) South-west of Stirchley Centre is **Brookside**, a large residential area of little character. The community centre has only one access point and is difficult to find. Possibly this is deliberate for there are no pleasures to be found here. A 'built on the cheap' row of 5 flat-roofed shops faces the pastoral centre over a very dirty covered way. There is a fish and chip shop, and a Spar store, 2 schools and a Probation Office. Nearby are blocks of flat-roofed flats and groups of houses with very low pitched roofs, some of which are not unpleasing. East of Brookside is **Halmer Lake** where there are 2 more schools, The Lakeside pub, and a green area that is being developed as an upmarket village. Indeed, the cottage-style houses of brick, stone and stucco are both varied and delightful. The illusion of a village is shattered, though, by the roar of the Queensway which sweeps by over a causeway dam 300 yards away.

STOKE SAINT MILBOROUGH *7m. NE of Ludlow (off the B4364 to Bridgnorth road).*

An attractive village in a valley on the south-eastern slopes of Brown Clee Hill. It was here that St. Milburga, pursued by bloodhounds for 2 days and nights, fell exhausted from her steed. She lay dying of thirst but was saved when her horse kicked the ground and a spring burst forth. The saint was able to refresh herself and make good her escape. The spring, one of several in the area, still exists but has been treated very badly by the Severn Trent Water Authority. They draw water from here and have built a rough, ugly, concrete platform and inserted a black iron pipe. No attempt whatever has been made to landscape the area or keep it tidy despite the fact that it gets many visitors. St. Milburga was the first Prioress of Wenlock Abbey and the grand-daughter of King Penda, Lord of Anglo-Saxon Mercia. There is a 7th Century land Charter giving her 63 'manetes' which included the land around what we now call Stoke St. Milborough. The village probably originated as a 'daughter settlement' of Much Wenlock. The Domesday record shows that the old Anglo-Saxon name of Stoke St. Milborough was Godestoc. 'Stoc' is Old English for 'daughter settlement'. In its turn Stoke St. Milborough had 'daughter settlements' of its own. The inhabitants of such homesteads and hamlets were to all intents and purposes colonists. They were sent out to clear the woods and heathland and make the ground productive. Such clearances were especially common in the 9th and 10th Centuries. Initially, the new settlements were entirely dependent on the 'mother-town' for religious, financial and administrative support, but in time many became established villages in their own right. In 1086 Stoke St. Milborough had 6 dependent 'daughter settlements'. The church is named after the saint and was built in the 13th and 14th Centuries with a strong, barn-like roof and has a timber-framed porch of 1707. Nearby is a stone-built National School of 1856 and Moor Farm, a timber-framed house. In 1809 an Act of Parliament was passed enabling the common land in Abdon and Stoke St. Milborough to be enclosed. Private ownership resulted in the land being greatly improved. It was hedged and walled, trees for shelter were planted and good pasture created up to a height of 1300 ft. The adjoining parish of Clee St. Margaret was never enclosed, possibly because there were too many smallholders and they were unable to reach an agreement. The land there is still only rough pasture.

STOCKTON *5m. N of Bridgnorth on the A442.*

The name Stockton can either mean 'a settlement belonging to a monastery', or 'a house built of logs'. The church of St. Chad has a Norman chancel but has been much restored at various times. The Rectory is red brick of 1702. There is also a cottage. These constitute the village and stand at one of the entrances to **Apley** Hall, set in Apley Park. Until recently the Hall was used as a boys school. At the time of writing it is up for sale and is likely to become an old people's home. It is a large 3-storeyed mansion of Grinshill stone built for Thomas Whitmore in 1811. It incorporates an older Georgian house, which in its turn replaced an even older fortified house of 1308. Apley Hall is most easily approached from Cheswardine Lane at Norton. If one bears left just before the Hall the track leads down the valley to wooded cliffs and caves by the riverside. Another track from the Hall leads to the bridge across the River Severn which is mentioned later. The valley in which the Hall lies is extremely beautiful with areas of parkland, woods, pastures with grazing sheep, and yellow corn fields. The ground is undulating in some places and flat in others. The whole effect is of a medieval estate, cultivated and lived in but unspoilt. There is an atmosphere here that is rarely found, even in parks that have been lavishly landscaped at great cost. When we visited it was damp and misty. In the sunshine it must be glorious. Across the river is **Apley Forge** and all around are coppiced woodlands. Trees were often planted near ironworks for until the 18th Century they were the only source of fuel for the furnaces. When coal – as coke – became useable as a fuel many coppices were left uncut but were kept as sporting estates. (Coppicing involves cutting a tree to its stump, thus encouraging the growth of many small side shoots. These are 'cropped' every 10 years or so. When a tree is then left it grows many trunks from the same bole, or stump.) The settlement of Apley Forge consists of a block of 5 Victorian red brick cottages and a block of 6 white painted brick cottages; a railway station (for Linley) and a disused forge pool which is now dry and planted with tall trees; and, remarkably, a private metal suspension bridge across the Severn (erected by David Rowell Ltd., of Westminster) which carries vehicular traffic between Apley Park and Apley Forge. This bridge is very little known, even to local people, and we came upon it with surprise. Some 100 yds. downstream from the bridge are the remains of wharves on the banks of the River Severn. There is a disused railway line along the W bank of the river which can be negotiated by a motor car. We joined it at Severn Hall, 2m. N of Bridgnorth, just past the golf course which is on Stanley Lane. It takes you to Apley Forge and on through **Hifnal**, which now consists only of one

In the grounds of Apley Park, near Stockton.

The suspension bridge over the River Severn between Apley Park and Apley Forge, near Stockton.

The gorge of the River Severn at Apley Park.

Stokesay Castle.

derelict brick house, the ruins of another, some disturbed ground and a conical spoil heap on the banks of the Severn. There are several exits from the track. One passes through a farm and leads up the hill to the site of the old **Caughley** Pottery, of which there is now no trace but a field name, Factory Field, and a large, still working, clay quarry. (Caughley is pronounced locally as Corfley.) Caughley Quarry, as it is called, is best approached from Broseley, down Pound Lane next to the Forester's pub. Opposite the entrance to the quarry are spoil heaps from the old coal mines. The whole area between Apley Hall and Caughley is quite fascinating. Nowhere could be more peaceful and beautiful, the epitomy of rural England and yet here were once the flames and thumps of forges, a passenger railway line and a world famous pottery. Caughley ware is very expensive and much sought after. Even very small, but perfect, pieces can be worth a thousand pounds or more. Treasure seekers are not unusual here. They search for reject pottery. Broken pieces are quite commonly found in the fields but these are only of historical interest and are not of great commercial value.

STOKESAY *¾m. SSE of Craven Arms.*
Stokesay is famous for its castle and is much visited by tourists. The settlement was originally called simply Stoke, then South Stoke and finally Stoke de Say which was abbreviated to Stokesay. The Norman de Say family were tenants of the manor. Shortly after 1068 Picot de Say built a castle and a church. The church was dependent on that of the neighbouring manor of Aldon, but within a few years the position was reversed. (A dependent church can only hold services and sometimes baptisms; the mother church retains the rights of marriage and burial.) The fortified manor house we see today was built about 1280 by Laurence de Ludlow, a wealthy wool merchant. It is not actually a castle, but a fortified house, one of the very ealiest in the whole country. The picturesque Gatehouse has stone lower parts and 16th Century half timber on the upper. Opposite the Gatehouse is the Hall Range flanked by 2 towers. The N tower is older than the Hall and contains the well. The Hall is huge and impressive. The Solar, the main private living room, can only be approached from outside, no doubt for security. The buildings are joined by walls to make a fully enclosed courtyard and the whole complex is surrounded by a moat. It is often remarked that the huge Hall windows constitute a chink in the armour of the house were it to have been attacked. The implication is that the Lord of the Manor felt that this was an unlikely eventuality. Most of the timber-framed structures date from Elizabethan times, as does the wall panelling. During the Civil War the castle surrendered and so escaped undamaged. However, the church of St. John the Baptist did not escape Cromwell's heavy hand and was largely rebuilt between 1654 and 1664. The result is a Commonwealth church. Such churches are, in fact, very uncommon. The cast-iron single-span bridge across the Onny was built by Thomas Telford in 1832. The countryside around the castle is most attractive, and a drive W from Craven Arms through the Clun Valley (the B4368) to Aston on Clun, Clunton, Clun and Newcastle is to be recommended. There is a folk tale associated with Stokesay to the effect that many years ago the lands here belonged to 2 giants who kept a vast treasure in vaults beneath the castle which was guarded by a raven. However, one of the giants accidentally dropped the key to the vaults into the moat and it has never been found to this day though many have searched for it. Neither, one must add, have the vaults themselves been located.

STOKE-UPON-TERN *5½m. SW of Market Drayton.*
The original name of the village was simply Stoke from *stoc*, Old English for 'a place', 'a religeous place', and quite commonly 'a daughter settlement' of a larger town on whom it originally depended. The Normans called it North Stoke to distinguish it from another Stoke in the S of the county which they called South Stoke. This later became Stokesay and is famous for its castle. The de Say family were tenants of both manors. The flat, drained land around Stoke upon Tern is used for both arable and pastoral farming. The sandstone church of St. Peter (1874-5) lies apart from the village. It stands on the site of the previous Norman church, a walled mound that juts into the flood plain of the River Tern, and has a striking turret attached to the tower. There is a collection of Norman coffin lids in the churchyard and inside are alabaster effigies of Sir Reginald Corbet, d.1566, and his wife. The Manor House lies on a track to the E of the village and just to the S of the Manor House are the remains of a rectangular moat that marks the site of Stoke Castle, once the home of the Corbets of Adderley and fortified for Parliament during the Civil War. Over the river, ¼m. SW, is the haunted house of Petsey, a black and white farmhouse of some character. It is dated 1634 and an unusual feature of the main gables is that the diagonal strutting turns outward from the post, not inward as is the norm. There is a school of 1870 and a travelling shop calls twice a week; of social facilities that is all. In a cottage at Manor Gate the Shropshire Giant, Thomas Dutton, was born in 1853. He stood 7'3", weighed 23 stones and was renowned for his great strength. Dutton later lived at 'The Poorhouse', a stone cottage in Stoke Heath, and is buried in Stoke-upon-Tern churchyard. Most of his descendants live in the Manchester area. Half-a-mile to the W of the village is Stoke Park a symmetrical red brick Georgian house of about 1750. It has 3 storeys, 5 bays, a hipped roof, stone dressings W irregular quoins, a string course, and a pedimented doorway. In 1987 it was divorced from its farm and park. North of Stoke-upon-Tern is Tern Hill Airfield, an RAF base. To the NW is **Stoke Heath**, a strange area of old wartime huts used as industrial buildings, a Borstal, and modern bungalows with smallholdings attached. It has a frontier atmosphere about it. On the A41 is Rosehill, a house with an old water mill and a large mill pond.

STONEY STRETTON *7m. WSW of Shrewsbury.*
Between Yockleton and Westbury the B4386 follows the route of a Roman road. Lying just off this ancient highway is Stoney Stretton. It is a typical small country hamlet with no shop, no pub and no church, just a handful of houses and farms. We saw not a soul. Ducks and geese lay in the sun on the banks of the brook. Opposite was a farm full of old machines and down the road was a large Georgian-style stuccoed house hidden in a wooded garden. The name Stretton is from the Old English *straet-tun* meaning 'a settlement on a Roman road'.

STOTTESDON *6m. N of Cleobury Mortimer.*
An attractive village lost in the lanes of the quiet, rolling country S of Bridgnorth. There was a large Saxon manor based on Stottesdon and in medieval times it possessed 7 villages. All of them are now reduced to single farms. At Pickthorn Farm, near Stottesdon, is a field containing the earthworks of a deserted manor of that name. As late as 1571 there were 8 tenements. Now there is just one farm. Stottesdon church is essentially Norman Decorated but there is a tympanum carving believed to be Anglo-Saxon. This is inside the tower and above the former W doorway, which is approached through a door at the back of the organ. The tympanum and lintel have carved figures of a bearded head and animals, 2 of which are upside down. The Norman font is very fine, and with little doubt is the best in Shropshire. It dates from about 1160 and is probably by craftsmen of the famous Hereford School. There is a Jacobean pulpit, a neo-Gothic screen of 1901 and a collection of gold thread garments. In the village there are some old cottages and farmhouses; new bungalows and houses; a Post Office in a private dwelling; a doctor's surgery; a Methodist Chapel; 2 pubs; and a school of 1872. Just N of Prescott, which is 1½m. SW of Stottesdon, is a bridge (SO.662.815) over the River Rea that is believed to be of Roman origin. It has 2 spans with a central cutwater. The arch with dressed stone voussoirs could be Roman, but the other span with 4 arch ribs of similar voussiors but with gaps between them covered by flat stones, is probably the work of masons employed by a medieval monastery. A disused railway track follows the river valley for several miles hereabouts. The name Stottesdon probably means 'the hill of the herd of horses'.

STOW *1½m. ENE of Knighton (Powys).*
It lies just off the narrow lane that follows the northern bank of the River Teme, an unspoilt hamlet in a splendid little secluded valley on the hillside below Holloway Rocks. The church of St.

Michael is framed by fir trees and has 13th Century masonry with new stone dressings, a good 17th Century roof, a weather-boarded belfry, 19th Century stained glass and 2 beautiful Art Nouveau (c.1900) mosaics of coloured stones, copper and mother of pearl. The church stands on a hillside, above the handsome stone-built Old Rectory, and has wide views across the valley. The name Stow is Old English and can mean either 'a place', or, more commonly, 'a religious place', often 'a church' but sometimes 'a monastery' or 'a hermitage'. One mile E is **Weston** (near Bucknell) with a motte and bailey castle facing it on the opposite bank of the River Teme. The Teme forms the boundary between Shropshire and Wales for several miles to the W. The hedges obscure some lovely riverside scenery but there are places where a brave man can park and take in the view.

STREFFORD *2¼m. NNE of Craven Arms.*
The ford is on the fast flowing Byne Brook. Beside it is a wooden footbridge supported by stone piers with cutwaters. The settlement lies between the busy A49 and the tree-clad slopes of Wenlock Edge. It is a tiny but charming little place with cottages of half-timber, brick, stone, and stucco and a few substantial houses such as Strefford Lodge and a farm with a lantern-like skylight that plies the bed and breakfast trade.

STRETTON WESTWOOD *2½m. SW of Much Wenlock.*
The settlement lies on the B4371 which runs along the escarpment of Wenlock Edge. Limestone mines old and new dominate both the landscape and the economy. Beyond the large quarries that adjoin the road is a wood called Blakeway Coppice. Here, above Blakeway Farm, is a precipice called the Major's Leap where Major Thomas Smallman of Wilderhope, whilst being pursued by Parliamentarians, forced his horse to jump to its death. The major landed in a crab-apple tree and lived to tell the tale.

SUTTON *2m. SSE of Shrewsbury.*
It is now a suburb of Shrewsbury and is not recognisable as an old community with an entry in Domesday Book. It was formerly a parish but is now a part of Meole Brace. All that is left of the church of St. John is the ruined chancel in a farmyard. Today, the few old buildings – a cottage, a mill, 2 farms and the Grange – are dominated by new houses. The redirected A4 by-passes Shrewsbury and this ring road has become the limit of estate housing. On the country-side are a row of early 20th Century houses with hipped roofs and the late Percy Thrower's excellent Garden Centre. Opposite is the Grange Nursery, where there are 2 small wooden windmills. Adjacent to the Grange is the Municipal Golf Course. At one time Sutton had a small reputation as a spa.

SUTTON MADDOCK *4½m. SSW of Shifnal.*
The country here is flatish and the farming mainly arable. The settlement is in 2 parts. The village lies on the busy A442, the Shifnal to Bridgnorth road, and the Hall and the church stand together ¼m. W in landlordly isolation. The village has an Old Schoolhouse and a timber-framed dwelling but is mostly a more recent red brick main road development. At the crossroads there is a Jubilee Fountain erected in memory of Queen Victoria. This was once the only source of water in the village but it is now dry. The Hall is an irregular brick Georgian house of 2 storeys with a hipped roof, a pedimented doorway and sash windows. The gardens are mainly laid to lawn but there are numerous trees and shrubs, all immaculately tended. What is more they are neither hedged nor fenced about but left open to the road. This provides a most pleasant prospect. To the side of the Hall is a farm. The substantial church of St. Mary is built on a low, circular, mound. The tower is dated at 1579 and the lower parts may be even earlier. The rest belongs to the rebuilding of 1888. The bell openings are of a Tudor design – 3 tall openings with a straight-headed lintel. In Anglo-Saxon times the settlement was simply called Sutton (meaning 'Southern settlement') and in 1066 was held by Earl Morcar. Henry II gave the manor to Iorwerth Goch for his services as an interpreter during the Anglo-Welsh border troubles. Goch was the son of the ruler of the Welsh princedom of Powys. The estate passed from Goch to his son Madoc, then to his brother Gruffydd who married Matilda le Strange, and then to their son Madoc. It was he who added the Celtic suffix, Madoc, to the Anglo-Saxon name, Sutton. Later lords of the manor were the Brooks family. Their crest depicts a badger and this symbol can be found on some of the houses in the village.

SWANCOTE *1½m. NE of Bridgnorth.*
A hamlet on the A454 Bridgnorth to Wolverhampton road. Swancote is known to many people in the Black Country for its blue and white painted open-air swimming pool. Sad to say this now lies abandoned; the timber-frames are rotting and vandals are no strangers here. Such places have a romantic, nostalgic atmosphere in death that they could never achieve in life – the empty ballroom syndrome. To the right of the swimming pool is Swancote Farm Restaurante, a late 18th Century brick house of 2½ storeys which has a summerhouse with a dovecote and a pyramid roof. The undulating country is most attractive hereabouts; in the season of ploughing plump red hills rear against the road like waves against a shore. The name Swancote probably means 'the cottage of the swineherd'. The word 'swan' originally meant either a swan, the bird, or a young man-servant, later a swineherd.

TASLEY *1m. NW of Bridgnorth.*
It lies just beyond the NW suburbs of Bridgnorth, a tiny place on high ground with good views. The church of St. Peter and St. Paul was built of hard, yellow brick in 1840 to a design by Josiah Griffith. Inside is a Jacobean pulpit and a Norman font. In the wood on the opposite side of the road to the church is a pond bay – a small dam that once formed a pool. This could possibly be connected with the iron-working industry that we know was well developed in the area. (Aldenham Park, once an old home of the Acton family, is 1½m. to the NW and was a medieval centre of the iron trade. Tasley was once a part of the Acton estate.) Racecourse Farm and Racecourse Drive commemmorate the old National Hunt course and Brickyard Cottages are on the site of a short-lifed brickworks – the bricks of the Village Hall were made here. There is no shop, pub, Post Office or school.

TELFORD (New Town) *7m. E of Shrewsbury.*
This is a collective name, similar to 'the Potteries' and 'the Black Country'. It consists of many old, small towns and villages linked by a motorway-like road system which are slowly becoming one as new industrial estates and housing developments 'fill the gaps in between'. The total area of town is 19,300 acres. Included within its boundaries is the Ironbridge Gorge, a beautiful and dramatic place, and Coalbrookdale where Abraham Darby first smelted iron with coke on a commercial scale. Together they are now an official World Heritage Site; that is, they are recognised as being in the same league as the Grand Canyon, the Pyramids of Egypt and the Statue of Liberty. Little wonder that Telford wanted the Ironbridge Gorge. It enabled their publicity people to make the exaggerated claim that Telford was 'the birth place of industry' and not look too shamefaced. Most national (as distinct from local) authorities point out that one technological advance is very little compared to the availability of capital, the development of a transport network, and above all the real revolution of the factory system which used mass-organized labour, first developed in Derbyshire and Lancashire. When the Development Corporation first started work here in 1963 much of the area was a derelict industrial wasteland, a place of old pit mounds and spoil heaps, uncapped mine shafts and ruined works all clad in naturally regenerating vegetation. It was not, though, an unattractive place. The corporation was originally called the Dawley Development Corporation and was armed with extensive powers under the New Towns Act. Its purpose was to develop the area as an overflow for the overcrowded Birmingham – Black Country Connurbation. Growth was slow, partly because of bad communications – the M54 was a long time coming – and the unemployment rate was one of the worst in the country. In 1968 the name was changed to Telford and in 1992 the Corporation is to be disbanded, its job completed. There are now 20,000 houses and 12,000 more are planned; the population is 111,000; there are nearly 100 shops in the Shopping Centre which receive 160,000 customers each week; and there

Telford Centre, Reynold's House.

Thomas Telford surveys the Telford Town Centre over the pool opposite the new Development Corporation office block.

are 3 major Industrial Estates, each the size of a small town. **Telford Town Centre** is a modernistic place, largely constructed of iron frames infilled and cladded with glass, concrete and plastic. It has a distinct air of impermanence. Here is the somewhat ill-lit undercover Shopping Centre in which most of the national chain stores represented; small, individual shops are conspicuously few and far between. Here, too, are the Wrekin District Council offices, the indoor sports arenas, office blocks and the little railway station. The station is separated from the town centre by the railway line and the ever noisy M54. A long, ungainly red bridge crosses this divide. The best group of buildings is what might be called the Civic Centre to the W of the road that encircles the Shopping Centre. Here are the Police Station, the County Court, the Magistrates Courts, the Probation Offices, and, most recently New Town House, the offices of the Telford Development Corporation when they leave Priorslee in 1989. In the central courtyard there is a cascade waterfall and a sculpture of Thomas Telford (1757-1834), casually leaning on the letters of his name writ large. Adjacent to this complex is the new A.M.C 10 screen cinema, a yellow barn of a place with a flat roof and a glass verandah. There is a hotel, The Moat House, and the TSB Training School but dominating all else here are the 4 buildings occupied by the Inland Revenue. These are Reynolds House, all glass and skinny red frames with a grey slate porch; Addenbrooke House, light buff-grey concrete; Boyd House, pebble dash panels, a blue porch, black railings and a bed of red roses; Matheson House in pale grey with what appears to be scaffolding incorporated into the design; and Darby House (1978), a 9-storey block with a higher octagonal tower and a large extension of dark glass to the front. The latter is shared with other occupants. These buildings are all secure places for in them are the Inland Revenue national computers. There is not a lot more to say about the Town Centre, it is, after all, a relatively small place. Adjoining the Town Centre to the W is **Hollinswood**, the site of an old ironworks. Unlike most of the new housing estates, which have orbital linking roads, the short residential roads of Hollinswood are connected to a meandering central thoroughfare called Dale Acre Way. There is a First School and a Middle School, a pub, the Woodcutter, a flat-roofed Community Centre and playing fields. The whole estate is well planted with small trees. To the E the houses come hard up against the Queensway on the other side of which is the industrial estate of **Stafford Park**. Hidden amongst the factories and warehouses is the red brick and dark glass Fire Brigade Headquarters. It is named Stafford after the Marquises of Stafford, the Leveson-Gower family (pronounced Lewson-Gore), who were major landowners in the area. Within the park is an office complex of little grey boxes with port hole windows and green and red trim; a row of new car sales rooms; and numerous and varied factories and warehouses, including the premises of: Mencon, Menk UK, Huck Trigon Packaging Systems, Magna (confectioners) Unimation, Telford Extrusions (plastics), Nortec Bodies, Mangum Fabrications, Whippendell Electrical Manufacturing Company, Samsung (TV's, videos, microwave ovens), Tatung, and the Airship and Baloon Company amongst many, many others. It is an attractive and well planned park but because local building materials and styles are not used it has no regional character at all. It could be anywhere. Only in the new office block called Stafford Court is there any sense of tradition and having built for the future. This is a yellow brick building with blue brick decorative arches and pediments under a green corrugated roof. One feels that this will be standing when all about it have long gone. The separate parts of the New Town are discussed in this book under their individual names – Stirchley, Madley, Ketley etc. Most are treated in considerable detail because in the past they have been somewhat neglected.

TERN HILL *2½m. SW of Market Drayton.*
Tern Hill is a relatively new settlement close to the junction of the A41 Newport to Whitchurch road and the A53 Shrewsbury to Newcastle-under-Lyme road. There used to be traffic lights to control the crossing but after several fatal accidents these were replaced, in 1988, by a roundabout. There are 2 garages, a Georgian farmhouse and a pet food store at the crossroads but half-a-mile SE on the A41 (which here follows the course of a Roman road) is the nucleus of a small town: a Post Office, a General Store and Cafe, a Service Station and a substantial brick built pub, the Stormy Petrel. This recent development has grown up to service the RAF station (established here in 1916), and the Army Camp. The airfield is used mainly by training 'planes and helicopters. In 1980 a major part of the excellent, purpose built accommodation was taken from the RAF, renamed the Clive Barracks, and put at the disposal of the Army. It is at present occupied by the 630 soldiers of the 2nd Battalion of the Parachute Regiment who are here on 'home defence duties'. The barracks, which were the subject of an IRA attack in 1989, cluster around the lower slopes of Tern Hill. This is a tree-covered rise just to the S of the River Tern, an unimpressive waterway which for most of its length is little more than a drainage ditch. Some 200 yds. S of the crossroads is Tern Hall Hotel, an irregular red brick house nicely sited amongst specimen trees with good views over an attractive landscape to the SW. Between the airfield and **Buntingsdale** Hall, 1m. NE, are the RAF living quarters, a reasonable sized village in its own right. The RAF used to occupy Buntingsdale itself but after having mutilated the Hall and all but destroyed the landscaped grounds they returned it to the private sector, selling it off in bits and pieces. The Hall was built about 1730, probably to a design by Francis Smith of Warwick. It is of brick with stone dressings and extends to 9 bays, 2½ storeys high, with giant pilasters and notable carved capitals. There is a lake and the Tern flows through the partly wooded park. The name Buntingsdale probably means 'the dale where buntings are found'. The bunting is a bird and the name is very old. (*See Little Drayton.*) At 3.00 a.m. on the 20th February 1989 one of the accommodation blocks was blown up by IRA terrorists. There was no loss of life because a sentry had spotted the intruders and raised the alarm in time for the dormitories to be cleared. Later that day we passed the camp which then was under siege from the national press. Complacent middle-aged male reporters and their hard-faced female colleagues sat in cars with engines running whilst they spoke into radio telephones; TV cameramen, some of the worst dressed men in England, and drunken press photographers lurched out of the pub, cast an eye in the direction of the billowing smoke and lurched back to wait for nothing to happen in comfort. In the car park despatch riders paced up and down beside their motorbikes waiting to take film and video tapes to Birmingham, and mobile hoists stood like giant birds of prey. Overhead a police helicopter whirred noisily. At the entrance to the camp fresh-faced paratroopers still in their 'teens stood guard with sub-machine guns and, not unnaturally, rejoiced in having their photographs taken. No one expressed fear or repugnance, far from it, there was an air of subdued excitement amongst some and amongst others expressions of boredom.

TETCHILL *1¼m. SSW of Ellesmere.*
A pleasant residential hamlet of mainly modern houses. 'Threeways' is a delightful black and white thatched cottage.

TIBBERTON CUM CHERRINGTON *4m. W of Newport.*
It lies in flattish lands, an attractive village despite much modern housing. The church of All Saints lies on a bank above the River Meese. It was built in 1842 of red sandstone ashlar to a Gothic design by T. Baddeley. It replaced an earlier, 12th Century, church. The Manor House lies ½m. away, close to the Newport road, the B5062. In the village there are a few black and white cottages; a pub, the Sutherland Arms (the village lay on the Duke of Sutherland's estate); a Church of England School of 1970; and a primitive Methodist Chapel of 1842. Just below the church, by the river, is a brick built house. This was formerly a paper mill, reputedly the last in Shropshire when it closed down in 1932. (It was also probably the iron slitting mill owned by Richard Ford II when he went bankrupt in 1756.) The name Tibberton is thought to mean 'the settlement of *Tibbeorht*'s people'. Two miles to the E, towards Newport, and clearly visible from the main road, is the Harper Adams Agricultural College, an imposing building of red brick. Thomas Harper Adams was a farmer from nearby Edgmond. In 1892 he died and left £37,000 and 178 acres as an endowment for an agricultural college. The College finally opened in

Stokesay Court, near Onibury.

The Clive Barracks, Tern Hill, on the day the IRA attacked it in 1989.

Buntingsdale Hall, Tern Hill, near Market Drayton.

1901, with support from the county councils of Shropshire and Staffordshire. One mile NNW of Tibberton is a Holy Well. The spring issues into a large concrete and brick basin. (Such basins were made so that the aflicted seeking a cure could bathe in the waters.) The well lies in a wooded hollow in a field close to the road. It is not well cared for. To get there take Mill Lane out of Tibberton, follow this road for about ¾m. and take the right-hand turn. About 100 yds. down this lane is a metal 5-bar gate opposite an oak tree. The well lies in the trees 20 yds. behind the gate. **Cherrington** lies ½m. W of Tibberton. It is a small, rambling village. Cherrington Manor (1635) is a superb black and white half-timbered house that lays claim to being the 'House that Jack Built'. There is a substantial moat close by. The village spreads between the Manor house and the main road. On the other side of the road, the B5062, is Cherrington Moor, part of the once wild Weald Moors, which were drained by Thomas Telford at the behest of the Duke of Sutherland. The name Cherrington means 'the settlement of *Ceorl*'s (or *Corra*'s) people'. There is an old rhyme that goes like this: 'Tibberton Tawnies and Cherrington Chats, Edgmond Bulldogs and Adeney Cats, Edgmond Bulldogs made up in a pen, Darna come out for Tibberton men.' Tibberton people were in times past noted for being dark complexioned – hence tawny; a chat is a gossip.

TILLEY *1m. SSW of Wem.*
The name is from the Old English *telga-leah*; *telga* in local dialect was 'tillow', and means 'a branch'. Tilley lies just beyond the suburban clutches of Wem. It is a small hamlet with a handful of Georgian buildings, a few modern dwellings, The Raven pub and several black and white houses 2 of which are most handsome. Oak Cottage has chevron strutting on both floors and Tilley Hall has projecting, gabled wings with varied decoration and a recessed centre with the date 1613 over the porch. Tilley Manor has a similar layout but has been partly rendered. The main line railway from Whitchurch to Shrewsbury passes along an embankment beside the access road. To the W the Sleap Brook, now deepened as a drainage ditch, joins the River Roden which has been similarly treated. The lands to both E and W are very flat and were once marshy and prone to flooding. **Tilley Green** lies ½m. SE of Tilley on the other side of the railway and the main road, the B5476. There are a few old cottages, a couple of new houses and Trench Hall which lies on a low hill. The Hall is of red brick with stone dressings and has a 3-bay front with a columned porch supporting an entablature. At roof level there are 2 pediments – most unusual. There are specimen trees in the garden and from the front of the house there are long views over the wide North Shropshire plain to the hazy hills of Wales. Behind the Hall is Trench Farm, a long, low part timber-framed house which faces a cobbled courtyard with red brick farm buildings to the other 3 sides. In the centre of the courtyard is a large lawn, and in the entrance range is a tall factory-like chimney, both unusual features. **Palms Hill** House lies ½m. E of the Hall on the B5063. It stands on the edge of a sandstone bluff, a fine example of a large Victorian country house. This mansion is one of those buildings that has a bit of everything: bright red brick walls with upper elevations clad with patterned tiles in some parts and black and white framing in others, stone dressings and bay windows, dormers and gables etc. Only the Victorians, with their exuberant confidence, would have dared to combine. These houses were well built and constructed of good materials; they will be with us for a long time yet. Palms Hill House is embowered in specimen trees and faces NW with views over the town of Wem just over a mile away.

TILSTOCK *1m. W of Prees Heath, which is 2½m. S of Whitchurch on the A41.*
A mainly red brick village with an attractive red brick church of 1835. Christ Church has a slender tower with a pyramid roof and cast-iron glazing bars in the windows. In the village there are 2 shops, a Post Office, a pub, a school, a children's playgroup, a Tennis and Bowling Club, a plant nursery and a few too many new houses. Agriculture in the fields about is largely arable: potatoes, carrots, wheat, sugar beet, peas and beans. Cattle and sheep from the Welsh hills are also fattened here for slaughter. Tilstock was called Tildestok in early medieval times. It is Old English and means 'the farm of *Tidhild*'. *Tidhild* was a woman's name. **Alkington** Hall lies 1¼m. NW of Tilstock. Only a small part of the original Elizabethan gabled red brick mansion of 1592 remains, but it is very striking. Two miles to the W of the Tilstock is the Shropshire Union Canal, crossed hereabouts by picturesque counterweight swing-up bridges. The canal makes a long detour between Blackoe and Bettisfield to skirt the great Fenn's Moss and Whixall Moss. These have been drained and are now farmed, but the drains and ditches need constant attention because the land easily becomes waterlogged. Indeed, small untended areas have once again become marshes. When passing another car in the country lanes around here beware of pulling on the grass verges. They are often very wet and very soft, even in summer.

TITTENLEY *2½m. NNW of Market Drayton.*
The name is from the Old English *Tytta*'s-*leah* and probably means '*Tytta*'s clearing in the forest'. Until 1895 Tittenley was in Cheshire. It lies on the high ground to the E of the broad valley of the River Duckow. Of buildings there are only 3: a gabled 2-storey farmhouse with traditional outbuildings all in red brick, and 2 square lodges flanking the entrance drive to Shavington Park. (*See Adderley.*) These have stone quoins and feature a Ventian window surmounted by a segmental 'eyebrow' window. There is a stand of tall trees and hollies in the hedgerows. Castle Hill stands 1m. SE. This tree-clad symmetrical rise has been researched (but not excavated) by Keele University. No evidence of a castle has been found. The red brick farm adjacent belongs to the Shavington estate and has some good traditional outbuildings. **Rhiews** lies ½m. SW of Tittenley to the W of the River Duckow. The name is probably from the Welsh *rhiw*, meaning a hill or a slope. There is little more here than a part timber-framed farm. In the valley below there used to be a square moat. This has been destroyed and is now barely discernible. It lies close to the river, to the S of and adjacent to the road, at SJ 643.375. The river is hardly a river anymore having been deepened so that it now lies about 8ft. below ground level. A once wandering and attractive waterway is now a deep, dangerous drainage ditch. **New Street Lane** lies 1m. W of Rhiews. Here is another lodge to Shavington Hall (now dismantled), a handsome red brick building with black and white upper elevations. But it stands empty with flaking paint and locked gates and it has been many years since carriages rolled down the hill through the long avenue of tall trees. Just along the road is the large and most handsome Shavington Grange Farm, mock Tudor, like the lodge, with gables, dormers, and many chimney stacks. Opposite is Moat Farm, brick with dormers and black and white end gables dated at 1870. The initials JPH stand for John Pemberton Heywood. A track leads past the farm buildings to a small wood in which is a square moat. To the N is the Big Wood of Shavington Park; to the W is the Big Wood of Cloverley Hall. (*See Calverhall.*) This is a gentle, quiet land of rolling green fields; only the ocassional fox hunt and shooting party disturb the peace. The most distasteful sight we witnessed here was a troup of about 50 hunters on 50 horses with a pack of about 50 hounds aided by several hundred 'onlookers', who in fact were often deliberately blocking escape routes, all in pursuit of one terrified fox. It might be said that the odds were a little stacked against him. We saw the fox in a field not knowing which way to turn, such a brave, spirited and attractive creature; how can people be so heartless? We talked to several of the participants and it was evident that despite their words they felt guilty.

TITTERSTONE CLEE HILL *6m. ENE of Ludlow.*
A large, sprawling hill (1,750ft. above sea level) composed of hard, dark, volcanic basalt lavas known locally as dhustone – *dhu* is Welsh for black. These are capped by coal bearing rocks. In the past both minerals have been quarried here; dhustone is used as a road metal, the 'grit' in 'Tarmac'. From most directions the hill has a distinctive outline – a sloping, scarped summit – and is easily distinguished today by the radar dishes of the Civil Aviation Authority. The whole of the flat top of Titterstone Clee is the site of a huge Iron Age settlement which covers 71 acres. The enclosure was origin-

The pyramid dovecote in a farmyard near Tong.

ally protected by a single earth and timber wall. Later, this was reinforced by secondary defences and finally by rough hewn blocks of dhustone. It must have been an impressive sight. At 533 metres above sea level Titterstone Clee Hill is not the highest hill in Shropshire, that distinction belongs to Brown Clee at 540 metres, but it seems to have been held in the highest esteem by local people. A vestige of its former religious importance was the Titterstone Wake which commenced on the last Sunday in August and often lasted for a week. It was always the last midsummer festival in the county. Sports and games of all kinds were played, much beer was drunk and the 'goings-on' were notorious. The proceedings began when the young men, who had assembled at the Forked Pole, were joined by the young women, all dressed in their best, and together walked in procession down Tea Kettle Alley – a passage made of dhustone – to drink tea made by the older women. From this decorous beginning things rapidly became Bacchanalian. The name Titterstone is said to be from Totterstone, a stone that rocked. One 19th Century antiquarian reported that he had seen such a stone. But we are not convinced. The names of many rivers and hills are Celtic and even pre-Celtic; their meanings lost in the mists of time. There are several modern settlements on the hill, but see in particular Dhustone and Bedlam; both have their own articles. South of and adjoining Titterstone Clee Hill is Clee Hill, the flanks of which are traversed by the Ludlow-Cleobury Mortimer road. Three miles to the N of Titterstone Clee Hill is Brown Clee Hill on which are 2 tall aerial masts.

TONG *3m. E of Shifnal.*

Tong is an old and most attractive village noted for its handsome church but with a number of other historically interesting sites. These include a Norman castle, a medieval chantry college, an exotic mansion by 'Capability' Brown, a 17th Century iron-working forge and a mysterious ruin. The village stands just off the A41 Newport to Wolverhampton road close to Junction 3 on the M54. The surrounding countryside is most pleasant and can be seen to good advantage from the A41 between Tong and Newport where the road follows a low ridge. The settlement was of some importance in Anglo-Saxon times and in 1066 was held by Earl Morcar. After the Norman Conquest all the lands of Shropshire were given to Earl Roger de Montgomery and he redistributed them amongst Norman families. Tong was one of the manors he kept for himself. In 1086 he had 8 slaves working the land for him and there were also 3 villagers and 2 smallholders on the manor. The first castle was probably built in about 1127 by Richard de Belmeis, Bishop of London and Viceroy of Salop. It consisted of an enclosure some 30 yds. by 40 yds. on a promontory at the confluence of 2 westward flowing streams (SJ.792.069). It is likely that the fork shape made by the 2 streams joining together is the origin of the name Tong. 'Fork' in Old English is *tang* (sometimes *twang*), and this is also the root of the current word 'tong', as in 'a pair of tongs'. In the mid-12th Century the enclosure was walled around by the La Zouche family who in the 13th Century added a large outer bailey to the E, complete with a new gatehouse. The property then passed to the Harcourts and then to the Pembruges who built a new manor house about 1300. Parts of the chapel still survive. Licence to crenellate was given in 1381 and in 1447 the castle passed to the Vernons who built a new house in about 1500. This was rebuilt by the Pierrepoints in the late 17th Century. In 1765 all the old buildings were cleared and a new extravagant house of Moorish-Gothic design was built for George Durant by 'Capability' Brown. It was probably at this time that the lake opposite the church was landscaped and various lodges and follys were erected. One of the best known of these is the pyramid dovecote in the farm at SJ.783.076 on the road to Shifnal. The Durant house was demolished in 1954 by the Earl of Bradford and the M54 now runs straight through the site where the mansion once stood. All that is left are the foundations of some outbuildings, which include a stable and smithy, on the N side of the motorway between the road and the lake. South of the road is the site of the earliest castle, most of which remains. There are foundations of a 13th Century kitchen, the chapel of about 1300, a 14th-15th Century well (40 ft. deep), the 17th Century garden walls (in places 15 ft. high) and cellars of outbuildings of the 18th Century Durant mansion. The position on the ground of the southern remains are, from W to E: garden retaining wall, kitchen, well and chapel. Note: In 1972, during the construction of the M54, a Bronze Age axe (c.700 B.C.) was found during excavations near Tong. The Durant family came here in 1760 when George Durant bought the estate with some of the fortune he had made in the West Indies as Clerk to the British Forces. His son was something of a ladies man. He had 12 children by his first wife, 10 by his second wife and 32 illegitimate children by various village maidens! This puts a second twist to the Durant family motto: 'Beati qui Durant', which means 'Blessed are those who endure' but which is easily misconstrued as 'Blessed are the Durants'. In the late Middle Ages the manor was owned by the Vernons (whose family seat was Haddon Hall in Derbyshire) and from them it passed by marriage to Sir Thomas Stanley (d.1576). In 1600 Venetia Stanley was born at Tong. She was one of the great beauties of her age. Amongst her admirers were the poet, Ben Jonson, the painter, Van Dyck and the nobleman, the Earl of Dorset. Her husband was Sir Kevelin Digby, son of Sir Everard Digby, who was executed for his part in the Gunpowder Plot. She had numerous children, not all by her husband, and died suddenly at the age of 33. The post-mortem revealed that her brain had shrunk to the size of a walnut. This was attributed to the lady having drunk viper wine which was believed to aid the preservation of a youthful complexion. Cynics say she was poisoned by the husband she had cuckolded. Another lady of renown with Tong connections was Mary Anne Smythe (of the Smythes of Acton Burnell), who, as Maria Fitzherbert became the mistress and later the unlawful wife of George IV. Her father was the agent in control of Tong Castle at the time of her birth. The village of Tong is most handsome and now that the old main road that ran through the settlement has been by-passed it can be enjoyed in peace. There are several black and white cottages and the timber-framed Church Farm, now a riding establishment, and some good brick houses such as the curious Old Post Office. This has 3 bays with a central Venetian window and pediment on the upper floor and side bays with lean-to roofs. Adjacent is the elegant Red House of Georgian red brick and next to that are the 18th Century Almshouses, 4 small one storey dwellings facing a quadrangle open to the road. The substantial brick built former Vicarage is also 18th Century and has 5 bays with a Tuscan pilastered doorway surmounted by a segmented pediment. The old 'school on stilts' in which in 1868 the American traveller Elihu Burritt found children 'rural, ruddy and happy as birds' is long gone. It stood at the back of the church on the mound in the grassy paddock where horses now graze. Abounding this field is a sandstone ruin, a section of thick walling with a round-headed window, a pointed arched window and small pointed arch doorway. This was obviously a building of some substance and yet its origin and use are unknown. This is quite remarkable. What is more, the guide books and reference works simply ignore it, not even acknowledging that it exists. The building might have had something to do with the church of St. Mary and St. Bartholomew and in particular with its founder, Elizabeth, widow of Sir Fulke de Pembruge. There was an early Norman church on the site but of this only a part of S arcade has survived. The church was rebuilt after 1410 when Elizabeth Pembruge founded a Chantry College of one warden, 4 clergy and 2 clerks to pray daily for the souls of herself and her 3 deceased husbands. The college buildings were in the field in front of the church (to the S) and medieval tiles have been found near the road, though nothing remains of the buildings above ground level. The college also had to care for 13 paupers and it could be that the mysterious ruin was a part of their accommodation. The priests had to eat and drink modestly and refrain from hunting and hawking. The college was seized by the Royal Commission of 1546 as part of the Dissolution of the Monasteries by King Henry VIII. The next year the college and the church were sold to Sir Richard Manners who had married into the great Vernon family. The church is a striking Gothic building constructed of stone in Perpendicular style. It is cruciform, though the transepts do not extend beyond the aisle walls, and it has a square tower that becomes octagonal and is topped by a spire. The church is decorated

Tong, the M54 looking west. Tong castle once stood here.

with battlements, pinnacles and gargoyles and has been called 'the village Westminster Abbey'. Inside this nickname is further justified by the many monuments to deceased noblepeople, especially the Vernons. The alabaster effigies, in particular, are very fine. There can be no doubt that they dominate the interior to the detriment of the church as a whole but that can be forgiven because of their excellence. The monuments include memorials to: an unknown 13th Century priest; Sir Fulke and Lady de Pembruge; Sir Richard Vernon of Haddon Hall, d.1451, and wife; Sir William Vernon, Lord High Constable of Henry VI, d.1467, and wife (the Constable led the English army in the absence of the king); Ralph Elcott, d.1510; Sir Henry Vernon, d.1515, the founder of the Vernon Chapel; Arthur Vernon, d.1517; Richard Vernon, d.1517; Humphrey Vernon, d.1542 and wife; Mrs. Wylde, d.1624; Sir Edward Stanley, d.1632, and his parents, Elizabeth Pierrepoint, d.1696; and George Durant, d.1780. The Vernon Chapel, sometimes called the 'Golden Chapel', is a small but beautiful chantry with a fan-vaulted stone ceiling on which are traces of the original gilding and stencilling. The rest of the church walls were scraped during the restoration of 1892 by Ewan Christian. Amongst the furnishings are a Jacobean pulpit of 1629; an original door from the chancel to the vestry; a Royal Coat of Arms of 1814; misericords on the chancel seats; an embroidery vestment of about 1600; stained glass in the W window with 15th Century figures and E window glass by William Kempe of 1900; and a most beautiful chalice cup of about 1545 in silver gilt about 11 inches high. Outside the church, by the entrance porch is 'The Reputed Grave of Little Nell'. Charles Dickens did in fact know Tong and set the church porch scene of The Old Curiosity Shop here. The grave, though, is the work of a verger who found that showing tourists Nell's last resting place was a lucrative sideline. He even forged an entry in the burial register! The old main road that runs at the back of the church forms a dam and held back the waters of a long narrow lake that is now silted up and marshy. This stretches northwards to **Tong Norton** and Norton Mere which lies below Tong Knoll. A rent from the Norton manor was provided by Henry Vernon to pay for the ringing of the Great Bell whenever a Vernon came to Tong. To the W, accessed off the road to Shifnal, is **Tong Forge** (SJ.784.083). If one leaves the church at Tong, turns right onto the main road, first left onto the Shifnal road and then first right down a rough lane one passes through the yards of an earth moving equipment firm and down a steep little hill to a most delightful little valley. The stream, a tributary of the River Worfe, has been dammed and a sequence of reservoir pools created. The valley is well wooded with both conifers and broadleafed trees. Jacob sheep and the occasional horse graze the pastures and at low water levels huge glacial boulders of grey granite can be seen in the bed of the main pool. There are 2 Georgian-Victorian dwellings, a modern dormer bungalow and a brick farm building. The lane crosses the stream on a sandstone bridge and climbs the far bank through a rock cutting. The furnace-forge buildings are no longer to be seen. This is probably the site of the ironworks sometimes called the Lizard Forge. Wrought iron was made here by beating out the impurities of re-heated cast-iron with great hammers. It was owned by the Slaney family of Hatton Grange and is known to have been operating in the late 17th Century. The Slaneys also owned a charcoal blast furnace at Kemberton and an iron slitting mill at Ryton. These last 2 were leased out by 1714 but the Ryton mill processed iron made at the Lizard Forge. Lizard Farm lies ¼m. NW on the lower slopes of the low-domed and conifer clad Lizard Hill. North of Tong Forge by one mile is Lizard Mill, a red brick building now converted to a dwelling. Here, the stream is crossed by a ford. The banks of the brook are wooded for over 3 miles hereabouts. Just S of Junction 3 on the M54 is the striking New Buildings Farm. It is built of red brick with a 3-bay, 2½-storey centre and 1-bay slightly projecting wings. To the left is a large walled garden and to the front is a paddock, frequented by geese, that borders a small stream. On the opposite side of the road is what appears to be one of the lodges to Tong Castle. It has unusual stepped walls and a variety of windows; round-headed, pointed-arched, and rectangular. War games, organised by Task Force, are played in the adjoining woods.

TREFLACH *4m. SW of Oswestry.*
A small, scattered, village with old, grassed over quarries beside the main road. There are houses and cottages of stone, modern bungalows, the stuccoed Royal Oak and a Post Office cum general store. Treflach Hall is a good stone-built house of about 1700 with a recessed centre of 3 bays and 2-bay wings capped by a hipped roof. Inside is some Queen Anne panelling and a small, possibly Jacobean, staircase. The countryside seems ordinary enough, but take the road to Porth-y-waen and as one descends into the valley a huge landscape unfolds – mysterious mountains in never ending rows, beautiful woods and valleys, and the great white blob of the ARC Western Blodwel Quarry.

TREFONEN *3m. SW of Oswestry.*
It lies on high ground in the line of Offa's Dyke, a village that was once small but which in recent years has been greatly extended. Modern red brick sits uneasily beside old grey stone, and fast growing conifers are looked upon with disdain by ancient holly bushes. Even the house names are incompatible; Casa Vago stands opposite Brynhyfryd. Nevertheless the place has some charm, most of which it owes to the way the houses are scattered about all higgledy piggledy. Most of the dwellings lie on a hill, off the main road. Here, too, are the Post Office, the Carneddau Chapel and a general store. Below, are the pubs, the Barley Mow and the Efel Inn, and 2 more churches. All Saints was built of grey stone in 1821-8 and the windows were Gothic and the chancel and apse added in 1876. The Church of England school stands beside it. Trenant United Reform is of 1832, also in stone with Gothic pointed-arched windows. On the main road is a hairdresser's shop and the horrendous modern shed of Martin's Cash and Carry. The country hereabouts is unspectacular, but not so to the S and W where there is some splendid mountain scenery.

TUCK HILL *6m. SE of Bridgnorth.*
A hamlet of houses and farms that lies just off the A458 Bridgnorth to Stourbridge road. The church of Holy Innocents lies embowered amongst mature, handsome beeches and Spanish chestnuts, a most beautiful setting. It was built in 1865 to a design by J.P.Saint Aubyn and has stained glass windows by Kempe and a timber bellcote with a spire. The name Tuck Hill may mean 'the hill, (or settlement on the hill) of tuckers (or fullers) of cloth'.

TUGFORD *2m., as the crow flies, E of Munslow, which is about 10m. SW of Much Wenlock on the B4378-B4368.*
Tugford lies on the lower slopes of Brown Clee Hill, sheltered in the valley of the Tugford Brook in Corvedale. It is a very pleasant spot spoiled only by inconsiderate farmers and their black barns. There are stone-built houses and a Norman church, St. Catherine's, which has a 13th Century tower and chancel. Inside the church on either side of the entrance door arch, placed where they are least likely to be noticed, are two small Sheela-na-gigs. These indecorous sculptures of pagan women are very common in Ireland but very rare in England. In the whole country there are only 18, and 4 of them are in Shropshire: 2 at Tugford, one at Church Stretton and one at Holdgate. They are presumed to be pre-Christian fertility symbols, and the fact that they have survived for so long is quite remarkable. The church also has clear glass windows, box pews and a gallery. The village green at Tugford is now partly incorporated in the churchyard. The houses of the settlement were originally built around the green and the villagers had common rights upon it. The low, earthen platforms on this communal area are the remains of early 17th Century squatter houses, the destruction of which was ordered by the Manor Court. Common rights on the green were abolished in 1815. Tugford Mill stands opposite the church across the Stream. The leats and pools are deep and well established and it is likely that this is the site of the mill mentioned in Domesday Book. Between 1801 and 1961 the population of Tugford fell from 168 to 86, a decline not untypical of many Shropshire villages. The first element of the name Tugford is probably from the Anglo-Saxon personal name *Tucga*. Two miles NE of Tugford is New Earnstrey Park. The house here was encased in brick in Elizabethan times. It represents, along with Plaish Hall, the earliest use of brick

Tong Castle.

Blockley's brickworks, Trench Lock.

The old Court Steelworks, Tweedale, being demolished.

in the county. Brick was then a more expensive material than stone or timber. Earnstrey Park was originally a part of Clee Forest. (Clee comes from the Old English 'Clag' meaning clay.) The Park is one of the oldest in Shropshire. The Norman deer park, or haye, largely coincides with a much earlier Saxon hunting estate. This is interesting because Saxon parks are difficult to trace because they were not enclosed. The Norman parks were enclosed by an earth bank with a ditch inside. This was topped with a wooden paling or fence, which was sometimes replaced by a hedge or stone wall. These boundaries can still be traced at Earnstrey. The name Earnstrey is from the Old English *earnes treo* meaning 'the eagle's tree'.

TWEEDALE *2½m. S of Telford Town Centre.*
The name is probably from the Indo-European root *teva* meaning 'powerful'. What was powerful we do not know, a stream, perhaps, or a religious centre. The dale stretches NW and is followed by the old Bridgnorth Road. To the S the boundary of Tweedale is Kemberton Road. Below the tree-clad spoil mound at Cuckoo Island is the Cuckoo pub, a handsome, modern, pagoda-style structure that faces the fire and ambulance stations. In Prince Street is a nice row of yellow brick houses and a delightful little red and black painted dwelling called Foundry Cottage. It stands beside a brick, barn-like building that is, in fact, the old foundry. This used to have a tall chimney but it was pulled down and the bricks used to build the cottage. The foundry is now a car repair workshop. North of this small residential area the land is almost entirely devoted to industry. Large steel-framed buildings clad in grey sheeting and black and white bricks house a variety of businesses: Raleigh Caravan Awnings, Bitec (motor components), Tool Hire, GKN Sankey (Wheels) Division, Coventry Hood and Seating and Telford Marble, etc. On the opposite side of the road, amongst the trees on the spoil heaps, is the Severn Gorge Caravan Park, an enterprise that has strayed somewhat far from home. Uphill, opposite the hipped-roofed Three Furnaces pub, is the entrance to the famous Court Steelworks. The last of the old buildings is a huge brown and black shed in the process of being demolished. New, blue and yellow units stand about, like brash young dogs turning on their old mother. Most of the new firms installed here are also in the metal-working business. Just W of the old works, hidden from view by trees and spoil mounds, is the elegant and ancient Madeley Court from which the company took its name. Just S of the Court is the Madeley Recreation Centre where there are tennis courts, bowling greens, sports grounds, a theatre and an artificial ski-slope.

TWITCHEN *1½m. S of Clunbury.*
Once, this was a flourishing little village with a chapel, pub, toll-gate, rope-maker, shop, malthouse, herbalist, cobbler, tailor and 20 or more houses clinging to the side of Clunbury Hill. Today there is but a handful of grey stone cottages. For many centuries this was an important crossing place. The old (possibly Roman) road to Clunbury cuts across the ancient Clee-Clun Ridgeway at Twitchen. The name is from the Old English *twicen*, meaning 'a fork in the road'. In the valley there are dark coniferous forests, sheep, grey squirrels and the occasional black and white cow. Twitchen is one of the places mentioned by Ida Gandy in her charming book, 'An Idler on the Shropshire Borders'. Her husband had a medical practice at Clunbury from 1930 to 1945.

TYN-Y-COED *2½m. WSW of Oswestry.*
It lies deep in the valley of the Morda Brook just N of the Oswestry to Croesau Bach road off which the Old Mill inn is signposted. In the woods and meadows are a few farmhouses and the Morda Brook is a delightfull, fast flowing stream of bright, clear water. The Old Llanforda Mill has been restored so that it no longer looks old and is guarded by several noisy dogs and a small army of hens many of which are bent on committing suicide beneath the wheels of passing motor vehicles. Offa's Dyke crosses the Morda just to the E of the mill and continues northwards through the oak and beech woods of the Llanforda estate. If the traveller turns left at the mill, and then forks right he can follow the stream northwards past an untidy farm, a small conifer plantation, and a romantically derelict stone cottage, ivy-clad and in ruins but with carefully tended front garden in which grow a variety of vegetables. The road bears left and up a steep hill. Near the summit is the white and black 2½ storey farmhouse of Nant-y-gollen complete with a clutch of new, shiny, corrugated-iron roofed sheds. The soil cover is very thin here and rocks regularly protrude from beneath the pastures where sheep graze. At the top of the hill we are on the top of the raised plateau of old rocks that have subsequently been eroded by streams to form valleys and mountains. One mile to the W is the hilltop lake of Rhuddwyn – Llyn Rhuddwyn (SJ.232.288) – and the pre-historic fort of Coed-y-gaer. (*See Croesau Bach.*)

UFFINGTON *3m. ENE of Shrewsbury.*
A linear village adjacent to the River Severn ½m. S off the B5062. To the W is 'yon bosky hill' – Haughmond Hill – at the base of which is Queen Eleanor's Bower, a prehistoric fort on a small concial foothill. The village has some character with a large, gaunt, empty farmhouse, some black and white cottages, the Corbet Arms pub, and the church of Holy Trinity built in 1856 to a design by S. P. Smith. Wilfrid Owen (1893-1918), the poet cut down whilst crossing the Sambre Canal at the end of the First World War, used to attend the church. He lived at 69, Monkmoor Road, Shrewsbury, but he and his family crossed the river by the rope-cable ferry on Sundays during the summer. The name Uffington means 'the settlement of Uffa's people'. (*See Haughmond.*)

UPPER AFFCOT *5m. S of Church Stretton on the A49.*
A main road hamlet with a substantial pub, the Traveller's Rest, and a restaurant, the White House. The Limes has pointed-arched windows on the ground floor, unusual in a house. On the road to Alcaston is a large, modern factory building noticeably not declaring its trade by means of a sign. It is, in fact, a hatchery where chicks are produced for poultry farms. All the males are killed. Where the lane crosses the Quinny Brook a track leads S to Affcot Manor Farm. The Manor House is an irregular building of brick with sash windows, a timber-framed gable, and a rendered facade painted pink. It is a friendly place, delightfully set amongst flowery lawns and an old orchard. A few yards S is the Byne Brook, and beyond that are the steep, wooded slopes of Wenlock Edge. The early medieval name of Affcot was Effechota, which probably means '*Aeffa*'s cottage'. The manor belonged to the church of St. Alkmund in Shrewsbury before 1066. St. Alkmund's was a royal foundation, probably made by Lady Aethelfleda of Mercia (daughter of King Alfred) who died in 975.

UPPINGTON *7m. ESE of Shrewsbury.*
The village is approached through a long straight avenue of mature trees down a lane off the A5. There are several large, ivy-clad houses, large red brick farms, a beautiful black and white cottage and a church with a Roman altar by the porch. The whole is well-wooded with some splendid horse chestnuts. It is altogether a charming place with the Wrekin, looming close by in the E, providing a grand setting. In the churchyard is a huge, ancient yew tree which measures 29 ft. around the girth; experts say it must be just over 1,000 years old. It is completely hollow and is currently being used to store planks. The church of Holy Trinity stands on a raised mound. It was heavily restored in 1885 but there is a Norman window in the N wall of the nave and a doorway with a dragon sculpture in the tympanum which is probably Anglo-Saxon. The Roman altar by the porch (1678) was dug up from the ground near the spot where it now stands. Opposite is a fine Tudor timber-framed house with good carving on the porch. Richard Allestree (1619-1681), the theologian, was born at Uppington and Goronwy Owen (1723-1769?), a highly respected Welsh language poet, was curate of the church from 1748 to 1753. On the A5, between Uppington and Atcham, are two small, thatched black and white cottages, one on either side of the road. Near to them, opposite the turn to Ironbridge, off the A5, is a stone-faced toll house. The name Uppington means 'the home of *Uppinga*'s tribe'.

UPTON CRESSET *4½m. W of Bridgnorth down lanes off the A458.*
Upton, an Anglo-Saxon name which means 'higher settlement', was given the suffix Cresset, the name of its Norman lord, after the Conquest to avoid confusion with other Uptons. Upton Cresset

consists of the former Hall, with its splendid brick gate-house, the church, a farm, and the site of a deserted village. The village was in decline as early as 1341 when its tax assessment was severely reduced because 'the tenants . . . have withdrawn because of penury' and the land lay uncultivated. By the mid-16th Century the little that was left was removed when the park of the Hall was extended. The earthworks can still be seen. In this century the Hall was empty and about to become ruinous but it has now been excellently restored. The church of St. Michael is a Norman building with some original parts, specifically the entrance doorway and the chancel windows. The N aisle was added in the 13th Century. It has a timber-framed turret belfry with a short spire and lies embowered in trees. The Hall was built about 1540, in brick which encased much of an earlier building to which belongs the fine 'great hall'. There is blue brick diaper work and rare mullioned and transomed windows of brick. The handsome, turreted and gabled Gatehouse was constructed 40 years later. The Hall was probably built by Thomas Cresset who, in 1517, was alleged to have emparked 40 acres of arable land at Upton without permission. When the Park was extended in the 16th Century, and the village was demolished, all but one of the tracks that led to neighbouring villages were abandoned and closed off. They exist today only as public footpaths. The only road to Upton Cresset is now the lane off the A448 east of Morville. After crossing the bridge over the stream keep straight on. The lane twists its way up a long, very steep and heavily-wooded hill called Meadowley Bank; drivers must find it a nightmare to negotiate in the ice and snow of winter.

UPTON MAGNA *3m. E of Shrewsbury.*
Up until the 18th Century it is believed that Welsh was the main language in Shrewsbury and indeed for about 3m. E, the line being drawn at Upton Magna. East of here English predominated. Haughmond Hill, 1m. NW, has many prehistoric earthworks. It is quite possible that Upton Magna was a lowland farm supplying the people of the Hill. There is evidence, mainly connected with old parish boundaries, to indicate that there was a large Romano-British agrarian estate centred on Upton Magna and Wroxeter which lasted well into the Anglo-Saxon period. This was later divided and in 1086 Upton Magna was a large manor in its own right. The Norman church of St. Lucy has a Perpendicular tower and was well restored and added to in 1856 by G. E. Street. Close to the church are several good timber-framed houses, including one of cruck construction. Behind the high brick wall lurks the Rectory. There is a pub, the Corbet Arms, a Post Office-cum-General Store, and a modern school. It is altogether a handsome village. In the late 17th Century there was an ironworks near to Upton Magna, one of several in the Tern valley. The others were at Tern Hall (which was later incorporated into Attingham, Hall) Withington, Wytheford and Moreton Corbet, and there may be others yet to be found. R. Chaplin suggests that between them they produced between 800 and 900 tons of iron a year. The Upton Magna works stood by the bridge over the Tern at **Upton Forge** ¾m. SSW of Upton Magna. The links for Thomas Telford's Menai Bridge were forged here. Today there is little left to see of the works but the water channels of the forge mill. A mundane concrete bridge crosses the river and there is a handful of brick cottages and an octagonal brick machine-gun bunker built during the war to defend the bridge. Tall pylons carrying mains electricity traverse the site with a noisesome hum. Despite such disadvantages Upton Forge is a pleasant enough spot. There is access to the river and when we were there we saw swans on the water and lolloping hares in the meadows. E. M. Almedingen (1898-1971), the prolific novelist, children's author and biographer, who fled Russia in the 1920's, settled in Shropshire. She had homes, at various times, at Worfield, Church Stretton and Frogmore (SJ.553.116), which lies between Upton Magna and Atcham. Frogmore is a typical small Victorian country house of brick with stone dressings which stands in charming gardens with lawns and bushes and mature trees. In fact this site is very ancient. Aerial photographs taken in 1974 show 2 rectangular marks in the crops in the field immediately to the S of Frogmore. They adjoin the road and almost touch the southern garden boundary hedge of the house. These marks have been interpreted as signs of Dark Age settlement, an aisled hall of the 7th Century.

VENNINGTON *1¼m. W of Westbury which is 9m. WSW of Shrewsbury.*
A pleasant little place with a stone-built windmill, now converted to a dwelling; cottages of stone, brick and half-timber; modern houses; a large 4-bay house of 2½ storeys; and a farm with black barns. The hamlet lies on a slope in undulating country. Half-a-mile W is Vron Gate, an even smaller little place. As a matter of interest there are very few settlement names beginning with the letter 'V' in Shropshire. Most gazetteers do not list any; here we have 2 adjacent to each other, surely not a coincidence.

WAGBEACH *2m. SSW of Minsterley on the A488.*
The main road winds gently uphill towards Hope. Alongside it stand the stone and brick cottages of Wagbeach, a hamlet with only one noteworthy building, namely **Hogstow** Mill. This is fronted by a timber-framed structure with red brick infill, but beyond are some most handsome and substantial stone-built ranges. Hogstow Hall itself lies 1m. S, down a track, and Hogstow itself lies further S again. The suffix 'beach' occurs in several place names in the area. No one seems to know for certain what it means. A beach, of course, normally means 'a pebbly or sandy shore'. But it can also be a corruption of *bec*, meaning a brook, or simply be a description meaning 'beach-like', that is, a level area of ground, or 'a shelf'. The hamlet called **The Waterwheel** lies ¼m. S of Wagbeach. Here the stream that carved out the Hope Valley runs in a steep-sided channel and is crossed by a bridge. The settlement consists of a handful of improved and extended cottages with terraced gardens and the ruins of the old watermill. The stone-built mill stood beside the most northerly of the dwellings, now an 'L' shaped, stuccoed house. The water was channelled along a leat that passes along the bank behind the house. On top of this bank is a tree-clad mound of indeterminate origin. On the road side of the stream are some ponds and irregular ground which presumably had a connection with the mill, but, as they lie above stream level, need some explanation. The Waterwheel is a strange little place but with its wide lawns, stream and bridge will, when some of the clutter is cleared away and the building works are completed, become a most attractive area. In the woods on the hill above the main road is a black timber shack barely worthy of the name The Waterwheel Institute.

WALCOT *7m. E of Shrewsbury on the B4394.*
A hamlet of red brick houses and farms in flat country. There is a substantial complex of weirs and races associated with an old mill and the confluence of the River Tern and the River Roden. Swans grace the waters and trees line the banks but for animals and children this could be a dangerous place. In the early Autumn large lorries trundle along the lanes carrying sugar beet to the large, gleaming processing factory at Allscott, 1m. to the NE. **Charlton**, a red brick hamlet with stands of pine trees lies ½m. S of Walcot. Amongst the traditional farm buildings are unlovely battery chicken sheds and close by are the remains of the castle of the Charlton family. Licence to crenellate the castle was given in 1317 by Edward II to John de Charlton, Lord of Powys. Today what can be seen is this: a rectangular platform about 73 yds. by 52 yds. and 7 ft. high from the bottom of the water-filled moat (dry only on the NW side); a small section of sandstone wall 9 ft. high and 3 ft. thick, probably a part of the Hall; buried foundations at the N and S corners and fallen fragments of the E corner of the curtain wall; and a causeway across the moat, adjacent to which would have been the gatehouse. The Wrekin looms large in this flat, wet country. As to the place names mentioned here Walcot means 'the cottage of the Celt' from *walh*, an Anglo-Saxon term of abuse, from which Welsh and Wales was derived (the Welsh call their country Cymru); Allscott means 'the cottage of Aldred' (the medieval name was Alderescote); and Charlton means 'the settlement of the free peasants (or villeins)'.

WALFORD *7m. NW of Shrewsbury on the B5067.*
The name is from the Old English *woelleford*, meaning 'the ford over the river'.

Today the Walford College of Agriculture has taken over this trim little village. The substantial main building is of brick and has 10 bays and a projecting porch. A little to the S, hidden from the road, is Walford Manor which is also a part of the college. It is large, of red brick with stone dressings, gables, mullioned and transomed windows, a stone porch, and for all that is singularly unimpressive. The country here is flat and the fields tend to be large. A mile to the SE is **Walford Heath**, a pleasant enough little place with a variety of red brick, sandstone, and white-painted cottages, a garden centre, 2 garages and a stuccoed Primitive Methodist Chapel of 1841 with round-headed windows opposite which stands the village pump.

WALTON *½m. N of High Ercall which is 8m. ENE of Shrewsbury.*
A hamlet of red brick and timber-framed cottages and farms dominated by several wartime aeroplane hangars now used as warehouses. The land is flat and sugar beet is an important crop hereabouts. The name Walton is from the Old English *Walh-tun*, 'the settlement of the Welsh (or the serfs)'. To call someone a *walh* was to abuse them. (The Welsh call their country Cymru; Wales is the English name for that country, though it has now lost its derrogative overtones.)

WALTON *1m. ENE of Onibury which is 3½m. SSE of Craven Arms.*
The name Walton can mean 'a settlement by either a stream, or in a wood, or by a wall, or of the *Wahls* (the British Celts)'. The settlement lies on a rise between 2 streams in undulating country. There is a tall stone-built barn with graceful iron columns supporting the roof; a red brick farmhouse; stone-built cottages; and arable fields with red soil **Vernolds Common**, 1m. NNE, is a scatter of stone and red brick cottages in similar undulating country. The lanes here are 'enclosure roads', quite wide with a narrow metalled track flanked by broad grass verges, ditches and hedges. **Norton**, 1m. WNW of Vernolds Common, consists of 2 substantial stone-built farmhouses and a red brick cottage dated at 1877. To the N is Wenlock Edge and to the NE the delights of Corvedale.

WAPPENSHALL *2m. NNE of Wellington.*
Wappenshall was called *Whatmundeshal* in the early 13th Century. The name probably means 'the secluded place where *Hwaetmund* lived'. The hamlet lies on the southern margins of the Weald Moors – the Wild Moors – now drained and a land of large, arable fields and rich pastures. The modern hamlet consists of a handful of red brick cottages and a farm, but a ¼m. NE is something rather more interesting. At **Wappenshall Junction** are the substantial remains of the warehouses, cottages, offices, stables, bridges and basin of what was once a thriving and very busy canal junction. The main warehouse, now used by a motor transport company, is a large building of some character built over a barge dock. The beautiful, carving white stone bridge adjoining it is quite superb. The bridge was about to be demolished in the early 1970's but the County Council moved in, carried out restorative work and scheduled it as an Ancient Monument. The Canal Basin has survived also, deep and overgrown but still partly water-filled. The great stone blocks of the Wharf walls are still in position and most of the other buildings have been saved from dereliction by being converted to dwellings. Wappenshall Junction was a transhipment point that handled coal, iron, bricks and general goods. It was here that in 1833 a branch of the Birmingham-Liverpool Junction Canal (authorised in 1826) from Norbury via Newport linked with the Shrewsbury Canal (completed in 1797). Such handsome and complete buildings must surely be preserved, not necessarily renovated or re-newed, but certainly protected. The ditch of the Shrewsbury Canal still exists and even carries water, a small stream flowing through tall rushes. There is still something to see of the Liverpool Junction Canal, though this has been largely filled in.

WATER'S UPTON *5m. N of Wellington on the A442.*
This attractive village, sometimes called Upton Parva, is now by-passed by the newly improved A422. The Water's in Water's Upton is a corruption of Walter, the Christian name of Walter Fitz-John, lord of the manor in 1200. At the centre of the village is a mound, without any doubt the centre of the Anglo-Saxon settlement. This has been cut through by a later road; to one side is the church, to the other a lawned area, formerly the village green, now the front garden of the Hall. The present rather ordinary red sandstone church was built in 1864 by G. E. Street on the site of an 11th Century chapel. The Hall was formerly the Manor House and parts of the old, cruck-framed house remain behind the Queen Anne facade of 1703 which was constructed by John Wase, a Shrewsbury haberdasher. There are some good farm buildings, a cider press in the cellar, and a 40ft. deep stone-lined well. Opposite the church is the old priest's house, where medieval monks from Shrewsbury stayed on Saturday nights to be available for the early Sunday morning mass. There are some new houses, both Council and privately owned and 2 main road pubs, The Lion and The Swan. The River Meese and the River Tern meet 1m. NNE of the village. Today there is a bridge but formerly the waters were crossed by stepping stones and the name Nobridge survives. The Tern flows ½m. W of the village but is now more like a deep ditch having been artificially lowered as part of the Weald Moors drainage scheme. Parallel with the river runs the line of the now dismantled Market Drayton-Wellington railway line which crosses the low lands hereabouts on a raised embankment. Meeson Hall lies 1m. NE of Water's Upton; it is a 3-gabled, red sandstone house dated at 1640.

WATTLESBOROUGH HEATH *5m. W of Ford and 11m. W of Shrewsbury on the A458.*
To the E of the settlement a dead end track leads northwards off the main road for ½m. to Wattlesborough Hall (SJ.355.127), a 5-bay, stone-built Georgian farmhouse. Adjoining it is the impressive 50 ft. high square keep of a Norman castle built by the Corbets in about 1200. It was one of a chain of strongholds built by the Marcher-Lords to contain the Welsh. The castle at Wattlesborough has a later medieval wing built by the Leighton family who lived here until about 1711 when Sir Edward Leighton moved to Loton Hall. In the main road village there is a brick-built church dedicated to St. Margaret of Scotland (1931), a Methodist church (1893), a Church of England school (1871), a village hall (1920), a garage, and cottages of red brick, stucco and stone. The name Wattlesborough is thought to be Old English, derived from *Wacol's-burg*, meaning '*Wacol*'s homestead by the old fort'.

WELLINGTON *9m. E of Shrewsbury on the A5.*
The name is from the Old English and probably means 'the settlement by the temple-grove'. Wellington is an old market town on the eastern edge of the Shropshire Coalfield. Since ancient times it has been a centre of communications. Here the road from Chester, via Whitchurch to Bridgnorth and down to Worcester, crosses the old Roman road of Watling St., now the A5. The main line railway from Shrewsbury to Wolverhampton and Birmingham passes through the town which has one of only 3 stations serving the whole of the Shropshire coalfield. In the past several branch lines also terminated here. The town also had good access to the canal system. Today Wellington is a reluctant consituent member of Telford New Town. Unlike most of the other towns in this 30 square-mile-sprawl Wellington has always prospered and feels she has nothing to gain but everything to lose by the association. A factor in the development of towns that is often ignored is 'ambience'. Some places, like some houses, are friendly and appealing. Others, for whatever reason, are not. That is not to say that Wellington has anything particularly special to offer. It has few buildings of interest, no historical attractions or connections of any import, and the countryside around is either reclaimed moorland or scarred with pit heaps. However, the town centre was always very pleasant and is even more so now that it has been pedestrianised. The streets are narrow and twisting, and there are some shops with traditional facades and a handful of timber-framed black and white buildings. Wellington has 2 bustling market days and, all in all, feels more like a country market town than an industrial centre. For industry a plenty there is in the area about: Brewing, agricultural machinery production, furniture making and sugar beet processing etc. In the past it had a flourishing bell making foundry, nailmakers and glass factories. As one

Old windmill, now a house, at Vennington.

The elegant stone canal bridge at Wappenshall.

Wellington Town centre.

industry died another took its place. There are two Anglican churches. All Saints is a smoke blackened building which many think quite highly of but which looks out of character with its surroundings. In a central London Square it would look fine. The architect was George Steuart and it was built in 1790. Christ Church is hidden in the suburbs, not too far from the centre, but unless you are a native almost impossible to find. The tall tower appears and disappears as the motorist explores a multitude of cul-de-sacs before finding that the entrance road leads off the A5. Christ Church was built of yellow brick in 1838 by Thomas Smith of Madeley. Back on the A5, the Holyhead road, is the disappointing Old Hall, an average though early timber-framed building, the lower part of which is hidden by a brick wall and the upper part spoiled by a large red sign advertising that the property is now a preparatory school. It was built in the 15th Century as a house for the Keepers of the Wrekin. Wellington has 2 famous sons, or rather a son and a daughter. Sarah Smith was born to a local postmaster. She adopted the pseudonym of Hesba Stretton and wrote a series of best selling novels. Her first novel, 'Jessica's First Prayer' was published in 1866. It sold a phenomenal 1½ million copies, and that in the days before cheap paperbacks. Very, very few books today sell that number in hardback. It was even translated and used as a set work in Russian Schools, apparently on the orders of the Czar. Sarah Smith used her wealth to travel and do good works. One of which was the foundation of the London Society for Prevention of Cruelty to Children, which was later expanded to become the NSPCC. William Withering was an 18th Century scientist of high standing; he was one of the first to speak out against the slave trade. Philip Larkin (1922-1985), one of Britain's leading modern poets, was librarian at Wellington Library for 3 years. **Dothill** is now a suburb of Wellington lying ¾m. NNW of the town centre. Here are a mixed bag of 20th Century residential developments with a small shopping centre linked by a little inner ring road called Severn Drive. This almost encircles the green acres and wood fringed pool of Dothill Park. There was once a big house here, and the estate had Norman roots. The remains of a moat lie just N of Dothill County School. West of the park, near Morville Drive, is The Lake. There are 2 grassy hillocks, one on each side of the pool, which were raised in 1734 probably as stands for spectators to watch water jousting and such like frivolities. Two and a half miles SW of Wellington is the summit of the legend-riddled **Wrekin**. The lower slopes of its foothill, the Ercall, come within ½m. of the town. The Wrekin was the densely populated capital of the Cornovii, the Celtic tribe who ruled the Welsh borderlands from Chester in the N to Hereford in the S. They were subjugated by the Romans and later moved their capital to Wroxeter (Uriconeum). When the Romans left, 350 years later, many of the Cornovii abandoned the city and went back to the Wrekin. Wellington, lying at the start of the easiest access route to the Wrekin, may well have been first settled as a farm to supply the hill town. Today the M54 slices between them, though one direct road, Ercall Road, between the town and the hill has been kept open by means of a bridge over the motorway. The Wrekin is a landmark for much of the county. It is not as high as some other hills in Shropshire, but because it rises alone out of a flat plain is the most distinctive single natural feature, not just in the county but in the whole of the West Midlands. Its NW flank is composed of hard, very old Pre-Cambrian rocks and the SE flank is composed of younger Cambrian rocks. (The Long Mynd and the Caradoc Hills of Church Stretton are also composed of Pre-Cambrian rocks. The Stiperstones are Ordovician Quartzites.) The summit can be reached quite easily on foot by way of the gentle slope from the NE, off the road running between the Ercall and the Wrekin itself. The path runs up the shoulder of the hill, through a prehistoric earthwork at Hell Gate and through the narrower Heaven Gate. The rocky outcrop of the Bladder Stone with its Raven's Bowl is near the summit to the E, and the narrow cleft of Needle's Eye is just to the S. One legend has it that the Wrekin was formed as the spoil heap when 2 giants dug out the bed of the River Severn. The story goes on to say that the 2 giants fell out and during the quarrel one giant threw his spade at the other, but missed him. The spade struck the hill and made a gash – the Needle's Eye. They continued to fight until one killed the other. A raven then swooped down and pecked at the victor's eyes which made them water and a huge tear fell on to the rocks of the Bladder Stone causing the bowl-shaped depression called the Raven's Bowl. There are splendid views in all directions from the summit. (*See Cluddley*.) Orleton lies 1m. W of Wellington. (*See Wrockwardine*.) Arleston Manor lies 1m. SW of Wellington. (*See Arleston*.)

WELSHAMPTON *2¼m. E of Ellesmere on the A495.*

A main road village with a pub, the Sun Inn, and 2 garages close to the mere country. Its name bears out modern research, which is providing reasons to believe that up to the 18th Century Welsh was the first language in areas far further E than was previously thought. Welshampton is about 11m. from the Welsh border. For such a small village a surprising number of lanes converge here. In all likelihood it was more important in the past than it is today. The present church was built in 1863 by Sir Gilbert Scott and is mock 13th Century Gothic with a patterned roof. A touching curiosity is the grave of Moshueshue, the son of an African chief who died at the age of 24 and was buried here in 1863. His father was the King of Basutoland and Moshueshue was here to study at the theological college in Canterbury. At the time of his death he was staying at the vicarage at Welshampton. That year the new church was completed and a stained glass window showing Moshueshue in a white robe being baptised in a river was incorporated into a chancel window.

WELSH FRANKTON *2½m. WSW of Ellesmere on the A495.*

Welsh Frankton, sometimes simply called Frankton, is a main road village of red brick houses on the A495 between Ellesmere and Oswestry. There is a garage, a Methodist Chapel and the rock-faced Anglican Church of St. Andrew. To the E of the village, on the main road, there is a cottage which has a small private collection of articles relating to local country life. There is no charge for admission. We were told that even in this century a man had to be able to play cricket reasonably well if he was to stand much of a chance of gaining employment on local estates. Inter-estate cricket matches were matters of importance. The Frankton village cricket ground is, and always has been, on a meadow in the grounds of the nearby Hardwick Hall. To this day the local squire, a Colonel, is keenly interested in the team which until recently he captained. Hardwick Hall was built in brick for John Kynaston in 1693. The forward projecting wings were added circa 1720, and are not an ideal complement to the house. Later again a terrace was built to the front. The Hardwick estate is 1m. E of Welsh Frankton and 1m. W of Ellesmere, and adjoins the N flank of the main road (the A495). Opposite the church at Welsh Frankton is a lane that leads SSE to **Lower Frankton**. This was an important junction of 2 branches of the Shropshire Union Canal. There are 4 locks which have recently been restored. They lift the southern branch of the canal by 30 feet to the level of the main canal but the southern waterway itself still needs to be restored. It has been dry since it burst its banks in 1936. There was once a flourishing community here. The white house on the W bank was once a pub and there used to be a boat builder's yard and dry dock by the bottom lock. Some of the cottages have been made into 'desireable residences' whilst at least one stands derelict. There are shady places to park the car and verdant tow paths on which to walk the dog. When we were there the local wild ducks were waging war upon one another. Altogether there are about a dozen cottages, a ruined red brick chapel and farms in meadows bedecked with buttercups and daisies. The name Frankton is from the Old English *Franka's-tun*, Franka being a personal name and *tun* meaning 'a homestead, or a settlement'.

WEM *10m. N of Shrewsbury.*

Wem is a small country town, with a full range of shops, banks and professional services. Seven roads meet here, there is a main line railway station and the town lies on the banks of the River Roden which in the past has provided power for its mills. The name is problematical. It could be from the Old English '*wamm*', meaning 'a stain' hence marshy ground; or, from *wenn* meaning 'a swelling', 'a wart' and hence a hill; or, from *hwen*, or *hwemm*, meaning 'a corner'. Whatever the old

Welsh Frankton, a restored canal lock.

name may mean the town we see today is descended from the Norman 'plantation' made when the castle was built. This was a substantial stone-built structure and the mound and some of the masonry still stand next to the church. The castle was the stronghold of William Pandulph, Baron of Wem, a major landowner in the county. The name of Pandulph died in 1233 when the heiress, Matilda Pandulph, married Ralph Botiler. The castle was destroyed largely by 'quarrying', the taking of the stone to use in the construction of new buildings, walls, bridges and roadways. In medieval times the settlement had town walls. The country around Wem was still heavily forested at the beginning of the 16th Century. Between 1550 and 1650 this was cleared to provide fuel, firstly for the glass industry and then for the iron furnaces. The cleared land was then used to grow hay and corn. The marshes and mosses around the town were exploited for their peat, the cutting of which began about 1560, and later, after drainage, used as agricultural land. In North Shropshire there were an unusually large number of moated farm houses. Many have survived but some have lost their moats. The moating of a house was usually only necessary if it was a single isolated farm some distance away from the village. Settlement of this kind is typically early medieval. Northwood and Highfield are moated houses near Wem. In the 18th Century the Enclosure of the common lands and the increased use of improved methods used by landowners on their estates led to a reduction in the number of men needed on the land. Villages were reduced to hamlets and hamlets to single farms. Examples of the latter are the farms at Lowe, Trench and Newton, all once hamlets. Despite agricultural improvement the roads remained virtually impassable in winter. As late as 1762 John Wesley on a journey from Shrewsbury to Wem had to turn back when his chaise became stuck in the mud. During the 17th Century Wem market began to specialize in livestock, as did those at Newport and Ellesmere. Market towns are usually well endowed with inns, and one of the local breweries that supplied those of Wem, Wem Breweries, went on to become a substantial enterprise serving a wide area. Sadly it was forced to close a few years ago but the old buildings have been preserved. During the Civil War Wem took the side of Parliament and repulsed a Royalist attack led by Lord Capel; the leading Puritan, Richard Baxter, took part in the defence. In 1677 a great fire destroyed much of the old town and the timber-framed buildings were replaced in brick. Lowe Hall was the home of the notorious Judge Jeffreys. In 1685 he bought the barony of Wem but he only enjoyed the title for 2 years; he died in 1688. People of note who were born in Wem include John Ireland, the biographer of William Hogarth; and John Astley, portrait painter and friend of Joshua Reynolds. William Henry Betty, the child actor, lived here for a time as did the essayist, William Hazlitt (in Noble St., where his house still stands). The sweet pea was developed by Henry Eckford who settled at Wem in the 1880's. In 1988 the National Sweet Pea Society held a celebration in the town. The Grammar School was founded in 1650 by Sir Thomas Adams, a Lord Mayor of London but a native of Wem. The church of St. Peter and St. Paul has an early 14th Century doorway of unusual design; the rest has been rebuilt or altered at various times between 1667 and 1886. There is a former market hall with columns and arcades; a red brick Town Hall of 1905; a Cheese Hall of 1928; Wem Hall, a good Georgian red brick house of 5 bays with an Ionic porch; and next to it a fine black and white house. The town is not a bustling place despite the presence of the North Shropshire District Council in their large, new premises. Around the town are some good houses: Tilley Hall, ¾m. S is a modest but attractive black and white building; The Ditches, 1½m. NE, is a handsome half-timbered house of 1612; Lowe Hall, 1½m. NNW, has a 19th Century brick facade to a house of 1666 and inside there is a fine Jacobean staircase; and Northwood Hall, 2m. NNW, is a small early 18th Century brick house with 5 bays, 2 storeys and a hipped roof. On the outskirts of Wem on the road to Ellesmere is Horton. No.2 Horton Villas is a delightful and very ancient little black and white cottage with bright green window shutters. The houses of humble folk so rarely survive.

WENTNOR *6m. NE of Bishop's Castle.*
The village lies high on a ridge between the River East Onny and the Criftin Brook. To the W is Norbury Hill and to the E is the Long Mynd on which villagers have the right to cut peat. It must be bitter here in the winter. The stone houses and farms look frozen to their bones even in the summer. If the T.V. aerials and cars were removed this could become a medieval hill town overnight. There are wide views in all directions. The Norman church of St. Michael was rebuilt in 1885 by Henry Curzon. He re-used much of the original Norman masonry and the Perpendicular roof. The pulpit is Jacobean. There is a pub and a Post Office-General Store and to the W is a campsite with its own facilities. A field called Black Graves lies off Adstone Lane; it is believed to be a Black Death burial site. The name Wentnor is probably derived from the Old English *Want-ofer* meaning '*Want*'s settlement on the slope'. At the bottom of the hill, to the W, is the delightful, wooded hamlet called **Walkmill**. The fulling mill, where urine soaked wool was literally walked on to remove excess oils and to flatten the fibres, has become ruinous but several of the stone built cottages remain on the banks of the River East Onny.

WESTBURY *9m. WSW of Shrewsbury on the B4386.*
The name is Anglo-Saxon and means 'western fort'. Today Westbury consists of little more than a garage, some modern red brick houses, a handful of stone-built cottages, a school, a Post Office and a pub. The main line railway passes by the village 1m. to the NW. There is a level crossing and a medium sized Station Hotel but there is no longer a station and the Hotel is now a house. There used to be a cattle market, held in a field at the back of the Hotel, and there was a brick works nearby. This must once have been a busy little place. However, the main attraction here lies 1½m. to the SW, where **Caus** Castle Farm marks the hill on which once stood a stone-built Norman castle and a village surrounded by massive defences. The earthworks can still be seen from the road, but the best view is from the track to Knapps Farm on the other side of the valley. Only a little of the masonry remains. Caux is a district of Normandy, the home of Roger Fitz Corbet who built Caus Castle. The fortress is situated at the eastern end of Long Mountain and commands the road from Shrewsbury to Montgomery. The town of Caus was created deliberately, that is it was 'planted' by Robert Corbet in 1198. By 1349 there were 58 burgesses and their families living here. Throughout the 13th Century Thomas Corbet of Caus was engaged in almost continuous warfare with Griffith ap Gwenwynwyn. Owen Glendower burned the town down and as a consequence the settlement was excused taxation in 1405-6. (Glendower also burned down the villages of Marsh, Vennington, Westbury and Yockleton.) The whole area around Caus was really Welsh. The castle was an English island in a Welsh sea. Even the taxes collected by the Lord of Caus were of the Welsh kind and special courts were held that judged according to Welsh Law. In the mid-15th Century the town was burned again, during the rebellion of Sir Griffith Vaughan, and in 1521 it was described as being in great ruin and decay. As times became more peaceful the castle and the town ceased to have a purpose. The last occupants of Caus were the Thynne family who moved to Minsterley when the castle was finally dismantled during the Civil War. The hamlet of Marche (or Marshe) lies 1¼m. NW of Westbury. The Manor House is a most attractive timber-framed building of 1604 with decorative lozenges in the gables. (*See Marche.*) Whitton Hall, a most attractive and beautifully situated large Georgian house lies ½m. W of Westbury. (*See Whitton.*)

WEST FELTON *4m. SE of Oswestry on the A5.*
The old village developed around the Norman castle in a rough horse-shoe shape, but the A5 has attracted the more recent development. The church of St. Michael is 12th Century with a Norman chancel arcade, a Georgian tower, Victorian restoration and a 20th Century window by William Kempe. Just to the W of the church is the mound and moat of the castle. John Freeman Dovaston was born at 'The Nursery' in 1782. He was a musician, poet and naturalist. Amongst his friends was the great engraver, Thomas Bewick. In the village there are 2 pubs (beside the Punch Bowl Inn an old wool packer's stone can be seen); a school; a Post Office; a military museum; and developments of expensive modern

St. Winifred's Well, near West Felton.

houses. The name Felton is from the Old English *feld-ton*, meaning 'a settlement on a fell', that is, open, moorland country with few or no trees. Just over 1m. SE is Tedsmore Hall, an 18th Century house in brick with a Park. One mile to the SE of West Felton, on the road to Woolstan and just past a double sharp bend, is one of the delights of Shropshire. **St. Winifred's Well** lies at the end of a short footpath which commences at a house called Fir Tree Villa. Against the bank and above the spring there is a small black and white, timber-framed cottage. The waters run into a series of grey-green stone basins in which afflicted persons bathed in the hope of gaining a cure. The cottage and the well are in a small wooded dingle with a grassy glade, flowers and greenery drooping from the trees. On a summer's day it is truly delightful and should not be missed. The well is very ancient and was almost certainly a pre-Christian holy well before its dedication to St. Winifred. This dedication arose because the body of the dead saint was rested here on its journey from North Wales to Shrewsbury Abbey. The waters of the well are reputed to be especially effective in the treatment of cuts, bruises and broken bones; those of a small spring below the main well are said to aid in the cure of eye complaints. The cottage above St. Winifred's Well was built circa 1600 as the court-house of the manor and used as such until about 1824. By 1883 it was occupied as a dwelling. In the 18th Century feasting and drinking took place around the well from the middle of May to the Harvest Home. There were as many as 5 temporary ale-houses in booths. These frolicks ceased in about 1755.

WESTHOPE *3¼m. NE of Craven Arms.*
It lies in Hope Dale on Wenlock Edge, a scattered village of cottages and farms with a church, a much used meeting room and an old moated site. The local landowners are the Swynnerton Dyer family. The little church was built about 1650 and stands in a daffodil filled orchard. The yew tree by the church entrance is reputed to be some 800 years old. Westhope Manor was constructed in 1901 to a neo-Tudor design by Sir Guy Dawber. It is a handsome, gabled house clad in roughcast render and tiles. Westhope used to be in Diddlebury parish but is now a parish in its own right. The last Anglo-Saxon to hold the manor was Almund. In 1086 it was held by Picot, the nickname of Robert de Say (from Sai, in the Department of Orne, France). There were 4 villagers and 6 slaves.

WESTON LULLINGFIELDS *1½m. N of Baschurch, which is 8m. NE of Shrewsbury.*
An old village lying on the edge of the vast acres of Baggy Moor, once a morass but now drained and under grass. The drainage works were some of the last in the county, not completed until after 1861 when an Act of Parliament allowed the improvement scheme to go ahead. An arm of the Ellesmere canal from Frankton ended at a wharf in Weston Lullingfields. The canal ditch is still there but the basin has been commandeered by a local farmer who now stores enormous numbers of large, black plastic bags full of hay in a makeshift shelter. The canal came here to service the lime kilns, to bring in coal and to take away the burnt lime which was used for agricultural, industrial and building purposes. It was originally intended to take the canal on to Shrewsbury but it was diverted via Ellesmere and Whitchurch to join the Shropshire Union at Hurleston Junction. At Weston Lullingfields the canal company built "a wharf, four lime kilns, a public house, stables, a clerk's house and weighing machine". These were opened in 1797 and operated until 1917 when the canal burst. The company had been in financial difficulties for some time and this was the last straw. The operation was abandoned but most of the buildings still stand. The village has a meeting hall but the shop and Post Office closed in the 1970's. The church and vicarage, both of 1857, lie amongst trees away from the village. This is, presumably, the old centre of the settlement. Two miles N, at **Stanwardine in the Wood**, is Stanwardine Hall, an Elizabethan mansion of about 1560, constructed in brick with stone facings to the front. It is irregular and was not all built at the same time. The house is believed to have been raised for Robert Corbet, brother of Sir Andrew Corbet of Moreton Corbet (d.1578), and enlarged by Robert Corbet's son. Adjacent to the house is a circular moat, probably the site of the first settlement here. The name of the hamlet, Stanwardine in the Wood, and the moat indicate that this was an isolated medieval farm planted in a clearing in virgin forest. Just over 1m. NE is **Petton** Hall. It also has a now abandoned moat, just N of Petton Grange Farm. A grange was a monastic farm on the edge of cultivation. This area was a desolate place until comparatively late. The Baggy Moors have been mentioned, but to the E the country was also waterlogged and heavily wooded. Names such as The Wood, Woodgate, Boulton Grange and Brandwood House indicate that this was frontier territory at a late date. Two-thirds of a mile SW of Weston Lullingfield is **Stanwardine in the Fields**, a hamlet with a few brick cottages, a black and white farm with a weather-boarded barn, a Primitive Methodist Chapel of 1869, some walling of massive red sandstone blocks, and a bridge over the railway line from Gobowen to Shrewsbury.

WESTON RHYN *4m. N of Oswestry.*
A small village with the many attractions of the Ceiriog Valley and Chirk Castle a little to the N, but with few of its own. The Quinta is in disappointing Gothic of 1850-60 with many and various modern satellite buildings attached and a model of Stonehenge on a rise in the NE of the grounds. It was built for Thomas Barnes, the Chairman of the Lancashire and Yorkshire Railway and a cotton mill owner. The house is approached by curving driveways designed to make the grounds appear larger than they actually are. In Victorian times the property had its own fire station and gas making plant. Today the Quinta is a Christian Study Centre. ('Quin' in a place name is usually a corruption of Queen.) In High Street is the Victorian Sunday School, turreted and elaborately decorated with ceramics, a gift of Mr. Barnes. There is a bakery, a Post Office and general stores, a newsagents, a fish and chip shop a hairdressers, several pubs and a choir. Moreton Hall, 1m. E of the village, is a 7-bay brick house of the 17th Century with a hipped roof on carved brackets. The church of St. John the Divine is of 1878. One mile NW is **Pentre-newydd** to the W of which are some good sections of Offa's Dyke. Adjoining Pentre-newydd is Bronygarth where there are well-preserved lime-kilns on the hill alongside the road. Tyn-y-Rhos Hall lies ¾m. W of Weston Rhyn. It is an irregular, gabled Tudor house parts of which are even older. In 1165 Owain Gwynedd, Prince of Wales, is said to have stayed here. The exterior facade is covered in rough, white stucco and the carved barge boards are broken in places; a care-worn place in dignified decline. In the Summer the house is open to the public at least one of whom, a German gentleman, has seen the ghost of the son of a wealthy industrialist who died on the day-bed in the dining room during the First World War. There are several other ghosts in occupation and the clairvoyant Gipsy Smith said that the property was the most haunted house this side of the Welsh border.

WESTON UNDER REDCASTLE *3½m. E of Wem.*
The Anglo-Saxon village of Weston had its qualification added by the Normans. This refers to the 12th-13th Century red sandstone castle of Hawkstone ½m. NE. Before the Norman Conquest Weston was held by Wild Edric, the Saxon lord who fought the Normans and who has since entered into the realms of myth and legend. After the Conquest the lands of Shropshire were redistributed by Earl Roger de Montgomery and he gave Weston to Ranulf Peverel. In time it passed to the Audley family and it was they who are believed to have built the castle on the 'Redcliffe'. The church of St. Luke is of 1791 in Gothic style except for the Georgian tower. The chancel is of 1879. In the churchyard are the village stocks. Weston is a most attractive little place with a genuinely old village green (a rarity in Shropshire) and dwellings built in a variety of styles. There is a rough track that leads from the Lodge, past the scanty remains of the Red Castle and through a short tunnel to Hawkstone Hall, over a mile away to the NE and hidden from view by the ridge. Beware: if you find the gate open and drive in it may well be locked when you return. The sights of Hawkstone can be viewed on foot, having parked in the grounds of the Golf Club. The Golf Club has taken over the Hawkstone Park Hotel but it remains open to the public. There are two 18 hole courses here. The hostelry was built in the late 18th Century to cater for the many hun-

The High Street, Whitchurch.

opposite, left: Wooden sculpture on house at Stanley Green Whixall.

below: A Bangalore Bomber bird scarer gas gun near Willaston.

Winsley Hall, near Westbury.

dreds, if not thousands, of tourists who came to marvel at the attractions, both natural and man-made, that were famous throughout the country. Only a few remain but it is still a splendid spot. The hotel has a 7-bay Georgian facade, which has been rendered, and one-bay wings with Venetian windows. *(See Hawkstone.)* Sandy Lyle, the international golf celebrity, learned his craft on the golf courses at Hawkstone and until recently his home was in the village. Note: There is another castle at Weston, the 15 ft. high mound of an early Norman castle on the hill just W of the golf club car park. Wixall lies just to the SW of Weston. *(See Wixall.)*

WHEATHILL *3m. S of Burwarton, which is 8m. SW of Bridgnorth on the B4364.*
A tiny hamlet amongst small hills beneath the southern summit of Brown Clee Hill. The Norman church was much restored in Victorian times and all that remains of the original is the entrance doorway and tympanum, the chancel arch and a chancel window. There is a good 17th Century roof. Three quarters of a mile W is a Youth Hostel. As to the name: Wheathill probably means what it says, 'the settlement on (or by) the hill on which wheat is grown'. The last Anglo-Saxon holder of the manor was Almund. In 1086 it was held by Roger Lacy who as part of his rent to Earl Roger paid one hawk. Lacy had 10 slaves to work his land here, an unusually high number.

WHITCHURCH *19m. NNE Shrewsbury.*
Whitchurch, meaning White Church, has been identified as Mediolanum. This Roman town is something of a mystery. It was mentioned in ancient writings but never given an exact location; the authors assumed that their readers knew its whereabouts. However, few places have such good claim as Whitchurch. Certainly there was a Roman fort here in the First Century AD, on the road from Uriconeum (Wroxeter) to Chester. Later, a small town developed around the fort. In the countryside nearby a number of Romano-British villa sites have been found. The modern High Street of Whitchurch follows the Old Roman road and the basic plan of the town is Roman. Whitchurch lies in the moss country. Glacial clays, impervious to water, have caused the land to become waterlogged. Not until the 16th Century did drainage work commence. The land is now mostly pasture, with some arable farming on the better drained sands and gravels which usually occur on slight rises. The soil is very peaty and has, indeed, been cut for fuel. Bronze Age artefacts and remains have been recovered from the peat bogs. The whole area was originally covered in a dense woodland with bog oaks, alder, birch and thickets of furze. Before being drained it could only be used as very rough grazing for a few weeks in the summer. Today Whitchurch is a busy and attractive market town, not seen by most motorists who are, quite rightly, sent on a detour of the town centre. It has good shops, pubs and all the usual services of a regional social and business centre. The sandstone church of St. Alkmund (1712) sits atop the hill. Higginson's Almshouses, in Georgian brick and the 19th Century neo-Elizabethan Old Grammar School lie on the downside of the hill to the N, and the High Street shops on the downside to the S. The church site is obviously old and the modern building replaces the previous, white limestone church that simply collapsed one night after evensong in the year 1711. Inside it is most impressive. There is a monument to John Talbot, 1st Earl of Shrewsbury and 'The Scourge of France' of Shakespeare's Henry VI, who was born hereabouts. He died in his 80's at Chatillon near Bordeaux in 1453 but his bones and heart lie here. (His full titles were: Marshal of France, Earl of Shrewsbury, Lord Lieutenant of Ireland, Earl of Waterford and Wexford and Lord Lieutenant of Aquitaine.) During the restoration of the church in 1874 the tomb of the Earl was opened and a church mouse was found to have made its nest in his skull. His heart is buried in a silver chalice in the church porch. One of the rectors of the church was Francis Egerton. He became the 8th and last Earl of Bridgewater and spent the last years of his life in Paris. He was an eccentric who dressed his pet dogs in formal attire and had his servants wait upon them at dinner. Whitchurch is a cosy little town and though not known for its architecture there are some buildings which, in addition to those already mentioned, should be noted. The National Westminster Bank in High Street, is a timber-framed former inn which has the old church doors; Barclay's Bank, which now occupies the old Town Hall; the Mansion House. probably the best town house in the best street, namely Dodington; and the fine Congregational church 1846 in Clasical Greek style. Whitchurch is surrounded by dairy farms and was once famous for its cheese: St. Ivel have a creamery about 1m. out of town on the road to Ossmere. Perhaps the best known manufacturer in the town, however, so is J.B.Joyce & Co. who are one of the few turret-clock makers in Britain. Sir Edward German (real name E.G.Jones) was born at a pub near the Trustee Savings Bank. His most famous work was the operetta 'Merrie England'. Randolph Caldecott, the illustrator of children's nursery rhymes was employed as a clerk at the Whitchurch and Ellesmere Bank. Dearnford Hall (of about 1840) lies 1m. out of the town. It is constructed of brick and looks a bit like a civic building. On the western fringe of Whitchurch, on the A525, is the residential suburb of **Chemistry**. This unusual place name may well be derived from a personal name the root of which is 'Chem' and 'tree' which can mean 'cross' in the same way that Oswestry means 'Oswald's Tree (or cross)'. At Blake Mere, 1m. NW of the town, in Black Park. On the S side of the lake is a moated mound. This marks the site of the manor house of John le Strange. His daughter, Ankaret, was the heiress to his titles and fortune. She married the great John Talbot (b.about 1390) already mentioned. The name Blake Mere is derived from *blake* which is Old English for 'bleached' or 'white', and *mere* means 'lake' – hence the White Lake. This gives an alternative derivation of the name Whitchurch, normally taken to mean White Church because the old church was made of limestone; it could, in fact, mean 'the church by the white lake'. The Black Park countryside is most attractive. There are many small hills and dales mostly laid to pasture grazed by cattle, and many small lakes. As well as Blake Mere there is Oss Mere and just over the border, in Cheshire, are numerous pools including Comber Mere on the shores of which stands the famous Combermere Abbey (3m. ENE of Whitchurch). The Abbot of this monastery was a major medieval landowner. Many of the pools are used as sport fisheries. One mile SW of Whitchurch is Pan Castle. Only the earthworks remain. A natural bank projects into a marsh (now drained) and on this bank a mound was raised. This is 15 feet high and covers a large area. To the SW is a bailey protected by a ditch and also in places by an earthen wall. It is presumed to be of Norman origin. **Hinton**, 1m. NNW of Whitchurch, is not a village but there is an Old Hall of red brick with stone quoins and a row of timber-framed cottages set on a slope. On the other side of the main road, the A49, a little further N and set on a wooded knoll is the new Hinton Hall, also in red brick and built to a design by the architect Pountney Smith. The Llangollen branch of the Shropshire Union Canal passes by the Hall just to the W and here forms the border with Cheshire for a short distance.

WHITCOTT KEYSETT *See Cefn Einion.*

WHITTINGTON *2¼m. NE of Oswestry.*
Whittington is a sizeable village on the A5, Holyhead Road, and its picturesque castle, complete with water-filled moat, is a familiar sight to many. It looks like a stage set but is real enough. There has been a castle here since 845 when Ynyr ap Cadfarch built the first fort of which we have definite knowledge. This was captured by Roger de Montgomery and later passed to Sir William Peverel. The substantial stone castle the remains of which we see today was built by Fulke Fitz Warin about 1221 in the reign of Henry III. The Anglo-Norman prose romance 'Foulques Fitz-Warin' relates the struggles of this Marcher-Lord family. The Gatehouse of the castle has 2 large, circular towers and between them is the entrance arch. The keep stood on a mound 30ft. high and was an average 125ft. in diameter at the top. There were at least 4 more towers and 3 additional raised platforms. To the S and W was a system of defensive ditches, and to the N and E a marsh provided protection. The present linear shape of the village was established when the Normans 'planted' the town. It is strung out along the A495 to the S of the castle. The irregular group of buildings opposite the fortress and behind the church may well be the site of the original Anglo-Saxon village. There is surprisingly little linear growth along the A5. The

Whitchurch, Oss Mere.

church of St. John the Baptist stands on the site of a medieval chapel. It has a tower of 1747, a nave of 1804 and was much restored in 1894. The E window is by William Morris and Burne-Jones. A former vicar here was William Walsham How who became the Bishop of Wakefield. He wrote many hymns including "Summer Suns are Glowing", "Soldiers of the Cross, Arise" and "For All the Saints who from their Labours Rest". Facing the tree-clad village green is the Big House, a timber-framed dwelling of 1631 with a right-hand extension of 1924. Just over the road is a row of Tudor-style bungalows erected in 1986. In the grounds of the old Park Hall estate are a sports complex and a concealed gipsy camping ground. The name Whittington is probably from the Old English personal name *Hwita* and means 'the settlement of *Hwita*'s people'. **Halston** Hall lies 1m. W of Whittington. Here lived one of Shropshire's best known characters, Mad Jack Mytton. He devoted his life to squandering the family fortune in gambling and general loose living. In between times he was a Member of Parliament and a hunter of foxes. He died in 1835, aged 38, whilst being held as a debtor at King's Bench Prison, London. The red brick Hall was built in 1690 with additions in 1766. The park is an agricultural estate and in a field about ¼m. to the rear of the Hall is one of only 2 timber-framed churches in the county. It is raised up on a mound and surrounded by yew trees in flat country. The tower is of brick and the framing of the nave-chancel is post-and-pan. It has a gallery and clear glass windows with outside opening shutters. The body of the church is probably of the early 16th Century. All the furnishings are period pieces – panelled walls, box pews, christening pew, altar rail and two-decker pulpit etc., mostly belonging to the early 17th Century, that is, pre-Reformation. In the last century there lived in a cottage near the Rectory a witch called Kitty Williams who had the power to cause horses to go lame and cattle to die. One mile SSE of Whittington is the village of **Babbinswood**. This is said to have originally been called Babies' Wood and is reputed to stand on the site where the Babes in the Wood died.

WHITTON *5m. SE of Ludlow.*
This is red sandstone country, a land of little hills and many coloured fields cut through by winding, high-banked lanes. Whitton lies on the lower, SE, slopes of Titterstone Clee Hill, 1m. S of Knowbury. The houses of the village are a mixture of black and white, stone and red brick. The church of St. Mary stands on a slope at the N end of the settlement. The nave and chancel are Norman, the tower is 14th Century, the extension to the chancel was made in 1891 and there is a William Morris – Burne-Jones E window of 1893. Whitton Court lies 1½m. N of the church. It is an attractive and intrigueing house. The core is a stone hall of the 14th Century with 15th Century ogee-leaded lights in the straight-headed windows. In Tudor times the stone was faced with brick and the building was altered again in 1621 and 1682. It is now an 'E' shaped house with bay windows and an embattled, off-centre porch. At the rear there is an enclosed courtyard; the W range of this is a fine black and white gable with lozenge decoration. Inside there is some good fresco wall painting of 1682 and a hunting scene of the 18th Century over the fireplace. The Hollins is a black and white building that is thought to have been a dower house to the Court. In early medieval times Whitton was an outlier of the Manor of Burford. The name is from the Old English and means either 'the homestead of *Hwitta*', or 'the white house'.

WHITTON *¾m. WSW of Westbury which is 9m. WSW of Shrewsbury.*
It lies, unsignposted, on low ground just off the lane between Westbury and Vennington. About a mile to the SW is Caus Castle and so it comes as no surprise to learn that in 1086 the manor was held by Roger, son of Corbet of Caus Castle. (*See Westbury.*) Before the Norman Conquest 3 Anglo-Saxon freemen, Leofnoth, Ledmer and Ulfketel, had held the land from King Edward. Today Whitton is a scattered farming community – Whitton Farm, Whitton Grange and a few cottages – centred on the Hall. An avenue of trees leads off the main road to a most attractive wooded area watered by a tiny stream. It is a delightful spot and a fitting setting for the substantial and well-matured red brick Hall. The house has not been firmly dated but is probably of the early 18th Century. It has gabled wings and a pedimented doorway. To the front left and front right are detached service wings. That to the right is older than the present house. In the grounds there is a circular dovecote and on the hill opposite the Hall is a folly. The estate is currently in the possession of the Halliday family; the previous owner was Lady Childs. The name Whitton is from the Old English and means either '*Hwitta*'s settlement', or 'the white settlement'.

WHIXALL *4m. N of Wem.*
The early forms of the name suggest that it means *Hwituc*'s-*halh*. Hwitic is a personal name and in the Midlands *halh* usually means 'a nook', a hidden place often in a valley. All the houses line the lanes; there are no clusters of buildings. The whole area, from Northwood in the W to Wem in the S and Tilstock in the N, is like this – a tangle of lanes with houses strung along them and no nuclear settlements at all. Equally noticeable is the tangle of footpaths. This is Moss country – old marshes and wet ground which at one time were densely wooded. The settlement pattern is typical of that which developed during late medieval forest clearance when individual 'colonists' built a cottage and cleared an acre or two. The unremarkable parish church of St. Mary (1897) was built in red brick to a design by G.E.Street. It lies half way between Whixall and **Stanley Green** which lies ½m. NE. Green in a place name usually refers to the clearing hacked out of virgin woodland in which the early settlement was made, not a village green. The name Stanley Green probably means 'the stoney clearing'; 'ley', from *leah* can mean either 'a wood', or 'a clearing in a wood'. There are 4 other churches in Whixall parish: Primitive Methodist, Congregational, United Reform and Methodist. The Old Smithy is now an agricultural store but the blacksmith's forge is still in place. There are a fair number of timber-framed buildings in the area. Willow Cottage, near Lower House in Stanley Green, is a black and white cottage with a most benign carved head nailed to the side wall facing the road. Bostock Hall lies to the E of Whixall at a crossroads. It is a small 17th Century farmhouse with well-kept gardens. At **Dobson's Bridge** is a substantial canal marina complex where narrowboats can be hired from the Black Prince Line. This stands on a spur of the Shropshire Union Canal (Llangollen Branch, opened 1797-1806) which follows the southern boundary of the extensive Whixall Moss and is crossed by quaint counter-weight bridges. On the Moss the peat can be up to 8ft. thick and very boggy. In the past people have ventured out and never been seen again. In places the peat is cropped, that is to say it is cut and dried, and sold mainly for horticultural useage. Much of the Whixall Moss has been drained and is now under pasture but there are still vast areas that are a virtual wilderness of marshy ground where the vegetation consists of tough grasses, bracken, thorns and scrub birch. The landscape is quite superb and remarkably similar to an African Savannah in general appearance. The wilder parts are not always easy to see. One of the best sites to visit is at SJ.492.358 which is down the earth track opposite Whixall Moss canal bridge. It is not wise to take a car down this 'lane'. Do beware, the ground can be treacherous.

WILCOTT *9m. NW of Shrewsbury.*
Wilcott (meaning *Winela*'s Cottage) lies just off the A5 near Nesscliffe. It is a curious mixture of smart homes, ancient and modern; the Grange, the Manor and the Hall; a pig farm; weather-boarded barns; sandstone boundary walls; the cream and green Nissen huts of a fair sized army camp; stands of mature, broadleafed trees; and the motte of a (presumably) Norman castle at the crossroads by the Hall. Wilcott Hall is a substantial black and white house that looks Victorian. It stands in attractive gardens that lead down to a stream. This has been dammed to make a pool, and was possibly used as a moat in medieval times. The tree-clad castle mound rears up beside it and on the opposite side of the road is an attractive cottage. Between Wilcott and the River Severn, 2m. S, are some 2,000 acres of sparsely populated, badly drained agricultural land. During the Second World War the Central Ammunition Storage Depot was sited here. The brick built storage sheds with their steel doors and high earthen explosion baffle mounds still stand. The area is now used by the military for training purposes though most of the land is either under cultivation or laid to pasture. The long, straight and

Whittington Castle.

dusty track that leads from **Pentre** to the River Severn near Shrawardine is the course of the now dismantled Shropshire and Montgomery Railway that went to Shrewsbury. It crossed the river near the motte and bailey castle at Little Shrawardine. The village of Pentre – the name is Welsh and simply means 'village' – is an airy little place 1m. SW of Wilcott. There are houses and cottages in stone and brick; some timber-framed dwellings; a pub called the Grove; a Post Office – General Store; a manufacturer of portable office accommodation; a scrap-metal merchant; a factory farm with a yellow painted roof housed in 2 of the old army buildings; and a few orthodox farms, one with a small orchard. The Celtic connection is strengthened with property names such as Pen-y-wern and Ty Bwchan. Half-a-mile W is a pub called the Royal Hill. It stands on the banks of the River Severn and is frequented by fishermen.

WILDERHOPE *See Rushbury.*

WILDERLEY *4½m. SE of Pontesbury.*
It lies lost in the lanes of the quiet country between the A488 and the A49 near Church Pulverbatch. The small Hall now looks very modern with its rendered walls and new window frames. Of the old house only one timber-framed gable end remains externally. To the front of the Hall is a range of farm buildings and to the rear is a derelict cottage. A very rough and stoney lane leads past this cottage up a gentle hill to a clump of trees. Amidst this little wood is the 17 foot high mound of a Norman Castle partly surrounded by a ditch that varies from between 6ft to 3ft deep, and the remains of 2 quadrangular courts, or baileys, protected by banks and ditches on the down slope to the E. In the lane is a magnificent old oak tree. Two-thirds-of-a-mile SW of Wilderley is Coppice Farm, a timber-framed house with 2 gables and brick chimneys. In the nearby valleys, too steep to graze cattle, are several coniferous plantations. The name Wilderley probably means 'the wood or clearing belonging to *Wilred*.'

WILLASTON *3m. NE of Prees.*
The first element of the name could be from the Old English *Willa*, a personal name, or from *willig*, meaning 'a willow tree'. The second element, 'aston', means 'eastern'. Willaston was part of the ancient Anglo-Saxon manor of Prees which was part of the Eccleshall estate of the Bishop of Lichfield. Today the area is remarkable for the number of oak trees, not just in the hedgerows, but free-standing in the fields. At first one thinks it is a parkland, but the area covered is too large; it is more like a forest that has been thinned. The country here is in fact part of the Cloverley Estate of the Heywood-Lonsdale family who also own the adjoining lands of Shavington. The present owner 'Mr. Tim' Heywood-Lonsdale lives at the Laundry House, Shavington. It was his predecessors, especially John Pemberton Heywood and his wife Anna Marie, who created and conserved this landscape and who built many of the cottages and farmhouses in the area. Most of these dwellings still belong to the estate. Willaston is a trim little place with 2 pairs of semi-detached houses (one pair dated 1937, the other, with black and white gables and dormers, dated 1875) and 2 farms, Wistaston Farm and Higher Kempley Farm, dated 1878. 'Kemp' is probably from the Old English *Cempa* a personal name from *cempa*, meaning 'a warrior'. At the sharp bend in the road just to the W of Kempley Farm is a public footpath that leads northwards past a circular, tree-clad mound (SJ.597.359). This lies on a slight rise and is about 6ft. high and 46ft. in diameter at the top. It is 'scheduled' as the motte of a Norman castle but local tradition has it that it is a burial mound within which lie the bodies of cattle that were destroyed after becoming infected with Rhinder-pest in the 1860's. That could, of course, be either proved or disproved by a simple excavation. Less than 100 yds to the NE of the mound is a small, broadleaf wood within which are the water-filled remains of a moat. There are game birds in this copse and here we chanced upon Mr. Lonsdale-Heywood and a shooting party. The lane that runs eastwards is an 'enclosure road', straight with hedges and deep ditches. We turned S down the drive to The Lawn Farm and stopped to photograph an especially handsome oak tree. My goodness what a shock we had – first a short piercing siren call and then a deafening explosion. Unknowingly we were standing about 4ft. away from an automatic gas operated bird scarer colourfully called a Bangalore Bomber. The gently rolling lands of the Cloverley estate ought to be more widely known. They are at their best on a misty winter's morning, the old oaks standing stark against a silver sky in what is everyone's idea of rich Shropshire farming country. In the fields hereabouts are numerous small ponds, called 'kettle holes', which were formed as the glacier's moved N after the last Ice Age. Blocks of ice broke away from the retreating ice front and were surrounded by the clays, sands and gravels deposited by the melt waters. When the blocks finally melted they formed the pools we see today. In some hollows and dips where the land has not been drained there are marshy areas covered in scrub woodland. (*See Calverhall.*)

WILLEY *5m. NW of Bridgnorth.*
Willey (meaning 'willow wood') is most easily approached by a lane off the B4373 Bridgnorth to Broseley road. There was a settlement here prior to 1086 when the medieval hay, or deer park, was recorded in Domesday Book as King's Wood, which then belonged to Much Wenlock. Willey was in Shirlett Forest and there is much woodland here today, though this is largely as a result of "controlled coppicing", the planned growing and cutting of small timber, which began in 1550 to supply fuel for the glass and ironworks of the area. Ironworking especially flourished at Willey. It is hard to imagine today but the whole area once throbbed to the roar of furnaces and the thumping of forges. Now there are only a few small, overgrown slag heaps and furnace pools to bear witness. Industry came early to Willey primarily through the efforts of John Weld, the son of a wealthy London merchant. In 1618 he bought the manor of Willey and the following year purchased the adjacent manor of Marsh. He either built or rebuilt the Old Willey blast furnace and exploited the coal reserves on his estates. By 1757 the Forester family owned Willey. In that year the New Willey Company was formed and a new furnace was constructed. One of the partners, with a twelfth share, was John Wilkinson. In 1774 he became sole lessee of the works and the old furnace was closed down. Within 2 years he was to introduce several innovations that were to make him without any doubt the leading ironmaster of his age. It was here, in 1776 at the New Willey Furnace, that Wilkinson made the first application of Boulton and Watts new steam engine to an ironworks. The engine powered the bellows that blasted the air into the furnace. This was an important development because it meant that an ironworks need no longer be dependant on the clumsy system of pools and water-wheels traditionally used as a source of power. (Many works had to close down for several months during the Summer because water levels were too low to operate the machinery.) It was at Willey that Wilkinson made the first use of steam forge hammers and in 1786 the first steam-powered rolling and slitting mill. He built the first iron boat here and launched it at Willey Wharf on the Severn in 1787. It was here, too, that Wilkinson developed his process for boring canon which was to be the basis of his fortune when he later moved to the Black Country. There has been a Hall at Willey since Anglo-Saxon times and it is possible that the medieval fishponds were used as furnace pools during the industrial period. The present Hall is very recent, being designed by Lewis Wyatt and built in 1812-15. It is an elegant building with a lovely setting. It lies in an extensive Park in which there are many handsome ornamental trees and a large, commercially operated forest, with herds of sheep and deer in the meadows. The old fishponds lie in a landscaped valley in front of the Hall. The house has an enormous portico of 4 Corinthian columns supporting a pediment and the interior is as fine as its facade with a splendid galleried and domed hall and many original draperies, wallpapers and chandeliers. The Old Hall lay to the E next to the church. It was stone built but later encased in brick. One side of it faced into the large courtyard, around which still stands the handsome stone stable block with its gables and mullioned windows. The right-hand half of this is currently occupied as a dower house. The brick coach-house block has been converted to cottages, as has a building on the other side of the entrance to the yard, which is stone with mullioned windows in its lower parts and is probably a remnant of the Old Hall. Tom Moody (d.1796) worked at the Willey Stables. He was a horseman of great renown and

Whixall Moss, Bridge No. 3, Morrise's Bridge.

'Whipper-in' of the local hunt. His ghost is often seen in the country hereabouts, and by his side always is his favourite hound. He has been the subject of many paintings and poems; his grave is in the churchyard at nearby Barrow. The church of St. John at Willey has a Norman nave and chancel, an 18th Century tower and aisles and a family chapel of 1880. The village of Willey stood on the slope NE of the Old Hall. It was demolished when the New Hall was constructed to give greater privacy to the new owners, the Forester family, and was never rebuilt. However, the conversion of the old courtyard buildings into homes has effectively reinstated it. Today Lord Forrester still occupies the New Hall and manages the Willey Estate. One mile SSW, in Shirlett High Park, is a monument to a dog, a retreiver who fell down the shaft of a coal mine.

WILLOWMOOR *3/4m. NNW of Little Wenlock.*
This green and pleasant valley is flanked by wooded hills and lies on the SE slopes of The Wrekin. There is a tangible atmosphere here, a feeling of being remote from the world despite the presence of a modernised cottage. The valley is cut through by the Little Wenlock to Wellington road that traverses the pass between The Wrekin and The Ercall. Near the cottage 6 prehistoric burial mounds have been identified; the road actually cuts through one of them (SJ.641.643). There are 3 more close to the road on the W, and 2 more to the E. Almost certainly there were others, perhaps many more, a kind of Valley of the Kings. (The Victorians plundered and destroyed many tumulii and the smaller ones are easily ploughed out.) The Wrekin was the centre of the pre-Roman kingdom of the Celtic Cornovii for several hundred years.

WILLSTONE *2 1/4m. NE of Church Stretton.*
The name might mean 'the settlement by the stone in the willows'. Willstone is a superbly positioned hamlet on a slope in the Stretton hills with Caer Caradoc to its back and long views downhill to Cardington in the W. The settlement centres on a large 5-bay stone house which is in a state of disrepair. There are stone steps to the front door above which is a cartouche with the date 1738. The old stone barns have been augmented with a variety of modern corrugated iron sheds and a giant silo which together have ruined the prospect of the place. On the road between Cardington and Comley is an excellent half-timbered farmhouse. The looming hill behind it is Caer Caradoc. (Caradoc is the Welsh name of Caractacus, who led the Britons in their fight against the Romans.) His last stand was on a hill, probably one of those now called Caer Caradoc, but there are several of these and no one really knows the whereabouts of the actual site described by Tacitus.

WISTANSTOW *1 1/2m. N of Craven Arms.*
This Anglo-Saxon settlement developed at an ancient crossroads and the village is actually built along the line of the old Roman road to Church Stretton and Uriconeum (Wroxeter), now by-passed by the A49. One mile S the River Onny is joined by the Byrne Brook. The village takes its name from St. Wystan (or Wigstan). Wystan was a Mercian prince who refused the crown to devote his life to Christ. He was murdered in 849 by his uncle and was later canonised. Shortly after his death the church was built. The present cruciform church is Norman, of about 1200, with some late 13th and 14th Century windows. The roof of the nave is judged to be of 1630 and that of the N transept circa 1200. The village is a pretty little place with some black and white thatched cottages but is largely built of stone. Behind the church is a handsome 19th Century, grey sandstone school despoiled by an almost unbelievably awful modern extension. Whoever designed this can only be called an architectural hooligan. There is a Post Office and General Store; a pub, the Plough Inn; and a black and white Village Hall, the gift of Mrs. Harriet Greene of the Grove Estate in 1925.

WISTANSWICK *3 1/2m. S of Market Drayton.*
The name probably means '*Wigstan*'s dairy farm, and it is still very much an agricultural community. The village lies just to the W of the busy A41 in flattish country where the land is laid mainly to pasture. There is a pub, the Red Lion clad in pale green stucco, a green and red corrugated-iron village hall, and a sandstone United Reformed Church now converted to a dwelling with a small graveyard adjacent. The main village lies just to the N, an unspoilt little place with a variety of buildings: the Manor Farm House of sandstone and stucco; a brick house painted to look like a black and white cottage; several red brick farms with sandstone outbuildings; cottages with dormer windows; and a handful of modern bungalows. The atmosphere can have changed little since Anglo-Saxon times. **Heathcote** lies in flat country 3/4m. WSW, a crossroads hamlet of red brick houses and an old chapel now used as a school which has attracted the usual awful modern extension. To the E of Wistanswick there are 2 charmingly named main road developments. **Crickmery** consists of a few stone cottages and houses of red brick and stucco and the Tern Garage. Just S of Crickmery is **Sweet Appletree**, a crossroads marked by a collection of very scruffy sheds amongst which a girl with long blonde hair feeds geese. The dwellings include the red brick Mona House and the white stuccoed Eidda Cottage. However, the building of greatest interest hereabouts lies 1m. SSW of Wistanswick. This is **Hurst Farm** situated at SJ.661.277 in gently rolling country. In the late 17th Century and through most of the 18th Century this was a Particular Baptist Meeting House and in the early years had larger congregations than the parish church at Stoke upon Tern, much to the chagrin of the vicar there. The Lowe family, owners of Hurst Farm, were the leaders of the community and on several occasions members of the family were excommunicated from the established church. Thomas Lowe was preaching in Cheshire, Lancashire, Shropshire, Denbighshire and Flintshire from the 1660's until his death in 1696 at Hill Cliffe in Cheshire, the centre of the Particular Baptists. Members of the Hurst Farm congregation were buried in unknown places until 1715 when a burial ground was established in Clover Field (now called Graveyard Field) where some 60 bodies were interred. This field lies between the road and the farmhouse at SJ.601.274. A few years ago it was ploughed and human bones came to the surface of the light grey soil. Fodder beet is grown there now. Allowing for the fact that we visited on a dull December day this is a melancholy, bleak and relatively isolated place. Hurst Farm is a modest 2-storey building of cream painted brick with 3 bays and a central porch. To the rear is an extension part of sandstone and part of brick; to the front is a most handsome Scots fir. Amongst the farm buildings there are some red painted barns. Young cattle are brought here to be fattened. The hedgerows to the E are ancient and contain oak trees, hollies, ivy and ferns.

WITHINGTON *6m. E of Shrewsbury.*
It lies in flat land near the confluence of the River Roden and the River Tern. The now abandoned Shrewsbury Canal (opened in 1797, closed in 1944) passed through the village. G. E. Street rebuilt the church of John the Baptist in 1874 using local rock-faced red sandstone. Adam Grafton, d.1530, was rector here and later became chaplain to Edward V and Prince Arthur. There was an iron forge in the area during the early 18th Century. At Allscott, 2m. E, there is a sugar beet processing plant, a gleaming factory that is a landmark for many miles around. In 1066 Withington was held as 2 manors by Wulfin and Wulfric. In 1086 it was held by Fulcwy. It is a pity that names such as these are no longer current. The name Withington is Anglo-Saxon and could mean either 'the settlement amongst the willows' or '*Widia*'s place', a homestead or settlement dedicated to the legendary German hero. Today there is a pub, the Hare and Hounds, some new houses and bungalows and 6 distinctive white farmhouses, built in 1918 by the council when the old manor estate was divided into small-holdings. The red brick Manor House is dated at 1710.

WIXHALL *1/4m. WSW of Weston-under-Redcastle which is 3 1/2m. E of Wem.*
The name is from the Old English and means 'the hall, or homestead, of Wittuc'. There are only 5 houses in the hamlet. One is stone-built and the others are red brick with parts of stone and parts of half-timber. The farm buildings are substantial and the fields laid mainly to pasture. The place feels old. There is holly and ivy in the hedgerows and between the settlement and the main road the lane passes through a sandstone cutting. Just N of the buildings is a disused quarry and just to the W are old mine shafts. Without being

Wistanstow, the church of Holy Trinity.

picturesque Wixhall is most attractive.

WOLFS HEAD *1½m. NW of Nesscliffe.*
At the junction of the B4396 and the A5 is a farmhouse that was formerly a pub, the Wolf's Head Inn. This had a great reputation as a resort of highwaymen. The section of the A5 from here southwards, past Nesscliffe Hill, was so dangerous that in 1583 the Draper's Company ordained that no draper should set out for Oswestry (from Shrewsbury) on Monday before 6 o'clock in the morning and that he should be armed and travel on in the company of others. If he did otherwise he would be fined 6s. 8d.

WOODSEAVES *2½m. S of Market Drayton on the A529.*
A straggling hamlet of red brick cottages and farms some of which have been rendered. The soil is red and clayey and the fields mostly laid to pasture. There is a small Methodist Chapel and at the N end of the settlement is the Four Alls pub. West of the pub is a lane that leads past the modern **Tyreley** church, which looks a bit like a bungalow, to Tyreley Locks. Alongside the Shropshire Union Canal there is a handful of cottages, a craft shop and a stand of splendid, tall trees which alltogether make an attractive ensemble. The canal forms the boundary between Shropshire and Staffordshire for a couple of miles hereabouts. A lane just S of the Four Alls leads to Sutton. Two hundred yards along this lane is The Dairy house Farm which lies adjacent to the now derelict Woodseaves Farm (SJ.678.318), a substantial yeoman farmer's black and white house. Queen Anne is reputed to have stayed here. It is sad to see it so neglected. At the end of a rough track to the SW of the village is The Sydnall, a substantial brick house of 2 broad bays flanked by pavilions with 2 storey blank arches. The lower parts of the outbuildings are sandstone. The name Woodseaves probably means 'the boundary wood', or possibly 'the wood beyond the boundary'. Eaves is Old English and can either mean 'boundary' or 'beyond'. (The eaves of a house are the part of the roof that extend beyond the vertical walls.)

WOLLASTON *1m. NW of Half Way House, which is 10m. W of Shrewsbury on the A458.*
There is little more here than a handful of houses and farms and the grey stone church of St. John which was rebuilt in 1788 and restored in 1885 when the windows with Venetian tracery were inserted. Just to the SW of the church is the overgrown 15 ft. high mound of a Norman castle. It stands on the E side of a bailey platform. Of the ringwork marked on the Ordnance Survey map at SJ.324.120 nothing now remains. In the church is a plaque to Old Parr who lived at **Glyn Common**, 2m. SW. Old Parr was reputed to be 152 years old when he died and is buried in Westminster Abbey. He first married at the age of 80 and fathered 2 children. He married again at 120 and fathered another. His diet consisted of coarse bread, milk, rancid cheese and whey. At 145 he was still doing everyday farm work. He was taken to London, met the king, became the talk of the town, and died there. His cottage has just been rebuilt in exact replica by a girl called Jenny. To get to the cottage leave the A458 at the sign for Winnington at the Winnington-Wollaston crossroads. Follow this road for 1¼m. Opposite the entrance to a farm with white gates called Lower Trafnant turn right into a little unmade lane. Follow this for ⅓m. and you will find Old Parr's cottage (SJ.306.112). It is truly delightful and has lovely views over Welshpool and into the mountains of Wales. One mile N of Wollaston is Braggington Hall Farm a 17th Century red brick house with red sandstone dressings, gables, a 5-bay centre with 2-bay projecting wings, and oval windows at the back. The date above the columned doorway is 1675; oval windows were in fashion at that time as in no other.

WOLLERTON *4m. SW of Market Drayton.*
The village lies on the A53 in the valley of the River Tern and was served by the Market Drayton – Wellington railway until it became a victim of Dr. Beeching's axe. The older part of the settlement lies around the Old Hall which stands ¼m. E of the main road. Close by there are several black and white cottages; the brick built mill, now occupied by a craft furniture maker; Riverside, a substantial house with a timber-framed barn; and a brick built Animal Pound (SJ.623.297) in which stray animals were placed and released on payment of a fine when claimed by their owners. Between the main road and the old settlement, and along the main road itself, there are substantial Georgian and Victorian houses with, inevitably, some modern infill. The local hostelry is The Squirrel. To the E of the A53 there is a 3-bay Georgian house called Brookside and a United Reform Church of 1867 with lancet windows. These stand beside the lane to the hamlet of Wollerton Wood which lies 1m. NW of Wollerton. **Wollerton Wood** has no nucleus but consists of a few farms and a brick built Primitive Methodist church of 1865. The moat of Moat Farm lies isolated in fields to the S of the homestead. We spotted a wild dog rose in a hedge near here, an unusual sight these days. There is another moated site at Manor Farm, **Lostford**, 1m. NNE of Wollerton. Fords are not easily lost so perhaps the name was originally Lastford, or possibly the ford came to be abandoned when the River Tern was lowered as part of drainage works. A network of strines drain the land and empties into the river. In 1066 the manor of Wollerton (or Ulvretone) was held by the Anglo-Saxon, Askell, but by 1086 it had passed to Gerard who held it from Earl Roger of Montgomery. Gerard had 7 serfs, or slaves, rather more than was usual. Later, the manor became the property of the Abbot of Shrewsbury Abbey. The name Wollerton is from the Old English *Wulfrun*'s-*tun*.

WOMBRIDGE *2¼m. E of Wellington.*
Today Wombridge is a northern suburb of Oakengates. It is a place of 20th Century dwellings with green areas protecting them from the noisy dangers of the Wombridge Way and Queensway (A442). There is very little left to remind one of its long and varied industrial history. In the 16th Century the monks of Wombridge Priory were digging coal and smelting iron. The bloomery, where the iron was refined, was adjacent to the monastery. The Foley family had charcoal burning furnaces in the area and in the early 18th Century the Charlton family of Apley Castle were mining coal here. At the end of the 18th Century William Reynolds constructed a tunnel 1¼m. long from the Wombridge mines to his blast furnaces at Donnington Wood. Sand was quarried here and sent to the Horsehay ironworks. In the 1790's there was a chemical factory that produced sulphuric acid from iron pyrites found in the coal seams; in about 1799 the plant changed to alkali production and made soap, dyes and sodium sulphate before closing in 1803. In 1818 James Foster built 2 iron smelting furnaces and another in 1834 before he closed down and moved to Madeley Court in 1847 – taking his skilled men with him. In that year almost half of the parish of Wombridge was derelict industrial wasteland. The old Wombridge church that had witnessed this depravation was itself torn down and rebuilt in 1869. The new church of St. Mary and St. Leonard is a sturdy building constructed of rock-faced rusticated stone. Below the churchyard are the remains of the Augustinian priory, founded circa 1140. There is little else at Wombridge today other than large numbers of modern houses many of which can be blamed on the council. The name Wombridge probably means 'the bridge by the lake'; 'wom' is from the Old English *wamb* meaning 'womb' but often means, symbolically, 'a lake'.

WOODCOTE *2½m. SSE of Newport, just off the A41.*
Woodcote Hall is a Victorian house of 1875 with an exotic front facade which is a mixture of Jacobean and Georgian with Eastern style ogee-shaped turrets to a design by F. P. Cockerell. In the last few years the hall has been put to many uses including a Youth Centre, a Christian Training College, a Country Club and currently an Old People's Home. It has lost its park but has retained its charming, red sandstone Norman church. The doorway and most of the fabric is original but the windows are of a later date. It lies to the side and rear of the house on lower ground hidden by trees. The Hall has been rebuilt several times. On each rebuild it has been moved in the direction of the road. When we visited trial pits in the floor of the modern Roman Catholic chapel had exposed the foundations of a previous Hall about 3ft. down. (The chapel is going to be converted into an indoor swimming pool.) There is a very old sandstone cottage at the back of the present hall. The stonework has weathered in almost identical fashion to that of the church. Stones and bricks from dismantled buildings are found in several

Hurst Farm, near Wistanswick, once a Particular Baptist Meeting House.

Old Parr's Cottage (reconstructed) near Wollaston.

The old steeplechase course stables near Woore.

places in the grounds including the car park to the front of the Hall. It seems extremely likely that there was a village here in Norman times and that it was removed on one of the occasions when the Hall was rebuilt. The church is not a private chapel and a service is still held here once a month which is attended by people from the surrounding area. The Hall can be glimpsed from the road. The name Woodcote means 'the cottage in the wood'.

WOOLSTASTON *1½m. W of Leebotwood, which is 9m. S of Shrewsbury on the A49.*
A delightful village on the northern slopes of the Long Mynd. It has a castle, a church, a Hall, black and white cottages and red brick farms set amongst trees at the end of deep lanes. The lane up from the main road at Leebotwood is lined with hedges which contain bracken, a sure sign that this was the natural cover of the land before man cleared it for grazing. The little village green is not ancient. It was created in 1865 on the site of a dung heap. The Norman motte and bailey castle lies at the uphill end of the village next to a farm. The earthworks are considerable and are best seen from a field at the back of the farm. Excavations here produced no signs of permanent occupation and there was no evidence of stone on the site. It was not unusual for the Normans to abandon castles after only a very short time and move on elsewhere. The church of St. Michael is 13th Century with a restoration in 1864-6. There are two 12th Century fonts. The woodwork was carved by William Hill, a local carpenter, in the 1860's. The church lies next to the Hall. This is late 17th Century. What we see is only one wing of a much larger 'H' shaped mansion; the rest was demolished. A rector of this church wrote an account of a terrible ordeal he experienced one winter's night in 1865. He was returning from Ratlinghope, which lies over the Long Mynd, and was caught in a snowstorm. He lost his way and for 27 hours stumbled about the mountain. He very nearly died but lived to write 'A Night in the Snow', the proceeds of which paid for the carving in the church. The Walk Mills formerly ground corn and fulled wool but now house a metal working business. The name Wolstaston means '*Wulstan*'s homestead'.

WOOLSTON *See West Felton.*

WOORE *7m. NE of Market Drayton.*
Dick francis, the best selling author of racing novels, rode his first race here on the old National Hunt course, about which more is said later. Woore is an attractive and well matured brick-built place on the old London to Chester coaching road. It is a large village which has developed partly as a dormitory town for the nearby connurbation of Stoke-on-Trent. The undulating country around is essentially pastoral and to the NE gives way to the Cheshire plain. The white stuccoed Classical church of St. Leonard is quite charming. It was built in 1830 by G. E. Hamilton, though the chancel was rebuilt in 1887. Behind the church is a modern housing estate; in front of it is the handsome facade of the red and orange painted Swan Hotel, an old coaching inn that has 15th Century origins. The North Staffordshire Hunt meets here on Boxing Day, a spectacle that attracts many onlookers. There are 2 other pubs: the Cooper's Arms and the Falcon Inn. The ivy-clad Manor House stands at the crossroads embowered in tall trees. It is of red brick on a stone plinth and has 5 bays and a hipped roof. On the opposite side of the road is a tiny but well cared for grey rendered Methodist church. In the village there are several shops, an antique-seller, an undertaker, a petrol station, a garden nursery and the black and white Tudor Tea Rooms. There is a long-standing tradition that there should always be an ash tree in the village, especially on the shooting butts, high ground where it could have acted as a drover's landmark. The name Woore (Wavre in Domesday Book) is probably derived from the Old English, which might mean 'brushwood'. However, some authorities think that the name might be related to a word meaning 'boundary'. Woore is very close to the borders of Cheshire and Staffordshire. The village was once well known for its Steeplechase meetings. The old course used to be W of the Manor House but a new course was made later at **Pipe Gate**, 1¼m. SW on the A51. The old, quadrangular stables (SJ.731.412) still stand, somewhat forlorn, in the middle of a field. They are constructed of old railway sleepers and have rusty-red corrugated iron roofs. Racing ceased here about 25 years ago. There used to be a railway station at Pipe Gate, but the line fell a victim to Dr. Beeching's axe. The yellow and red brick factory which dominates the hamlet used to be the Creamery but is now occupied by Phoenix Rubber Ltd. The original creamery was on the opposite side of the road and has now been converted into dwellings. The name Pipe Gate could mean several things: 'pipe' could be either a stream or a water drainage culvert or a personal name; 'gate' can either be a gap (a feature of the landscape) or the gate to an estate or a road toll-gate. There are many small pools in the fields hereabouts. They lie in hollows in the clayey soil and are called 'kettle-holes'. They are believed to have originated during the retreat of the last ice-sheet, some 20,000 years ago. Blocks of ice are thought to have broken off and then been surrounded by deposits of clay, silt and sand as the ice sheet melted. When the blocks finally melted themselves a water-filled hollow was produced. The hamlet of **Gravenhunger** Moss lies ¼m. E of Woore. It consists of a handful of well cared for cottages; the moss, or marsh, was drained long ago. A track leads N through an immense, mile-long field to the isolated Phynsons Hayes Farm. This is a homely place, well stocked with dogs, by a stream and a small wood and lies within yards of the Cheshire border. The name Gravenhunger is from the Old English *graf-hanger* which means 'the slope covered in brushwood'.

WORFIELD *3½m. NE of Bridgnorth, off the A454 Wolverhampton – Bridgnorth road.*
Worfield is one of the prettiest villages in Shropshire. It nestles between the River Worfe and a steep sandstone cliff. The main street is lined with delightful cottages of half-timber, stone, brick and stucco all jostled contentedly together. Much of this property is still owned by the Davenport estate and twice a year the tenants resort to the Davenport Arms (known locally as The Dog) to pay their rent. The sandstone church of St. Peter is set on a slight rise below a steep, wooded slope. It was built in the 13th Century and has 14th Century additions, which include the tall elegant grey spire, but there are architectural puzzles a-plenty here. Inside is a good alabaster monument to Sir George Bromley (d.1588). There was a much earlier church on the same site of which nothing remains; indeed the parish and manor were both large and in times past important. At the time of Domesday Worfield lay in Staffordshire and was held by Hugh de Montgomery. Before the Conquest it had been held by Earl Algar, son of the King of Mercia. By the entrance to St. Peter's is the tiny black and white former Worfield Grammar School. On the other side of the road is the superb Lower House. A high stone wall unfortunately masks much of this building from the road. The house has two 3-storey ranges; the back range is the earlier. To the side are several additions around the stone and brick chimneys. It is a delightfully irregular dwelling with simple, unadorned framing. In the parish of Worfield have been found some of the earliest remains of man in the county, namely scrapers and other implements from the Mesolithic (Middle Stone Age circa 4000 BC). The name Worfield is from the Old English and means 'open land by the River Worfe'. The name Worfe is thought to mean 'wandering, or winding'. Davenport House lies ½m. SW of Worfield. It has a Dovecote which can be seen from the road near the entrance drive. The substantial red brick house was built in 1726 by Francis Smith of Warwick and stands in a large landscaped park. It is 9 bays wide and has a later porch with fluted Ionic columns. Inside there is a good carved staircase and a feature fireplace in the White Drawing Room. Chesterton Farm lies 2m. NE of Worfield. It is a fine, large, early 18th Century brick house with a hipped roof and wooden window casement crosses. All that remains of the chapel are a Perpendicular doorway and some windows. The stones were probably used to build the nearby cottages. (The stone of dismantled churches was often used for house building.) About 300yds. S of the farm is the huge Iron Age settlement called The Walls. Trees surround it, but the enclosure is clear. Ewdness manor house, 2¼m. NW of Worfield, is a 17th Century symmetrical red sandstone house with projecting gables and large, star-shaped chimneys. One mile N of Worfield is **The Bog**. This is a large, oval pool in

The cast-iron Trellis porch of the Old Vicarage, Wrockwardine.

The Gatehouse, Orleton Hall, Wrockwardine.

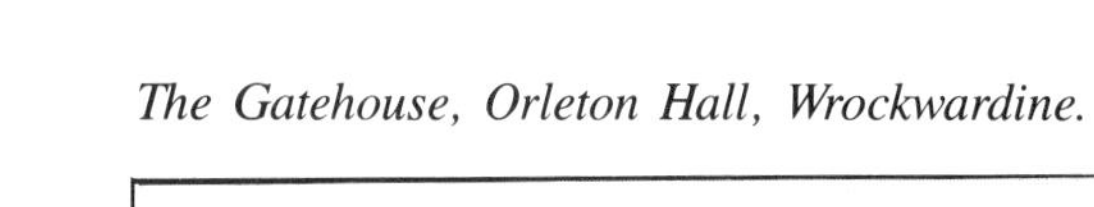

John Talbot, 'The scourge of France.'

Woodseaves Farm, Woodseaves, near Market Drayton.

which is a large, wooded island of peat. Birch predominate on the island and the northern shore of the pool is fringed with willow trees. This is a natural bog, a peat swamp which was once drained by a circular ditch and the water carried away to the S by an underground conduit. The land was then used as meadow, but, about 30 years ago, large machines were brought in to cut and dig out the peat for use as a horticultural fertilizer. These machines broke the conduit and after the cutting had stopped the area filled up with water and formed the lake we see today. The lake is fed by a natural spring and is stocked with carp. Ducks, geese and other wild fowl also frequent these waters. It is a somewhat strange place and not a little eerie.

WORTHEN *13m. WSW of Shrewsbury on the b4386 to Montgomery.*
It lies in the Rea Valley. The Welsh drovers came this way and the name of Cowlane is a reminder that cattle were kept off the main highways because their sharp hooves and great weight could destroy even metalled surfaces. At the time of Domesday Worthen was an established town with 13 berewicks (dependent villages or hamlets). The buildings of this substantial village are largely of stone. The church of All Saints is set back from the road. The tower is 12th Century at the bottom, Perpendicular in the upper and 18th Century in the topmost part; the nave is Early English; the brick chancel is of 1761; the benches are medieval, possibly reconstructed in the 17th Century; and the pews are Jacobean. There was once a mill, on Worthen Brook, but this was demolished in 1973 and a modern housing estate now stands on the site. There were once 6 pubs, now there are none; but there is a Post Office and general store, a Village Hall and a new Primary School. Hampton Hall, 1½m. NW of Worthen, is a 17th Century red brick house with a porch of 1749. It incorporates parts of an older building. The hall of the house still lies in the medieval position, that is to the side of the porch, not immediately behind it. The house is well sited and has views towards the Stiperstones. Half-a-mile to the W are the earthworks of a pre-historic fort. Just over 1m. NE of Worthen is **Aston Rogers** which has a Hall, a pub, some cottages and a moat. Just N of Aston Rogers is the curiously named Mondaytown. The name Worthen is thought to be derived from the Old English *worpign* which can mean several things: homestead, enclosure, open place in a village, and the yard of a house.

WOTHERTON *4m. NE of Montgomery just off the B4386.*
The name means 'the settlement by a ford by a wood'. The woods are still here but the stream is now crossed by a bridge. There is a Norman motte, a multilated mound by the stream; a substantial farmhouse with pedimented 3-bay front and stone and brick outbuildings guarded by a querulous goose; stone and brick cottages; a curious little tower-house of stone with round-headed windows that stands by disturbed ground opposite an overgrown quarry; a weather-boarded barn; a stone-built farm; and on the main road a handsome stone-built house called Woodmore. The village stands on rising ground above the flood plain ofthe Rea Brook and is clearly visible from the busy B4386. Rorrington lies 1m. E and the access lane from Wotherton passes through a delightful valley with a stream, stands of conifers, a gorse covered knoll, broadleaf trees, meadows and a mysterious little wooden chalet with a verandah almost hidden in woods by the road.

THE WREKIN *Just S of Wellington.*
The geologists will tell you that the Wrekin, and its foothill the Ercall, are composed of Pre-Cambrian volcanic rocks that resisted the elements which eroded the softer rocks around them. Not so; this is really what happened. A giant bore a grudge against the people of Shrewsbury and set off from his home in Wales with a huge spadeful of earth with which to dam the River Severn and so flood the town. However, he lost his way and meeting a cobbler carrying a bag of old boots asked him how far it was to Shrewsbury. The cobbler asked what he wanted there and on being told was horrified, after all he would lose many of his customers. Thinking quickly he said "Shrewsbury! You'll never get to Shrewsbury, neither today nor tomorrow. Why, look at me! I've just come from there and I've worn out all these boots, it's so far". The giant was tired and hungry. He threw down his load of earth, so making the Wrekin, scrapped his boots on the spade, so making the Ercall, and strode home back to Wales. There is more about the Wrekin in the articles on Wellington and Cluddeley.

WROCKWARDINE *2m. W of Wellington.*
Wrockwardine – the name means 'the homestead by the Wrekin' – is an old village of some character that has attracted an unfortunate development of bland modern houses. However, it must be said that most of these have been kept away from the ancient centre. There is a Post Office and general store, now fighting for its life; a Church of England Primary School of 1837; and a church hall. The church itself is dedicated to St. Peter and in early Norman times belonged to Shrewsbury Abbey, the gift of Roger de Montgomery. The present church is still substantially Norman. It has a nave, chancel and transepts, and a crossing tower. Inside there are 2 chapels in the broad chancel, a finely carved Jacobean pulpit, several stained glass windows by William Kempe and a notable tablet memorial to William Chudde, d.1765. A later Chudde, Edward, of Orleton has an even grander memorial, namely the handsome, brick and stone Almshouses of 1841. Above the pointed-arch doorway is an inscription that tells us that he '. . . was an eminent example of pure and undefiled religion visiting the fatherless and widows and keeping himself unspotted from the world'. Indeed, it is said that the reason for the absence of a pub in the village is that the local squires were inclined to Puritanism. There is, however, a village smithy, though nowadays he is an agricultural welder. Opposite the smithy, across the small traingular green, is a part-timbered cottage with a beautifully kept roadside rockery. There are wide views northwards over the Weald Moors from here. Wrockwardine is not short of large red brick farms but the 'big house' is Wrockwardine Hall; it is half of 1628, to the right, half Georgian of 1750, to the left, and was built for William Pemberton. The Georgian facade has 7 bays with a segmental pediment of fluted pilasters over the doorway and a hipped roof. Since 1948 the Hall has been the official residence of the Commander of the Western District (Army). The Old Vicarage is of mellowed brick with a good, white-painted cast-iron trellis porch. Under Anglo-Saxon rule Wrockwardine was held by King Edward as a Royal Manor. In 1086 it was the *caput*, the head, of the Hundred of Recordine (Wrockwardine), had 7½ outlying dependant settlements and was part of the personal estate of the great Earl Roger de Montgomery who in Shropshire acted on behalf of the king. There were 13 villagers, 4 smallholders, and 8 ploughmen and their families and a priest, a not inconsiderable community for the time. **Orleton** lies ½m. SE of Wrockwardine but is approached off the Holyhead Road, to the S. The name Orleton is Anglo-Saxon and means 'the settlement of the Earl'. It certainly was held by the Norman, Earl Roger, but the name probably refers to an earlier Mercian nobleman. There is no village, just one large Georgian style stuccoed house which has a main front of 9 bays with a 3-bay pediment and porch columns of Grinshill stone. The building is of several periods but especially of 1766 though something of a much earlier timber-framed manor house still survives in the fabric of the central part. Facing the entrance is a handsome, timber-framed Gatehouse, complete with a stone bridge over a water-filled section of the old moat. The Gatehouse is dated at 1588 but has been altered in later times – a Gothic dormer, an oval window, a Georgian lantern and renewed brickwork. The moat is popular with wild ducks several of whom zoomed in and made a very splashy landing whilst we were taking photographs. In the walled garden is a splendid gazebo (circa 1730) with an ogee roof and unusual windows. This is locally called The Bothy, and indeed was used as such in Victorian times when the gardener's assistants slept there. The present incumbents of the house are the Holt family. Mrs Holt is deputy chairman of the Telford Development Corporation. Her husband is V. M. E. Holt, nephew of the late E. R. H. Herbert, 5th earl of Powys who came to the property in 1952. Orleton is accepted as being the manor house, the head, of the large Domesday manor. It has even been suggested that the estate dates back to the 5th Century and that it took over as the centre of administration when the Romans left. It would, if that were true, be the site of the mysterious hall of Cynddylan, the long lost place called Pengwern which was burnt by the

Lower House, Worfield.

St. Andrew's, Wroxeter.

The bailey of the Norman castle at Woolstaston.

Mercians in about 660. The modern estate is only a fraction of its original size.

WROCKWARDINE WOOD *1m. NNW of Oakengates.*
Wrockwardine Wood is not near Wrockwardine; it lies 4m. E near Oakengates and became a separate parish in 1834. The settlement stands around a hill, a hill which from the early 14th Century to the mid-19th Century was torn apart as men worked the rich seams of coal and iron ore. In early medieval times it was heavily wooded and the central area was (and still is) called Cockshutt Piece, a reference to the practice of catching wild birds by hanging nets across clearings. Today it is wooded once more and is a most attractive place with stands of mature beech and sweet chestnut and areas of moor under a cover of rough grass, gorse and birch. Nevertheless, the old spoil mounds are still quite evident. The wood lies on the Silkin Way footpath and a good access point is off Lincoln Road on the W flank of the hill. Old cottages and small terraces of 19th Century houses abut the woods though modern bungalows have infilled to the point of becoming predominant. It is noticeable that as the houses get younger so they become grander; people are beginning to appreciate that the area has much to commend it. There are 2 pubs, the Red Lion and the Peasant, a Post Office and general store, and the Old Beehive Bakery. This little enclave is now cut off from the old centre of Wrockwardine Wood by the new, and busy Wrockwardine Wood Way. Down the hill, on the other side of the road, are: the attractive brick-built parish church of Holy Trinity (1833) by Samuel Smith of Madeley, restored by Gilbert Scott (1889); the White House pub (1734); the Victorian green and brown tile clad Bull's Head which is a listed building; the modest red brick Old Rectory in grounds with specimen trees; the Donnington Wood Mill (opened 1818, closed 1970), a steam powered corn mill complete with granary and baker's ovens; the yellow brick County Infants School (adjacent to the mill); the Social Centre for the Unemployed; the clumsy looking modern building that houses the substantial Oakengates Leisure Centre; the Methodist Chapel; and the Church of Jehovah's Witnesses. Wrockwardine Wood was best known for its coal and iron ore but there were other industries. Amongst these was a glassworks in which the iron master William Reynolds (1758-1803) had an interest. This works stood some 200 yds. SE of the parish church and produced a variety of products – buttons, striped rolling pins, door stoppers, table ware – but mostly they made bottles for the French wine trade. The present Old Rectory was the factory manager's house – the works was established about 1792, more than 40 years before the church was built. The basic raw material was slag from the Donnington Wood blast furnaces and Black Rock from Little Dawley. The works changed hands several times and, as Biddle and Mountford, closed in 1841. To the E and S of the central hill are modern housing estates. Adjoining Wrockwardine Wood to the NW is **Trench**; just where the boundary lies seems to be anybody's guess. Most of the older houses and shops lie along Trench Road, the A518. Here are Flanagan and Allen's pub, and a firm called Kiss Kiss. South of Trench Road are large modern housing estates and large modern schools. **Trench Lock** adjoins the Queensway (A442), a motorway in all but name. On its northern flank is a slope. This is the famous Trench Incline (of about 1797) which joined canals at 2 levels. It was 223 yds. long and had a fall of 75ft. and was the last working Incline in Britain when it closed in 1921. There is little to see today but the Blue Pig pub stands at the foot of the Incline, clearly marking the site. Blue Pig was the name given to slag skimmed off the molton iron in the blast furnaces. It is a hard, shiny, glass like material often used for road making and is sometimes mixed with sand and cement to form a hard concrete used for kerbstones and paving slabs. It could be green in colour as well as blue. To the left of the pub is Trench Pool, a canal feeder reservoir for the Shrewsbury Canal. This canal lost a lot of water because the lining was frequently cracked by settlement caused by mining. Today there are boats on the waters of the pool and fishermen frequent its shores. On the other side of Queensway, a little to the S, is Middle Pool. Both lakes have a good and varied population of wild water birds, including swans. Middle Pool now lies almost surrounded by high road embankments. It was originally 2 pools which were drained during construction work and the ground remodelled to make one pool. This was at the same time greatly extended northwards. There is a car park next to the pool with an access off Somerfield Road. The big pale green sheds on the other side of the highweay belong to Blockley's, manufacturers of bricks, tiles and chimney pots. The embankment between the highway and the factory conceals an enormous clay pit part of which is being back-filled. Blockley's use a lot of water. The waste is full of red clay particles and before the water enters the Middle Pool it is sent along an 'S' shaped settlement canal where most of the suspended clay is caught. This 'canal' is adjacent to the car park by the pool.

WROXETER *5m. SE of Shrewsbury.*
Wroxeter is a tiny hamlet with a noteworthy church, St. Andrew's. A section of the church's N wall is Anglo-Saxon, the stones being taken from Uriconeum, the nearby deserted Roman town; the chancel is Norman, as are the lower part of the tower and a small window in the gable; the S wall was rebuilt in the 18th Century when an aisle was removed; the porch is Victorian; the font was probably carved out of the base of a Roman column; there is a 13th Century chest; the box pews and pulpit are Jacobean; and the front entrance gate posts are pillaged Roman pillars. The church is redundant but is now in the care of English Heritage. Adjacent to the church is the stuccoed, cream painted Grange and opposite is the drive to the Boathouse, a cottage by the River Severn. In the field to the left of the Boathouse are considerable earthworks and a stream. These were part of the southern defences of Uriconeum. The river is relatively shallow here and there are several small islands and mud flats, a natural crossing place. To the left of the church is the Wroxeter Hotel. There is a house called The Cottage, which it is not, and a black and white dwelling which really is. Three hundred yards NE of Wroxeter are the ruins of Uriconeum, (*not* Viriconeum, the Roman symbol for U was a V). This was the fourth largest city in Roman Britain and its defences enclose an area of 180 acres. It was the capital of the Roman province called Britannia Secunda. Uriconeum Cornovoirum, to give the town its full name, is on a low plateau overlooking the Severn in territory which belonged to the Celtic-British Cornovii tribe who had ruled much of central Mercia from their stronghold on the Wrekin before the Roman conquest. Watling St., one of the great Roman roads, crossed the Severn here; traces of a timber bridge have been found on the mid-stream island near the Boathouse. Uriconeum was established as a base fort for the conquest of Wales. When Chester assumed command of this task in AD.78 Uriconeum was developed as a civil town. The Cornovii were assimilated and used it as their capital. The town had a regular street grid, a public bath house, a water supply and a drainage system. This was a civilised, cultivated place and for over 300 years the Roman administrators and the local people lived ordered, settled lives. The large upstanding part of the ruins, called Old Work, was part of a large 4-roomed building the purpose of which is unknown. Most of the rest of the uncovered foundations are part of the 2nd Century Baths. On the opposite side of the road from the exposed and enclosed site is a row of 16 column bases. They lie in a trench below the present ground level and are the remains of the colonnade that fronted the W entrance of the City Forum. Over the portico of this entrance was a stone tablet with a most beautifully carved inscription. This tablet is now at the Victoria and Albert Museum in London. Aerial photography has established the line of the earthwork defences which totally surround the town, even on the river side. (Wroxeter and its church lie within these boundaries.) The flat lands around Uriconeum were probably already under cultivation by the Cornovii before the Romans arrived, a factor no doubt taken into consideration when choosing this site in the first place. The Legionaries had to be fed. After the Romans left there is some evidence that the town continued to be at least partially occupied: the tomb of a late 5th Century chieftain called Cunarix; a Frankish throwing axe of the same date; and an area of timber buildings to the NE of the site, identified from crop marks. It would seem, however, that the stone city itself was abandoned. Quite probably the Cornovii simply did not feel at ease there and reverted to their previous life style in the interests of safety

Wroxeter, the Severn near the Roman crossing point.

when the Roman legions and law were no longer there to protect them. Aligned with the Roman defensive ditches are many pits, some rectangular in shape. These could very well be the below-ground areas of early Anglo-Saxon houses, or huts. They would certainly have been attracted to this flat, already cultivated land. The abandoned city became ruined and in time was "quarried" by local people for building material – Atcham and Wroxeter churches both contain Roman stone – and simply cleared away by farmers wishing to cultivate the land. Even when the remains of the city were mostly buried under the soil farmers would dig up stone when required. They noticed that in times of drought the crops or grass above an area of stone quickly became parched and turned a lighter colour, thus indicating where to dig. It was Thomas Telford who made the first serious archaeological excavations at Wroxeter and it was probably he who first discovered the baths. Since then the site has been extensively studied and is still being investigated. The excavations are open to the public and there is a small museum. Note: There is sometimes confusion over the names Cornovii and Wrocensaetna. The Cornovii were the Celtic-British tribe who the Romans found in occupation of what we today call Shropshire. They remained after the Romans left. The term Wrocensaetna was first used to describe the people of the area by Anglo-Saxon scribes when the Tribal Hidage was compiled in 660.AD. The name could be derived from either Wroxeter or the Wrekin. We do not know. The majority of the inhabitants were still the Cornovii, but the term Wrocensaetna means what it says – the people of Wroxeter, or the Wrekin, regardless of bloodstock – and could include other peoples, such as Latins and Anglo-Saxons, not only Cornovii.

WYKE, THE *1m. SW of Shifnal.*
A red brick hamlet that lies in a gently rolling, wooded, landscape off the A4169. There is an Equestrian Centre attached to a farm and some ramshackle barns. Cows and corn country. The name Wyke is probably from the Old English *wic*, meaning 'a dairy-farm'. **Nedge Hill** lies 1m. WNW at SJ.717.073. There is a picnic place in the woods here and good views to both the E and the W.

WYRE FOREST *13m. E of Ludlow.*
The ancient Wyre Forest stretched from Bewdley in Worcestershire to Cleobury Mortimer. There is still a large wooded area within the county, along the border. Most of it is in the hands of the Forestry Commission and much of it is coniferous desert, but not all; there are still some mixed deciduous woods and areas of heath and moor to be enjoyed. The majority of the forest is only accessible to the walker. The name Wyre is, of course, from the River Wyre, which may mean 'winding'.

WYTHEFORD *See Great Wytheford.*

YEATON *6½m. NW of Shrewsbury.*
The name means 'the settlement on the banks of the river'. Yeaton lies by the River Perry, a pleasant hamlet with cottages some of half-timber and some of red brick and boundary walls of sandstone. A mile to the SW, set amidst its landscaped park, is the spectacular mansion of Yeaton Pevery. It was built in 1890-2, an irregular, neo-Jacobean fantasy by Aston Webb and very well done. It looks good from any angle and has not yet attracted unsympathetic modern satellite buildings. It was last used as a girl's school but has been empty for 18 years and is now up for sale. The present owner is Sir David Wakeman who still lives on the estate in a much smaller, modern house. Webb's big house has been sadly neglected but is still most handsome. It is build of sandstone, with a finely carved first storey half-timbered central section on the garden side; mullioned and transomed windows; an entrance tower; a tall turret; gables; a raised patio; a walled garden and 2, domed summer-houses. Inside, several rooms are wood panelled. Four hundred yards to the NE of the house are moated earthworks, presumably the site of the medieval farmstead.

YEYE *Somewhere near Stanton Lacy.*
Yeye was a small village last recorded in 1350 when it was then in ruins. It lay in the important manor of Stanton Lacy, 3m. NW of Ludlow.

YOCKLETON *7m. WSW of Shrewsbury.*
A small village on the B4386 Shrewsbury to Montgomery road. Exotic butterflies from all over the world, not to mention tarantulas and scorpions, water fowl and a children's pets corner can be found at Country World which lies, well signposted, just off the main road. In 1066 the manor of Yockleton was held by the Anglo-Saxon Edric. By 1068 it had passed to Roger, son of Corbet. It was a substantial manor with at least 33 families resident here, a mill 'which pays one packload of malt' and woodland for fattening 100 pigs. The Domesday Book name is Ioclehuile which is from the Old English *iocled*, meaning 'a small manor', and *hyll*, meaning 'a hill'. The 'hill' ending was later altered to 'ton'. The present church of Holy Trinity was built in 1861 to a design by Edward Haycock Junior. It is constructed of stone of three different colours and has dormer windows in the spire which are not much admired. There is a Church of England School, a new housing estate at Brookside Gardens, a Post Office, a garage, a weather-boarded barn, a black and white cottage, a stone bridge over the railway, a garden machine centre, a stream, a Norman motte (a mound 12 ft. high, 36 ft. wide, and 77 ft. long, NE of the church), and that rarity, a village blacksmith, Mr. Eddie Price. The Grange, a substantial house with gables, bay windows and dormers, is now an old persons home. The fields are mainly laid to pasture, grazed by dairy cows, but there is some arable farming and pigs and chickens are no strangers hereabouts.

Glossary of Architectural Terms

Apse A semi-circular or polygonal end to the chancel and or a chapel of a church. It is usually vaulted.

Arcade A range of arches supported by either columns or piers and usually of structural importance. *(See Blind Arcade.)*

Ashlar Blocks of squared stone, smooth faced, even in size and usually laid in courses like bricks. The blocks can either be structural or merely a facing to a brick built building.

Bailey An enclosed and protected open space within the walls of a castle.

Bargeboards Boards which hide the join of the roof and the wall on a gable and which are often ornamented.

Battlement The alternately raised (merlons) and lowered (embrasures) configuration of a parapet, usually associated with the tops of the walls and towers of a castle but also used as a decoration on civilian buildings. An alternative word for battlemented is crenellated.

Bay A bay is a vertical division of a building either by internal or external features. The most common feature used to define the number of bays is the arrangement of the windows as seen from the outside. *(See the illustration of 'Queen Anne Style' house. It has 7 Bays.)* The building in the drawing has 5 bays.

Beaker People Late Neolithic (New Stone Age) people who came to Britain from Europe. They introduced weapons and tools made from metal and buried their dead in round barrows.

Belfrey It can be either a specially built bell tower or the upper floor of the main church tower where the bells are hung. In the Middle Ages a belfry was a moveable seige tower and later came to mean a a watch tower from which the alarm was raised by ringing a bell.

Bellcote A simple framework from which bells are hung. It is often an extension of a gable end but can mean anything that literally 'houses a bell'.

Broach Spire When an octagonal tapering spire meets the square tower on which it stands, sloping half-pyramids of wood, brick or stone are positioned at the base of the four oblique faces of the spire to 'square the octagon'.

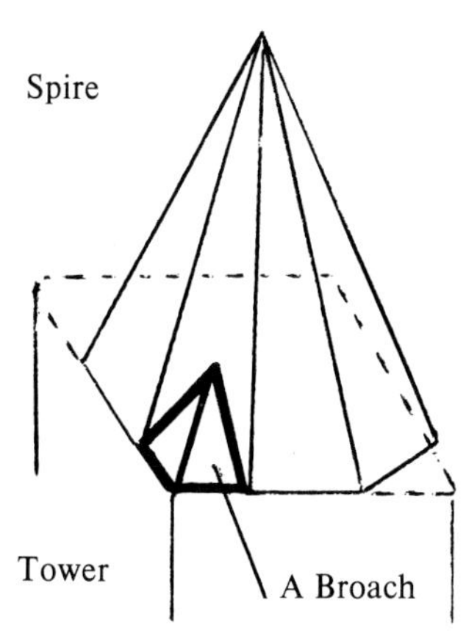

Blind Arcade A series of arches supported by columns or pilasters arranged against a wall almost always as a decorative, not structural, feature. The Normans were very fond of it.

Broken Pediment A pediment in which the horizontal base line has a gap at the centre.

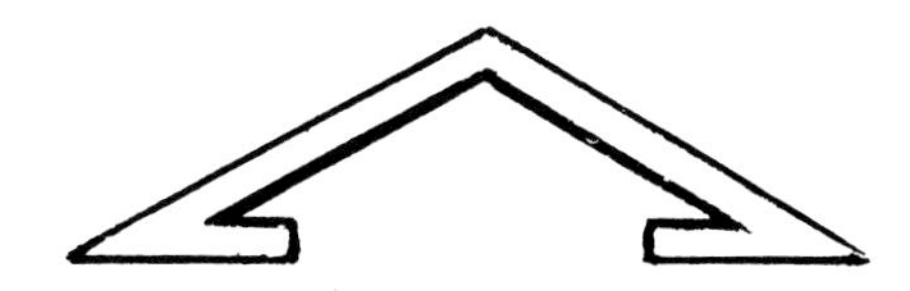

Bronze Age Bronze is a mixture of tin and copper. It was the first metal that man was able to work with effectively. In Britain the Bronze Age lasted from about 2000 B.C. (the end of the Stone Age) to about 600 B.C. (the beginning of the Iron Age).

Canted Bay Window A protruding window the front of which is straight and parallel to the main building, and which is joined to the main building by side walls set at an angle of more than 90 degrees.

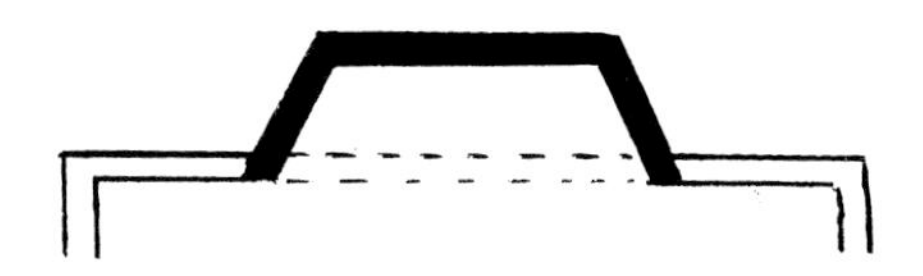

Cable Moulding A decorative band usually carved in stone or wood imitating a twisted rope or cable and much used by the Normans.

Capital The top part of a column or pilaster usually highly decorated. It tapers outwards from the column beneath to provide a platform from which an arch may spring or upon which an entablature may rest.

Cartouche An ornament in the form of a tablet or shield with an ornate frame, often in the form of scrolls made to look like paper, and which usually bears either an inscription or a coat of arms.

Castellated Identical in meaning to 'battlemented'.

Chancel The part of the church, usually the east end, where is placed the altar. In a cruciform church it is the whole of the area east of the crossing. The sanctuary is the area immediately around the altar.

Chantry Chapel A chapel either inside or attached to a church or other religious foundation in which prayers were said, and the mass was chanted, for the founder or other particular person (often a relative of the founder). The founder usually built the chapel himself and endowed (i.e. left money or land) to the church to pay for the clergy to officiate after his death.

Chevet A French word meaning the east end of the church.

Classical A building or style based on Roman or Greek architecture.

Clerestory Derived from clear-storey. The upper part of the main walls of a church, above the aisle roofs, which are pierced by windows. *(See Triforium diagram.)*

Cob Walls constructed of mud or clay mixed with straw and, or, cow dung. The exterior surface of the cob was white-washed to protect it from rain and properly maintained is a durable material. The houses of the poor in Britain were commonly made of cob, and some still exist in the south-west. However, if neglected and exposed to water they almost literally melt away. Whole villages, once abandoned, could disappear in a matter of years.

Corinthian See Order.

Cornice A projecting ornamental moulding around the top of a building, arch or wall. *(See illustration to 'Queen Anne Style' house.)*

Crenellation Means battlemented.

Crossing The area where nave, chancel and transepts meet.

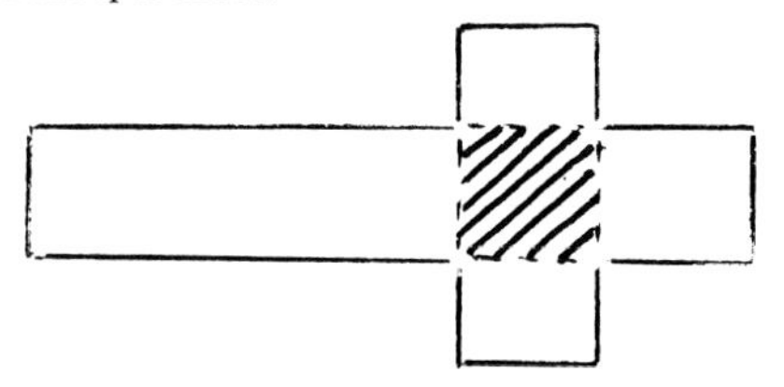

In a cruciform church the tower is usually built above the crossing.

Cruck A curved beam that supports both the walls and the roof of a timber-framed building.

It was usually obtained from a tree that had a major branch leaving the trunk at an angle of about 45 degrees. The tree was then cut down and the main branch and the trunk trimmed of side branches. This curved timber was then spilt in two. The two parts were then erected opposite each other to form an arch. Several such arches were constructed and these formed the basic structure of a 'cruck house'. The area between the arches were called bays.

A two-bay cruck house.

Cruciform In the shape of a cross, usually meaning a Christian cross.

Crypt A room below floor level and usually at least partly below ground level, most commonly positioned at the east end of a church underneath the chancel. The crypt normally contained graves and religious relics.

Cupola A small dome, usually circular but sometimes polygonal, that either crowns the roof of a building or a tower or caps a turret.

Decorated (Sometimes called 'Middle Pointed'.) A style of English Gothic current in the period 1300-1350 or thereabouts. As the name suggests ornamentation was rich in this period. *(See Gothic.)*

Doric See Order.

Dormer A window placed vertically in the sloping plane of a roof. If it lay in the plane of the roof it would be a skylight. *(See illustration of 'Queen Anne Style' house.)*

Early English Sometimes abbreviated to 'E.E.' and sometimes called 'Lancet' or 'First Pointed'. Early English is the earliest style of English Gothic and was current between 1200 and 1300. It is characterised by a simplicity of line; by tall, pointed, untraceried, lancet windows; and by slender spires and tall piers. *(See Gothic.)*

Easter Sepulchre A recess in a wall (usually the N wall) of the chancel in which is a tomb chest. An effigy of Christ is placed here during the Easter ceremonies.

Embattled It means there are battlements. *(See Battlement.)*

Fan Vault A vault is an arched interior roof of stone or brick. A fan vault is a late medieval vault where all the ribs from the same springer are of the same length, the same distance from each other and the same curvature.

Fluting Vertical grooves or channels in the shaft of a column.

Foliated As of foliage; the materials are ornamented with carvings of plant foliage, especially leaves.

Gable The triangular upper part of a wall that rises to the slopes of a pitched roof. The gable at the side of a building is called the End Gable. Gables facing to the front or back of a building are simply called gables.

Gallery In a church a gallery is an upper floor over an aisle which is open to the nave. There is also sometimes a gallery at the west end to house the organ.

Gazebo A small pavilion or summerhouse in a garden or park. When placed on the roof of a building it is called a Belvedere.

Georgian Buildings constructed in England during the reigns of of the four King Georges, 1714-1830, in a style influenced by Greek and Roman (Classical) ideas are said to be Georgian. Neatness, formality and symmetry are to the fore. Famous architects associated with the Georgian styles include Robert Adam, John Nash, James 'Athenian' Stuart and James Wyatt. Many of the great English country houses were either built or were given new facades in the 18th Century.

Gothic Gothic is a style that can only be defined by listing its characteristic features: the ribbed vault, the pointed arch, the flying buttress, the traceried window, the lofty steeple, the panelled stonework, the triforium, spacious clerestory windows etc. When all or many of these elements are present in a building it can be described as Gothic. The word 'Gothic' was not used during the period called Gothic, that is between the later 12th Century and the mid-16th Century. It was first used in the late 16th Century to describe the architecture of the previous centuries. It was a term of abuse. To say the style was Gothic meant that it was barbaric and uncivilised, as were the ancient German tribes called Goths. Gothic architecture had absolutely no connection with the Gothic tribes themselves. The style, though always recognizable as one style, underwent changes over the years. The main historical sub-divisions in England were:

Early English	1200-1300
Decorated	1300-1350
Perpendicular	1350-1550

These dates, of course, overlap and are only a rough guide. It should be remembered that individuai elements of the Gothic style can be present in buildings that are not Gothic. It is the coming together of several elements that defines the style. Gothic evolved out of the Norman style and so some later Norman churches might best be described as Transitional.

Hammerbeam Horizontal timbers 'projecting' out from opposite sides of the walls supported by brackets. From these, vertical timbers called hammer-posts rise to support the purlins.

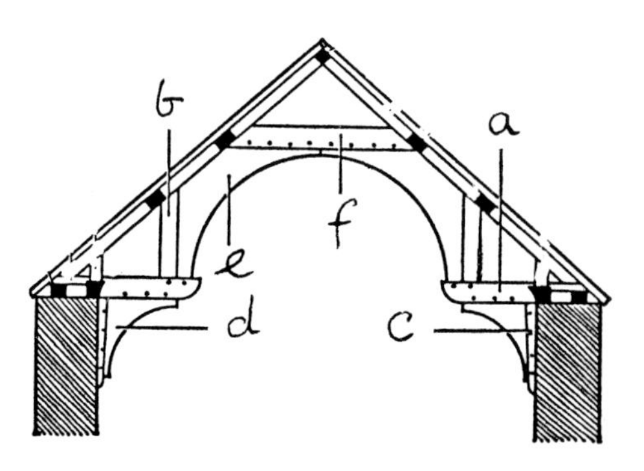

a – hammer beam
b – hammer post
c – bracket
d – brace
e – arched brace
f – collar beam

Herringbone A design created by laying bricks, stones or tiles so as to make a zig-zag pattern.

Hipped Roof A roof which has sloping ends, not vertical ends with gables as is usual. *(See illustration of 'Queen Anne Style' house.)*

Ionic See Order.

Iron Age The Iron Age began when iron working was introduced into Britain by invading Celtic tribes who arrived on these shores in about 600 B.C. The Iron Age lasted until the coming of the Romans which was effectively A.D.43. However, in some remote areas the Iron Age continued until the coming of the Anglo Saxons in the 6th Century A.D.

Linenfold Tudor panelling, usually of wood to an interior wall, decorated with a stylized representation of a piece of linen laid in vertical folds. The pattern is contained within one panel and repeated in every other panel.

Lower Palaeolithic See Palaeolithic.

Lancet A tall, slender, one-light window with a sharply pointed arch, much used in early Gothic 13th Century architecture. They can be either single or arranged in groups, usually of 3, 5, or 7.

Mansard A pitched roof with two slopes, the lower being longer and steeper than the other. Named after Francois Mansart (sic).

Mesolithic Means Middle Stone Age, the period of man as hunter and fisher from about 8,000 B.C. to 3,500 B.C.

Megalith A large stone or block of stone, often irregular and either rough hewn or left as found. Such stones were erected by prehistoric man as monuments between about 4,000 B.C. and 1,000 B.C.

Middle Pointed Another name for the period of English Gothic more commonly called Decorated. *(See Decorated and also Gothic.)*

Misericord Sometimes called a Miserere. A bracket attached to the underside of a hinged choir stall seat, which, when the seat is turned upright, provide support for a person in the standing position. They are frequently carved and because they are not seen by the congregation the subjects are sometimes non-religious and on occasion even verge on the profane.

Motte Means a mound, usually of earth, on which the main fort of a motte and bailey castle was built.

Mullion A vertical post that divides a window into seperate lights. The term is usually reserved for substantial uprights of brick or stone. *(See Transom).*

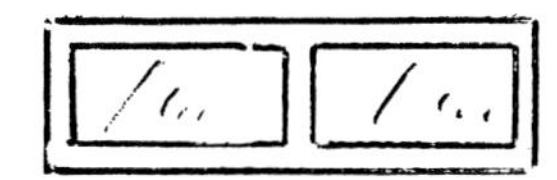

Nave The western part of a church, often with aisles to the sides which are occupied by the congregation.

Neolithic Means New Stone Age, the period of pre-history in Britain from about 3,500 B.C., when settled farming communities were established, to the emergence of the Bronze Age about 1,800 B.C.

Norman What in England is called Norman is really Romanesque. The Norman architects and builders did not have a style particular to themselves. They built in the manner current on the Continent at that time. Indeed, there are in England examples of 'Norman' work – such as Westminster Abbey – that were constructed before the Conquest. *(See Romanesque.)*

Obelisk A pillar of square section that tapers to the top and ends with a pyramid. It can be of any size but is usually of stone, large and erected as a monument.

Ogee An arch introduced into Britain about 1300 and very popular during the 14th Century. It has a characteristic double curve, one concave and the other convex, and is best described by a drawing:

Order In Classical architecture an Order is a complete system which comprises a column and its associated base shaft, capital and entablature (a horizontal table or lintel supported by the columns). The number of inventions and permutations is endless but a mere handful achieved popular acceptance in Rome and Greece and are still used. They are Greek Doric, Roman Doric, Tuscan Doric, Ionian and Corinthian.

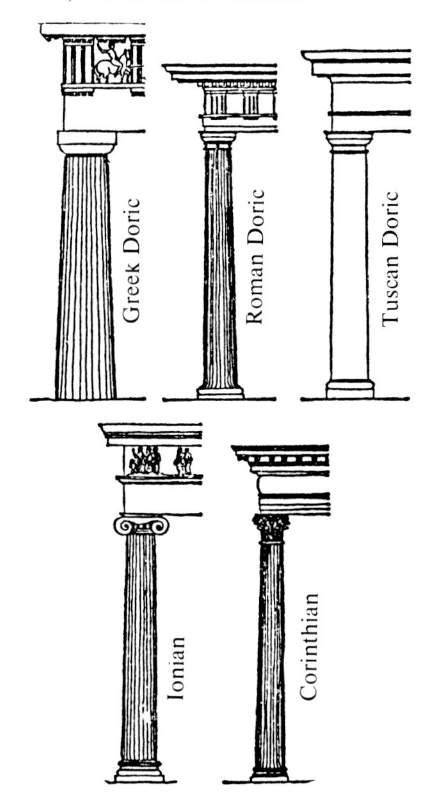

Palaeolithic Means Old Stone Age and lasted from about 26,000 B.C. to the start of the Middle Stone Age (the Mesolithic) about 8,000 B.C. As it is such a long period it is sometimes divided into Lower (the early part) and Upper (the later part).

Palimpest When a metal plate, usually of brass, is re-used by engraving on the back the plate is said to be a palimpest; likewise a parchment re-used after removing the original writing; and a wall painting where one painting overlaps and obscures an earlier work.

Pantile A tile with an 'S' configuration. When laid such tiles give a roof a corrugated surface.

Pediment A low-pitched gable much used in Classical architecture and essentially of an ornamental nature. Pediments can be found not only at roof level but above windows, doors and porticos. *(See illustration of 'Queen Anne Style' house.)*

Perpendicular A style of Early English Gothic current in the period 1350-1550. It is characterised by the vertical lines in the window tracery and the vertical articulation of the panelling in the stonework – hence Perpendicular. *(See Gothic.)*

Pier A pillar-like support but square (or composite) in section, usually of masonry. The brickwork or stonework between windows and doorways in a building can be described as piers.

Pilasters A shallow pier or rectangular column projecting only slightly from a wall.

Piscina A stone basin, complete with drain, in which the Communion or the Mass vessels are washed. It is usually set in or against the wall south of the altar.

Porte Corchere A coach porch.

Portico A porch with columns and a pediment decorating the main entrance to a building. If it projects forward, as is commonly the case, it is said to be 'prostyle'. If it recedes into the building (with the columns arranged in line with the front wall) it is said to be 'in antis'.

Presbytery Has two meanings. It is part of the church east of the choir where the high altar is placed. It is also the name given to the house of a Roman Catholic priest.

Priory A monastery whose head is a Prior or Prioress, as distinct from the Abbot or Abbess of an Abbey.

'Queen Anne Style' A style developed in England by Eden Nesfield (1835-88) and Richard Norman Shaw (1831-1912) and very much influenced by 17th Century Dutch brick architecture and the William and Mary style (1689-1702). *(See the illustration)* Queen Ann reigned from 1702-1714.

Queen Anne house

Quoins Stones, usually dressed, at the corners of a building. Sometimes they are of equal size; more commonly they alternate long and short. *(See illustration of 'Queen Ann Style' house. It has long and short quoins.)*

Recusant A person who refuses to submit or comply. Most commonly means a practising Catholic in the period when that religion was suppressed, that is, between the Reformation and Catholic Emancipation in the early 19th Century.

Reredos A wall or screen, usually well decorated, that lies behind the altar in a church.

Rib-vault A vault with diagonal ribs projecting along the groins. 'A framework of diagonal arched ribs carrying the cells which cover the spaces between them.'

Rock-Faced Masonry cut to regular blocks but deliberately roughened on the exposed surface to look rough-hewn. The aim is to look natural, but usually the result is most unnatural.

Roll Moulding A decorative moulding of semi-circular and more than semi-circular (but less than circular) cross section.

Romanesque A Continental term for what in England is called Norman, and which covers architecture of the 11th and 12th Centuries. The style is characterised by the round arch, thick walling, small windows, bays clearly marked internally by vertical shafts from floor to ceiling, arcading, tunnel vaults and apses. Ornamentation was vigorous and depicted foliage, birds, animals and monsters, and utilised various bold geometric designs such as zig-zags and chevrons.

Rood A cross or crucifix (Saxon).

Rood Screen A screen which separates the chancel from the nave. Above, or fixed to, it is often a rood (a crucifix or cross).

Saddle-Back Roof This is a normal pitched roof, as on an ordinary house, but placed above a tower.

Saltire Cross A cross with four equal limbs laid diagonally i.e. an X shaped cross.

Sarcophagus A coffin, usually of stone and ornamented with carvings.

Screen A porclose screen separates a chapel from the rest of the church.

Screens Passage The passage between the kitchen and other work places, and the screen that protected the privacy of the occupants of the great hall of a medieval house.

Sedilia Masonry seats for use by priests on the south side of the chancel of a church. There are usually three in number.

Segmental Arch An arch with a profile that is part of a circle in which the centre of the circle is below the springing line; that means that it will always be less than a semi-circle.

Solar An upstairs living room in a medieval house.

Spandrel The triangular space between the side of an arch and the horizontal line drawn through its apex and the vertical line drawn from the side of the opening below the arch.

Spire A tall cone, pyramid or polygon that is placed on top of a tower. it has no structural purpose and its function is primarily to act as a landmark.

Steeple A term that means the tower and spire of a church taken together.

Strapwork Decoration consisting of interlaced bands similar to leather straps. It can be open (like fretwork), as in a vertical partition or screen, or closed as in plasterwork on ceilings. It was especially popular during the Renaissance.

String Course A horizontal band projecting from an exterior wall.

Stucco Plaster or cement rendering to walls or ceilings. The term is most commonly used to describe external wall rendering which is usually given a smooth finish.

Three-Decker Pulpit A tall pulpit with three seats, one above the other: a reading desk, the clerk's stall, and the preacher's stand.

Tomb Chest A stone coffin shaped like a chest, often with an effigy of the deceased placed on top. Some times the side walls are elaborately carved. It was commonly used in medieval times.

Tracery In the Middle Ages it was called form-pieces or forms. It is the intersecting ornamental work in the upper part of a window, a screen, a panel, a blank arch or a vault.

Transept The transverse arm of a cruciform church which usually projects out from the junction of the nave and the chancel.

Transom A horizontal bar across a window opening which divides the window into separate lights. The term is usually only applied to a substantial bar, one commonly made of stone.

Triforium A galleried arcade facing on to the nave between the wall arcade (below) and the clerestory (above).

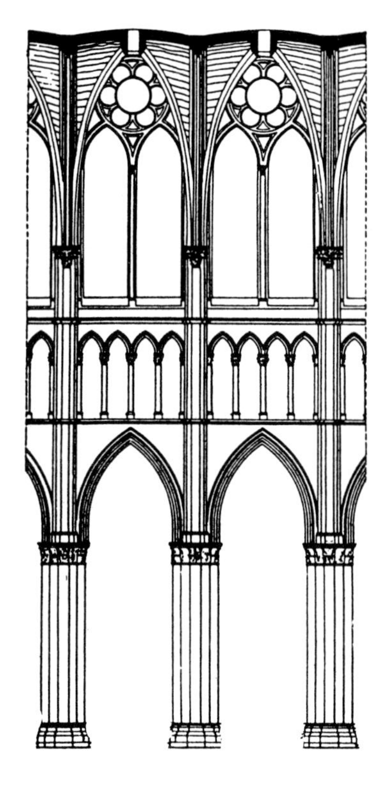

Turret a small tower, round or polygonal in shape.

Tuscan See Order.

Undercroft A vaulted room under a church or other building, sometimes below ground level.

Upper Palaeolithic See Palaeolithic.

Venetian Window A tripartite window. The central part is arched and wider than the square-topped parts to either side. It was developed from the Italian Serliana.

Victorian Architecture and general style associated with the period during which Queen Victoria reigned in Britain, 1837-1901.

Weather-Boarding Horizontal exterior overlapping boards, usually of a timber-framed building but sometimes used decoratively on a stone or brick structure.

A Brief Guide to Shropshire

GEOLOGY OF THE SOUTH SHROPSHIRE HILLS

A very brief guide.

Clee Hills (Brown Clee and Titterstone Clee). Volcanic rocks capped by Coal Measures lifted up by volcanic action.

Clun Hills Old red sandstone and Silurian limestone. Clun itself is built on Old Red Sandstone.

Long Mynd Pre-Cambrian sedimentary rocks.

Pontesford Hill Pre-Cambrian volcanic rocks.

Stiperstones Ordovician quartzites and shales; lead, zinc, barytes and fluorspar deposited in fissures during later volcanic action.

Stretton Hills (Ragleth, Helmeth, Caer Caradoc and The Lawley). Pre-Cambrian volcanic rocks overlain by later Cambrian rocks on lower eastern flanks. These hills lie along the E side of the Church Stretton fault.

Wenlock Edge Silurian limestone, probably formed as a huge reef in a shallow sea, like the present Great Barrier Reef in Australia.

The Wrekin Pre-Cambrian volcanic rocks overlain on eastern flank by Cambrian rocks.

Note: Sir Roderick Murchison, one of the founding fathers of modern geology, did much work in the area during the 1830's, especially on the Silurian system.

MAJOR PREHISTORIC SITES

Many sites are now occupied by towns, castles and moated mansions. Listed here are the most important sites not developed in later ages.

Baschurch, The Berth, a lowland hill fort, possible site of Pengwern, SJ.429.236.

Bishop's Castle, Bury Ditches, 7 acre hill fort, SO.327.837.

Bromfield, Bronze Age burial mounds, SO.496.776.

Brown Clee Hill, Nordy Bank, 4 acre hill fort, SO.576.847.

Caynham Camp, 8 acre hill fort SO.545.736.

Chapel Lawn, Caer Caradoc, 3 acre hill fort, SO.310.758.

Chirbury, Mitchells Fold, Bronze Age stone circle, SO.304.983.

Church Stretton, Bronze Age barrows along The Portway on The Long Mynd.

Church Stretton, Caer Caradoc, 8 acre hill fort, SO.545.736.

Craven Arms, Norton Camp, 10 acre hill fort, SO.447.820.

Offa's Dyke, King Offa's 8th Century Anglo-Welsh boundary bank and ditch that stretched from Prestatyn to Chepstow. Best seen in Shropshire from Ceiriog Valley SJ.264.375 to SJ.253.324; from Oswestry SJ.295.300 to SJ.256.283; from Chirbury SO.233.988 to SO.246.947; from Clun SO.258.898 to SO.256.827 and south of Lower Spoad, near Newcastle.

Old Oswestry, 15 acre lowland hill fort, SJ.296.310; also, sections of Wat's Dyke, the mysterious predecessor to Offa's Dyke can be seen by Old Oswestry and the west side of Oswestry cattle market.

Titterstone Clee Hill, small remains of 70 acre hill fort, SO.592.779.

The Wrekin, hill fort capital of the Cornovii, SJ.629.083.

Wroxeter, Uriconium, Roman city, SJ.565.086.

MOATED SITES

If a moat encircles a substantial building of stone it will be listed as a castle. The Ordnance Survey map references give the exact location of each site.

Acton Burnell	SJ.529.019
Alberbury Priory	SJ.375.151
Alcaston	SO.459.870
Albright Hussey	SJ.523.176
Battlefield Priory	SJ.375.151
Bearstone	SJ.724.394
Belswardine	SJ.609.025
Blakemere	SJ.560.425
Bower	SO.555.722
Bromfield	SO.479.769
Calverhall	SJ.603.377
Cherrington	SJ.666.202
Cleeton St. Mary	SO.608.791
Cloverley Hall	SJ.612.372
Fauls	SJ.586.327
Gadlas	SJ.373.371
Hadnall	SJ.522.199
Hanwood	SJ.447.094
Harlescott Grange	SJ.502.160
High Ercall Hall	SJ.594.173
Highfields	SJ.510.309
Humphreston Hall	SJ.818.050
Humphreston	SJ.814.050
Hunkington	SJ.565.141
Hurst	SJ.355.073
Ightfield Hall	SJ.600.394
The Isle	SJ.457.167
Langley Hall	SJ.540.002
Leafields	SO.646.725
Lea Hall	SJ.583.385
Leahead	SJ.760.422
The Lees	SJ.668.263
Longnor, The Moat House	SJ.494.002
Lower Newton	SJ.487.313
Ludstone Hall	SO.800.945
Middle Morrey	SJ.624.403
Middleton	SO.296.987
New Marton	SJ.340.345
Newstreet Lane	SJ.627.373
Northwood	SJ.493.311
Old Park Farm	SJ.714.005
Pool Hall	SO.768.838
Shawbury	SJ.561.212
Shifnal	SJ.746.074
Snitton	SO.557.754
Soudley	SJ.731.297
Soulton Hall	SJ.545.304
Stanwardine	SJ.427.277
Stapleton, Moat Farm	SJ.457.035
Startlewood	SJ.387.206
Syllenhurst	SJ.725.427
Thonglands	SO.549.891
Uppington	SJ.592.087
Upton Cressett	SO.656.924
Watling Street Grange	SJ.722.113
Wem	SJ.599.305
Westhope	SO.467.859
Whitley Grange	SJ.453.096
Whixall	SJ.504.337
Willaston	SJ.597.360
Wistanstow	SO.422.861
Wollerton Wood	SJ.609.311
Woodhouse (Priory)	SO.647.771

MONASTERIES AND OTHER MEDIEVAL RELIGIOUS HOUSES

An entry here does not necessarily mean that there are any physical remains.

Alberbury Priory, Grandmontine monks.

Battlefield, chantry college of St. Mary Magdalene.

Brewood Priory, Augustinian canonesses.

Bridgnorth, chantry college of St. Mary Magdalen; Franciscan friars; Hospital of St. John; Hospital of St. James.

Bromfield Priory, Benedictine monks.

Buildwas Abbey, Cistercian monks.

Chirbury Priory, Augustinian canons.

Halston Preceptory, Knights Hospitallers.

Haughmond Abbey, Augustinian canons.

Lilleshall Abbey, Augustinian canons.

Lydley Preceptory, Knights Templars.

Ludford, Hospital of St. Giles.

Ludlow, Augustinian friars; Carmelite friars; Palmers' Guild; Hosier's Almshouses; Hospital of St. John.

Morville Priory, Benedictine monks.

Newport, Town Almshouses; chantry college of St. Mary.

Oswestry, Hospital of St. John.

Preen Priory, Cluniac monks.

Ratlinghope Priory, Augustinian canons.

Shrewsbury Abbey, Benedictine monks; Franciscan friars; Dominican friars; Augustinian friars; Hospital of St. George; Hospital of St. Giles; Hospital of St. John; St. Chad's Almshouses; Drapers' Almshouses; chantry college of St. Chad; chantry college of St. Mary.

Tong, chantry college of St. Bartholomew.

Wenlock Priory, Cluniac monks.

Wombridge Priory (Telford), Augustinian canons. (The site is opposite the church.)

Woodhouse, Augustinian friars. (Woodhouse is a remote spot on the lower E slopes of Titterstone Clee Hill near Hopton Wafers.) The hermit friars arrived here in 1250.

CASTLES

This list includes small Norman mottes (indicated by N in parenthesis), substantial fortified houses and ruins of castles however slight; it does not include pre-historic sites, unless used as forts at a later period, nor moated mansions, which are listed separately. Most of the castles listed here are described in the gazetteer. Richard's Castle and the 2 castles at Montgomery are not listed here though they are described in the gazetteer.

Acton Burnell Castle	SJ.534.019
Adderley Castle	SJ.665.404
Alberbury Castle	SJ.357.144
Apley Castle	SJ.656.132
Belan Bank	SJ.342.220
Beguildy, The Moat	SO.188.805
Bicton	SO.289.826
Binweston (N)	SO.302.041
Bishop's Castle	SO.393.891
Bishop's Moat	SO.291.896
Bretchel (N)	SJ.302.041
Bridgnorth Castle	SO.717.927
Broadward (N)	SO.394.766
Brockhurst	SO.446.925
Brockton (N)	SO.580.933
Brogyntyn	SJ.274.314
Bromlow (N)	SJ.320.024
Brompton Motte	SO.251.932
Broncroft Castle	SO.545.867
Bryn Amlwg Castle	SO.167.846
Bucknell (N)	SO.356.739
Burford	SO.594.686
Caus Castle	SJ.337.078
Charlton Castle	SJ.597.112
Cheney Longville Castle	SO.417.848
Cheswardine Castle	SJ.719.301
Cleobury Mortimer Castle (site)	SO.674.757
Cleobury Mortimer, Castle Toot	SO.682.758
Clun Castle	SO.299.809
Clungunford (N)	SO.395.788
Colebatch (N)	SO.320.871
Corfham Castle	SO.525.850
Corfton (N)	SO.496.847
Culmington (N)	SO.497.822
Dawley Castle (site)	SJ.687.063
Dudston	SO.245.974
Ellesmere Castle	SJ.404.346
Gwarthlow (N)	SO.252.955
Hardwick (N)	SO.368.906
Hawcocks Farm Ring	SJ.349.078
Heath Farm, Ring and Bailey	SJ.379.113
Hisland (N)	SJ.317.275
Hockleton (N)	SO.270.999
Holdgate Castle	SO.562.897
Hodnet Castle	SJ.614.284
Hope (N)	SJ.344.023
Hopton Castle	SO.367.779
Knockin Castle	SJ.336.224
Lea Castle	SO.351.892
Leebotwood (N)	SO.446.991
Lee Brockhurst (N)	SJ.546.273
Leigh Hall	SJ.333.037
Little Ness (N)	SJ.407.197
Little Shrawardine (N)	SJ.393.151
Lower Down (N)	SO.336.846
Ludlow Castle	SO.508.746
Lydham Castle	SO.334.910
Marshbrook (local tradition)	SO.445.897
Marton	SJ.290.026
Meole Brace	SO.334.910
Middlehope (N)	SO.499.886
Minton (N)	SO.432.906
More Castle	SO.339.914
Moreton Corbet Castle	SJ.561.231
Myddle Castle	SJ.468.236
Newcastle (N)	SJ.244.821
Oswestry Castle	SJ.292.298
Pan Castle	SJ.526.404
Pennerley (N)	SO.351.994
Petton (N)	SJ.441.262
Pontesbury Castle	SJ.401.058
Pulverbatch Castle	SJ.423.022
Quatford Castle	SO.739.908
Red Castle	SJ.572.295
Rorrington (N)	SJ.303.004
Rowton Castle	SJ.379.128
Ruyton Castle	SJ.393.223
Ryton (N)	SJ.761.029
Sandford (N)	SJ.581.344
Shrawardine Castle	SJ.401.184
Shrewsbury Castle	SJ.388.206
Smethcott (N)	SO.448.994
Stapleton (N)	SJ.581.344
Stoke Upon Tern Castle	SJ.646.276
Stokesay Castle	SO.436.817
Tong Castle	SJ.792.069
Tyreley Castle	SJ.678.330
Upper Millichope Lodge	SO.522.892
Wattlesborough Castle	SJ.355.126
Wem Castle	SJ.504.288
West Felton (N)	SJ.340.252
Weston Under Redcastle (N)	SJ.564.292
Whittington Castle	SJ.326.312
Wilcott (N)	SJ.379.185
Wilderley (N)	SJ.433.017
Willaston (N)	SJ.597.359
Wilmington (N)	SJ.433.017
Wollaston (N)	SJ.328.123
Woolstaston (N)	SO.450.985
Wotherton (N)	SJ.280.007
Yockleton (N)	SJ.396.103

LOST VILLAGES

Listed here are settlements that have their own entry in Domesday Book but which no longer exist. Some are known by a later name now fossilized as the name of a wood, a hill, a field, etc., and some are mentioned in later medieval documents by a different name, e.g. Munton was Muletune in Domesday Book. The Ordnance Survey reference is given in a shortened 4 figure form which gives the square but not the exact position of the site within it.

Charlton	SJ.56.22
Chatsall	SJ.63.31
Corfham	SO.52.85
Little Eaton	SJ.52.04
Fouswardine	SO.67.85
Hawkesley	SJ.53.00
Horsewell (Wales)	SO.20.98
Lydley Hayes	SO.39.65
Marston	SO.52.85
Munton	Witterley Hundred
Slackbury	Baschurch Hundred
Stanway (Herefordshire)	SO.40.70
Starcote (Wales)	SO.21.98
Thornbury (Wales) called The Gaer	SO.20.99
Tibeton	Oswestry Area
Yagdons	SJ.59.19

The following list is of lost villages the location of which is either unknown or unsure. Some have disappeared but others may still exist under different, later, names with which they cannot be connected.

Bolebec
Bosle
Buchehale
Burtone
Courtone
Cheneltone
Chinbaldescote
Cleu
Estone
Fech
Goseford
Humet
Lel
Newetone
Stantune
Sudtelch
Tumbelawe
Tuneston
Udeford
Wiferes Forde.

There are also many small lost settlements not named in Domesday Book but which are known to have existed from references to them in later medieval documents.

BUILDINGS MOST VISITED BY TOURISTS

All sites are open to the public though some have restricted visiting hours. For further information on these properties see 'A Shropshire Gazetteer', by M. Raven.

Acton Burnell Castle. A 13th Century fortified manor house built by the King's Treasurer.

Adcote. A large Victorian mansion with a superb baronial entrance hall. Now a school but open to visitors in the summer.

Aston Eyre Church. The Anglo-Saxon carving on the tympanum is one of the county's treasures.

Acton Round Hall. A handsome 18th Century dower house to Aldenham Hall.

Aston Munslow, The White House. A curious dwelling with a collection of farm implements.

Attingham Park. The noblest country house in Shropshire built for Noel Hill, First Lord Berwick.

Benthall Hall, near Broseley. A fine 16th Century stone house with a collection of Caughley china.

Boscobel House. Famous as hiding place of Charles II, and site of The Royal Oak. Nearby is White Ladies, the only nunnery in Shropshire.

Bridgnorth, Bishop Percy's House. The famous collector of old poetry was born here on the steep Cartway.

Bridgnorth Castle. The Norman keep lies at a crazy angle after having been blown up by Cromwell's men.

Bromfield Church And Gatehouse, near Ludlow. Remains of a 13th Century Benedictine Priory.

Broseley, The Lawns. Near the Church, John Wilkinson and John Rose both lived here. There is a collection of pottery.

Buildwas Abbey, near Ironbridge. Substantial and very good remains of Norman abbey of about 1200. Magnificent arcades in the nave.

Chirbury, Mitchell's Fold Stone Circle. Prehistoric religious site with splendid views over mysterious hills.

Claverley Church. Has rare medieval wall paintings and an ancient yew tree. A beautiful village.

Cleobury Mortimer, Mawley Hall. An 18th Century mansion. The bland exterior belies beautiful interior decoration.

Clun Castle. A home of the FitzAlan family. Striking ruins with huge earthworks. Charming, quiet town with a rare medieval bridge.

CONDOVER HALL. The best Elizabethan mansion in Shropshire. Now a school for the blind.

DUDMASTON HALL, Quatt, near Bridgnorth. A 17th Century house with collections of furniture and Dutch flower paintings.

EARDINGTON, DANIEL'S MILL. Near Bridgnorth. Has largest working water-wheel in Britain.

HAUGHMOND ABBEY. Ruins of a 12th Century Augustinian monastery.

HEATH CHAPEL. An almost unaltered Norman church. It stands alone on the site of a deserted village.

HOPTON CASTLE. A Norman castle with a 14th Century stone keep. The Civil War defenders were battered to death in a muddy pit.

LANGLEY CHAPEL. A 16th Century chapel with early 17th Century furniture and layout.

LILLESHALL ABBEY. The ruins of a 12th Century monastery. It has a ghostly monk.

LLAN-Y-BLODWELL CHURCH. It has a unique pencil tower. The beautiful river is crossed by a handsome stone bridge.

LUDLOW CASTLE. Shropshire's largest castle. Old home of the powerful Mortimer family. The Princes of the Tower, Catherine of Aragon and Mary Tudor were guests here.

MORETON CORBET CASTLE. Spectacular ruins of one of the most perfect buildings in England, an Elizabethan mansion in the French style built beside a Norman keep.

MUCH WENLOCK, THE PRIORY. A Norman monastery on an ancient Anglo-Saxon site. The Prior's Lodge is held in high regard. Attractive small town.

OLD OSWESTRY HILL FORT. The most impressive (and easily accessible) of the many Bronze Age-Iron Age forts in Shropshire.

SHIPTON HALL. A handsome Elizabethan mansion built of stone with a good dovecote, barn, stables and secret tunnel.

SHREWSBURY ABBEY. The great Roger de Montgomery was buried here but his tomb is lost. Only the church remains, and that has been reduced. The Council wanted the road widened so down came the monastic buildings.

SHREWSBURY CASTLE. Parts of the perimeter wall and the gateway are 11th Century but much rebuilt in the late 13th Century by Edward I.

SNAILBEACH LEAD MINES. A romantic place. Old mine buildings, chimneys, rail tracks, sealed shafts and mountains of white spoil.

STOKESAY CASTLE. Internationally famous and rightly so. Mostly the work of Lawrence de Ludlow, a wool merchant, in the 13th Century. The gatehouse is Elizabethan.

STOTTESDON CHURCH. It has the finest carved stone font in the County and an Anglo-Saxon tympanum behind the organ.

TONG CHURCH. "The Village Westminster Abbey." Noble monuments, the Golden Chapel and the Grave of Little Nell.

TITTERSTONE CLEE HILL FORT. The summit of this hill was the site of the most important prehistoric religious centre in Shropshire. The huge fort had stone ramparts but has been largely destroyed by roadstone quarrying.

TYN-Y-RHOS HALL, WESTON RHYN. A Tudor house with some older parts. Owain Gwynedd, a Welsh Prince, stayed here. Has a ghost.

UPTON CRESSETT HALL. Long forgotten, recently restored Elizabethan mansion and gatehouse with medieval great hall. Nearby is the site of a deserted village.

WEST FELTON, ST. WINIFRED'S WELL. A true delight. A watery dell you will never forget, but not easy to find. (See A Shropshire Gazetteer by M. Raven.)

WESTON PARK, Staffordshire. The Earls of Bradford no longer live here. Very much a tourist place.

WHITTINGTON CASTLE. Spectacular ruins and roadside moat of once massive fortress of the FitzWarin family in medieval times.

WILDERHOPE MANOR, near Easthope. A gaunt stone mansion of the 16th Century in a secluded valley. Now a Youth Hostel.

WROXETER, THE ROMAN CITY OF URICONIUM. Extensive ruins of the 4th largest town in Roman Britain. Coins are found regularly in the fields about.

WATER MILLS OPEN TO THE PUBLIC

CHADWELL MILL, near Newport, Sundays April to end of October.

CLEOBURY MORTIMER, Claybury Mill, Pinkham, irregular opening times.

EARDINGTON, near Bridgnorth, Saturday and Sunday from Easter to end of September.

WINDMILLS

All the buildings listed are tower mills

ALBRIGHTON, site of	SJ.818.042
ALBRIGHTON, house	SJ.802.039
ASTERLEY, Pontesbury, being renovated	SJ.372.075
BISHOP'S CASTLE, site of	SO.327.897
CLUDDELEY, Wrockwardine, derelict	SJ.630.104
COTONWOOD, Tilstock, house	SJ.542.351
DITTON PRIORS (Hillside), derelict	593.877
ELLESMERE, site of	SJ.406.342
HADLEY PARK, Telford, derelict	SJ.657.115
HADNALL, house	SJ.523.210
HARLEY, site of	SJ.597.018
HAWKSTONE, Weston-u-Redcastle, derelict	SJ.566.297
HOWL, Sambrook, derelict	SJ.695.235
LONGFORD, Newport, derelict	SJ.718.181
LOPPINGTON, site of	SJ.463.301
LYTH HILL, Bayston Hill, derelict	SJ.469.067
MADELEY Court, Telford, derelict	SJ.695.053
MUCH WENLOCK, derelict	SJ.624.008
RODINGTON, remains only	SJ.590.144
ROWTON, Cardeston, derelict	SJ.365.129
SHIFNAL (Upton), remains only	SJ.756.067
VENNINGTON, Westbury, house	SJ.337.096
WHIXALL, site of	SJ.518.352
WROXETER, site of	SJ.588.075

MUSEUMS

ACTON SCOTT Farm Museum, end March to end October.

ASTON MUNSLOW, White House Museum of Buildings and Country Life, April to October.

BRIDGNORTH Museum, Northgate, April to end September. Postern Gate, Museum of Childhood and Costumes, open all year. Midland Motor Museum, Stanmore, open all year.

CLUN TOWN Trust Museum, Easter to end October.

COSFORD Aerospace Museum, open all year except weekends December to February.

IRONBRIDGE Gorge Museum, on several sites, most open all year: Coalbrookdale Museum of Iron, Jackfield Tile Museum, Coalport China Museum, Blists Hill Open Air Museum, The Tar Tunnel, Bedlam Furnaces and the Iron Bridge itself.

LONGDEN, Coleham, Pumping Station, appointment only.

LUDLOW MUSEUM, Butter Cross, mid-March to end September except Sundays.

LUDLOW Museum, Old Street, all year, weekdays for study and research.

MUCH WENLOCK Museum, April to September except Sunday.

ONIBURY, the Wernlas Collection of Rare Poultry, Easter to end of October.

OSWESTRY Bicycle Museum, all year except Sunday and Thursday afternoon.

SHREWSBURY, Clive House, College Hill, open all year except Sunday and Monday morning; Reabrook Centre of Catering and Management Studies, Radbrook Road, all year weekdays; Rowley's House Museum, open all year except Sundays mid-September to March; Shrewsbury Castle, Shropshire Regimental Museum, open all year.

WEST FELTON, the Oswestry Military Museum, open all year.

FAMOUS PEOPLE ASSOCIATED WITH SHROPSHIRE

SIR HAROLD ACTON. Several Acton's (once of Adenham Hall, Morville), achieved high social standing but Sir Harold achieved immortality when he was caricatured by Evelyn Waugh in 'Brideshead Revisited.'

RICHARD BAXTER. Puritan divine born at Eaton Constantine where his house still stands.

ADMIRAL JOHN BENBOW (1653–1702). Fought pirates in the Caribbean. Born in Shrewsbury.

WILLIAM BETTY. Theatrical child prodigy born at Wem.

WILLIAM BROOKES. Founded first modern Olympic Games at Much Wenlock in 1850.

CHARLOTTE BURNE. The editor of the internationally acclaimed Shropshire Folk Lore (1883), based on the field work of Georgina Jackson. Miss Burne lived at Pyebirch Manor, near

Eccleshall, Staffordshire.

SAMUEL BUTLER (1835–1902). Novelist and author of 'Erewhon', lived in Ludlow for a time.

RANDOLPH CALDECOTT (1846–1886). Illustrator of children's books worked as a bank clerk at Whitchurch for 6 years.

ROSAMUND CLIFFORD. 'Fair Rosamund', mistress of Henry II may have been born at Corfham Castle, near Diddlebury.

ROBERT CLIVE (1723–1774). 'Clive of India', born at Styche Hall, near Market Drayton, buried in an unmarked grave at Moreton Say.

ABRAHAM DARBY. Came to Coalbrookdale in 1690 from Bristol. Developed process of smelting iron with coke.

CHARLES DARWIN. Born at The Mount, Frankwell, Shrewsbury. His statue sits outside the town library, the old Shrewsbury School.

HENRY WALFORD DAVIES. A minor composer and native of Oswestry.

BENJAMIN DISRAELI. Politician and confidante of Queen Victoria. He was MP for Shrewsbury from 1841 to 1847.

ARTHUR CONAN DOYLE (1859–1930). Creator of Sherlock Holmes. As a young man he was a doctor's assistant at Ruyton XI Towns.

SIR EDWARD GERMAN (1862–1936). Born at Whitchurch. Famous for light operas such as 'Merrie England.'

WILLIAM HAZLITT. Essayist. Lived in Noble Street, Whitchurch as a youth.

HENRY HILL HICKMAN (1800–1830). A pioneer of anaesthesia by inhalation. He was a native of Bromfield, near Ludlow.

GENERAL ROWLAND HILL. Wellington's right hand man at the Battle of Waterloo. Is perched on a column opposite the Shirehall.

SARAH HOGGINS. A maid of Great Bolas who married Lord Burleigh and whose sad story is told in 'The Cottage Countess'.

A. E. HOUSMAN (1859–1928). Through his poetry did much to popularize Shropshire. His remains are interred at St. Laurence's, Ludlow.

AGNES HUNT. Born at Boreatton Park, Baschurch. Founded the famous orthopaedic hospital at Park Hall, Oswestry, with the assistance of the surgeon Robert Jones.

LORD JOHN HUNT. Leader of the expedition that conquered Mount Everest. Has a home near Llanfair Waterdine.

EGLANTYNE JEBB. (1876–1928). Born at The Lyth, Ellesmere. Founded the Save the Children Fund.

JUDGE GEORGE JEFFREYS. Bought the barony of Wem and lived at Lowe Hall, Wem, for the last 2 years of his life.

HUMPHREY KYNASTON. A 15th Century nobleman turned robber. Kynaston's Cave, a rock dwelling, can be seen in the woods at Nesscliffe.

WILLIAM LANGLAND. 14th Century poet believed to have lived near Cleobury Mortimer. Famous for his 'The Vision of Piers Ploughman'.

THOMAS MINTON (1766–1836). A native of Shropshire worked at the Caughley Pottery before founding his own company at Stoke-on-Trent.

ROGER DE MONTGOMERY. William the Conqueror's best friend and one of the first three great Marcher Lords. Montgomery, once in Shropshire, is named after him.

FRANCIS MOORE (1657–1715). Astrologer and originator of the famous 'Old Moore's Almanac'. He was born at Bridgnorth.

'MAD JACK' MYTTON. Wild aristocrat who gambled away the family fortunes. Born at Halston Hall, near Whittington; died in a debtors prison in London.

WILFRID OWEN. Wilfrid not Wilfred. Born at Oswestry in 1893. His enormous poetic talent had barely begun to flower when he was cut down at the end of the First World War. He also lived at 69, Monkmoor Road, Shrewsbury.

THOMAS PARR (1483–1635). Lived to be 153 years old. Ate mainly dairy products. His cottage stands at SJ.306.112 near Winnington, near Halfway House. He is buried at Westminster Abbey.

BISHOP PERCY (1729–1811). Famous for his Relics of Ancient English Poetry. His handsome house stands in the Cartway, Bridgnorth.

SIR GORDON RICHARDS (1904–1987). Twenty times champion jockey. Born Oakengates and worked for the Lilleshall Company before turning to the track.

SIR PHILIP SIDNEY (1554–1586). Former pupil of Shrewsbury School. Soldier, poet, scholar and courtier – the archetypal 'Renaissance Man'.

HESBA STRETTON (1832–1911). Best selling authoress Sarah Smith was born at Wellington. Later she lived at Church Stretton (hence her pen name). 'Jessica's First Prayer' sold over a million copies. At the time this was phenomenal.

SIR JOHN TALBOT. The 'Scourge of France'. Killed in the battle of Castillon in 1453. Buried at Whitchurch. Amongst his titles was that of Earl of Shrewsbury. This is now the oldest earldom in England. The present earl lives near Uttoxeter in Staffordshire. As the holder of the oldest title he is 'the Premier Earl of England'.

RICHARD TARLETON. Queen Elizabeth I's Court jester. He was born at the Pyepits, Condover. A feared wit and a fine swordsman.

THOMAS TELFORD (1757–1834). A Scottish stonemason made good as County Surveyor of Shropshire. Telford is named after him. Probably the country's greatest civil engineer.

PERCY THROWER. The best known gardener in Britain, even though he is no longer with us. He was Parks Superintendent at Shrewsbury and lived at Merrington, near Bomere Heath.

JOHN WEAVER (1673–1760). A native of Shrewsbury. A dancing master who is believed to have brought the pantomime to England.

MARY WEBB (1881–1927). A regional novelist born at Leighton. 'Precious Bane' was dramatised on TV in 1988.

CAPTAIN MATHEW WEBB (1848–1883). The first man to swim the English Channel. Born at Dawley. Died attempting to swim the Niagara rapids.

STANLEY WEYMAN. A best-selling Victorian novelist.

JOHN 'IRON MAD' WILKINSON. The leading 18th Century producer of iron and iron goods in the West Midlands. Made many innovations at his works at Willey, near Broseley, before moving to the Black Country.

WILLIAM WITHERING (1741–1799). Born at Wellington. He developed the use of digitalis in the treatment of heart disease.

P. G. WODEHOUSE (1881–1975). As a young man Wodehouse lived at Hay's House, Stableford, near Bridgnorth. Many of his fictional settings are, in fact, descriptions of local houses and landscapes.

WILLIAM WYCHERLEY (1640–1716). The author of 'The Country Wife' and other rather coarse comedies, was born at Clive.

PRIVATE GARDENS OPEN TO THE PUBLIC

BURFORD House Gardens, end of March to end of October.

MARKET DRAYTON, Willoughbridge (Staffordshire), Dorothy Clive Memorial Garden, beginning of March to end of November.

HODNET HALL Gardens, April to end of September.

NESSCLIFFE, Oak Cottage Herb Farm, all year but 'phone first.

YOCKLETON, Shropshire Country World (Butterfly World), Easter to end of October.

PRIVATE RAILWAYS

BRIDGNORTH Castle Hill Railway, a 2 car funicular railway between High Town and Low Town, closed Christmas Day and Boxing Day.

OSWESTRY, the Cambrian Railway Society, Oswald Road, old station, rolling stock, engines, Sundays 10 a.m. –4 p.m.

SEVERN VALLEY Railway, steam trains run between Bridgnorth and Kidderminster weekends March to October and daily mid-May to early September.

HORSEHAY. The Telford Horsehay Steam Trust, loco shed, 1m. track, 4 steam engines, 2 diesels. Also the oldest auto-trailer in Britain.

OUTDOOR SHOWS

BISHOP'S CASTLE Traction Engine Rally, August Bank Holiday. Tel. Baschurch 260595.

BRIDGNORTH Festival, Spring Bank Holiday. Tel. Munslow 383.

BURWARTON SHOW, first Thursday in August. Tel. Stoke St. Milborough 309.

COSFORD Air Day, a Sunday in June. Tel. Albrighton 4872.

CHENEY LONGVILLE (Craven Arms), the Shropshire Game Fair, first weekend

after the May Bank Holiday. Tel. Craven Arms 2708.
CRUCKTON, the Horse and Tractor Ploughing Championship of the British Isles, last Saturday in September. Tel. Church Stretton 722701.
MARKET DRAYTON, the Tern Valley Vintage Machinery Club Rally, third weekend in September. Tel. Cheswardine 333.
MINSTERLEY Show, second Saturday in August. Tel. Shrewsbury 88446.
NEWPORT Show, third Saturday in July. Tel. Stafford 822532.
OSWESTRY Show, early August. Tel. Oswestry 654875.
SHREWSBURY Flower Show, middle of August. Tel. 64051; West Midlands Show, to be announced. Tel. 62824.
TELFORD Show, Town Park, Bank Holiday Monday. Tel. Telford 505370.

REGATTAS
BRIDGNORTH Regatta, second weekend in June. Tel. Bridgnorth 4101.
IRONBRIDGE Regatta, last weekend in June. Tel. Telford 613007.
SHREWSBURY Regatta, third weekend in May. Tel. Hadnall 442.
SHREWSBURY, the Severn Head of the River Race, third Saturday in October. Tel. Hadnall 442.

YOUTH HOSTELS
BRIDGES, Hugh Gibbins Memorial Hostel. Tel. Linley 656.
CLUN, The Mill. Tel. Craven Arms 582.
COALBROOKDALE, Ironbridge, the Coalbrookdale Institute. Tel. Ironbridge 3281.
EASTHOPE, Wilderhope Manor. Tel. Longville 363.
LUDLOW, Ludford Lodge. Tel. Ludlow 2472.
SHREWSBURY, The Woodlands, Abbey Foregate. Tel. Shrewsbury 260179.
WHEATHILL, Malthouse Farm. Tel. Burwarton 236.

COUNTRYSIDE LEISURE AREAS
The following sites are managed by the County Council and are open to the public.
BROWN MOSS (near Prees Heath) 80 acres heathland, scrub woodland, marsh and fishing pools.
CANTLOP BRIDGE, car park and picnic area.
CHURCH STRETTON, Old Rectory Wood and Rectory Field, 16 acres of woodland and 7 acres of pasture.
COLEMERE, the mere and 57 acres of woodland. Sailing.
ELLESMERE, The Mere and adjoining Cremorne Gardens. Like a seaside resort on sunny summer evenings.
GRINSHILL, Corbet Wood, 23 acres of woodland, including Scots pines, and old quarry workings.
LYTH HILL, 70 acres of hillside, good views.
NESSCLIFFE and Hopton Hill, 150 acres of woodland.
PONTESBURY, Poles Coppice, 47 acres of broadleaved woodland, 1m. S of village.
SHELVE, The Bog lead mine, old workings partly landscaped and information boards.
WELLINGTON, Ercall Wood, 38 acres of woodland.

The following are organized Forestry Commission Walks.
BURY DITCHES, near Clun. 3 walks. Temporary car park. From Clunton on the B4368 take the road to Brockton for 2.5 miles. The car park is on the left. SO.334.839.
HIGH VINALLS, near Ludlow. 4 walks. Car park from where the walks start can be found 4 miles from Ludlow, on the Wigmore road. SO.474.732.
HOLLY COPPICE, Haughmond Hill, near Shrewsbury. 2 walks. Take B5062 from Shrewsbury towards Haughmond Hill. Just after passing Haughmond Abbey turn right and the Forestry Commission car park is on the right. (Completion date for car park April 1989). SJ.545.148.
WHITCLIFFE, near Ludlow. 2 walks. Park in the Forest Office car park – weekdays and in the main gateway/bellmouth at weekends. SO.494.742.
WYRE FOREST, Hawkbatch. 2 walks. Car park 3 miles from Bewdley on B4194. SO.761.777.
WYRE FOREST, VISITOR CENTRE, Callow Hill, near Bewdley. 3 walks. Car park at Wyre Visitor Centre, 3 miles west of Bewdley on A456. The newly redesigned visitor centre and refreshment bar will be open in June 1989. SO.753.740.

The following are areas of either outstanding natural beauty or of scientific interest not included in the preceeding two lists.
BENTHALL EDGE. Coppiced woodland in the Ironbridge Gorge. Signposted trail. Car park is by the Iron Bridge, opposite the town.
BROWN CLEE HILL. Highest hill in Shropshire. Moor and woodland nature reserve.
CARDING MILL VALLEY. A narrow valley near Church Stretton. Very popular in summer with picnicers by the stream.
CEFN COCH. Forestry Commission picnic site at SJ.242.329, near Oswestry.
EARL'S HILL, PONTESFORD. Shropshire Conservation Trust pamphlet available.
EDGE WOOD, WENLOCK EDGE. In remote situation at SO.478.877. Pamphlet available for trail.
HOPESAY HILL. A National Trust heathland. Lovely views over the South Shropshire hills.
LLANYMYNECH. The hills here have been mined for copper and limestone since Roman times. Dramatic cliffs and Offa's Dyke.
THE LONG MYND. Ten miles long with numerous small hills and valleys on the eastern flank. The Portway, a prehistoric ridgeway, is a favourite walk. Grand moorlands and spectacular views.
OSWESTRY OLD RACECOURSE. Two miles West of Oswestry. A high ridge near Offa's Dyke. Good views.
THE STIPERSTONES. Legend ridden outcrop of quartzite rocks shattered in Ice Age frosts. Harsh, dramatic country with remains of old lead mines.
WENLOCK EDGE. A steep, tree-clad limestone slope that runs for 15 miles between Much Wenlock and Craven Arms.
WHITECLIFFE. An area of common land near Ludford. Superb view over Ludlow and the Clee Hills beyond.
THE WREKIN. The best known hill in the West Midlands. The hill fort here was the capital of the Cornovii, the pre-Roman tribe who ruled much of the border country. Best explored from the car park at Forest Glen, SJ.638.093.

MARKET DAYS OF PRINCIPAL TOWNS
EC = Early closing day.
BISHOP'S CASTLE, Friday.
BRIDGNORTH, Monday and Saturday (EC Thursday).
CHURCH STRETTON, Thursday (EC Wednesday).
CLEOBURY MORTIMER, alternate Wednesdays.
CLUN, Tuesday.
LUDLOW, Monday, Friday and Saturday (EC Thursday).
MARKET DRAYTON, Wednesday (EC Thursday).
MUCH WENLOCK (EC Wednesday).
NEWPORT, Friday and Saturday (EC Thursday).
OSWESTRY, Wednesday, and Saturday in Summer (EC Thursday).
SHIFNAL (EC Thursday).
SHREWSBURY, Tuesday and Friday (EC Thursday).
WHITCHURCH, Friday (EC Wednesday).

CINEMAS
BRIDGNORTH, the Majestic Cinema (2 screens).
OSWESTRY, the Regal Cinema (3 screens).
SHREWSBURY, the Empire Cinema (1 screen); the Music Hall Cinema (1 screen).
TELFORD, AMC Cinema Complex (10 screens).
WELLINGTON, Clifton Cinema (1 screen).

LEISURE CENTRES
BRIDGNORTH Sports and Leisure Centre, Northgate.
CLEOBURY MORTIMER, Lacon Childe Sports Centre.
DAWLEY, Phoenix Centre, College Road.
KETLEY Golf and Squash Centre.
LILLESHALL Hall, National Sports Centre.
MADELEY Court Sports Centre.
MUCH WENLOCK Sports Centre, Farley Road.
OSWESTRY Leisure Centre, College Road.
SHIFNAL, Idsall Sports Centre, Coppice Green Lane.
SHREWSBURY, London Road Sports Centre.
STOKE HEATH (nr. Market Drayton), Maurice Chandler Sports Centre.
STIRCHLEY Recreation Centre.
TELFORD Ice Rink, Telford Town Centre.

Racquet Centre, St. Quentin Gate.
WEM, Adams Sports Centre, Lowe Hill Road.
WROCKWARDINE WOOD, Oakengates Leisure Centre, New Road.

CRICKET AND FOOTBALL

CRICKET

Shropshire has a cricket team that plays in the Minor Counties League, Western Division. In the 1985 National Westminster Bank Trophy Competition Shropshire defeated the Yorkshire first class county side. For details of local clubs contact Shropshire Libraries' Central Information Service, 1A, Castle Gates, Shrewsbury. Tel. 0743 254506.

FOOTBALL

There are 2 professional football clubs in Shropshire: Shrewsbury F.C., Gay Meadow, Shrewsbury; and, Telford United F.C., Bucks Head, Watling Street, Wellington. For details of local clubs contact Shropshire Libraries' Central Information Service, 1A, Castle Gates, Shrewsbury. Tel. 0743 254506.

SWIMMING POOLS

BISHOP'S CASTLE High School (private).
BRIDGNORTH Sports and Leisure Centre, Northgate.
BRIDGNORTH, St. Mary's Junior School.
CHURCH STRETTON Junior School (private).
CLEOBURY MORTIMER, Lacon Childe School.
DONNINGTON (outdoor).
ELLESMERE Junior School.
HIGHLEY (outdoor).
KETLEY Recreation Centre, Holyhead Road (outdoor).
LUDLOW, Dinham Bridge.
MADELEY, Court Centre.
MARKET DRAYTON (outdoor).
MUCH WENLOCK (outdoor).
NEWPORT, Victoria Park.
OSWESTRY Leisure Centre.
PONTESBURY, the Mary Webb School (private).
SAINT MARTIN'S, near Oswestry, Rhyn Park School.
SHREWSBURY, Quarry Swimming Centre, Priory Road.
WELLINGTON, Walker Street.
WEM, Bowen's Field.
WHITCHURCH.
WROCKWARDINE WOOD, Oakengates Leisure Centre.

NIGHT CLUBS

LUDLOW, the Starline Club. Tel. 3358.
OSWESTRY, the Empire Night Club. Tel. 656117.
NEWPORT, Main Street Night Club. Tel. 811949.
SHREWSBURY, The Buttermarket, Castle Foregate. Tel. 241455; Cheers, Riverside. Tel. 50669; Park Lane Night Club, Raven Meadows. Tel. 58786; Mr. Pews, Abbey Foregate. Tel. 53357.
TELFORD, Cascades Discotheque. Tel. 502233.
WELLINGTON, Barons Club. Tel. 242243.
WHITCHURCH, the Deja Vu Night Club. Tel. 3367.

GOLF CLUBS

BRIDGNORTH Golf Club. Tel. 3315.
CHURCH STRETTON Golf Club. Tel. 722281.
HAWKSTONE Park Hotel Golf Club. Tel. Lee Brockhurst 209.
LILLESHALL Hall Golf Club. Tel. Telford 603840.
LLANYMYNECH Golf Club. Tel. 830983.
LUDLOW Golf Club. Tel. Bromfield 285 or 366.
MARKET DRAYTON Golf Club. Tel. 2266.
MEOLE BRACE Golf Club. Tel. Shrewsbury 64050.
OSWESTRY Golf Club. Tel. Queens Head 221.
SHIFNAL Golf Club. Tel. Telford 460467.
SHREWSBURY Golf Club. Tel. Bayston Hill 2977; Municipal Golf Club. Tel. 64050.
MADELEY, the Telford Hotel Golf and Country Club. Tel. Telford 585642.
WELLINGTON, The Wrekin Golf Club. Tel. Telford 44032.

AERO-SPORT

SLEAP Airfield (near Wem), Shropshire Aero Club.
ASTERTON, Long Mynd, Midland Gliding Club.
PREES HEATH Airfield, Sport parachute jumping school.

HORSE RACING AND ICE HOCKEY

HORSE RACING. Ludlow Racecourse, Bromfield. National Hunt only. Tel. Bromfield 221.
ICE HOCKEY. The Telford Tigers, Telford Town Centre. Tel. Telford 291551.

MOTOR SPORT

HAWKSTONE PARK, near Hodnet, is a world renowned venue for motor-cycle scrambling, moto-cross as it is now called. It has a terrifying hill climb and an even dizzier descent.
LOTON PARK, near Alberbury, is a venue for motor car tarmac racing.

TOURIST INFORMATION CENTRES

BRIDGNORTH The Library, Listley Street. Tel. 3358.
CHURCH STRETTON The Library, Church Street. Tel. 722535.
IRONBRIDGE Museum Visitor Centre. Tel. 2166.
LUDLOW Information Centre, Castle Street. Tel. 3857.
NEWPORT Information Centre, St. Mary's Street. Tel. 814109.
OSWESTRY Mile End Service Area. Tel. 662488; The Library, Arthur Street. Tel. 662753.
SHREWSBURY The Music Hall, The Square. Tel. 50761.
WELLINGTON Information Centre, Walker Street. Tel. Telford 48295.
WHITCHURCH Civic Centre, High Street. Tel. 4577.

NEWSPAPERS

LUDLOW Advertiser, Upper Galdeford, Ludlow.
MARKET DRAYTON Advertiser, 81 Shropshire Street, Market Drayton.
NEWPORT ADVERTISER, 32, St. Mary's Street, Newport.
OSWESTRY AND WELSH BORDER SENTINEL, English Walls, Oswestry.
SHREWSBURY ADMAG LTD., 39 Hills Lane, Shrewsbury.
SHREWSBURY CHRONICLE, Castle Foregate, Shrewsbury.
SHROPSHIRE MAGAZINE, 77 Wyle Cop, Shrewsbury.
SHROPSHIRE STAR, Waterloo Road, Ketley, Wellington.
WHITCHURCH HERALD, Whitchurch.
WHAT'S ON MAGAZINE, 5a, Shoplatch, Shrewsbury.

GOVERNMENT

There are 8 divisions of local government.
THE COUNTY COUNCIL, Shrewsbury.
TELFORD DEVELOPMENT CORPORATION, Telford 293131 and 6 District Councils:
BRIDGNORTH, Bridgnorth 5131.
NORTH SHROPSHIRE, Wem 32771.
OSWESTRY Borough, Oswestry 654411.
SHREWSBURY and Atcham Borough, Shrewsbury 232255.
SOUTH SHROPSHIRE, Ludlow 4941.
THE WREKIN, Telford 505051.
There are 4 parliamentary constituencies in the County:
SHREWSBURY & ATCHAM.
THE WREKIN.
NORTH SHROPSHIRE.
LUDLOW.

Index to the Photographs

Index to the Gazetteer

Ludstone.

Lee Hall, near Ellesmere.

Stokesay Castle.

Elsick.

Bromcroft Castle.

Boscobel.

The etchings are from *Shropshire* by A. J. C. Hare, published by George Allen, London 1898.

Road map of Shropshire, 4 miles to 1 inch.

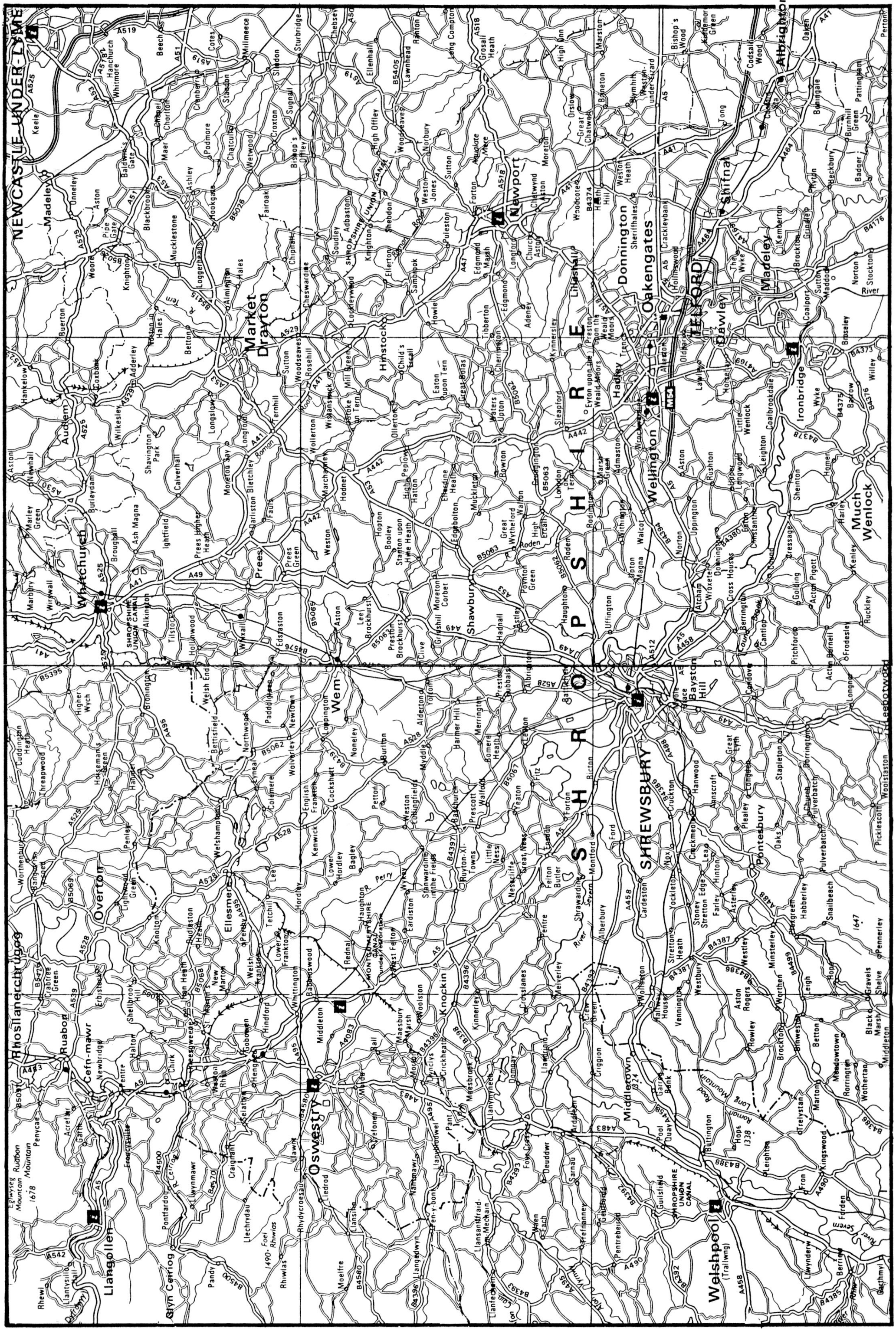

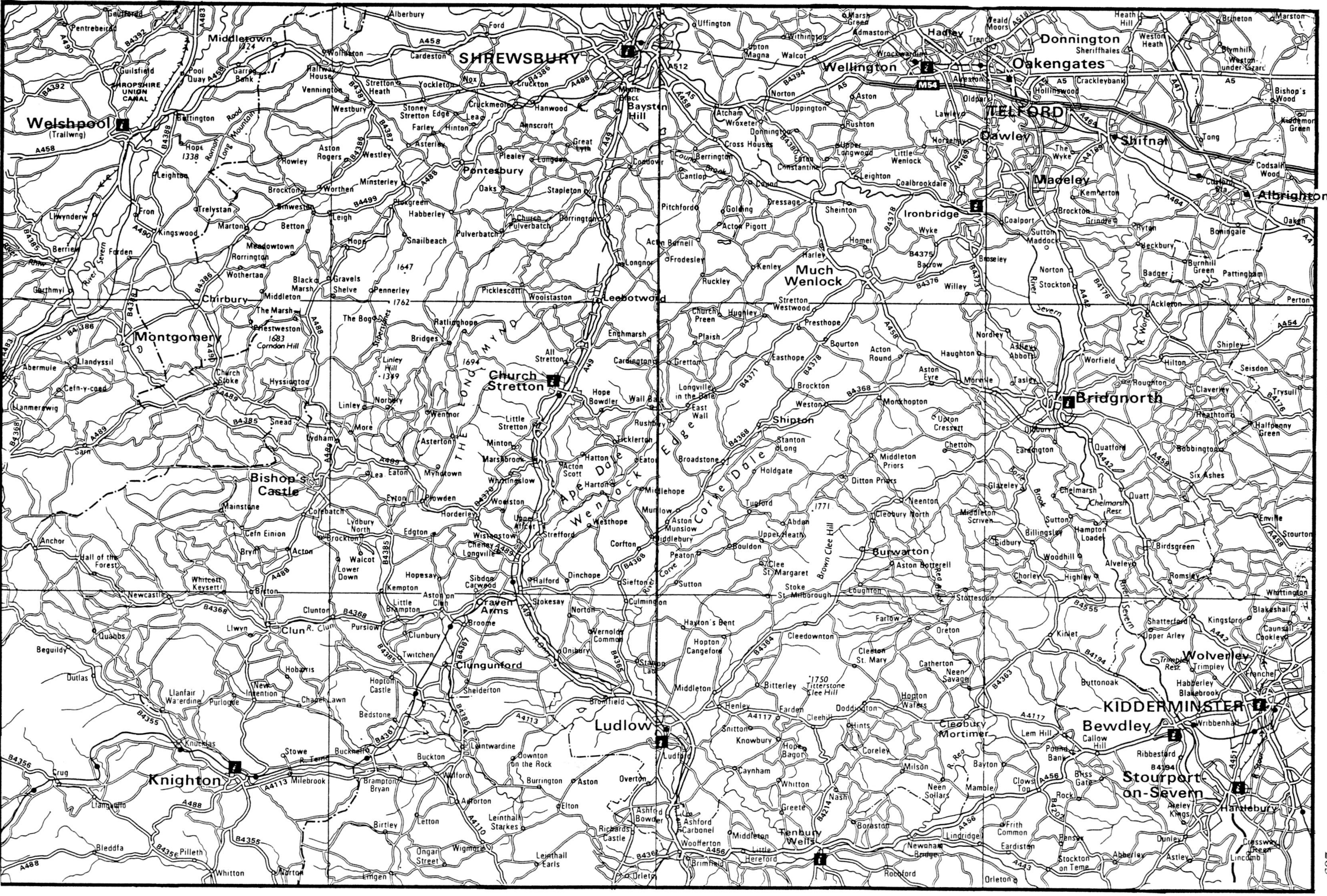

SHREWSBURY
Wellington
Oakengates
Donnington
TELFORD
Dawley
Madeley
Shifnal
Albrighton
Ironbridge
Much Wenlock
Bridgnorth
KIDDERMINSTER
Bewdley
Stourport-on-Severn
Cleobury Mortimer
Tenbury Wells
Ludlow
Craven Arms
Church Stretton
Pontesbury
Bishop's Castle
Clun
Knighton
Montgomery
Welshpool
Shipton
Burwarton
Clungunford

The Red Castle, Hawkestone.

Stokesay—The Gateway.

The Old Lodge of Frodesley.

Gateway, Langley Hall.

House where Queen Margaret slept, Betton.

Tong Church.

The etchings are from *Shropshire* by A. J. C. Hare, published by George Allen, London 1898.